A COMPARATIVE METHODS APPROACH TO THE STUDY OF OOCYTES AND EMBRYOS

Advances in Molecular Biology

Series Editor
Allan Jacobson, University of Massachusetts Medical Center

Editorial Advisory Board
Joan Brugge, ARIAD Pharmaceuticals, Inc.
Henry Erlich, Roche Molecular Systems
Stanley Fields, University of Washington
Larry Gold, NeXstar Pharmaceuticals, Inc.

The Yeast Two-Hybrid System, edited by Paul L. Bartel and Stanley Fields

A Comparative Methods Approach to the Study of Oocytes and Embryos, edited by Joel D. Richter

A COMPARATIVE METHODS APPROACH TO THE STUDY OF OOCYTES AND EMBRYOS

Edited by
Joel D. Richter

New York   Oxford • *Oxford University Press   1999*

Oxford University Press

Oxford   New York
Athens   Auckland   Bangkok   Bogotá   Buenos Aires
Calcutta   Cape Town   Chennai   Dar es Salaam   Delhi   Florence   Hong Kong
Istanbul   Karachi   Kuala Lumpur   Madrid   Melbourne
Mexico City   Mumbai   Nairobi   Paris   Singapore   Taipei   Tokyo   Toronto   Warsaw
and associated companies in
Berlin   Ibadan

Copyright © 1999 by Oxford University Press, Inc.

Published by Oxford University Press, Inc.
198 Madison Avenue, New York, New York 10016

Oxford is a registered trademark of Oxford University Press

All rights reserved. No part of this publication may be reproduced,
stored in a retrieval system, or transmitted, in any form or by any means,
electronic, mechanical, photocopying, recording, or otherwise,
without the prior permission of Oxford University Press.

Library of Congress Cataloging-in-Publication Data
A comparative methods approach to the study of oocytes and embryos/
   edited by Joel D. Richter.
      p. cm. — (Advances in molecular biology)
   Includes bibliographical references and index.
   ISBN 0-19-511594-5
   1. Embryology—Mammals.   2. Embryology—Amphibians.
3. Embryology—Insects.   4. Mice.   5. Frogs.   6. Fruit-flies.
I. Richter, Joel D.   II. Series.
QL959.C58   1998
571.8′61—dc21   98-13183

9 8 7 6 5 4 3 2 1

Printed in the United States of America
on acid-free paper

Preface

This volume of the *Advances in Molecular Biology* series takes a comparative approach to the study of oocytes and embryos of frogs, mice, and flies, which are among the most important organisms for the analysis of early development. The chapters are written by molecular and cellular developmental biologists who have compiled some of the most salient tried-and-true techniques they use, and in many cases developed, in their own laboratories.

In part I, nine chapters focus on mouse oocytes and embryos. Four of them (Eppig, Litscher and Wassarman, Pauken and Capco, and Balz et al.) deal with the in vitro culture of oocytes and embryos, the characterization of oocyte zona pellucida components, signal transduction analysis, and intracellular ion measurements. Five chapters (Stutz et al., Oh et al., Villar, Schultz, and Brewster et al.) deal with various aspects of gene regulation, from transcription and translation to genomic imprinting. The techniques include oocyte injection, antisense oligonucleotide mRNA ablation, PCR-based RFLP for genome analysis, and differential display in the search for unique gene products in early embryos and genital ridges.

The study of *Xenopus* oocytes and embryos is covered in the 10 chapters of part II. Four chapters (Carroll, Menut et al., Murray and Reinbold, and Wolffe) cover DNA and RNA metabolism in oocytes and embryos, including chromatin assembly, DNA recombination and transcription, mRNP isolation and characterization, and mRNA translation. Four chapters

(Kloc and Etkin, Holleman et al., Carl and Klymkowsky, and McGrew et al.) focus on mRNA and protein localization in oocytes and embryos by in situ hybridization and immunohistochemistry. One chapter, by Zuck et al., describes antisense technology for the analysis of maternal mRNA function in embryos. The final chapter in this part, by Moody, illustrates how the fates of single blastomeres are determined.

In part III, seven chapters are devoted to the analysis of *Drosophila* oogenesis and embryogenesis. The first four chapters (Waring, Montell, Kobayashi et al., and Glotzer and Ephrussi) cover oogenesis and include immunogold labeling of eggshell proteins, general methods such as ovary dissection, fixation procedures, and mosaic analysis, techniques for ultrastructural studies including immunoelectron microscopy, and the examination of mRNA distribution by fluorescent in situ hybridization. Finally, three chapters (Sallés and Strickland, Minden, and Van Vactor and Kopczynski) are devoted to embryogenesis and include the determination of poly(A) tail length of maternal mRNAs, time-lapse microscopy methods, and dissection and cell biological techniques for assessing the developing nervous system.

By bringing together some major techniques used by investigators devoted to three different organisms, I hope that this volume finds a wide audience among developmental biologists. This compilation of current but diverse methods may be used as a source not only by those who are already well acquainted with these three organisms, but also by those who are just entering the field. In addition, this comparative presentation should be useful to investigators who wish to adapt certain technologies in their examination of other tissues and animals.

Contents

Contributors xi

Part I Methods for Studying Mouse Oocytes and Embryos

1   Mouse Oocyte Maturation, Fertilization, and Preimplantation Development in Vitro   3
    John J. Eppig

2   Purification and Functional Analysis of Mouse Egg Zona Pellucida Glycoproteins   10
    Eveline S. Litscher & Paul M. Wassarman

3   Methods for Analysis of Cytoplasmic Signal Transduction in Mammalian Eggs and Embryos   23
    Christine M. Pauken & David G. Capco

4   Intracellular Ion Measurements in Single Eggs and Embryos Using Ion-Sensitive Fluorophores   39
    Jay M. Baltz & Karen P. Phillips

5   Approaches to Study Translational Control in Mouse Oocytes   83
    André Stutz, Joachim Huarte, Béatrice Conne, & Jean-Dominique Vassalli

6 Identification of Genes and Processes Guiding the Transition between the Mammalian Gamete and Embryo 101
   Bermseok Oh, Sue-Yun Hwang, Wilhelmine N. De Vries, Davor Solter, & Barbara B. Knowles

7 Approaches for the Analysis of Genomic Imprinting in Oocytes and Early Mouse Embryos 127
   Angela J. Villar

8 Gene Expression in Mouse Embryos: Use of mRNA Differential Display 148
   Richard M. Schultz

9 Use of Differential Display to Identify Gene Expression Differences in Male and Female Mouse Genital Ridges 157
   Jay Brewster, Jill O'Moore, & William Crain

Part II Methods for Studying *Xenopus* Oocytes and Embryos

10 DNA Recombination and Repair in *Xenopus* Oocytes and Eggs: Substrate Design, Direct Microinjection, and Extract Preparation 173
   Dana Carroll

11 DNA Replication and Chromatin Assembly Using *Xenopus* Eggs or Embryos 196
   Sophie Menut, Jean-Marc Lemaitre, Alan Hair, & Marcel Méchali

12 Isolation and Characterization of Masked mRNPs from *Xenopus* Oocytes 227
   Mary T. Murray & Rolland Reinbold

13 Complications and Problems in the Assay of Transcription in *Xenopus* Oocytes, Eggs, and Embryos 241
   Alan P. Wolffe

14 Analysis of Localized RNAs in *Xenopus* Oocytes 256
   Malgorzata Kloc & Laurence D. Etkin

15 In Situ Hybridization Techniques with *Xenopus* Embryos 279
   Thomas Hollemann, Frank Panitz, & Tomas Pieler

16 Visualizing Endogenous and Exogenous Proteins in *Xenopus laevis* (and Other Organisms) 291
   Timothy Carl & Michael W. Klymkowsky

17 Qualitative Analysis of Differential Gene Expression during Early *Xenopus* Embryogenesis by Whole-Mount In Situ Hybridization and RT–PCR 316
   L. Lynn McGrew, Stefan Hoppler, & Randall T. Moon

18 Studying the Function of Maternal mRNAs in *Xenopus* Embryos: An Antisense Approach 341
    M. V. Zuck, C. C. Wylie, & J. Heasman

19 Testing the Cell Fate Commitment of Single Blastomeres in *Xenopus laevis* 355
    Sally A. Moody

Part III Methods for Studying *Drosophila* Oocytes and Embryos

20 Eggshell Assembly in *Drosophila* 385
    Gail L. Waring

21 Cell Biological and Genetic Methods for Studying *Drosophila* Oogenesis 397
    Denise J. Montell

22 Techniques for Analyzing Protein and RNA Distribution in *Drosophila* Ovaries and Embryos at Structural and Ultrastructural Resolution 426
    Satoru Kobayashi, Reiko Amikura, Akira Nakamura, & Paul Lasko

23 Analysis of RNA Distribution during *Drosophila* Oogenesis Using Fluorescent In Situ Hybridization 446
    Jolanta B. Glotzer & Anne Ephrussi

24 Assessment of Poly(A) Tail Lengths on Specific mRNAs during Development 457
    Fernando J. Sallés & Sidney Strickland

25 Three-Dimensional, Time-Lapse Microscopy of *Drosophila* Embryogenesis 471
    Jonathan Minden

26 Anatomical Techniques for Analysis of Nervous System Development in the *Drosophila* Embryo 490
    David Van Vactor & Casey Kopczynski

Index 515

Contributors

Reiko Amikura
Institute of Biological Sciences
Gene Experiment Center and Center for
 Tsukuba Advanced Research Alliance
 (TARA)
University of Tsukuba
Tsukuba, Ibaraki 305
Japan

Jay M. Baltz
Loeb Medical Research Institute and
 Human IVF Laboratories
Ottawa Hospital
Departments of Obstetrics and
 Gynecology (Division of
 Reproductive Medicine) and Cellular
 and Molecular Medicine
University of Ottawa
Ottawa, Ontario K1Y 4E9
Canada

Jay Brewster
McLaughlin Research Institute
1520 23rd Street South
Great Falls, Montana 59405

David G. Capco
Molecular and Cellular Biology
 Program/Zoology
Arizona State University
Tempe, Arizona 85287-1501

Timothy Carl
Environmental, Organismic and
 Population Biology
University of Colorado, Boulder
Boulder, Colorado 80309

Dana Carroll
Department of Biochemistry
University of Utah School of Medicine
Salt Lake City, Utah 84132

Contributors

Béatrice Conne
Department of Morphology
University of Geneva Medical School
1211 Geneva 4
Switzerland

William Crain
McLaughlin Research Institute
1520 23rd Street South
Great Falls, Montana 59405

Wilhelmine N. De Vries
The Jackson Laboratory
600 Main Street
Bar Habor, Maine 04609

Anne Ephrussi
European Molecular Biology
  Laboratory
Developmental Biology Programme
Meyerhofstrasse 1
69117 Heidelberg
Germany

John J. Eppig
The Jackson Laboratory
Bar Harbor, Maine 04609

Laurence D. Etkin
Department of Molecular Genetics
The University of Texas
M.D. Anderson Cancer Center
1515 Holcombe Blvd.
Houston, Texas 77030

Jolanta B. Glotzer
Research Institute of Molecular
  Pathology
Dr. Bohr-Gasse 7
A-1030 Vienna
Austria

Alan Hair
Institut Jacques Monod—CNRS
2 place Jussieu
75251 Paris Cedex 05
France

J. Heasman
Institute of Human Genetics
Department of Cell Biology and
  Neuroanatomy, and Department of
  Pediatrics
University of Minnesota
Minneapolis, Minnesota 55455

Thomas Hollemann
Georg August Universität Göttingen
Institut für Biochemie und Molekulare
  Zellbiologie
Humboldtallee 23
37073 Göttingen
Germany

Stefan Hoppler
Department of Anatomy and Physiology
The Wellcome Trust Building
University of Dundee
Dundee DD1 4HN
Scotland, UK

Joachim Huarte
Department of Morphology
University of Geneva Medical School
1211 Geneva 4
Switzerland

Sue-Yun Hwang
The Jackson Laboratory
600 Main Street
Bar Harbor, Maine 04609

Malgorzata Kloc
Department of Molecular Genetics
The University of Texas
M.D. Anderson Cancer Center
1515 Holcombe Blvd.
Houston, Texas 77030

Michael W. Klymkowsky
Molecular, Cellular and Developmental
  Biology
University of Colorado, Boulder
Boulder, Colorado 80309

Barbara B. Knowles
The Jackson Laboratory
600 Main Street
Bar Harbor, Maine 04609

Satoru Kobayashi
Institute of Biological Sciences
Gene Experiment Center and Center for Tsukuba Advanced Research Alliance (TARA)
University of Tsukuba
Tsukuba, Ibaraki 305
Japan

Casey Kopczynski
Exelixis Pharmaceuticals, Inc.
260 Littlefield Ave.
S. San Francisco, California 94080

Paul Lasko
Department of Biology
McGill University
1205 Avenue Docteur Penfield
Montréal, Québec H3A 1B1
Canada

Jean-Marc Lemaitre
Institut Jacques Monod—CNRS
2 place Jussieu
75251 Paris Cedex 05
France

Eveline S. Litscher
Department of Cell Biology and Anatomy
Mount Sinai School of Medicine
One Gustave L. Levy Place
New York, New York 10029-6574

L. Lynn McGrew
Howard Hughes Medical Institute and Department of Pharmacology
K516 Health Sciences Building
University of Washington School of Medicine
Seattle, Washington 98195

Marcel Méchali
Institut de Génétique Humaine, CNRS
Dynamics of the Genome and Development
141 Rue de la Cardonille
34396 Montpellier Cedex 5
France

Sophie Menut
Institut Jacques Monod—CNRS
2 place Jussieu
75251 Paris Cedex 05
France

Jonathan Minden
Department of Biological Sciences
Carnegie Mellon University
Pittsburgh, Pennsylvania 15213

Denise J. Montell
Department of Biological Chemistry
Johns Hopkins School of Medicine
725 North Wolfe Street
Baltimore, Maryland 21205-2185

Sally A. Moody
Department of Anatomy and Cell Biology
The George Washington University Medical Center
2300 I Street, NW
Washington, District of Columbia 20037

Randall T. Moon
Howard Hughes Medical Institute and Department of Pharmacology
K516 Health Sciences Building
University of Washington School of Medicine
Seattle, Washington 98195

Mary T. Murray
Center for Molecular Medicine and Genetics
Wayne State University School of Medicine
5107 Biological Sciences Building
5047 Gullen Mall
Detroit, Michigan 48202

Akira Nakamura
Department of Biology
Mcill University
1205 Avenue Docteur Penfield
Montréal, Québec H3A 1B1
Canada

Jill O'Moore
McLaughlin Research Institute
1520 23rd Street South
Great Falls, Montana 59405

Bermseok Oh
The Jackson Laboratory
600 Main Street
Bar Harbor, Maine 04609

Frank Panitz
Georg August Universität Göttingen
Institut für Biochemie und Molekulare Zellbiologie
Humboldtallee 23
37073 Göttingen
Germany

Christine M. Pauken
Molecular and Cellular Biology Program/Zoology
Arizona State University
Tempe, Arizona 85287-1501

Karen P. Phillips
Loeb Medical Research Institute and Human IVF Laboratories
Ottawa Hospital
Department of Obstetrics and Gynecology (Division of Reproductive Medicine) and Cellular and Molecular Medicine
University of Ottawa
Ottawa, Ontario K1Y 4E9
Canada

Tomas Pieler
Georg August Universität Göttingen
Institut für Biochemie und Molekulare Zellbiologie
Humboldtallee 23
37073 Göttingen
Germany

Rolland Reinbold
Center for Molecular Medicine and Genetics
Wayne State University School of Medicine
5107 Biological Sciences Building
5047 Gullen Mall
Detroit, Michigan 48202

Fernando J. Sallés
Department of Cellular and Molecular Pharmacology
University Medical Center at Stony Brook
Stony Brook, New York 11794-8651

Richard M. Schultz
Department of Biology
University of Pennsylvania
Philadelphia, Pennsylvania 19104-6018

Davor Solter
The Jackson Laboratory
600 Main Street
Bar Harbor, Maine 04609

Sidney Strickland
Department of Cellular and Molecular Pharmacology
University Medical Center at Stony Brook
Stony Brook, New York 11794-8651

André Stutz
Department of Morphology
University of Geneva Medical School
1211 Geneva 4
Switzerland

David Van Vactor
Department of Cell Biology and The
 Program in Neuroscience
Harvard Medical School
240 Longwood Avenue
Boston, Massachusetts 02115

Jean-Dominique Vassalli
Department of Morphology
University of Geneva Medical School
1211 Geneva 4
Switzerland

Angela J. Villar
Porter Beach Foundation
2269 Chestnut Street #876
San Francisco, California 94123

Gail L. Waring
Department of Biology
Marquette University
WLS 109, PO Box 1881
Milwaukee, Wisconsin 53201-1881

Paul M. Wassarman
Department of Cell Biology and
 Anatomy
Mount Sinai School of Medicine
One Gustave L. Levy Place
New York, New York 10029-6574

Alan P. Wolffe
Laboratory of Molecular Embryology
National Institute of Child Health and
 Human Development
NIH, Bldg. 18T, Rm. 106
Bethesda, Maryland 20892-5431

C. C. Wylie
Institute of Human Genetics
Department of Cell Biology and
 Neuroanatomy, and Department of
 Pediatrics
University of Minnesota
Minneapolis, Minnesota 55455

M. V. Zuck
Institute of Human Genetics
Department of Cell Biology and
 Neuroanatomy, and Department of
 Pediatrics
University of Minnesota
Minneapolis, Minnesota 55455

Part I

METHODS FOR STUDYING MOUSE OOCYTES AND EMBRYOS

1

Mouse Oocyte Maturation, Fertilization, and Preimplantation Development In Vitro

John J. Eppig

Development of mouse oocytes and preimplantation embryos in vitro offers excellent opportunities for studying the mechanisms of oocyte and embryo development and also new approaches to rescue valuable genetic resources. Although complete growth and development of mouse oocytes from primordial follicles in vitro has been achieved (Eppig and O'Brien 1996), the frequency of success is still very low and cannot be considered a routine procedure; therefore, it will not be described here. In contrast, the embryonic developmental competence of oocytes matured in vitro after isolation at the germinal vesicle (GV) stage is highly successful and often equivalent to that obtained by exogenous gonadotropin-induced maturation in vivo. In fact, under some circumstances, such as aged mice (Eppig and O'Brien 1995) or unusual mouse strains having atypical oocyte development (Eppig and Wigglesworth 1994), maturation in vitro produces higher quality embryos than maturation induced by exogenous gonadotropins in vivo. Likewise, the media for preimplantation development of mouse embryos have dramatically improved in recent years (Gardner and Lane 1993; Gardner and Sakkas 1993; Lawitts and Biggers 1993; Erbach et al. 1994; Gardner 1994). Culture of preimplantation embryos in one of these media, KSOM/AA, produces blastocysts having a pattern of gene expression that is very similar to that of embryos grown in vivo (Ho et al. 1995). This chapter will describe methods used in the Jackson laboratory for oocyte maturation, fertilization, and preimplantation development in vitro.

In Vitro Maturation of Cumulus-Cell-Enclosed Oocytes

Oocyte–cumulus-cell complexes are liberated from large antral follicles of mice primed 44 hours previously by intraperitoneal injection of 5 IU equine chorionic gonadotropin (eCG). Mice should be injected with eCG either between the ages of 22 and 24 days or older than 6 weeks. Injection should be avoided in 4–5-week-old mice since the yield of high-quality oocytes from these mice is very low. The cumulus-cell-enclosed oocytes are liberated by immersing the ovaries in 2.5 ml of maturation medium (Table 1-1), in 35-mm polystyrene Petri dishes, and puncturing the large antral follicles with a 26–30G syringe needle. The oocyte–cumulus-cell complexes are then collected using a drawn-glass micropipette and washed by serially transferring them through three dishes of medium, avoiding cellular debris with each collection. Dishes used for collecting and washing the complexes should not be tissue-culture treated to avoid having the complexes adhere to the dishes, thus making manipulations very difficult and frustrating. Medium should be kept at approximately 37°C. The gas mixture for the maturation of cumulus-cell-enclosed oocytes is 5% CO_2 with 5–20% O_2, and the balance N_2 (Eppig and Wigglesworth 1995). Collection and manipulation of complexes should be carried out as quickly as possible to avoid dramatic changes in pH of the medium. If necessary, 25 mM if HEPES buffer can be added to the medium used for the collection of complexes to reduce pH changes, but the medium used for maturation should be HEPES-free.

Cumulus-cell-enclosed oocytes are allowed to undergo maturation for 15–17 hours in medium containing follicle-stimulating hormone (FSH). The amount of FSH depends upon the potency of the FSH preparation. In general, the concentration used should maximally stimulate the mucification and expansion of the cumulus oophorus. After maturation, the complexes are washed three times with fertilization medium (modified MEM; Table 1-2) immediately before placing them in drops containing sperm for fertilization.

In Vitro Maturation of Cumulus-Cell-Denuded Oocytes

Oocytes that mature in vitro with their companion cumulus cells are more likely to be competent to undergo fertilization and embryo development than oocytes that mature denuded of cumulus cells. Nevertheless, it is sometimes necessary to mature denuded oocytes when other aspects of the experimental protocol make the retention of cumulus cells difficult. For example, if it is necessary to microinject GV-stage oocytes, it is much easier to inject denuded rather than cumulus-cell-enclosed oocytes since cumulus cell debris quickly renders the microinjection pipette unusable. To mature cumulus-

Table 1-1 Composition of Waymouth medium for oocyte maturation

Component	g/liter	ml/liter
Group A		
Powdered Waymouth 752/1 medium (1-liter package)	—	—
Group B		
NaHCO$_3$	2.24	—
Sodium pyruvate	0.025	—
Streptomycin sulfate	0.05	—
Penicillin G	0.075	—
Group C		
Fetal bovine serum	—	50

cell-denuded oocytes with maximum potential for fertilization and embryogenesis, the following protocol is recommended. Denuded oocytes are matured for 15–17 hours in modified MEM (Table 1-2) supplemented with 10–20% fetal bovine serum (FBS), rather than bovine serum albumin (BSA), using an atmosphere of 5% O_2, 5% CO_2, and 90% N_2 (hereafter referred to as 5–5–90 gas); FSH is not required. After maturation, oocytes are washed three times with fertilization medium (modified MEM; Table 1-2) immediately before placing them in drops containing sperm for fertilization.

Fertilization In Vitro

Fertilization is carried out in 0.5-ml drops of fertilization medium under washed paraffin oil. Sperm are prepared for fertilization by removing cauda epididymides from mice caged individually for at least a month before use.

Table 1-2 Composition of MEM for fertilization

Component	g/liter
Group A	
Powdered MEM with Earle's salts and glutamine (1-liter package)	—
Group B	
NaHCO$_3$	2.2
EDTA (tetrasodium salt)	0.0038
Sodium pyruvate	0.022
Streptomycin sulfate	0.05
Penicillin G	0.075
Group C	
BSA[a]	3.0

[a]Crystallized, lyophilized.

A pair of epididymides are first blotted on an absorbent wipe and then immersed in 0.9 ml of fertilization medium under washed paraffin oil. The epididymides are cut 5–10 times with sterile scissors and sperm are allowed to swim out for about 15 minutes at 37°C in an atmosphere of 5-5-90 gas. This preparation, 10 µl, is added to 0.5 ml of fertilization medium under washed paraffin oil. The drops are now ready to receive the washed matured oocytes. It is important to move mature oocytes as quickly as possible from the maturation medium, through washes in the fertilization medium, and into the fertilization drops. This prevents changes in the zona pellucida, causing zona "hardening" that could significantly reduce the frequency of fertilization.

Eggs are inseminated for 4 hours at 37°C under 5-5-90 gas. This is best accomplished using modular incubation chambers thoroughly perfused with gas mixture before sealing. The chambers are then placed in a 37°C water-jacketed incubator. After insemination, eggs are removed from fertilization drops, washed twice in fresh medium, transferred to 12 × 75-mm borosilicate tubes containing 1 ml of fertilization medium, gassed with 5-5-90 gas, sealed with a rubber (bromo butyl) stopper, and placed in a 37°C water bath. Alternatively, washed inseminated eggs can be transferred to 50-µl drops of fertilization medium under paraffin oil and cultured in modular incubation chambers as described above. In either case, eggs are cultured for 15 hours and two-cell-stage embryos are collected and washed in embryo culture medium (KSOM/AA; Table 1-3).

Culture of Preimplantation Embryos

After washing, two-cell-stage embryos are transferred to tubes containing 1 ml of KSOM/AA and cultured for 4 more days to produce blastocysts. Tubes are gassed with 5-5-90 gas, sealed tightly with rubber stoppers, and placed in a 37°C water bath. As an alternative to culture in tubes, the embryos can be cultured in drops (10–50 µl) under washed paraffin oil. Dishes containing the drops with embryos are placed in modular incubation chambers, gassed with 5-5-90 gas that is then sealed and placed in a 37°C water-jacketed incubator.

Miscellaneous Methods

Washing Paraffin Oil

Light paraffin oil is washed by placing 500 ml of oil in a polystyrene bottle, used to store tissue culture medium, with 50 ml of fertilization medium. This is stirred vigorously with a magnetic stirrer for 24 hours and then allowed to separate into aqueous and oil phases. This is repeated twice. Oil is then decanted into a fresh bottle and is ready for use after passing through a 0.8-µm sterilization membrane.

Table 1-3 Composition of KSOM/AA for preimplantation embryos

Component	mM	g/liter	ml/liter
Group A			
NaCl	95	5.55	—
KCl	2.5	0.185	—
KH_2PO_4	0.35	0.0476	—
Glucose	0.2	0.036	—
Sodium pyruvate	0.2	0.022	—
L-Glutamine	1	0.146	—
Streptomycin sulfate	—	0.05	—
Penicillin G	—	0.063	—
EDTA (tetrasodium salt)	0.01	0.0038	—
Group B			
$MgSO_4 \cdot 7H_2O$	0.2	0.0493	—
$CaCl_2 \cdot 2H_2O$	1.71	0.251	—
Na lactate (syrup)	11.8	—	1.74
$NaHCO_3$	25	2.1	—
Essential amino acids (50× concentrate)	—	—	10.0
Nonessential amino acids (100× concentrate)	—	—	5.0
Phenol red[a]	—	—	1.0
Group C			
BSA[b]	—	1.0	—

[a]Prepare stock by dissolving 0.11 g of water-soluble phenol red in 10 ml of distilled water. Keep refrigerated and shake it just before use.
[b]Crystallized, lyophilized.

Preparation of Embryo Culture Tubes

Borosilicate tubes, 12 × 75 mm, are rinsed five times with culture-media-quality water and then baked, inverted, at 140°C for 2 hours. Rubber stoppers are soaked overnight in 95% ethanol, rinsed twice in media- quality water, and spread to dry overnight on paper towels. Stoppers are inserted into the tubes immediately after the baking oven is turned off and the combined tubes and stoppers allowed to cool with the oven. The rack containing the stoppered tubes is stored in a plastic bag until use.

Water

It is critically important that high-quality water be used in the preparation of media. In this laboratory, water from a remote spring is collected and transported in plastic bottles. It is distilled once and stored in glass bottles before use. Excellent results can also be achieved using 18+ megaohm-resistance water from a well-maintained purification apparatus.

General Procedures for the Preparation of Media

All media are prepared by weight using polystyrene plastic culture-medium storage bottles. This avoids the use of volumetric flasks cleaned with potentially harmful chemicals. The bottle is placed on a tared balance, components are added as described below, and then water is added until the total weight is 1 kg. Appropriate gas is bubbled through the media, using a sterile serological pipette, for 10 minutes before macromolecules, such as BSA, are added to the media. After addition of macromolecules, the media are sterilized by filtration using 0.45 μm, or smaller, pore-size membranes. Bottles are then gassed with the appropriate gas mixture and sealed tightly, and gassed again whenever medium is removed. Media should not be stored for more than 2 weeks.

Oocyte Maturation Medium and Fertilization Medium

The composition of oocyte maturation medium is presented in Table 1-1 and that for fertilization medium in Table 1-2. Tare an empty 1-liter plastic tissue culture storage bottle on a balance and add approximately 700 g of water. Powdered Waymouth 752/1 and MEM with Earle's salts are available commercially in 1-liter packages. Add a package to the bottle and dissolve by swirling. Then add the components listed in group B individually, dissolving each by swirling before adding the next. Add in the order shown in the tables and then add water to 1 kg. Bubble 5-5-90 gas through the medium and sterilize, as described above, then cap tightly, and refrigerate. When ready to use the oocyte maturation medium, add fetal bovine serum to a final concentration of 5% (v/v). Add BSA to the fertilization medium immediately after gas is bubbled through and allow to dissolve into the medium without agitation, then sterilize again by filtration.

Medium for Preimplantation Embryo Development

The composition of KSOM/AA embryo culture medium is shown in Table 1-3. Tare an empty 1-liter plastic tissue culture storage bottle on a balance and add the components listed in group A. Then add approximately 700 g of water and dissolve group A by swirling. Add the components in group B individually, according to the order indicated in the list, dissolving each solid component by swirling before adding the next. Add water to to final weight of 1 kg. Bubble 5-5-90 gas through the medium, add the BSA, and sterilize, as described above, then cap tightly, and refrigerate.

ACKNOWLEDGMENTS This research was supported as part of the National Cooperative Program on Non-Human In Vitro Fertilization and Preimplantation Development and was funded by the National Institute of Child Health and Human

Development (NICHD), NIH, through Cooperative Agreement HD21970. My thanks to Drs. Keith Latham, Larry Mobraaten, Randy Prather, and Mark Westhusin for suggestions in the preparation of this chapter.

References

Eppig, J. J., and O'Brien, M. (1995). In vitro maturation and fertilization of oocytes isolated from aged mice: a strategy to rescue valuable genetic resources. J. Assist. Reprod. Genet. 12:269–273.

Eppig, J. J., and O'Brien, M. J. (1996). Development in vitro of mouse oocytes from primordial follicles. Biol. Reprod. 54:197–207.

Eppig, J. J., and Wigglesworth, K. (1994). Atypical maturation of oocytes of strain I/LnJ mice. Human Reprod. 9:1136–1142.

Eppig, J. J., and Wigglesworth, K. (1995). Factors affecting the developmental competence of mouse oocytes grown in vitro: oxygen concentration. Mol. Reprod. Dev. 42:447–456.

Erbach, G.T., Lawitts, J. A., Papaioannou, V. E., and Biggers, J. D. (1994). Differential growth of the mouse preimplantation embryo in chemically defined media. Biol. Reprod. 50(5):1027–1033.

Gardner, D. K. (1994). Mammalian embryo culture in the absence of serum or somatic cell support. Cell Biol. Int. 18(12):1163–1179.

Gardner, D. K., and Lane, M. (1993). Amino acids and ammonium regulate mouse embryo development in culture. Biol. Reprod. 48 (2):377–385.

Gardner, D. K., and Sakkas, D. (1993). Mouse embryo cleavage, metabolism and viability: role of medium composition. Hum. Reprod. 8(2):288–295.

Ho, Y., Wigglesworth, K., Eppig, J. J., and Schultz, R. M. (1995). Preimplantation development of mouse embryos in KSOM: augmentation by amino acids and analysis of gene expression. Mol. Reprod. Dev. 41:232–238.

Lawitts, J. A., and Biggers, J. D. (1993). Culture of preimplantation embryos. In *Methods in Enzymology. Vol. 225. Guide to Techniques in Mouse Development*, Wassarman, P. M., and DePamphilis, M. L., eds. San Diego, Academic Press, 153–164.

2

Purification and Functional Analysis of Mouse Egg Zona Pellucida Glycoproteins

Eveline S. Litscher
Paul M. Wassarman

The plasma membrane of all mammalian eggs is surrounded by a relatively thick extracellular coat, called the *zona pellucida* (ZP; Figure 2-1) (Gwatkin 1977; Dietl 1989; Yanagimachi 1994). The ZP forms a viscous border between the oolemma and the innermost layer of follicle cells *(corona radiata)*. Cellular processes extend from the innermost follicle cells, through the ZP, and form gap junctions with egg plasma membrane. The ZP performs functions identical to those performed by the jelly coat and vitelline envelope of eggs from nonmammalian species (e.g., echinoderms and amphibia). In particular, it is the site of species-specific sperm receptors and acrosome reaction-inducers that function in vivo during fertilization of the egg (Wassarman 1990, 1995; Snell and White 1996). Removal of the ZP eliminates the barrier to fertilization of eggs in vitro by sperm from different species.

The mouse egg ZP is about 6.2 (±1.9) μm thick, contains about 3.5 (±0.5) ng of protein, and stains positively for glycoprotein (Wassarman 1988, 1998). It consists of three glycoproteins, mZP1 (~200 kDa M_r; dimer), mZP2 (~120 kDa M_r; monomer), and mZP3 (~83 kDa M_r; monomer), that combine to form an extensive network of long, cross-linked filaments by using noncovalent bonds. Each of the glycoproteins is heterogeneous with respect to M_r due to heterogeneous glycosylation of a unique polypeptide (polypeptide M_r—mZP1, 68 kDa; mZP2, 77 kDa; mZP3, 44 kDa) with

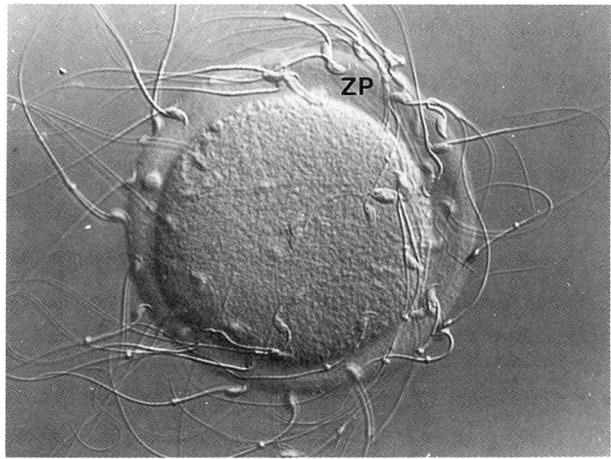

Figure 2-1 Light photomicrograph of free-swimming mouse sperm bound by their heads to the ZP of an ovulated mouse egg in vitro. The micrograph was taken by using Nomarski differential interference contrast (DIC) microscopy. ZP, zona pellucida. Magnification 550×.

both asparagine- (N-) linked (complex-type) and serine/threonine- (O-) linked oligosaccharides. The oligosaccharides are both sulfated and sialylated, which contributes to the M_r heterogeneity and makes all three glycoproteins relatively acidic. Targeted mutagenesis of the *mZP3* gene, by homologous recombination in embryonic stem (ES) cells, results in production of homozygous null mutant females ($mZP3^{-/-}$) that are infertile and whose eggs lack a ZP (Liu et al. 1996; Rankin et al. 1996; Wassarman et al. 1996, 1997).

In mice, free-swimming, acrosome-intact sperm recognize and bind to specific O-linked oligosaccharides located on the carboxy-terminal portion of mZP3 (a region encoded by *mZP3* exon-7) (Bleil and Wassarman 1980; Florman and Wassarman 1985; Rosiere and Wassarman 1992; Kinloch et al. 1995; Litscher and Wassarman 1996). Each acrosome-intact sperm head possesses tens-of-thousands of sites that are capable of binding to mZP3 in the ZP (Bleil and Wassarman 1986; Mortillo and Wassarman 1991). Once bound to mZP3 by plasma membrane overlying their head, sperm undergo the acrosome reaction (cellular exocytosis), lose plasma membrane and outer acrosomal membrane, and remain associated with the ZP by binding to mZP2 by their inner acrosomal membrane. Thus, during fertilization, mZP3 serves as a primary sperm receptor and acrosome reaction-inducer, while mZP2 serves as a secondary sperm receptor (Wassarman 1990).

Here, we describe procedures used routinely in our laboratory to purify ZP glycoproteins mZP1–3 from mouse ovaries and to assay ZP glycoproteins for sperm receptor and acrosome reaction-inducing activity in vitro.

These procedures have evolved over the years and have been used extensively in our laboratory. By addition of small modifications, these procedures may also be used to purify ZP glycoproteins from the ovaries of other mammals and to assay ZP glycoprotein biochemical characteristics and biological functions for other mammalian species. For supplementary information about this and related methodology on mammalian eggs and sperm see Rafferty (1970), Wassarman and DePamphilis (1993), Hogan et al. (1994), and Wassarman (1997).

Purification of Mouse Zona Pellucida Glycoproteins

Our initial work with ZP glycoproteins began by physically removing the ZP from eggs, one at a time, by using mouth-operated glass micropipettes, purifying the individual ZP glycoproteins by sodium dodecylsulfate–polyacrylamide gel electrophoresis (SDS-PAGE), and then electroeluting the three glycoproteins (mZP1–3) from the gels (Bleil and Wassarman 1980; Florman and Wassarman 1985). We no longer use this tedious procedure. Instead, we rely on preparation of ZP glycoproteins from ovarian homogenates by centrifugation of the homogenates through a Percoll gradient, recovery of a band of oocyte ZP, and fractionation of the ZP into individual glycoproteins by high-performance liquid chromatography (HPLC) on a size-exclusion column, as described below.

Excision and Homogenization of Ovaries

Generally, 100–150 ovaries, excised from 6–8-week-old, randomly bred, Swiss albino mice (CD-1; Charles River Labs), are suspended in 8 ml of homogenization buffer (25 mM triethanolamine, pH 8.5, 150 mM NaCl, 1 mM $MgCl_2$, 1 mM $CaCl_2$, 0.02% sodium azide, 1% NP-40) supplemented with 1 mg/ml turkey egg-white trypsin inhibitor (Sigma; type II), 0.1 mg/ml hyaluronidase (Sigma), and 0.1 mg/ml DNase I (Sigma). It should be noted that excised ovaries can be stored frozen at $-80°C$ for several months prior to use. The suspension is gently homogenized on ice with a VirTishear (Virtis) for about 1 minute at a setting of 30 until the ovaries are particulate, but not liquified. The homogenate is transferred to a glass homogenizer, 0.8 ml of 10% deoxycholate is added (1% final concentration), and the solution is then homogenized by hand on ice with several strokes (about five or six) until it becomes opalescent.

Centrifugation of Ovarian Homogenates

Two tubes are prepared for centrifugation by placing 4 ml of a 90% Percoll (Sigma) solution, made up in 10× homogenization buffer (90 ml of 100% Percoll and 10 ml of 10× homogenization buffer), into ultraclear, 16 × 76-mm Quick-Seal (Beckman) centrifuge tubes. Ovarian homogenate, 4 ml, is layered on top and the tubes are then filled to the top with 50% Percoll [50 ml of 100% Percoll, 10 ml of 10× homogenization buffer, and 40 ml of distilled water (DW)]. Sealed tubes are centrifuged in a 50 Ti Beckman rotor at 30,000 rpm, at 18°C for 30 minutes. At the conclusion of centrifugation, an opalescent band containing intact and fragmented ZP is present approximately one-third of the way down the tubes (Figure 2-2). The band is removed from each sealed centrifuge tube by piercing it with a 5-ml syringe prefilled with 0.5–1 ml of 50% Percoll and the bands are pooled. The centrifugation step (30,000 rpm, 18°C, 30 minutes) is repeated by placing the ZP on the bottom of a fresh centrifuge tube and filling the tube to the top with 50% Percoll. Once again, the band containing ZP (about one-half the way down the tube) is withdrawn with a 5-ml syringe as before. The ZP are then washed with 25 mM triethanolamine, pH 8.5, 150 mM NaCl, in a 15-ml conical polypropylene tube (Corning) by centrifugation at about 3000 rpm for 10 minutes at room temperature (RT) in a table-top centrifuge. The resulting pellet is transferred to an Eppendorf tube and is washed with phosphate-buffered saline (PBS; 30 mM sodium phosphate, pH 7.4, 150 mM NaCl) by centrifugation at about 10,000 rpm for 5 minutes at RT in a microfuge. The final pellet of ZP is resuspended in 50% glycerol/ 50% PBS and stored at −20°C for up to several months. This procedure yields about 0.5 µg of ZP glycoprotein per ovary or about 1 µg per mouse.

HPLC Fractionation of Isolated ZP

Isolated ZP, suspended in 50% glycerol/50% PBS, are washed in PBS, pH 6.8, by centrifugation at 10,000 rpm for 5 minutes at RT in a microfuge and then resuspended in 100–200 µl of 200 mM phosphate buffer, pH 6.8, containing 0.1% SDS (HPLC buffer) by vortexing. The ZP are solubilized by heating at 65°C for 10 minutes, cooled to RT, and centrifuged at full speed for 1 minute in a microfuge to remove any particulate matter. The supernatant, containing solubilized ZP, is then fractionated by HPLC (Hewlett Packard, Series II 1090) at RT on a size-exclusion column (BioRad; BioSil SEC-250) and eluted with HPLC buffer at a flow rate of 0.1 ml/minute. Absorbance at 280 nm is monitored continuously (Figure 2-3). By this procedure, the supernatant is fractionated into three peaks of absorbance that correspond to mZP1 (~200 kDa M_r), mZP2 (~120 kDa M_r), and mZP3 (~83 kDa M_r). The three peaks elute at about 55–60 minutes (mZP1), about 61–66 minutes (mZP2), and about 68–73 minutes (mZP3). Column fractions (100 µl each) are stored at 4°C prior to use.

Electrophoretic Analysis of HPLC Fractions

To determine the distribution of ZP glycoproteins, aliquots (1–2 µl) of the peak fractions are solubilized in SDS sample buffer (no reducing agent) and subjected to SDS–PAGE on a 7.5% minigel (Figure 2-4). Following electro-

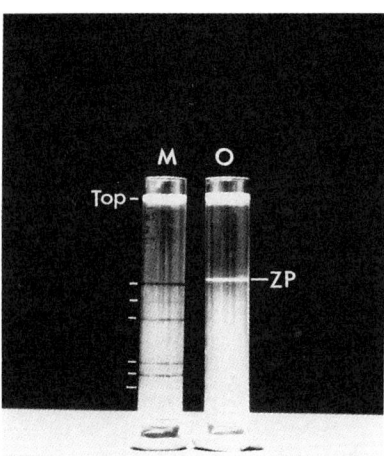

Figure 2-2 Purification of mouse egg ZP by centrifugation of an ovarian homogenate on a Percoll gradient. Shown are two Percoll gradients following centrifugation, essentially as described here. On the left (labeled M) is a gradient with colored M_r marker beads and on the right (labeled O) is a gradient with an ovarian homogenate. Note the opaque band of oocyte ZP located about one-third of the way down the gradient. This band is removed by puncturing the tube with a syringe and prepared for HPLC fractionation, essentially as described here. Several of the M_r marker bands (M tube) are indicated by white hash marks.

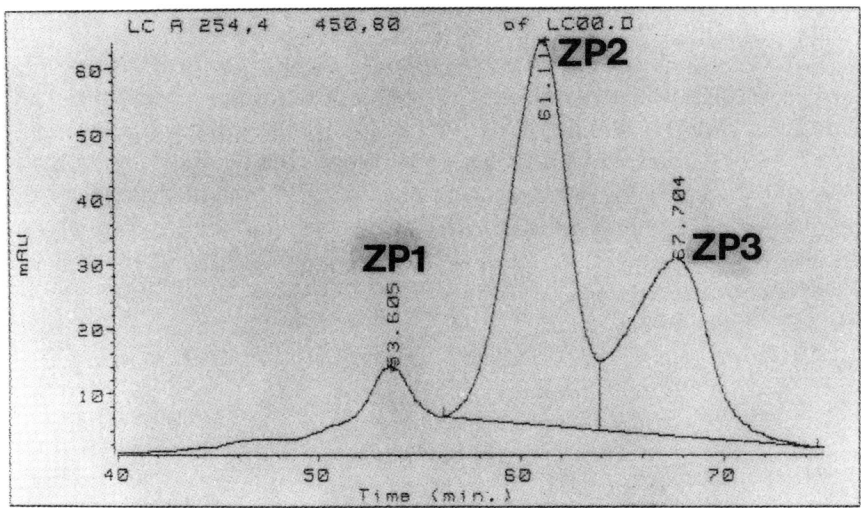

Figure 2-3 High-performance liquid chromatography (HPLC) of mouse ZP glycoproteins. Shown is an absorption profile at 280 nm following HPLC fractionation of a band of ZP purified by centrifugation of an ovarian homogenate on a Percoll gradient, essentially as described here. The band of oocyte ZP was transferred to HPLC buffer, heated to solubilize the ZP, and fractionated on a BioSil SEC-250 size-exclusion column, essentially as described here. The material was separated into three distinct peaks (retention time) that correspond to mZP1 (53.6 minutes), mZP2 (61.1 minutes), and mZP3 (67.7 minutes).

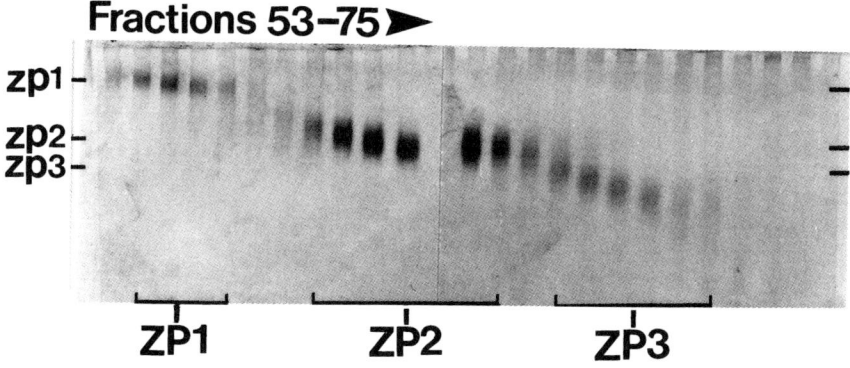

Figure 2-4 Electrophoretic analysis of HPLC-fractionated mouse ZP glycoproteins. Shown is a silver-stained SDS–polyacrylamide gel containing fractions obtained from the HPLC run presented in Figure 2-3. Fractions 53–57 (mZP1), 60–65 (mZP2), and 67–75 (mZP3) were each pooled and then rerun by SDS–PAGE. The yield of mZP3 in this experiment was about 300 ng per mouse (~150 ng per ovary; ~50 ZP).

phoresis, the minigel is stained with silver. Briefly, the minigel is fixed with 30% 2-propanol/10% acetic acid for 30 minutes (or for up to 16 hours), washed three times with 10% 2-propanol (7 minutes each), and rinsed extensively with DW. The gel is bathed for 15 minutes in 100 ml of DW containing 4 µl of 1 M dithiothreitol (DTT) and rinsed extensively with DW. The gel is then stained for 15 minutes with a solution made up of 1 ml of 10% silver nitrate in 100 ml DW, rinsed extensively with DW, and developed in a solution containing 75 µl of 37% formaldehyde in 150 ml of 3% sodium carbonate (gel is rinsed once and then incubated in the solution until stained). Development is terminated by addition of 1% acetic acid.

Recovery and Dialysis of ZP Glycoproteins

The HPLC fractions containing mZP1, mZP2, and mZP3 are pooled and the amount of each glycoprotein is determined spectrophotometrically at 280 nm. In general, 100 ovaries (50 mice) will yield ~5 µg of mZP1, ~25 µg of mZP2, and ~15 µg of mZP3. The individual glycoproteins are then dialyzed in precut, sterile dialysis bags (Spectrum; Spectra/Por CE) having a molecular weight cutoff of about 30 kDa. The first dialysis is against 8 M urea for 24 hours at RT, followed by extensive dialysis (72–96 hours) against DW, day 1 at RT and thereafter at 4°C. Finally, the samples are lyophilized and stored at −80°C prior to use.

In Vitro Assay of Sperm Receptor Activity

The sperm receptor activity of mZP3 is assessed by its ability to inhibit binding of capacitated sperm to ovulated eggs in vitro; that is, sperm receptor activity is assessed in a so-called 'competition assay' (Bleil and Wassarman 1980; Wassarman 1990). Briefly, sperm are incubated in the presence or absence of mZP3, and then eggs and embryos (either fixed or unfixed) are added to the culture medium. After a suitable time, eggs and embryos with ZP-associated sperm are removed from the medium, fixed, and examined under an inverted microscope. The number of sperm associated with each egg in the largest focal plane is determined for each set of conditions and the percent inhibition of sperm binding is calculated. The presence of embryos is essential in this assay since they serve as a negative control; sperm can attach but not bind to the ZP of fertilized eggs or embryos due to completion of the zona reaction.

Conditions of Cell Culture

All cell culture (sperm, oocytes, eggs, and embryos) is carried out in sterile-filtered M199-M [Earle's medium 199 (Gibco BRL; M199) containing L-glutamine, 2.2 mg/liter sodium bicarbonate, 25 mM HEPES, and supplemented with 4 mg/ml bovine serum albumin (BSA; Sigma; fraction V, A-4503) and 30 µg/ml pyruvate (Gibco BRL)] at 37°C in an incubator in a humidified atmosphere of 5% CO_2 in air, unless otherwise indicated. The M199-M should be stored no longer than 7–10 days and then a fresh batch should be used. Drops of medium are covered with oil (225 ml light mineral oil combined with 25 ml Earle's medium 199, autoclaved, and stored in the incubator at least 2 days prior to use) in sterile tissue culture dishes (Corning; 6 mm diameter) and maintained in the incubator throughout the experiment, unless otherwise indicated. Oocytes, eggs, and embryos are transferred from one droplet to another under a dissecting microscope (we use Wild stereo microscopes) by using mouth-operated, medium-bore (drawn to about 150 µm internal diameter) Pasteur pipettes connected to a syringe filter (Gelman Sciences). Sperm are transferred by using an automatic micropipette.

Isolation, Capacitation, and Culture of Sperm

Cauda epididymes and vasa deferentia are excised from sexually mature (≥ 3 months old) male mice and placed into a tissue culture dish containing 5 ml of M199-M/4 mM ethylene glycol tetraacetic acid (EGTA), pH 7.4. Sperm from the vasa deferentia are squeezed into the medium by using a sterile, bent steel needle and sperm from the cauda epididymes are released into the medium by puncturing and gently squeezing the tissue with a sterile steel needle (Becton Dickinson; Precision Glide Needle, 18G $\times \frac{1}{2}''$). Sperm are combined in a 10-ml conical polypropylene tube and pelleted by low-speed centrifugation for 4–5 minutes at RT. The supernatant is discarded and 5 ml of fresh M199-M, without EGTA, is added carefully on top of the sperm pellet without disturbing the pellet. Sperm are capacitated for 1 hour in the incubator at 37°C. During this time, the live sperm become increasingly motile and will "swim up" into the medium ("swim-up" sperm). Final sperm concentrations will vary depending upon the particular male mouse, but on average it is about 10^5–10^6 sperm per milliliter (based on 5 ml of supernatant from one male).

Isolation and Culture of Oocytes, Ovulated Eggs, and Two-Cell Embryos

Oocytes To obtain growing and fully grown oocytes, ovaries are excised from juvenile (≤ 3 weeks old; for growing oocytes) or adult (≥ 6 weeks old; for fully grown oocytes) female mice, placed in a tissue culture dish containing 5 ml of M199-M, and fat and connective tissue are removed manually. The ovaries are poked extensively with sterile steel needles (Becton Dickinson; 30G × ½″) under a dissecting microscope. Released oocytes are then washed through three 100-μl drops of M199-M. Typically, 10–15 fully grown oocytes can be recovered from a single ovary.

Ovulated Eggs To obtain ovulated eggs, 5 IU (international unit) of pregnant mare's serum (PMS; Sigma) is injected intraperitoneally, preferably in the evening of day 1, followed 46–48 hours later by the injection of 5 IU of human chorionic gonadotropin (hCG; day 3; Sigma). On the morning of day 4 (16–18 hours after injection of hCG), oviducts are excised and placed in a tissue culture dish containing 5 ml of M199-M. The upper portion of excised oviducts (where ovulated eggs are located) is gently torn apart by using two pairs of fine forceps (watchmaker's forceps; Peer no. 5) under a dissecting microscope. This results in the release of eggs surrounded by cumulus cells (cumulus mass). Cumulus masses are collected by using a mouth-operated, wide-bore Pasteur pipette (not drawn) and are treated with hyaluronidase (Sigma; 1 mg/ml) in a 200 μl drop of M199-M for several minutes. Generally, such treatment results in removal of cumulus cells from the ovulated eggs, but some up-and-down pipetting may be required in addition. The eggs are washed through three 100-μl drops of M199-M to remove any hyaluronidase. Approximately 250–350 ovulated eggs can be recovered from 10 superovulated, 6–8-week old female CD-1 mice (~25–35 eggs/mouse).

Two-Cell Embryos To obtain two-cell embryos, 5 IU of PMS is injected intraperitoneally, preferably in the afternoon of day 1, followed 46–48 hours later by the injection of 5 IU of hCG (day 3). At the time of hCG injection, the female is caged overnight with a proven male breeder and checked the following morning (day 4) for a copulation plug (indicative of a successful mating). On the morning of day 5, oviducts are excised and placed in M199-M, as described above. We use two methods to remove embryos from excised oviducts. In one procedure, a 1-ml syringe equipped with a sterile steel needle (Becton Dickinson; 30G × ½″) with a blunt tip is inserted into the oviduct opening and embryos are flushed out with M199-M under a dissecting microscope. In the other procedure, oviducts are torn at several points along their length with watchmaker's forceps and embryos are released into the medium under a dissecting microscope. Isolated two-cell embryos are then washed through three 100-μl drops of M199-M. Approximately 15–30 two-cell embryos can be obtained from one superovulated, 6–8-week old, female CD-1 mouse.

Storage of Oocytes, Ovulated Eggs, and Two-Cell Embryos

For storage for up to 2 weeks, oocytes, eggs, and embryos are washed through three 100-μl drops of phosphate-buffered saline/polyvinylpyrrolidone (PBS/PVP 30 mM sodium phosphate, pH 7.2, 150 mM NaCl, 4 mg/ml PVP-40, 0.02% sodium azide) at RT prior to fixation for 1 hour in a 100-μl drop of PBS/PVP/1% paraformaldehyde at RT. Following fixation, the cells are washed through three 100-μl drops of PBS/PVP at RT and are transferred to a 100-μl drop of storage buffer (M199 containing 50 mM Tris-Cl, pH 7.5, 4 mg/ml PVP-40, 0.02% sodium azide) for storage at 4°C.

Incubation of Sperm with Ovulated Eggs and Two-Cell Embryos

Ovulated eggs (or fully grown oocytes) and two-cell embryos, stored in storage buffer at 4°C, are washed through three 100-μl drops of M199-M at RT and transferred to the incubator (37°C, 5% CO_2 in air, humidified) for at least 30 minutes before being added to capacitated sperm. Ten-microliter drops of M199-M are pipetted into a tissue culture dish and covered with oil. Alternatively, if a test substance is being evaluated, the sample is dissolved in 2 μl of DW to which 8 μl of M199-M is added, and it is placed as a drop in a tissue culture dish and covered with oil. Then 10 μl of capacitated "swim-up" sperm is added to the drop (1:1 dilution) and the sample is incubated for 15 minutes. Following this period, eggs and embryos are added to the drop, with as little M199-M as possible, and the sample is incubated for an additional 35–40 minutes. Sperm motility is monitored continuously during this period and preparations with less than 70% motile sperm should be discarded. At the end of the incubation, eggs and embryos with associated sperm are removed from the drops with a mouth-operated, wide-bore Pasteur pipette (150–200 μm internal diameter) and are washed through three 100-μl drops of M199-M. The wash should involve extensive up-and-down pipetting, until no more than 1 or 2 sperm remain attached to the two-cell embryos. Under these conditions, sperm bound to the egg ZP remain bound, whereas sperm nonspecifically attached to egg and embryo ZP are removed. Eggs and embryos with bound sperm are then transferred to 10-μl drops of M199-M containing PBS/PVP/3% glutaraldehyde (1:1 by volume) covered with oil.

Analysis of Inhibition of Sperm Binding to Eggs ("Receptor Activity")

The number of sperm bound per egg is determined by counting sperm tails in one plane of focus using dark-field microscopy and 200× magnification (we use a Zeiss IM35 inverted microscope) (Figure 2-5). Typically, there are 12 ovulated eggs (or fully grown oocytes) and 2 two-cell embryos present per sample. After determining the number of sperm bound to each egg, the highest and lowest values are discarded, and 10 eggs are used to calculate the average number of sperm bound per egg (±SD). In control samples (sperm exposed to M199-M alone), at sperm concentrations of about 5×10^5 sperm per milliliter, there are 30–50 sperm bound per egg at the largest focal plane of the egg.

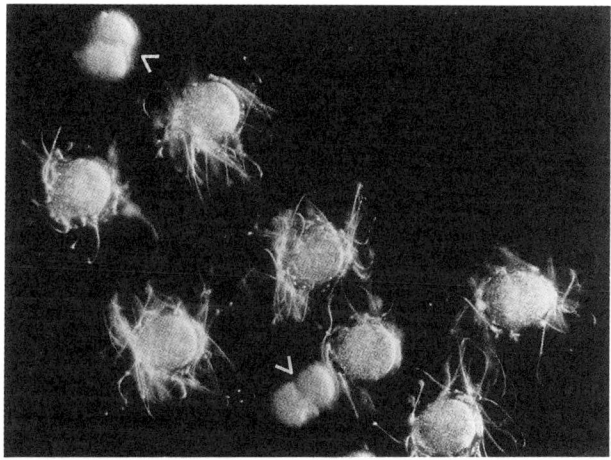

Figure 2-5 Light photomicrograph (dark-field) of sperm, eggs, and embryos at the end of a competition assay. Shown are sperm bound to 7 ovulated eggs, but not bound to 2 two-cell embryos (indicated by white arrowheads). The number of sperm bound to each egg is determined for the largest focal plane of the egg in order to calculate the percent inhibition of sperm binding by a test substance (e.g., mZP3).

In Vitro Assay of Acrosome Reaction-Inducing Activity

The acrosome is a large lysosome-like vesicle overlying the sperm nucleus. Capacitated mouse sperm undergo the acrosome reaction (AR) shortly after binding to the egg ZP or to mZP3 purified from oocytes or unfertilized eggs (Wassarman 1990; Yanagimachi 1994). The AR is a specialized form of cellular exocytosis restricted to sperm. The reaction involves multiple fusions between plasma membrane overlying the anterior region of the sperm head and the outer acrosomal membrane lying just beneath the plasma membrane. Small hybrid membrane vesicles form and the inner acrosomal membrane, overlying the sperm nucleus, is exposed to the egg ZP. Only acrosome-reacted sperm can penetrate the ZP, reach the plasma membrane, and fuse with (fertilize) the egg.

Analysis of Acrosome Reaction-Inducing Activity

To assess the state of the sperm acrosome, capacitated "swim-up" sperm (see above) are incubated, either in 20–50-μl drops under oil or in 0.5-ml Eppendorf tubes, for 1 hour in M199-M in the presence or absence of a test substance (e.g., mZP3). Calcium ionophore A23187 [Calbiochem; 40 mM stock solution made up in dimethylsulfoxide (DMSO) and diluted 1:40 with PBS], made up to 10 μM final concentration in M199-M, is used as a positive control since it induces sperm to undergo the AR in vitro. In this context, it should be noted that sperm incubated in A23187 become entirely immotile in a short time. Sperm are collected by centrifugation at about 5000 rpm for 5 minutes in a microfuge and are fixed with 5% formaldehyde/PBS overnight at 4°C. The following day, sperm are washed by centrifugation (as before) in 0.1 M ammonium acetate, pH 9.0, resuspended in 50 μl of 0.1 M ammonium acetate by gentle vortexing, and air-dried onto gelatin-coated (0.3% gelatin/0.01% chromium potassium sulfate) glass slides. The slides are washed with DW, methanol, and then DW for 5 minutes each, and are stained with 0.04% Coomassie brilliant blue G-250, made up in 3.5% perchloric acid, for 5 minutes and washed extensively with DW. Sperm can be observed directly or under a coverslip, using 30% glycerol/PBS as mounting medium, and are scored for the presence or absence of an acrosome by bright-field microscopy at 400–1000× magnification (we use a Zeiss IM35 and a Zeiss Universal microscope). Sperm retaining an intact acrosome will display a continuous blue-stained ridge overlying the sperm head (Figure 2-6). Sperm that have undergone the AR will display either a "patchy" blue pattern, indicative of a partial AR, or no staining at all, indicative of a completed AR (Figure 2-6). Typically, A23187-treated sperm samples will be 60–80% acrosome-reacted compared with control sperm, exposed to M199-M alone, that will be 15–25% acrosome-reacted (so-called "spontaneous AR").

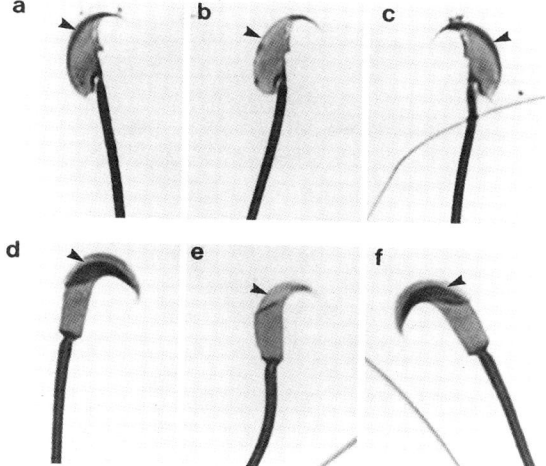

Figure 2-6 Light photomicrographs of acrosome-intact and acrosome-reacted mouse and hamster sperm. Capacitated (a–c) mouse and (d–f) hamster sperm were stained with Coomassie brilliant blue G-250, fixed, and examined by light microscopy, essentially as described here. Shown are photomicrographs of representative examples of (a, c, d, and f) acrosome-intact and (b and e) acrosome-reacted sperm. Magnification 750×.

ACKNOWLEDGMENTS The methodology described here either evolved over the years in our laboratory or was adopted, with and without modifications, from the literature. In our own laboratory, several talented investigators contributed to the development of this methodology; these include Drs. Jeffrey Bleil, Harvey Florman, Jeffrey Greve, Ross Kinloch, Christopher Moller, and Thomas Rosiere, and Mr. Steven Mortillo. We are grateful to them all for helping to make this chapter possible.

References

Bleil, J. D., and Wassarman, P. M. (1980). Mammalian sperm–egg interaction: identification of a glycoprotein in mouse egg zonae pellucidae possessing receptor activity for sperm. Cell 20:873 882.

Bleil, J. D., and Wassarman, P.M. (1986). Autoradiographic visualization of the mouse egg's sperm receptor bound to sperm. J. Cell Biol. 102:1363–1369.

Dietl, J., ed. (1989). *The Mammalian Egg Coat: Structure and Function.* Berlin, Springer-Verlag.

Florman, H. M., and Wassarman, P. M. (1985). O-Linked oligosaccharides of mouse egg ZP3 account for its sperm receptor activity. Cell 41:313–324.

Gwatkin, R. B. L. (1977). *Fertilization Mechanisms in Man and Mammals.* New York, Plenum Press.

Hogan, B., Beddington, R., Costantini, F., and Lacy, E. (1994). *Manipulating the Mouse Embryo: A Laboratory Manual,* 2nd ed. Cold Spring Harbor, NY, Cold Spring Harbor Laboratory Press.

Kinloch, R. A., Sakai, Y., and Wassarman, P. M. (1995). Mapping the mouse ZP3 combining site for sperm by exon swapping and site-directed mutagenesis. Proc. Natl. Acad. Sci. USA 92:263–267.

Litscher, E. S., and Wassarman, P. M. (1996). Characterization of a mouse ZP3-derived glycopeptide, gp55, that exhibits sperm receptor and acrosome reaction-inducing activity *in vitro*. Biochemistry 35:3980–3986.

Liu, C., Litscher, E. S., Mortillo, S., Sakai, Y., Kinloch, R. A., Stewart, C. L., and Wassarman, P.M. (1996). Targeted disruption of the *mZP3* gene results in production of eggs lacking a zona pellucida and infertility in female mice. Proc. Natl. Acad. Sci., USA 93:5431–5436.

Moller, C. C., Bleil, J. D., Kinloch, R. A., and Wassarman, P. M. (1990). Structural and functional relationships between mouse and hamster zona pellucida glycoproteins. Dev. Biol. 137:276–288.

Mortillo, S., and Wassarman, P.M. (1991). Differential binding of gold-labeled zona pellucida glycoproteins mZP2 and mZP3 to mouse sperm membrane compartments. Development 113:141–149.

Rafferty, K. A., Jr. (1970). *Methods in Experimental Embryology of the Mouse*. Baltimore, Johns Hopkins Press.

Rankin, T., Familiari, M., Lee, E., Ginsberg, A., Dwyer, N., Blanchette-Mackie, J., Drago, J., Westphal, H., and Dean, J. (1996). Mice homozygous for an insertional mutation in the *ZP3* gene lack a zona pellucida and are infertile. Development 122:2903–2910.

Rosiere, T. K., and Wassarman, P. M. (1992). Identification of a region of mouse zona pellucida glycoprotein mZP3 that possesses sperm receptor activity. Dev. Biol. 154:937–944.

Snell, W. J., and White, J. M. (1996). The molecules of mammalian fertilization. Cell 85:629–637.

Wassarman, P. M. (1988). Zona pellucida glycoproteins. Annu. Rev. Biochem. 57:415–442.

Wassarman, P. M. (1990). Profile of a mammalian sperm receptor. Development 108:1–17.

Wassarman, P. M. (1995). Towards molecular mechanisms for gamete adhesion and fusion during mammalian fertilization. Curr. Opin. Cell Biol. 7:658–664.

Wassarman, P. M. (1998). Egg zona pellucida glycoproteins. In *Guidebook to the Extracellular Matrix and Adhesion Proteins*, 2nd ed., Kreis, T., and Vale, R., eds. Oxford, Oxford University Press, in press.

Wassarman, P. M., and DePamphilis, M. L., eds. (1993). Guide to techniques in mouse development. Methods Enzymol. 125:1–1021.

Wassarman, P. M., Liu, C., and Litscher, E.S. (1996). Constructing the mammalian egg zona pellucida: some new pieces of an old puzzle. J. Cell Sci. 109:2001–2004.

Wassarman, P. M., Qi, H., and E. S. Litscher (1997). Female mice carry a single *mZP3* allele produce eggs with a thin zona pellucida, bur reproduce normally. Proc. R. Soc., Biol. Sci. 264:323–328.

Yanagimachi, R. (1994). Mammalian fertilization. In *The Physiology of Reproduction*, Vol. 1, Knobil, E., and Neill, J. D., eds. New York, Raven Press, pp. 189–317.

3

Methods for Analysis of Cytoplasmic Signal Transduction in Mammalian Eggs and Embryos

Christine M. Pauken
David G. Capco

Fertilization-competent eggs represent a special type of cell that has no comparable cellular counterpart in the full-grown organism. Eggs are cells that upon fertilization by the penetrating sperm, undergo an extensive and orderly change in their engineering. This results in the rescue of the cell from certain death by initiating a pathway of biochemical, molecular, and structural changes that create an entirely new individual. These changes lead to the process of embryogenesis which itself represents a special state as the cells of the embryo, referred to as blastomeres, turn on (and off) genes that may not again be engaged during the life of the organism. Additionally, the cells exhibit functions and activities related to the polarity of the organism that sets up the body plan of the organism. Throughout this time span, the egg and blastomeres of the embryos must function at a cellular level, as would any somatic cell, but these cells must also surmount the additional challenges imposed on normal cellular function by the requirement to unfold the developmental program of fertilization and embryogenesis. Specializations of cellular function exist in eggs and blastomeres to permit them to accomplish the additional challenges presented by the developmental program. These specializations highlight the structure–function (i.e., engineering) relationships present in any cell, but the details of the engineering, as well as the biochemical events regulating the changes in engineering, are highly choreographed in both time and space.

For reasons such as those discussed above, eggs and early embryos are often used for studies of complex interactions, e.g., cell-cycle regulation, mechanisms of calcium wave propagation (Schultz and Kopf 1995; Swann and Lai 1997), signal transduction via various kinases (Gallicano et al. 1997), initiation of cellular polarity (Collins and Fleming 1995), cell–cell contact and adhesion (Collins and Fleming 1995; Fagotto and Gumbiner 1996; Sastry and Horwitz, 1996), and gap-junctional communication (Lo 1996).

Eggs of a variety of different species have been employed for such studies with those from different classes of organisms having different advantages and disadvantages. For example, amphibia provide very large eggs (i.e., 1.4 mm diameter in the case of *Xenopus laevis*) and, as such, a great deal of cytoplasmic mass can be provided for study from a single egg. In contrast, invertebrate eggs, e.g., those of a sea urchin, individually provide a small amount of cytoplasmic mass per egg, but large numbers of eggs can be harvested from a single sea urchin. Mammalian eggs and embryos are technically difficult to work with for two reasons: first, the eggs are small, about 80 μm in diameter, and second, only 20–30 eggs can be obtained per female under optimal conditions. Moreover, development occurs internally in mammals, further complicating analysis. Consequently, there is little material for analysis, and the loss of even a small number of eggs during handling can significantly decrease the amount of material for study. Yet mammalian eggs serve as the best model for analysis of human development since only in mammals are seen the developmental transitions of compaction, development of inside–outside polarity, the embryonic blastocyst, uterine implantation, egg cylinder development, and placental development. This chapter will present methods for analysis of mammalian eggs and embryos and provide a means of examining the structure–function relationship in these specialized cells.

General Handling Procedures

Manipulation of eggs or embryos surrounded by the zona pellucida (an extracellular matrix around mammalian eggs and preimplantation embryos) can be carried out by pipetting them using a finely pulled pipette that is attached to a vapor filter that is connected by a piece of flexible tubing to a mouthpiece through which suction or pressure is applied (see Hogan et al. 1986). However, mouth pipetting would be hazardous when employing pharmacological agents, radioactive materials, or when contamination is a concern. Consequently, in place of the mouthpiece, we use a tuberculin syringe attached to the flexible tubing. Pasteur pipettes are pulled to the proper diameter in the flame of an ethanol lamp or Bunsen burner. Careful consideration of, and full knowledge of, the constitution of the media used for harvesting the eggs and embryos cannot be overstated. The goal of the media is to mimic as closely as possible oviductal or uterine conditions, especially with regard to ionic environment, osmolarity, and pH. We typically use two types of media for the culture of mouse embryos: 2A-HEPES (McGaughey 1977) and KSOM (Lawitts and Biggers 1992; Erbach et al. 1995; Ho et al. 1995; Summers et al. 1995). The principle difference between the two is that 2A utilizes glucose as the energy source while KSOM utilizes L-glutamine as well as glucose, plus it has a lower sodium chloride concentration and is supplemented with amino acids (Lawitts and Biggers 1992; Erbach et al. 1995; Ho et al. 1995; Summers et al. 1995). Additionally, a media unique for work with hamster eggs and early embryos has been developed (McKiernan et al. 1991, 1995).

Procurement of Eggs and Embryos

An outbred mouse strain (for example, CD-1) is used routinely in this laboratory and is maintained under a 14-hour light/10-hour dark schedule. In order to increase the numbers of eggs or preimplantation embryos available for an experiment, gonadotropins are administered to the females to induce superovulation. Pregnant mare serum gonadotropin (PMSG), 5.0 IU in a 0.1-ml volume of 0.1 M sodium phosphate is injected intraperitoneally (ip) in the late afternoon and then, 42–48 hours later, 5.0 IU human chorionic gonadotropin (hCG) is injected.

Unfertilized eggs are obtained from the females the next morning, approximately 16–18 hours after hCG injection. To remove the oviduct from the female mouse, first gently cut away the mesometrium while holding the uterine horn near the oviduct and ovary, then remove the fat that allows the ovary to adhere to the abdominal viscera. Cut between the oviduct and ovary and then cut at the top of the uterine horn to remove the oviduct. The cumulus mass (the eggs and accompanying cumulus cells represent the cumulus mass) will be present in the region of the oviduct that was closest to the ovary expanding the oviduct noticeably if they are healthy and collected at a time that would permit successful fertilization. The masses can be removed by scoring the oviduct on one side of the mass with a syringe needle and gently teasing them out. Alternatively, a 30G or 26G needle (attached to a tuberculin syringe filled with media), can be carefully inserted in the ostium, a cuff-like structure of the oviduct usually located next to the ovary, then the media is flushed through the oviduct so that the cumulus masses will come out of the opposite end. Cumulus masses are collected by the use of an unpulled Pasteur pipette and placed into a fresh plate containing KSOM media with 300 µg/ml hyaluronidase. Incubation of the cumulus masses in the media with hyaluronidase for about 5 minutes will remove the cumulus cells, a process that should be monitored under the dissecting scope. The denuded eggs are then washed through several droplets of fresh KSOM before use. As the egg ages, the cumulus cells are shed; thus, eggs with only a slight association of cumulus cells may be too old and undergo parthenogenic activation on the way to cell death.

For fertilized eggs and embryos, a female is placed with a male for copulation after hCG injection. The next morning, the female is checked for a white copulation (vaginal) plug to indicate that copulation has occurred. Fertilization usually occurs about midnight and the day the plug is detected is referred to as 0.5 days post coitum (dpc). Two-cell embryos are collected about noon on 1.5 dpc and four-cell embryos about 8–12 hours later. Eight-cell embryos are retrieved on the morning of 2.5 dpc with morula being present in the evening. Two-cell through morula stage embryos are most efficiently collected by flushing the oviducts with approximately 0.1 ml of media, and then transferring them to a dish of fresh KSOM media using a Pasteur pipette pulled to the proper diameter. Flushing the oviducts is more efficient than pulling them apart since, by this time, the cumulus cells have sloughed off the embryos and, consequently, the presence of embryos is difficult to detect.

On the morning of 3.5 dpc, the blastocysts enter the uterine horns in order to implant, so at this stage the uterine horns are removed intact with the oviducts still attached. The 26G needle is inserted into a horn at the cervical end and media is injected to expand the horn completely. The end near the oviduct is then quickly cut so that the media rushes out of the horn, flushing out any blastocysts. In this procedure, the oviducts also are flushed since the blastocysts may be in either or both locations.

Cautions Embryos that have a very grainy texture, are pale, almost ghost-like, have unevenly sized blastomeres, or have an inappropriate number of blastomeres for the stage expected are abnormal and should not be used. Also, when working with unfertilized eggs, release of eggs from meiotic metaphase II arrest occurs in a higher percentage of eggs at 16–18 hours post-hCG than with eggs retrieved at less than 14 hours post-hCG (Moses and Kline 1995; Moses and Masui 1995; Winston and Maro 1995).

Immobilization of Specimens for Living or Fixed Eggs/ Embryos

Eggs surrounded by the zona pellucida can be immobilized by attaching them to slides or coverslips coated with polylysine.

1. For this procedure, the slide or coverslip is thoroughly cleaned (sonicated three times for 5 minutes each in acetone, then sonicated three times for 5 minutes each in distilled water, followed by air-drying). A droplet (30 ml) of 0.1% polylysine (MW 150,000–300,000) is allowed to dry on the slide/coverslip on a slide warmer (40°C), covered to protect it from the dust. Immediately before use, the slide/coverslip is rinsed in a medium that is free of protein and/or amino acids (e.g., PBS).

2. Next, the eggs/embryos, prewashed in a protein/amino-acid-free medium, are added to the slide/coverslip and permitted to sit for at least 30 seconds without movement. At this point, the eggs/embryos adhere to the slide/coverslip.
3. If the specimen is to be cytologically fixed it is done at this point in a two-step procedure. A pulled pipette containing fixative is placed immediately above the eggs/embryos in the medium and the fixative is allowed to exude from the pipette. The fixative has higher density than the surrounding medium so it will settle around the specimens and further adhere them to the polylysine-coated substrate. The medium on the slide can then be removed and replaced with more fixative for thorough preservation. This procedure is necessary to ensure tight adhesion, otherwise replacement of medium with a large volume of fixative might shear the eggs off of the substrate due to viscosity differences between the media.
4. As an alternative to cytological fixation, living eggs can be processed into different solutions while immobilized on the polylysine-coated coverslips.

Kinase Assay

Kinases act to alter the engineering of a cell in short-term response to a stimuli. To investigate the role of a kinase, one needs to determine when the kinase is active and where in the cell it is active. The development of reporter dyes for some kinases permit tracking of the kinase to different locations in the cell and can provide an indication of the state of activity of the kinase. For example, RIM-1 and FIM-1 (i.e., bisindolylmaleimide conjugated to rhodamine or fluorescein, respectively) permit detection of the location of protein kinase C (PKC), and, when the common isotypes of PKC translocate to membranes, this is an indication of the activation state of the kinase (Chen and Poenie 1993). In addition, a reporter dye of protein kinase A (PKA) is referred to as "FlCRhR" and can detect the position and state of activity of this kinase (Sammak et al. 1992). While such dyes are useful in providing information about kinase activity wherever possible, biochemical confirmation of the state of activity of the kinase should be conducted even when employing the small numbers of eggs obtained from mammals.

Procedure for Localizing PKC with RIM-1

Protein kinase Cs can be localized by the use of the fluorescent PKC reporter dye, RIM-1 (Chen and Poenie 1993; Gallicano et al. 1995), and this procedure is described here for the examination of unfertilized and activated eggs. This reporter dye binds at the ATP binding site in the kinase (Chen and Poenie 1993) and can be detected by observation with an epifluorescent microscope using rhodamine settings.

1. Isolate eggs as detailed in the earlier section. The following steps are accomplished most efficiently in a multiwell glass plate. Cytologically fix eggs/embryos in 0.05% glutaraldehyde/2% paraformaldehyde in PBS for 5 minutes

at room temperature. Pipette eggs/embryos rapidly up and down several times after transfer to ensure maximal exposure to the fixative.
2. The eggs are then permeabilized by placing them in 0.1% Tween 20/2% paraformaldehyde in PBS for 10 minutes at room temperature.
3. Wash the eggs in ICB (ICB; 100 mM KCl, 5 mM $MgCl_2$, 3 mM EGTA, 20 mM HEPES, pH 6.8) three times for 20 minutes each at room temperature.
4. Stain eggs in 200 nM RIM-1 (Molecular Probes, Eugene, OR) in ICB for 30 minutes at room temperature in the dark.
5. Wash in ICB 3 times for 30 minutes each. Then mount the eggs on a poly-lysine-coated slide and apply a coverslip as detailed above. Seal coverslip to slide with nail polish.
6. The RIM-1-stained eggs can be viewed on an epifluorescent or confocal microscope using the rhodamine settings to detect RIM-1 or the fluorescein settings to detect FIM-1. Figure 3-1 shows (A) an unactivated mouse egg and (B) an egg 20 minutes after activation with A23187, stained with RIM-1, and viewed in a confocal microscope to demonstrate PKC distribution. The un-activated egg exhibits a relatively uniform distribution of PKC while the activated egg shows an enrichment near the plasma membrane and in some internal sites.

Procedure for Biochemical Kinase Assay

The activity of PKC can be assayed at a biochemical level by providing both a substrate and radiolabeled ATP and measuring phosphorylation of this substrate.

1. Three eggs are lysed in 2 µl of collection buffer (phosphate-buffered saline containing 1 mg/ml polyvinyl alcohol, 5 mM EDTA, 10 mM Na_3VO_4, and 10 mM NaFl) in an Eppendorf tube and flash-frozen in liquid N_2 to prepare a cell lysate.
    The following reagents are added, in order, to an Eppendorf tube to give a final volume of 10 µl:
2. 100 mM KCl, 5 mM $MgCl_2$, 10 mM EGTA, 20 mM HEPES (pH 7.6) to adjust ionic conditions, pH, and chelate calcium;
3. 54 mM β-glycerophosphate and 14.5 mM *p*-nitrophenylphosphate as phosphatase inhibitors;
4. 1 µg/ml of each of the following protease inhibitors—pepstatin, aprotinin, chymostatin, leupeptin, and trypsin-chymotrypsin inhibitor;
5. The following kinase inhibitors—2.4 µM PKI, 10 µM ML-9, 75 µM Genistein, to inhibit protein kinase A, myosin light-chain kinase, and tyrosine kinase, respectively;
6. 5 µCi of $\delta\text{-}^{32}P$-ATP (S.A. 6000 Ci/mM, Amersham Corp., Redivue™);
7. 5 µg of MARCKS peptide (a.a. 151–175) as the peptide substrate;
8. 1.5 µl of cell lysate which contains the presumptive PKC.

The kinase assay is reacted for 30 minutes at 37°C and is stopped by the addition of an equal volume of 2× concentrated SDS–sample buffer for elec-

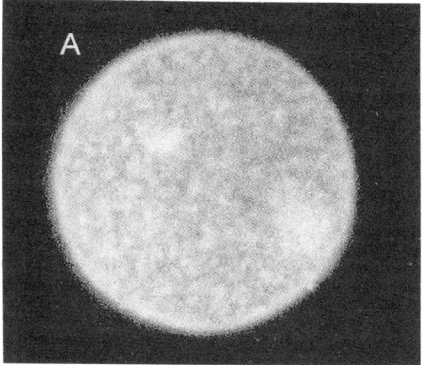 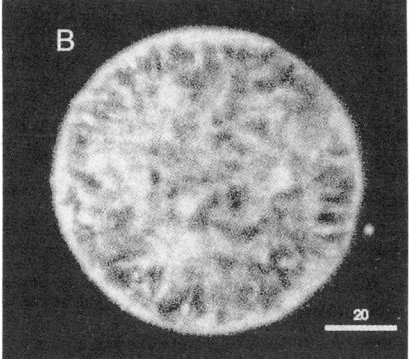

Figure 3-1 Redistribution of PKC after egg activation viewed with a confocal microscope: a stack of eight 0.5-μm optical sections of (A) an unfertilized egg and (B) an egg 20 minutes after activation with calcium ionophore (i.e., A23187). The unactivated egg contains a relatively uniform distribution of PKC, while the egg 20 minutes after activation contains an enrichment of PKC near the plasma membrane (unpublished results).

trophoresis. The MARCKS peptide can be viewed in 15% SDS–polyacrylamide gels after autoradiography. As a control, an aliquot of the same cell lysate is incubated in a reaction mixture identical to that described for the experimental mixture except that this reaction mixture is modified to contain both of the following PKC inhibitors: 200 μM PKCψ (a.a. 19–36) and 2 μM BIM. These two PKC inhibitors should suppress any phosphorylation due to PKC; consequently, if any phosphorylation of MARCKS peptide is present in the control lanes it suggests the activity of a kinase, other than PKC, that can phosphorylate MARCKS peptide. The two structurally unrelated PKC inhibitors PKCψ and BIM are highly specific for PKC. While PKCψ is a peptide inhibitor which contains the amino acids that compose the pseudosubstrate domain of PKC, BIM is a pharmacologic inhibitor that attaches to PKC at near the ATP binding site on the kinase (Davis et al. 1989). When these PKC inhibitors are used they should be preincubated with the lysate (or reaction mixture without substrate) for 15 minutes to permit binding of the inhibitor to PKC.

Cautions EGTA is present in the reaction mixture to chelate calcium when it is desired to measure the endogenous levels of PKC activity. This is because once PKC is activated it no longer requires calcium to remain active (Ashendel 1985). Since homogenization of the cells is likely to release calcium (albeit in varying levels in different homogenates) from sequestered stores, a lysate made in the absence of a calcium chelator would be likely to artifactually activate more PKC than was active just prior to homogenization. When it is desired to measure the total amount of PKC, then a specific amount of calcium in the form of calcium chloride can be added to the homogenate. Many of the commercially available kits are designed to quantify total PKC and recommend addition of calcium, rather than measuring the endogenous level of PKC activity. These commercial kits can be modified to measure endogenous PKC activity typically by removing from their protocol the addition of cofactors (usually calcium and lipid), and by the addition of a calcium chelator.

In our procedure, we employ MARCKS peptide (a.a.151–175) which is the most specific PKC substrate available (Blackshear 1993; Allen and Aderem 1995). More commonly, in the literature the reader will find the use of histone, myelin basic protein (MBP) or MARCKS protein as the substrate for PKC. These proteins have multiple phosphorylation sites on them and can be phosphorylated by a variety of kinases (Gallicano et al. 1997. Consequently, we have chosen MARCKS peptide which contains only the consensus substrate sequence for PKC and, in addition, our reaction mixture contains inhibitors to other kinases. Protease inhibitors such as AEBSF or PMSF are not recommended for these reactions as they can reduce enzymatic activity.

Substrates The use of specific substrates in the form of peptides containing the consensus phosphorylation site for different kinases and phosphatases is essential for the proper recognition of specific signaling agents. Many natural substrates contain phosphorylation sites for two or more kinases; consequently, phosphorylation of such an exogenous substrate, in a kinase assay conducted using a cell lysate, could provide misleading results. For example, most assays conducted on MPF activity monitor the phosphorylation of histone H1, while those conducted on MAP kinase employ MBP. While MAP kinase does not significantly phosphorylate histones (Payne et al. 1991), MPF does phosphorylate MBP (Shibuya et al., 1992). Significantly, both histone and MBP are substrates for PKC (Kishimoto et al. 1985: Walten et al. 1987). Consequently, some investigators may erroneously attribute kinase activity to MPF (or MAP kinase) when the activity really results from PKC. One way to circumvent this problem would be to determine the precise amino acid residue that is phosphorylated, but this can be time consuming. The peptide substrates (containing the consensus a.a. residues for specific kinases) provide a more rapid and precise measurement, particularly when the oligopeptide is large enough to remain in a separating gel or can be separated by some other mechanism. Even with the use of the peptide substrates, there is always the possibility that an as yet unidentified kinase is acting on the substrate. Consequently, an important control, which unfortunately is rarely employed by investigators, is to add a specific kinase inhibitor to a parallel reaction tube. Comparison of the intensity of signal between the experimental and control reaction mixture gives a measure of the amount of phosphorylation due to a specific kinase.

Kinase Inhibitors The recent development of peptide inhibitors allows much more specific inhibition of kinases than the more traditional drug inhibitors. Drugs often are used as inhibitors of various kinases, and clearly there is always the concern that drugs may have an unanticipated effect or act through an alternate pathway. With rare exceptions, few investigators have ever measured whether the kinase inhibitor drugs actually enter the cell and at what rate. A drug referred to as H-8 is an exception as antibodies to H-8 have been used to titer the intracellular concentration over time. It has been shown that an incubation time of 1 hour results in equilibration of the extracellular and intracellular concentration of H-8 (Hidaka et al. 1984). In an attempt to circumvent the problem of pleiotropic effects of drug inhibitors, investigators should employ two or more structurally different inhibitors for each kinase and assume that a common effect between different drugs is due to the kinase inhibited. Table 3-1 lists a variety of kinase inhibitors and concentrations reported for use in cell lysates or in vivo to inhibit the action of specific kinases. Ideally, investigators should carry this approach one step further and test in a biochemical assay whether the inhibitor inhibits the desired kinase. The peptide inhibitors are oligopeptides that fit into, and inhibit, the catalytic site of specific kinases. The amino acid residues comprising the peptides are based on the pseudosubstrate motif of the respective kinases and generally are selected so as not to interfere with the ATP binding site of the kinase (Combest et al. 1988; Eichholtz et al. 1993; Ward and O'Brian 1993) and have been shown not to interfere with the action of other kinases (Eichholtz et al. 1993). Such inhibitors can be used in cell lysates and other in vitro reactions and recently they have been made membrane-permeable by myristolation without loss of inhibitor specificity (Eichholtz et al. 1993; Ward and O'Brian 1993).

Permeabilized System

A permeabilized egg system has been developed that allows the introduction of various proteins, ions, inhibitors, and activators into the cell interior (Gallicano et al. 1995). This system can be used to access cell structures and functions. The following procedure indicates a mechanism that can be employed to flush in constitutively active PKC.

1. Eggs/embryos are permeabilized by placement in 1% Tween 20 in ICB (ICB; 100 mM KCl, 5 mM $MgCl_2$, 3 mM EGTA, 20 mM HEPES, pH 6.8) containing the following protease inhibitors at 1 mg/ml each—pepstatin, aprotinin, chymostatin, leupeptin, and trypsin-chymotrypsin inhibitor—for 3 minutes at room temperature.
2. Wash the eggs/embryos through three changes of ICB containing the protease cocktail to remove detergent.
3. The eggs/embryos are then transferred to droplets of temperature-equilibrated ICB with protease inhibitors that contain the conditions necessary for PKC action. That is, ATP (radiolabeled ATP can be used when phosphoproteins are

Table 3-1 In vivo (*) and cell lysate (†) analysis of kinase inhibitors

Target kinase	Inhibitors	PKC	PKA	MLCK	CaMKinase II	Tyrosine kinase
PKC	PKC pseudosubstrate (MYR)[a]	50% 8.0* 98% 100	Ineffective*	Ineffective	Ineffective*	Ineffective*
PKA	PKI pseudosubstrate (MYR)[b,c]	Ineffective†	100% 100†	Ineffective†	Ineffective†	Ineffective†
PKC	Calphostin C[d–f]	50% 0.1† 91% 1.0†	5% 10.0† 50% >50.9†	50% >5.0†	Undetermined	Ineffective†
PKC	Bisindolylmaleimide[g,h]	75% 5.0* 90% 10.0*	0% 5.0*	0% 5.0*	0% 5.0*	0% 1.0*
PKC	Sphingosine[i–l]	50% 2.0* 90% 10.0* Loaded for 30 minutes	Ineffective†	50% 100†	CaMKinase II, None CaMKinase III 20.0† Activates casein kinase II at 67.0†	Ineffective†
PKA	H-8[m,n]	50% 15.0†	50% 1.2* 80% 5.0*	0% 500* 22% 100*	Undetermined	Ineffective†
PKA	H-89[o,p]	10% 50.0† 0% 20.0*	80% 25.0* 0% 110*	Undetermined	8% 20† 0% 110*	Ineffective†
Tyr kinase	Genistein[q–s]	0% 110* 50% >350*	50% <350†	50% <350†	Undetermined	100% 100* 100% 100*
MLCK	ML09[t,u]	Undetermined	Undetermined	89% 30.0* 100% 100*	Undetermined	Ineffective
CaMKII	KN-93[u]	10*	Undetermined	Ineffective*	50.0% 0.3* 98.0% 10.0*	Undetermined

[a] J. Biol. Chem. 268:1982–1986; 1993.
[b] J. Biol. Chem. 264:8802–8810; 1988.
[c] J. Neurochem. 51:1581–1591; 1988.
[d] Pharmacology 46:181–192; 1993.
[e] Biochem. Biophys. Res. Commun. 159:548–553; 1989.
[f] FEBS Lett. 314:149–154; 1992.
[g] J. Biol. Chem. 266:15771–15781; 1991.
[h] FEBS Lett. 259:61–63; 1986.
[i] J. Biol. Chem. 261:13674–13681; 1987.
[j] J. Biol. Chem. 266:21773–21776; 1991.
[k] J. Biol. Chem. 266:15771–15781; 1991.
[l] Biochim. Biophys. Acta 1010:131–139; 1989.
[m] Biochemistry 23:5036–5041; 1984.
[n] Dev. Growth and Diff. 32:549–556; 1990.
[o] J. Biol. Chem. 267:14898–14906; 1993.
[p] J. Biol. Chem. 265:5276–5272; 1990.
[q] J. Biol. Chem. 262:5592–5595; 1987.
[r] Biochim. Biophys. Acta 1112:181–186; 1992.
[s] Nature (London) 353:558–560; 1991.
[t] Biol. Chem. 262:7796–7801; 1987.
[u] Biochem. Biophys. Res. Commun. 191:255–261; 1993.

to be detected), an ATP regenerating system (1 mM ATP, 10 mM phosphocreatine, 50 μg/ml creatine phosphokinase), and constitutively active PKC (referred to as PKM) as described in Gallicano et al. (1995), and incubated at 37°C for the desired time. The incubation should be conducted under oil to prevent evaporation.
4. At the end of the incubation, the eggs or embryos can be placed in 2× sample buffer for electrophoresis or cytologically fixed and transferred to polylysine-coated coverslips for visual examination.

Cautions The use of a permeabilized system has the advantages that changes can be induced inside cells that could lead to the identification of causal agents that mediate structural change. However, it is likely that the only changes that can be induced in the permeabilized cell are those that the cell was immediately engineered to undergo. Once the cell is permeabilized, the loss of some cellular components make it unlikely that the cell could properly assemble the structural changes that would occur hours into the future. To test such later changes, cells should be permeabilized at later points in the time course of events under consideration.

Preparation of the Detergent-Resistant Cytoskeleton

The cytoskeleton can be isolated by detergent extraction and this preparation can be analyzed at both a structural and a biochemical level. When prepared for biochemical analysis, the detergent-soluble fraction can be isolated from the detergent-resistant cytoskeleton and analyzed separately, e.g., by gel electrophoresis. In addition, the detergent-resistant cytoskeleton represents a structural component of the cell that can be examined by light microscopy (e.g., after binding with antibodies to specific antigens) or by electron microscopy.

Procedure to Prepare the Detergent-Resistant Cytoskeleton and Soluble Fraction

1. Eggs/embryos are washed through PBS and placed in detergent extraction media (ICB containing 200 mg/ml AEBSF and 1% Tween 20) and extracted for 3 minutes.
2. For analysis by gel electrophoresis, the eggs/embryos can be transferred into a SDS–sample buffer. The detergent-soluble components are precipitated from the media by the addition of 4 volumes of cold ethanol and incubation overnight at −20°C. The precipitated soluble proteins are pelleted and then solubilized by addition of SDS–sample buffer.
3. Alternatively, the detergent-resistant cytoskeleton of eggs/embryos is processed for observation by electron or light microscopy. For electron microscopy, the samples are cytologically fixed by addition of 2% glutaraldehyde and 1% paraformaldehyde and processed for electron microscopy (see below). If the specimens are to be viewed by a form of light microscopy, the choice of

fixative is modified for the form of analysis, e.g., to permit the efficient binding of antibodies. For immunocytochemistry, the detergent-extracted specimen is cytologically fixed by the addition of 0.05% glutaraldehyde and 2% paraformaldehyde made in PBS for 5 minutes at room temperature and then transferred to 2% paraformaldehyde for an additional 25 minutes. Any free aldehyde groups that remain can be inactivated by a subsequent wash in 5 mM glycine made in PBS for 1 hour. With this treatment, specimens can then be processed through antibody-containing solutions to view specific proteins, into RIM-1 to view the distribution of PKC, or into 1 µg/ml Hoechst 33342 for 30 minutes to view the DNA.

Cautions Detergent extraction is performed by immersing cells in an intracellular buffer containing protease inhibitors and a nonionic detergent. The most common nonionic detergent used with somatic cells is Triton X-100. However, it is important to note that Triton X-100 destabilizes the cytoskeleton of mouse eggs and embryos (Gallicano et al. 1991) and for that reason we do not recommend its use; instead use the nonionic detergent Tween 20 to isolate the cytoskeleton. In order to obtain a consistent preparation of the cytoskeleton, the nonionic detergent should be present above its critical micelle concentration and the detergent extraction should be performed for a consistent time to permit complete removal of all of the egg's soluble components. If the egg is left for long periods of time, the cytoskeleton will begin to disassemble since the detergent extraction results in a monomer concentration of actin and tubulin effectively at zero. For the reason just noted concerning consistency, detergent extraction is only effective when dealing with an embryo that is, at most, a few cell layers thick since the detergent will take much longer to reach cells at the interior of an embryo, while cells at the exterior of the embryo will be detergent extracted at a much earlier time point.

If eggs are transferred by pipette into detergent extraction medium they will float immediately to the surface and they may lyse or accumulate at the air–medium interface where it is difficult to retrieve them. One way to circumvent this problem is to pipette them quickly up and down in the detergent extraction medium upon initially transferring them. Alternatively, the eggs or embryos, surrounded by the zona pellucida, may be washed into a protein-free medium and then transferred onto a polylysine-coated coverslip in a small Petri dish. Once the eggs/embryos are adhered, an excess of detergent extraction medium can be added to the Petri dish. When this is done, subsequent handling steps can be performed by exchanging the medium in the petri plate rather than transferring the eggs/embryos by pipette.

Care must be exercised when selecting the protease inhibitor that will be used. Many individuals routinely use either **PMSF** or **AEBSF**; however, these inhibitors block proteases by sulfonating serine residues at the active site of the proteins. The addition of this sulfonating group clearly can reduce or block enzyme function. If the purpose of the experiment is to examine the structure of the cytoskeleton or apply cytoskeletal proteins to a separating gel, then either of these two inhibitors are adequate. However, if the purpose is to perform an enzyme assay or any other metabolic event, then neither PMSF or AEBSF should be used; here, a protease cocktail that contains 1 µg/ml of each of the following protease inhibitors—pepstatin, aprotinin, chymostatin, leupeptin, and trypsin-chymotrypsin inhibitor—and which acts by competitive inhibition should be employed.

Processing for Electron Microscopy

Mammalian eggs and embryos do not provide images of high contrast when processed for conventional, resin-embedded, transmission electron microscopy of thin sections. The level of contrast and cellular detail is greatly enhanced by the inclusion of one of the two following procedures: (1) addition of tannic acid (0.1%) to the glutaraldehyde fixation step to increase the electron density of the specimen or (2) En bloc staining with 2% aqueous uranyl acetate prior to embedding.

Eggs and embryos can be processed through the solutions necessary for embedding by using the eggs/embryos adhering to polylysine-coated coverslips as described above. Throughout the dehydration steps, solutions are exchanged by rapidly pouring off most of the contents of the plate into a waste beaker and applying new solution to the corner of the plate. In the final step, the eggs/embryos are transferred off of the coverslip by touching their zona pellucida with a blunt object (in a plate containing unpolymerized resin or molten wax) to release them from the coverslip and then transferring them into the embedding mold. Procedures for embedding in a removable medium and preparation of embedment-free sections are detailed in Capco (1993) and those for embedding in plastic resin in Capco et al. (1993).

Summary

Investigations employing mammalian eggs and embryos are technically difficult because of the small number of eggs/embryos obtained and the small size of the cells. However, technical advances in the assay systems for analysis of cytoplasmic signaling agents have made it possible to conduct investigations even on the small mammalian cells. These advances include the development of kinase reporter dyes, specific peptide substrates, specific peptide inhibitors that can be made membrane-permeable, and the development of permeabilized systems to name the most major advances.

The methods of analysis described here make it possible to link specific structural changes in the cell with functional changes and to analyze the agent(s) responsible for specific structural changes (and, consequently, functional changes) that are part of a cascade of signaling events that progressively remodel the cell. The fertilization-competent egg of any species is ideally suited for such investigations because, by its very nature, such an egg exhibits arrest points built into its developmental program, and because activated/fertilized eggs undergo extensive remodeling typically under the regulation of cytoplasmic signaling agents in order to become the zygote. Using the egg-to-zygote conversion as a starting point for analysis of the regulation of such structure–function changes should allow the extrapolation of approaches to investigate regulation during embryonic cleavage and in somatic cells.

ACKNOWLEDGMENTS This work was supported by NIH HD32621.

References

Allen, L. A., and Aderem, A. (1995). Protein kinase C regulates MARCKS cycling between the plasma membrane and lysosomes in fibroblasts. EMBO J. 14:1109–1121.

Ashendel, C. (1985). The phorbol ester receptor: a phospholipid-regulated protein kinase. Biochim. Biophys. Acta 822:219–242.

Blackshear, P. J. (1993). The MARCKS family of cellular protein kinase C substrates. J. Biol. Chem. 268:1501–1504.

Capco, D. G. (1993). Diethylene glycol distearate (DGD): an alternative to PEG as a removable embedment. In *Polyethylene Glycol as an Embedment for Microscopy and Histochemistry*, Gao, K., ed. Boca Raton, FL, CRC Press, pp. 97–111.

Capco, D. G., Gallicano, G. I., McGaughey, R. W., Downing, K. H., and Larabell, C. A. (1993).Cytoskeletal sheets of mammalian eggs and embryos: a lattice- like network of intermediate filaments. Cell Motil. Cytoskeleton 24:85–99.

Chen, C.-S., and Poenie, M. (1993). New fluorescent probes for protein kinase C. J. Biol. Chem. 268:15812–15822.

Collins, J. E., and Fleming, T. P. (1995).Epithelial differentiation in the mouse preimplantation embryo: Making adhesive cell contacts for the first time. TIBS 20:307–312.

Combest, W. L., Bloom, T. J., and Gilbert, L. I. (1988). Polyamines differentially inhibit cyclic AMP-dependent protein kinase mediated phosphorylation in the brain of the tobacco hornworm, *Manduca sexta*. J. Neurochem. 51:1581–1591.

Davis, P., Hill, C., Keech, E., Lawton, G., Nixon, J., Sedwick, A., Wadsworth, J., Westmacott, D., and Wikinson, S. (1989). Potent selective inhibitors of protein kinase C. FEBS Lett. 259:61–63.

Eichholtz, T., de Bont, D. B. A., Widt, J., Liskamp, R. M. J., and Ploegh, H. L. (1993). A myristolated pseudosubstrate peptide, a novel protein kinase C inhibitor. J. Biol. Chem. 268:1982–1986.

Erbach, G. T., Lawitts, J. A., Papaioannou, V. E., and Biggers, J. D. (1995). Differential growth of the mouse preimplantation embryo in chemically defined media. Biol. Reprod. 50:1027–1033.

Fagotto, F., and Gumbiner, B. M. (1996). Cell contact-dependent signaling. Dev. Biol. 180:445–454.

Gallicano, G. I., McGaughey, R. W., and Capco, D. G. (1991). Cytoskeleton of the mouse egg and embryo: Reorganization of planar elements. Cell Motil. Cytoskeleton 18:143–154.

Gallicano, G. I., McGaughey, R. W., and Capco, D. G. (1995). Protein kinase M, the cystolic counterpart of protein kinase C, remodels the internal cytoskeleton of the mammalian egg during activation. Dev. Biol. 167:482–501.

Gallicano, G. I., Yousef, M., and Capco, D. G. (1997). PKC—A pivotal regulator of early development. BioEssays 19:29–36.

Hidaka, H., Inagaki, M., Kawamoto, S., and Sasaki, Y. (1984). Isoquinolinesulfonamides, novel and potent inhibitors of cyclic nucleotide dependent protein kinase and protein kinase C. Biochemistry 23:5036–5041.

Ho, Y., Wigglesworth, K., Eppig, J. J., and Schultz, R. M. (1995). Preimplantation of mouse embryos in KSOM: Augmentation by amino acids and analysis of gene expression. Mol. Reprod. Dev. 41:232–238.

Hogan, B., Constantini, F., and Lacy, E. (1986). *Manipulating the Mouse Embryo*. Cold Spring Harbor, NY, Cold Spring Harbor Laboratory Press.

Kishimoto, A., Nishiyama, K., Nakanishi, H., Uratsuji, Y., Nomura, H., Takeyama, Y., and Nishizuka, Y. (1985). Studies on the phosphorylation of myelin basic protein by protein kinase C and adenosine $3'$: $5'$-monophosphate-dependent protein kinase C. J. Biol. Chem. 260:12492–12499.

Lawitts, J. A., and Biggers, J. D. (1992). Joint effects of sodium chloride, glutamine and glucose in mouse preimplantation embryo culture media. Mol. Reprod. Dev. 31:189–194.

Lo, C. W. The role of gap junction membrane channels in development. J. Biomembr. Bioenerg. 28:379–385.

McGaughey, R. W. (1977). The maturation of porcine oocytes in minimal, defined culture media with varied macromolecular supplements and varied osmolarity. Exp. Cell Res. 109:25–30.

McKiernan, S. H., Bavister, B. D., and Tasca, R. J. (1991). Energy requirements for in vitro development of hamster 1- and 2-cell embryos to the blastocyst stage. Human Reprod. 6:64–75.

McKiernan, S. H., Clayton, M. K., and Bavister, B. D. (1995). Analysis of stimulatory and inhibitory amino acids for development of hamster one-cell embryos in vitro. Mol. Reprod. Dev. 42:188–199.

Moses, R. M., and Kline, D. (1995). Calcium-independent, meiotic spindle-dependent metaphase-to-interphase transition in phorbol ester-treated mouse eggs. Dev. Biol. 171:111–122.

Moses, R. M., and Masui, Y. (1995). Metaphase arrest in newly matured or microtubule-depleted mouse eggs after calcium stimulation. Zygote 3:1–8.

Payne, D. M., Rossomando, A. J., Martino, P., Erikson, A. K., Her, J.-H., Shabanowitz, J., Hunt, D. F., Weber, M. J., and Sturgill, T. W. (1991). Identification of the regulatory phosphorylation sites in pp42/mitogen-activated protein kinase (MAP kinase). EMBO J. 10:885–892.

Sammak, P. J., Adams, S. R., Harootunian, A. T., Schliwa, M., and Tsien, R. Y. (1992). Intracellular cyclic AMP, not calcium, determines the direction of vesicle movement melanophores: direct measurement by fluorescence ratio imaging. J. Cell Biol. 117:57–72.

Sastry, S. K., and Horwitz, A. F. (1996). Adhesion-growth factor interactions during differentiation: an integrated biological response. Dev. Biol. 180:455–467.

Schultz, R. M., and Kopf, G. S. (1995). Molecular basis of mammalian egg activation. Current Topics in Developmental Biology 30:21–62.

Shibuya, E. K., Boulton, T. G., Cobb, M. H., and Ruderman, J. V. (1992). Activation of p42 MAP kinase and the release of oocytes from cell cycle arrest. EMBO J. 11:3963–3975.

Summers, M. C., Bhatnager, P. R., Lawitts, J. A., and Biggers, J. D. (1995). Fertilization in vitro of mouse ova from inbred and outbred strains: Complete preimplantation embryo development in glucose-supplemented KSOM. Biol. Reprod. 53:431–437.

Swann, K., and Lai, F. A. (1997). A novel signalling mechanism for generating Ca^{2+} oscillations at fertilization in mammals. BioEssays. 19:371–378.

Walten, G. M., Bertics, P. J., Hudson, L. G., Vedvick, T. S., and Gill, G. N. (1987). A three step purification process for protein kinase C: Characterization of the purified enzyme. Anal. Biochem. 161:425–437.

Ward, N. E., and O'Brian, C. A. (1993). Inhibition of protein kinase C by N-myristolated peptide substrate analogs. Biochemistry 32:11903–11909.

Winston, N. J., and Maro, B. (1995). Calmodulin-dependent protein kinase II is activated transiently in ethanol-stimulated mouse oocytes. Dev. Biol. 170:350–352.

4

Intracellular Ion Measurements in Single Eggs and Embryos Using Ion-Sensitive Fluorophores

Jay M. Baltz
Karen P. Phillips

Many cell functions rely on the maintenance or regulated alteration of a few ion species in the cytoplasm. The ability to measure ion concentrations in cells is therefore important when studying any of a large number of cellular processes, and the availability of relatively noninvasive methods to determine intracellular ion concentrations is of great utility. Techniques have been developed which allow the measurement of concentrations of a number of inorganic ions inside intact, living cells. These methods rely on the use of fluorescent compounds whose fluorescent properties are affected by the relevant ion. By quantitative measurement of the fluorescence arising from these fluorophores, it is possible to determine either the absolute ion concentration or relative changes in ion concentration. Such techniques have proven extremely valuable, especially in the study of intracellular signaling by Ca^{2+}, and in the study of intracellular pH (pH_i) regulation.

In mammalian eggs and early embryos, the majority of investigations using such techniques have involved measurements of changes in intracellular Ca^{2+} concentrations, especially around the time of fertilization and egg activation. Other studies have examined the regulation of pH_i and investigated whether there are pH_i changes at fertilization. In this chapter, we will describe the methods currently being used with mouse eggs and embryos for intracellular ion measurements. Because relatively few eggs or embryos can be obtained from mammals, techniques such as these, which can be used down to the single-cell level, are of great importance.

Mechanisms of Fluorophore-Based Ion Measurements

Fluorophores which are useful for determining ion concentrations are those whose fluorescence properties are altered in a specific, predictable way by one or a few species of ions, over the range of ion concentrations which are biologically relevant. The details of the physical chemistry governing such properties are beyond the scope of this chapter. However, it is important to understand the general principles behind fluorophore function when using them for ion measurements.

Most ion-sensitive fluorophores have a binding site for the ion. Upon binding, the fluorescence emitted by the fluorophore is altered. The effect of ion binding can be either to simply decrease (or increase) fluorescence intensity or, instead, to shift the spectrum of the fluorescence.

Those fluorophores whose spectra shift upon ion binding are particularly valuable for intracellular measurements. In this case, fluorescence at different wavelengths is affected differently by ion binding, while, in contrast, changes in fluorophore concentration, pathlength (i.e., cell thickness), or other nonspecific effects will have equal effect at all wavelengths (e.g., doubling the concentration of the fluorophore will usually double the fluorescence at all points in the spectrum). The advantage that such spectral shifts confer is that the ratio of fluorescence intensities at two chosen wavelengths with different dependencies on ion concentration will yield a ratio which is strongly dependent only on the ion concentration, but which is almost independent of other parameters such as fluorophore concentration, cell geometry, or excitation intensity. Such fluorophores which exhibit a spectral shift upon interacting with an ion are termed "ratiometric" fluorophores.

There are two types of ratiometric responses. The first are those which exhibit a spectral shift in their excitation spectrum when intensity of fluorescence emission is measured at a single wavelength. The second are those which exhibit a spectral shift in the emission spectrum when fluorescence is excited at a single wavelength. Most fluorophores are useful for excitation ratio measurements (two excitation wavelengths), while others can be used for emission ratio measurements (two emission wavelengths). Some, indeed, have properties which allow both types of ratios.

Ion-Sensitive Fluorophores

pH-Sensitive Fluorophores

A number of fluorescent indicators are available which are sensitive to changes in pH around the physiological range. One such indicator, BCECF [2',7'-bis(Carboxyethyl)-5(and -6)-carboxy fluorescein] was the first pH-sensitive fluorophore developed which had a pK_a near neutrality, and which was retained inside of cells for an appreciable length of time (Rink et al. 1982). BCECF has a pK_a of about 7.0, making it optimally useful in most cells, including embryos. It is a ratioable fluorophore, exhibiting a shift in its excitation spectrum upon protonation. A ratio of the fluorescence intensity excited at 490–500 nm to that excited at around 450 nm increases with increasing pH. Emission is maximal at about 530 nm.

More recently, pH-sensitive fluorophores have been developed which show a shift in both excitation and emission spectra. Carboxy-SNARF-1 (carboxy-seminaphthorhodafluor-1) exhibits both shifts, although generally only its usefulness as an emission ratioed fluorophore has been exploited (Bassnett et al. 1990). In this mode, excitation of carboxy-SNARF-1 is at around 535 nm, and the ratio of emission intensity at 640 nm to that at 600 nm increases with increasing pH. Related fluorophores, carboxy-SNAFL-1 and carboxy-SNAFL-2 (carboxy-seminaphthofluorescein-1 and -2, respectively) are also available; these are excited and fluoresce at longer wavelengths than carboxy-SNARF-1 (Haugland 1996).

Ca^{2+}-Sensitive Fluorophores

There are many different Ca^{2+}-sensitive fluorophores available. Indeed, a recent paper reports comparing properties of 13 different Ca^{2+} indicators loaded into frog muscle fibers (Zhao et al. 1997b). The best known and most widely used Ca^{2+}-sensitive fluorophore is fura-2, which is excited in the near-ultraviolet (UV) and has a fluorescence emission spectrum which peaks near 510 nm. Fura-2 is ratiometric, since its excitation spectrum shifts upon Ca^{2+} binding. Usually, fura-2 is excited at 340–350 nm, where increasing Ca^{2+} concentrations maximally increase fluorescence, and at around 380 nm, where the effect is opposite, and increasing Ca^{2+} concentrations decrease fluorescence. Thus, the ratio of intensities at 340:380 nm increases steeply with Ca^{2+} concentration. The dissociation constant of fura-2 for Ca^{2+} is around 0.2 µM (Grynkiewicz et al. 1985; Groden et al. 1991), making it very useful for measuring normal physiological levels of Ca^{2+} in cells.

Indo-1 is very similar to fura-2, both chemically and in its spectral response. However, its emission spectrum shifts in response to Ca^{2+} binding, and therefore the ratio of intensities at two different emission wavelengths can be used. Indo-1 is excited at around 340–350 nm, and its emission is measured at around 475 nm and 400 nm; the ratio of intensities at 400:475 nm increases with Ca^{2+} concentration.

The other common ratioable Ca^{2+} indicator is quin-2, which is of the first generation of such indicators developed. It is no longer used extensively because the high concentration of quin-2 which must be loaded buffers intracellular Ca^{2+}.

Several Ca^{2+}-sensitive fluorophores have been developed which have the advantage that they are excited by visible, rather than UV, light. Fluo-3, calcium green, calcium orange, and calcium crimson all exhibit large increases in fluorescence upon Ca^{2+} binding. However, these indicators do not undergo any spectral shift upon Ca^{2+} binding, and therefore cannot be ratioed. Of these, fluo-3, excited around 490 nm, has been used with germinal vesicle (GV)-intact oocytes, eggs, and morulae (Kline and Kline 1994; Stachecki et al. 1994; Jones et al. 1995b; Kono et al. 1995; Parrington et al. 1996).

One relatively new Ca^{2+} probe that may prove useful in eggs and embryos is C_{18}-fura-2, developed by Fay and colleagues (Etter et al. 1994). C_{18}-fura-2 carries a hydrophobic tail which associates with membranes. When injected into cells, C_{18}-fura-2 is retained at the inner face of the plasma membrane. Thus, it may be useful for probing Ca^{2+} concentrations during events such as cortical granule exocytosis.

Cl^--Sensitive Fluorophores

Several halide-sensitive fluorophores are available that are used for the measurement of Cl^-, the only halide normally present in large concentrations in or around cells (Krapf et al. 1988; Verkman 1990; Biwersi and Verkman 1991; Haugland 1996). Unlike the fluorescent indicators described above, these fluorophores do not bind Cl^- ions. Instead, their fluorescence is quenched by Cl^- and other halides in a nonradiative transfer of energy between the fluorophore in its excited state and the halide ion. Thus, as Cl^- concentration increases, fluorescence decreases, and vice versa. The Cl^--sensitive fluorophores include SPQ [6-methoxy-N-(3-sulfopropyl)quinolinium], ABQ [N-(4-aminobutyl)-6-methoxyquinolinium chloride], MEQ (6-methoxy-N-ethylquinolinium iodide) and MQAE [N-(ethoxycarbonylmethyl)-6-methoxyquinolinium bromide]. Because they are Cl^--sensitive due to quenching, these indicators are not ratiometric, and only exhibit a change in fluorescence intensity in the presence of Cl^-. All are excited in the UV range, around 350 nm, and emission is maximal at around 450 nm.

Na^+- and K^+-Sensitive Fluorophores

There are also fluorescent probes available for two other ions of biological importance — Na^+ and K^+. These are fluorescent benzofuranyl derivatives of crown ethers, whose fluorescence changes when the ions are bound in the crown ether's cavity. Probes SBFI and SBFO are selective for Na^+, while PBFI is selective for K^+ (Jezek et al. 1990; Wong and Foskett 1991; Haugland 1996). Selectivity is based largely on the size of the cavity. These compounds exhibit a shift in their excitation spectra upon ion binding, making them ratioable. Excitation is typically at 340–350 nm and 380 nm, with the ratio of intensity at 340–350 nm to intensity at 380 nm increasing with increased ion concentration. In addition, a Na^+-sensitive fluorophore which is excited in the visible range — sodium green — is available (Haugland 1996). It is not ratiometric, however. Sodium green is excited at around 490–510 nm, and peak emission is at around 535 nm.

Methods of Loading Ion-Sensitive Fluorophores into Cells

There are several methods of loading fluorescent compounds into cells. Direct methods include microinjection, hypotonic shock, scrape loading, and loading via liposomes. Microinjection has been used successfully with eggs and embryos (cf. Kline and Zagray 1995). We have used hypotonic conditions to load some fluorescent indicators into blastocoel cavities, as will be described below. Scrape loading is not an option for embryos, of course, and loading via liposomes has not been used with embryos to our knowledge.

Perhaps the single most important advance which has made the use of ion-sensitive fluorophores common for the determination of intracellular ion concentrations has been the development of relatively noninvasive methods of introducing these fluorophores into the cell (Thomas et al. 1979). Many fluorescent indicators are now available as precursors with hydrophobic groups conjugated to the fluorescent indicator, which allow them to permeate freely through the plasma membrane and reach the cytoplasm (Haugland 1996). Once in the cytoplasm, the hydrophobic groups, which are typically attached via ester bonds, are cleaved off by the action of endogenous esterases normally found in the cytoplasm. When cleaved, the fluorophore becomes hydrophilic and therefore is trapped in the cytoplasm (Thomas et al. 1979). By this means, a high concentration of fluorophore can be loaded into cells simply by incubating them for a period of time with the relatively hydrophobic precursor.

Although there are several variants on this scheme, the most common precursors are acetoxymethyl ester (AM) or diacetate derivatives of the fluorophore. The most commonly used fluorescent indicators, including fura-2 and most other Ca^{2+}-sensitive fluorophores, BCECF, and carboxy-SNARF-1, are available in the AM form. Esterase activity results in the removal of the AM group by hydrolysis at the ester bonds. Since there are typically 2–5 AM groups per molecule, this leaves a negatively charged moiety at each site of AM cleavage, rendering the fluorophore impermeant and trapped in the cytoplasm.

Hydrolysis of each AM group produces one molecule each of formaldehyde and acetic acid, which are toxic in large enough concentrations. Therefore, one potential drawback to this convenient method of loading cells with fluorescent indicators is the possible toxic effect of the loading procedure itself. Such potential toxicity, in relation to eggs and preimplantation embryos, will be discussed below.

Methods of Detection

A prerequisite for using ion-sensitive fluorophores is the ability to measure the fluorescence quantitatively. For mammalian eggs and preimplantation embryos, measurements of the fluorescence from single embryos or small groups of embryos is required, due to the very small amount of material available. Quantitative fluorescence measurements from single cells are routinely carried out using quantitative fluorescence microscopy. A fluorescence microscope with a suitably stable excitation source is used. The emitted fluorescence can be detected in several ways. The first is by employing a photomultiplier tube-based detector to collect the total fluorescence emission from a defined area of the microscopic field. Usually, a pinhole aperture is used in the microscope to restrict the collection of light to one cell or a portion of a cell in the visual field. All emitted fluorescence which passes through the aperture is detected by the photomultiplier, which gives a quantitative measure of the fluorescence. The advantages of this method are that photomultipliers are capable of detecting very low levels of light, and typically exhibit a low noise level. Thus, very little fluorophore and low levels of illumination are sufficient. The disadvantages are that any spatial information is lost, and that data can only be obtained from one cell (or embryo) at a time.

Another detection method is based on quantitative imaging. Here, a camera is employed to capture the fluorescence image of (usually) a group of cells. The camera must be capable of detecting very low light levels, and must have a linear response to intensity over a suitably large range of light intensities. Such cameras are widely available, and include digital cameras such as cooled, low-noise charge-coupled device (CCD) cameras, and modified video cameras such as CCD-based video cameras fitted with image intensifiers, or silicon intensification target (SIT) cameras. The advantage to imaging is that spatial information is retained, and a number of cells (or embryos) can be imaged at once, increasing the speed of data collection. The disadvantages are increased cost over a photomultiplier-based system, and the typically higher light levels (and hence higher fluorophore concentrations and excitation intensities) needed for imaging.

Such quantitative fluorescence microscopy systems are available commercially or can be constructed in the laboratory. Almost invariably, the data (either photomultiplier voltage output or images) are digitized and manipulated and stored digitally. This allows the computation of ratios for ratiometric measurements, and the display of the now-familiar pseudocolor images of Ca^{2+} concentrations or pH in cells.

Finally, confocal microscopes can be used for quantitative fluorescence imaging. The advantage of such systems is that they have good spatial resolution not only in the X–Y plane, but also in the vertical (Z) direction. The disadvantages are that the choice of excitation wavelengths is restricted to those available from the laser(s) in the system, and that confocal scanning is generally slower than other imaging methods.

Any detailed discussion of the various types of systems, their function, or their relative advantages and disadvantages is beyond the scope of this chapter. However, before attempting to use any system for quantitative fluorescence, it is vital that appropriate validation of the system be carried out. Such validation procedures have been described in detail elsewhere (Bright et al. 1987, 1989; Jericevic et al. 1989).

Microscope Stage Chambers for Fluorescence Measurements with Eggs and Embryos

There are probably almost as many schemes for controlling the environment of cells during microscopy as there are laboratories using such techniques. A chamber which is used to maintain eggs and embryos during microscopy should be optically clear over all relevant wavelengths, it should allow control over temperature and preferably gas phase, and it should allow for solution changes. In addition, the embryos should be immobilized in some way so that they do not move during measurements and are not lost during perfusion or solution changes. Finally, the chamber should not interfere with excitation or emission at the required wavelengths.

We use a temperature-controlled chamber whose bottom is a replaceable glass coverslip (Biophysica, Sparks, MD). It holds approximately 2.5 ml of liquid. Many similar temperature-controlled chambers are available. Our chamber has been modified to allow the gas phase over the liquid to be controlled by introducing a gas mixture (usually 5% CO_2/air) through an inlet tube in the cover. Gas is supplied to the approximately 5 cm^3 volume of the gas phase at about 30 cm^3/minute, which is sufficient to keep the medium gas-equilibrated. For longer term experiments (more than approximately 20 minutes), evaporation becomes significant when passing the dry gas mixture over the surface of the warm medium. In this case, the gas is first bubbled through distilled water at 60°C to raise the humidity to near 100%, with care taken that water does not condense in the tubing and enter the chamber.

The solution in the chamber can be changed in our setup using a syringe driver and vacuum aspiration to replace the solution in the chamber with a different, prewarmed and pre-equilibrated solution. Alternatively, we use a constant perfusion system to maintain a low rate of flow over the embryos, with solutions changed by switching between perfusates.

In order to prevent movement or loss of the eggs and embryos, particularly during solution changes, they must be immobilized. Polylysine coating of the coverslip immobilizes eggs and embryos fairly effectively, but we have found that there is frequently damage to two-cell-stage embryos in the presence of polylysine. Others, however, have successfully used polylysine with rat oocytes and eggs (Ben-Yousef et al. 1995). Wheat-germ agglutinin covalently linked to the glass, reported to effectively immobilize embryos by binding to the zona pellucida (Cherr and Cross 1986), was only partially effective at preventing loss of embryos in our hands. Finally, we found that the easiest and most consistent method was simply to allow the embryos to adhere to the clean glass surface. In the absence of any protein or other macromolecules in the medium, the zona will adhere tightly to glass and the adhesion will survive changes in the medium. Embryos are placed one by one onto the glass surface, in a tight group, using a standard mouth-operated pipette. Care must be taken not to allow anything to touch the area where the embryos are to adhere. In addition, the embryos must be well washed, and there can be no debris, red blood cells, or cumulus cells present. After placing the embryos, adhesion is checked by gently blowing a stream of medium onto them with the pipette. Zona-free eggs, zygotes, and blastocysts have also been similarly immobilized by adhesion to glass (Jones et al. 1995a,b; Y. Zhao, W.-L. Zhou, and J. M. Baltz, unpublished data).

Other laboratories have reported using various other means of successfully immobilizing embryos and eggs. These include the use of adhesion-promoting coatings such as the bio-adhesive, Cell-Tak (Collaborative Biosciences, Bedford, MA; Mehlmann and Kline 1994), or the lectin, concanavalin A (Cheek et al. 1993). House (1994) placed eggs into small (250-µm wide) slots milled into a silicon wafer which was laid on top of a coverslip, so that the eggs were held in place within the slots. Presumably, medium flowed over the top of the wafer during solution changes, and did not dislodge the eggs.

pH_i Measurements

pH_i measurements have been made in eggs and preimplantation embryos using the pH-sensitive fluorophores BCECF and carboxy-SNARF-1. We have loaded BCECF into eggs and zygotes, two-cell stage embryos, morulae, and blastocysts, as well as isolated inner cell masses, by incubation with either approximately 1.0 µM or 0.1 µM BCECF-AM at 37°C for 7–15 minutes (Baltz et al. 1990, 1991a,b; Zhao et al. 1997a). The exact molar concentration varies somewhat, since BCECF-AM is supplied as a mixture of at least two forms, one with two and the other with three AM groups, and the ratio varies from lot to lot (Haugland 1996). We used a 1 mg/ml stock solution in DMSO (dimethylsulfoxide; kept frozen), which was good for several months. In the past, we have monitored the extent of BCECF loading by periodically checking fluorescence. Since BCECF-AM is nonfluorescent, its presence in the medium does not interfere with intracellular fluorescence measurements. This has the advantage of assuring that similar levels of fluorophore are loaded each time. More recently, however, we have simply incubated the embryos for a fixed time of 10 minutes, with good results. Under a dissecting microscope, the embryos become very slightly orange in color after BCECF loading, although this can be difficult to see. We have found that, after loading, BCECF is retained for at least 1 hour at 37°C, although we have not assessed the loss of BCECF fluorescence quantitatively.

There are several other reports in the literature of the use of BCECF to measure intracellular pH in mouse embryos. LeClerc et al. (1994) loaded 8-cell and 16-cell stage embryos using 2.5 µM BCECF-AM for 5–10 minutes at 22–28°C. Kline and Zagray (1995) loaded unfertilized eggs using 1 µM BCECF-AM for 30 minutes, and also by microinjection of BCECF conjugated to dextran to a final concentration of 11 µM.

We have extensively used carboxy-SNARF-1 in eggs, zygotes, two-cell embryos, morulae and blastocysts. Embryos were loaded by incubation in 5 µM carboxy-SNARF-1-AM for 30 minutes at 37°C in HEPES-buffered medium. After loading, the embryos take on a dark pink color that is obvious under a dissecting microscope. We are aware of two other reports of the use of carboxy-SNARF-1 in eggs or embryos. House (1994) used an almost identical protocol to load unfertilized eggs with the fluorophore. Indeed, we adapted our loading protocol from his. Edwards et al. (1997) report using carboxy-SNARF-1 with zygotes and morulae.

Carboxy-SNARF-1 is retained well by embryos. Two-cell-stage embryos have been used for pH_i measurements 3–5 hours after loading, with good results. However, the length of time that the fluorophore is retained may be different for different stages of embryos. We were not able to detect much fluorescence in zygotes 3 hours after loading in preliminary experiments. This has not been investigated quantitatively, however.

Conversion of Ratio to pH

The relationship between pH and ratio is sigmoidal, since it arises from a titration curve of protonatable groups. Therefore, for maximum accuracy, ratios at a number of different pH values would be determined, and a sigmoidal fit to these data would be used for conversion. In practice, however, there is a relatively extensive range of pH values, centered on the pK_a of the fluorophore, over which the dependence of ratio on pH is nearly linear. For BCECF, the pK_a is approximately 7.0 (Rink et al. 1982), and for carboxy-SNARF-1, it is about 7.3–7.4 (Haugland 1996) at 37°C. Therefore, pH calibration is usually carried out by finding the ratios corresponding to three or four pH values within a suitable range, and fitting these to pH by linear regression. The usable range for BCECF is lower than that for carboxy-SNARF-1, owing to their different pK_a values; therefore, all things being equal, BCECF might be more useful for studies into pH_i regulation in the acid range, while carboxy-SNARF-1 might be more suitable for the alkaline range.

The simplest method of calibration is to use solutions of the fluorophore (not the AM form) with the pH adjusted to several appropriate values. The solutions used should approximate the composition of cytoplasm as much as possible. Practically, this entails using a high K^+, low Na^+ solution with added protein. Since the geometry of the fluorescent region can affect the measured ratio (depending upon the measurement system used), small droplets of the fluorescent solutions under oil, approximately the size of embryos, can be used.

This in vitro calibration method is far from optimal, however, and will yield only an approximate conversion to pH. The discrepancy will arise from the effect that the intracellular environment has on the fluorophore and its fluorescent properties. In practice, the pK_a and fluorescence spectrum of such fluorescent probes is significantly different in situ in cytoplasm than in solutions in vitro. The ratios are sensitive to viscosity, macromolecular components, and the variation of absorbance by the cell with wavelength, among other parameters which cannot be controlled. For accurate conversion of ratios to pH, an in situ conversion using fluorophores in the cell should be used.

The in situ conversion of pH-sensitive fluorophore ratio to pH exploits the properties of the K^+/H^+ exchanger, nigericin (Thomas et al. 1979). When nigericin is present, the 1:1 exchange of K^+ for H^+ across the plasma membrane results in an equalization of the gradients of these two ions across the membrane, given by

$$[H^+]_i/[H^+]_o = [K^+]_i/[K^+]_o \qquad (1)$$

where "i" and "o" represent intracellular and extracellular, respectively. Therefore, if $[K^+]_i = [K^+]_o$, then pH_i will have the same value as the pH of the external medium. To collapse the K^+ gradient, external solutions with high K^+ concentrations that approximate the intracellular K^+ concentration are used. The intracellular K^+ concentration is approximately 100–125 mM for embryos (Lee 1987; Baltz et al. 1997), so external solutions with such values are appropriate. In order to assure that the K^+ gradient is completely collapsed, the K^+-selective ionophore valinomycin is often used in conjunction with nigericin/high K^+.

We routinely convert ratios to pH_i using nigericin/high K^+ and valinomycin. The calibration solutions have 100 mM KCl, 25 mM NaCl, and 21 mM HEPES, with the pH adjusted using NaOH to several values between 6.8 and 8.0 at 37°C. The solutions also contain 75 mM sucrose to counter the swelling induced by exposure to high K^+/nigericin. Nigericin is stored as a 10 mg/ml stock in ethanol at −20°C, and valinomycin is stored as a 10 mg/ml stock in DMSO at 4°C. Just before calibrations are performed, nigericin is added to a final concentration of 10 µg/ml, and valinomycin to 5 µg/ml, in room temperature calibration solutions with extensive vortexing. The solutions are then warmed to 37°C before use. The drugs tend to precipitate out of solution if added too quickly, so they are added slowly and vortexed for 30–60 seconds. At 37°C the effect of nigericin introduction on pH_i is virtually instantaneous. A 5-minute exposure to each different pH solution before measuring the ratio is sufficient, although we allow a 10–15 minute exposure upon first introducing nigericin/high K^+ to ensure that the nigericin and valinomycin have equilibrated with the membrane, the K^+ gradient has collapsed, and the pH_i has reached equilibrium. Sample calibration

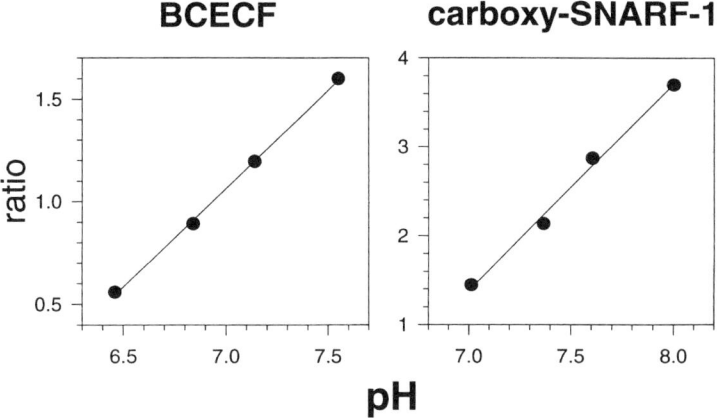

Figure 4-1 In situ calibration of BCECF and carboxy-SNARF-1. Sample calibrations of BCECF and carboxy-SNARF-1 in preimplantation mouse embryos (two-cell and zygote, respectively). Embryos loaded with the fluorophores were calibrated using the nigericin/high K^+ method as described in the text, with four different pH solutions. Note the approximately linear relation around the pK_a values of the fluorophores.

curves for BCECF and carboxy-SNARF-1 are shown in Figure 4-1. These curves are, of course, specific for our quantitative fluorescence imaging system.

Methods for Probing pH_i Regulation: Induced Alkalosis

In order to probe the regulation of pH_i in cells, it is necessary to perturb pH_i and examine the behavior during any recovery. Cells regulate pH_i, in general, by correcting deviations from the preferred "setpoint" pH_i using separate mechanisms for increasing or decreasing pH (Roos and Boron 1981; Boyarsky et al. 1988a,b; Vaughan-Jones, 1988; Alper 1994; Kim et al. 1996). To investigate the mechanisms used for correcting increases in pH_i, an intracellular alkalosis is induced.

To alkalinize the cytoplasm, a permeant weak base is introduced into the external medium. The most commonly used, and the one which we have used routinely for embryos, is ammonium. The introduction of a solution containing 25–35 mM NH_4Cl results in an immediate alkalinization of about 0.5–0.6 pH units (Figure 4-2; Baltz et al. 1991b). The mechanism behind this alkalinization is the rapid passive entry of membrane-permeable NH_3, which picks up H^+ in the cytoplasm until an equilibrium exists between NH_3 and NH_4^+ (Boron and DeWeer 1976; Boron 1977; Roos and Boron 1981). Other weak bases, such as triethylamine and procaine can also be used to produce the same effect (Aickin 1988; Szatkowski 1989; Szatkowski and Thomas 1989). Triethylamine (2–6 mM) has been

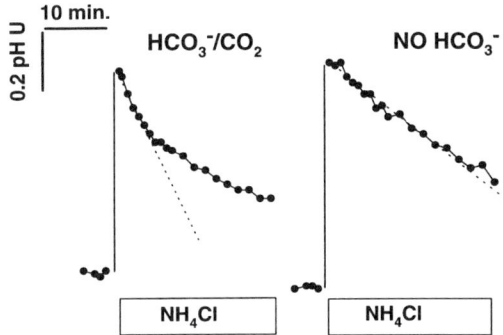

Figure 4-2 Alkaline loading with NH₄Cl. Two-cell stage embryos were alkaline loaded using the permeant weak base NH₄Cl (present as indicated by the box). The pH_i was monitored using BCECF. In the presence of HCO_3^-, the recovery from alkalosis is more rapid (compare slopes of dotted lines), due to activity of the pH_i-regulatory HCO_3^-/Cl^- exchanger. This figure was adapted from Baltz et al. (1991b).

used to produce large alkalinizations of egg cytoplasm (House 1994), and we have used procaine and tetracaine at millimolar levels for a similar level of alkalinization (J. M. Baltz, J. D. Biggers, and C. Lechene, unpublished data). In cells other than eggs or embryos, an array of such permeant weak bases has been used to induce cytoplasmic alkalosis, and these might be useful in eggs and embryos as well.

Induced alkalosis is followed by a recovery phase, in which pH_i returns to approximately the normal resting value. Part of this recovery is nonspecific. In the case of NH₄Cl-induced alkalosis, for example, relatively slow NH_4^+ entry into the cell, after the sharp NH_3-mediated alkalinization, serves to progressively lower pH_i even in the absence of specific pH_i-regulatory mechanisms (Figure 4-2; Boron and DeWeer 1976; Roos and Boron 1981). Thus, any specific mechanisms for recovery from alkalosis will be detected as increases over this nonspecific rate.

We have shown that the major specific mediator of recovery from alkalosis in mouse embryos is the bicarbonate/chloride (HCO_3^-/Cl^-) exchanger, which exports HCO_3^-, a weak base, in exchange for Cl^- running down its normal gradient into the cell, thus restoring pH_i (Baltz et al. 1991b; Zhao et al. 1995; Zhao and Baltz 1996; K. P. Phillips and J. M. Baltz, unpublished data). This same mechanism is employed by most other mammalian cells for the alleviation of alkalosis (Olsnes et al. 1986, 1987; Alper 1994). At least two members of the known HCO_3^-/Cl^- exchanger family of genes (termed AE for "anion exchanger"; Kopito 1990; Alper 1994) are expressed in preimplantation mouse embryos (Zhao et al. 1995; Zhao et al. 1997a). Our methods for investigating HCO_3^-/Cl^- exchanger activity in embryos, using pH-sensitive fluorophores, will be outlined here.

Analysis of HCO_3^-/Cl^- Exchanger Activity:
Recovery from Alkalosis

Because the HCO_3^-/Cl^- exchanger requires intracellular HCO_3^-, investigations into its function should be carried out in HCO_3^-/CO_2-buffered media. HCO_3^-/Cl^- exchange can be detected by its activity during recovery from an induced alkalosis. The HCO_3^-/Cl^- exchange-specific component of recovery is manifested as an increase in recovery rate over the baseline recovery rate. The HCO_3^-/Cl^- exchange component has several characteristics which are, together, diagnostic of pH_i regulation by HCO_3^-/Cl^- exchanger activity:-

1. Dependence on Cl^-: recovery from intracellular alkalosis requires an inwardly directed Cl^- gradient and therefore high extracellular Cl^-;
2. Requirement for HCO_3^-: there must be significant intracellular HCO_3^-, supplied by equilibrium with significant levels of CO_2;
3. Inhibition of recovery from alkalosis by stilbene-derivative inhibitors of anion transport and anion exchange.

Thus, a component of recovery from alkalosis which is equally inhibited by the absence of external Cl^-, HCO_3^-/CO_2-free conditions, or the presence of a stilbene inhibitor such as 4,4'-diisothiocyanatostilbene-2,2'-disulfonic acid (DIDS) is likely to be mediated by HCO_3^-/Cl^- exchange (Figure 4-2; Baltz et al. 1991b).

Refinements of Alkalinization Procedure

In order to follow the recovery from alkalosis, it is often desirable to produce the largest alkalinization feasible. First, the dependence of the recovery on pH_i (see below) can then be followed over the largest possible range of pH_i values. Second, we have found that there is a sharp drop in pH_i which immediately follows alkalinization, and which is not inhibitable by any of the manipulations that we have tried. This sharp drop is likely due to a slow component of intracellular buffering, which takes a minute or two to reach equilibrium, or, alternatively, to nonspecific effects of the solution change. Thus, we routinely disregard the first 2–3 minutes of recovery data.

In theory, raising the concentration of weak base should increase the extent of alkalosis. However, we have found that NH_4Cl concentrations above about 35 mM are not tolerated well by the embryos, and furthermore did not give very significant improvements in the extent of alkalinization. Therefore, to give the largest possible range, we exploit a property of the exchanger which will be described in detail below. The exchanger can be made to run backwards, importing HCO_3^- and exporting Cl^-, simply by removing external Cl^-. We have used this manipulation to improve our ability to follow recoveries from alkalosis. Instead of simply alkalinizing the embryos by exposure to NH_4Cl, we instead first expose them to Cl^--free medium for 10 minutes, and then to Cl^--containing NH_4Cl solution (Figure 4-3; Zhao and Baltz 1996). This confers two advantages. First, the

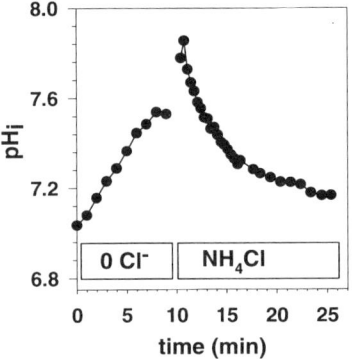

Figure 4-3 Alkaline loading with Cl⁻ removal followed by NH₄Cl. Zygotes were alkaline loaded by first removing external Cl⁻, which causes the HCO_3^-/Cl^- exchanger to run in reverse, alkalinizing the cell, and then introducing NH₄Cl. This maximizes alkalinization, and, by removing intracellular Cl⁻, maximizes the inward Cl⁻ gradient which energizes HCO_3^-/Cl^- exchange. The pH_i was measured using carboxy-SNARF-1. This figure was adapted from Zhao and Baltz (1996).

Cl⁻ removal phase results in an alkalinization of about 0.3–0.5 pH units, so that pH_i reaches a much higher level before recovery starts. Second, the depletion of intracellular Cl⁻ maximizes the inward Cl⁻ gradient, and hence maximizes the rate of HCO_3^-/Cl^- exchange activity. This two-step alkalinization has been used successfully with zygotes, two-cell embryos, morulae, and blastocysts (Zhao and Baltz 1996)

Determination of Dependence of Exchanger Activity on pH_i

pH_i-regulatory mechanisms are activated by deviations from the resting level, or setpoint, pH_i (Vaughan-Jones, 1988; Kim et al. 1996). In the case of the HCO_3^-/Cl^- exchanger, activity increases as pH_i increases, until it reaches the maximum elicitable level (Fineman et al. 1990; Jiang et al. 1994; Sekler et al. 1996; Zhang et al. 1996). The dependence of activity on pH_i is easily extracted from pH_i vs time data describing recovery from alkalosis (such as shown in Figures 4-2 and 4-3). An analysis of pH_i-dependence and setpoint requires determination of the rate of change of pH_i as a function of pH_i. In practice, it is simplest to fit a curve to the data (pH_i vs time), and then differentiate it to obtain the slope as a function of time (dpH_i/dt vs time) for each set of data. The two can be combined, and time eliminated as a variable, if the functions are in forms which can be solved for time. This can be done for simple exponentials (e.g., $pH_i = Be^{-Ct} + r$, where B, C, and r are constants). Therefore, fitting (by nonlinear regression) to an exponential and then differentiating allows the expression of the data in the form of dpH_i/dt as a function of pH_i: $dpH_i/dt = -CpH_i$.

Where this reaches zero, or the nonspecific background rate, is the setpoint of the exchanger (Zhao and Baltz 1996). We have performed this calculation for zygotes, two-cell embryos, morulae, and collapsed blastocysts, and have found that the setpoint is approximately 7.1–7.2 in our hands for all of these stages (Baltz et al. 1991b; Zhao and Baltz 1996).

Other Methods of Determining HCO_3^-/Cl^- Exchanger Activity

The methods outlined above for investigating HCO_3^-/Cl^- exchanger activity have the advantage that the exchanger is operating in a physiologically relevant manner, in which it is alleviating alkalosis. However, a major disadvantage is that there is always a background nonspecific recovery (Figure 4-2), and so detecting activity requires subtracting this variable rate from the total rate (Baltz et al. 1991b). This increases variability and decreases precision. Therefore, in many situations, the Cl^- removal technique (Nord et al. 1988) is used instead.

Removal of extracellular Cl^- causes the exchanger to import HCO_3^- and therefore results in a measurable alkalinization. Since the rate of alkalinization without Cl^- removal is essentially zero, this technique avoids the need to subtract a background rate of pH_i change. Cl^- free media are produced by substituting Cl^- salts with equimolar gluconate salts, or salts of other nonpermeant anions. Upon switching to a Cl^--free medium, pH_i immediately increases, with the increase reaching as much as 0.6 pH units or more in zygotes and two-cell embryos (Figure 4-4; Baltz et al. 1991b; Zhao et al. 1995; Zhao and Baltz 1996). The rate of increase or extent of the increase are measures of HCO_3^-/Cl^- exchanger activity. The increase is reversible upon reintroduction of Cl^-. We have used the Cl^- removal technique routinely to investigate HCO_3^-/Cl^- exchange activity. For example, this technique has been used to measure the K_i (inhibition constant) for inhibition of exchanger activity by DIDS at each stage of embryo development (Zhao and Baltz 1996).

Methods for Probing pH_i Regulation: Induced Acidosis

To investigate the mechanisms used for correcting decreases in pH_i, methods which induce intracellular acidosis are used. In a method analogous to the use of weak bases to induce alkalosis, permeant weak acids can be used to induce acidosis (DeHemptinne et al. 1983; Szatkowski and Thomas 1989). Butyrate (6–10 mM) has been used in eggs (House 1994) and 8–16 cell embryos (LeClerc et al. 1994), where it produces an acidification of about 0.4–0.6 pH units. We have tested the effect of acetate on two-cell-stage embryo pH_i; acetate up to about 20 mM had little effect, while 50 mM and 95 mM acetate caused acidifications of 0.2–0.3 pH units (J. M. Baltz, J. D. Biggers, and C. Lechene, unpublished data). The pH_i decrease was reversible upon subsequent removal of any of these weak acids.

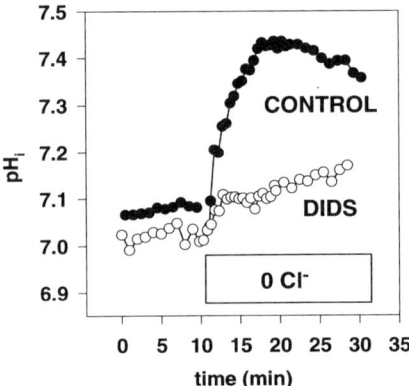

Figure 4-4 Assaying for HCO_3^-/Cl^- exchange activity by Cl^- removal. The HCO_3^-/Cl^- exchange activity in zygotes is revealed by an alkalinization upon removal of external Cl^-. The alkalinization is inhibited by the anion exchange inhibitor DIDS, as expected. The pH_i was measured using carboxy-SNARF-1.

The most common method for inducing acidosis is the ammonium pulse method. As detailed above, exposure of most cells to NH_4^+/NH_3 initially results in a sharp pH_i increase due to NH_3 entry, followed by a slower decrease in pH_i from the maximal level. The decrease is mediated, at least in part, by the entry of NH_4^+ into the cell. Upon removal of the external ammonium, all of the intracellular NH_4^+ and NH_3 exits as the more permeant NH_3, with all of the H^+ which was carried in as NH_4^+ left behind (Boron and DeWeer 1976; Roos and Boron 1981). This net excess in H^+ results in a significant acidification of the cytoplasm after a pulse of ammonium, which can be used as a method of acid loading the cell.

We have found that a 25 mM ammonium pulse which lasts for 15 minutes is sufficient to acidify the cytoplasm by about 0.3–0.5 pH units (Figure 4-5). This has been used successfully with zygotes, two-cell embryos, morulae, and blastocysts (Baltz et al. 1990, 1991a; J. M. Baltz, J. D. Biggers, and C. Lechene, unpublished data).

Investigating Recovery from Acidosis

The major mechanism mediating recovery from acidosis in mammalian cells is the Na^+/H^+ antiporters (Murer et al. 1976; Grinstein 1988; Kim et al. 1996), members of the *NHE* gene family (Fafournoux and Pouysségur 1996). The antiporter is inhibited by the diuretic drug amiloride, and more potently by several of its derivatives [e.g., 5-(*N-N*-dimethyl) amiloride, 5-(*N*-ethyl-*N*-isopropyl) amiloride]. In addition, the removal of external Na^+ prevents the functioning of the antiporter. In contrast, the antiporter does not require HCO_3^-, so investigations can be carried out under atmo-

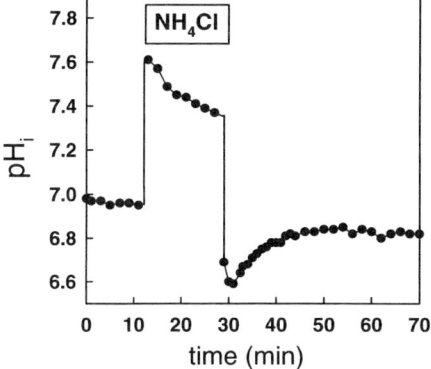

Figure 4-5 Acid loading by the NH$_4$Cl pulse method. Two-cell-stage embryos were acid loaded by a 15-minute pulse of NH$_4$Cl, as indicated by the box. Upon removal of NH$_4$Cl, a net acidification occurs relative to the original pH$_i$, the mechanism of which is described in the text. The pH$_i$ was monitored using BCECF. This figure was modified from Baltz (1993).

spheric levels of CO$_2$ and low HCO$_3^-$ concentrations, in solutions buffered by, for example, HEPES. However, it should be recognized that overall pH$_i$ regulation will be impaired under such conditions, since no HCO$_3^-$-dependent pH$_i$-regulatory mechanisms would be functional. Baseline pH$_i$, and net changes in pH$_i$, are functions of the coordinated action of all regulatory systems present (Ganz et al. 1989).

Another mechanism, the Na$^+$-dependent HCO$_3^-$/Cl$^-$ exchanger, as yet uncharacterized at the molecular level, has been demonstrated to mediate recovery from acidosis in mammalian cells (Boyarsky et al. 1988a,b; Kim et al. 1996). It functions by importing HCO$_3^-$ into the cytoplasm to neutralize excess H$^+$. Na$^+$ enters the cell while Cl$^-$ leaves the cell. For it to operate, there must be sufficient HCO$_3^-$ in the external medium. In addition, removal of external Na$^+$ will prevent pH$_i$ recovery. Furthermore, depletion of intracellular Cl$^-$ should inhibit this mechanism. This exchanger is also inhibited by anion exchange inhibitors such as DIDS.

We have previously investigated recovery from acid loading via ammonium pulses in the two-cell stage of embryos (Baltz et al. 1990, 1991a). A recovery occurred (Figure 4-5); however, none of the manipulations just listed had any measurable effects on the rate or extent of recovery. Therefore, we were unable to detect measurable Na$^+$/H$^+$ antiport activity or Na$^+$-dependent HCO$_3^-$/Cl$^-$ exchange activity during recovery from acidosis in two-cell mouse embryos.

We also found no recovery from acid loading by exposure to the weak acid, acetate (50 mM; J. M. Baltz, J. D. Biggers, and C. Lechene, unpublished data). This lack of significant recovery was also seen by LeClerc et al. (1994) upon exposure of 8–16-cell embryos to butyrate, by House (1994) upon similar butyrate exposure of eggs, and apparently by Edwards et al. (1997) upon lactate-induced acidification of zygotes.

In contrast, there is a report in which Na^+/H^+ antiporter-mediated recovery from acidosis has apparently been demonstrated in two-cell embryos (Gibb et al. 1995). Here, there was no recovery in the absence of external Na^+, nor in the presence of amiloride derivatives. In addition to the apparent Na^+/H^+ antiporter-mediated component, there was a small, slow component of the recovery which was due to lactate efflux from the cell via a lactate/H^+ cotransporter. In addition, House (1994) reported preliminary evidence for a HCO_3^--dependent mechanism for relieving acidosis in eggs, although not for Na^+/H^+ antiporter activity. The discrepancies between these reports have not been explained. Some differences may be due to the different mouse strains used. Indeed, in the report where apparent Na^+/H^+ antiporter activity was demonstrated, the resting pH_i of the embryos was higher (about 7.3) than in the other reports (about 7.0–7.2), possibly indicating a difference in pH_i-regulatory activity. Clearly, further investigations are needed.

Intracellular Ca^{2+} Measurements

There are a large number of reports in which Ca^{2+}-sensitive fluorophores have been used to measure intracellular Ca^{2+} concentrations and concentration changes in mouse eggs or embryos. The vast majority are investigations of the role of Ca^{2+}-mediated signaling at fertilization, including the repetitive Ca^{2+} transients which follow fertilization, signaling mechanisms, and sites of the intracellular Ca^{2+} pools involved. In addition, some work has been carried out to elucidate possible Ca^{2+}-mediated signaling involved in the formation of the blastocoel cavity in the blastocyst (Stachecki et al. 1994). Also, we have investigated the possible involvement of Ca^{2+}-mediated signaling in cell volume regulation in the zygote (Séguin and Baltz, 1997). It is beyond the scope of this chapter to review all of the findings concerning Ca^{2+} regulation in mouse eggs and embryos, or even to cite every research group which has used these techniques. Instead, this will be a brief review of the most recently reported methods used for measurement and calibration. The main Ca^{2+}-sensitive fluorophores in current use with eggs and embryos are fura-2, indo-1, and fluo-3.

Fura-2 is easily loaded using fura-2-AM. Low levels of fura-2-AM are sufficient for loading if the dispersing agent Pluronic F-127 is present. Kline and Kline (1994) used 0.33 µM fura-2-AM with 0.025% Pluronic for 30 minutes at 37°C. Use of Pluronic is not mandatory, however, as ovulated eggs can be loaded with fura-2 using somewhat higher levels of fura-2-AM (2.0–5.0 µM) for 20–60 min at 37° (Cheek et al. 1993; Masumoto et al. 1996). In the absence of Pluronic, the presence of bovine serum albumin (BSA) in the medium will help to carry the fairly hydrophobic fura-2-AM to the cell, but it is not essential.

Fura-2 loading of zygotes is accomplished by similar protocols as those used with eggs. We measured intracellular Ca^{2+} in zygotes after loading them with fura-2 using 2 µM of the AM form for 30 minutes at 37°C without Pluronic or BSA (Séguin and Baltz 1997).

Fura-2 conjugated to the polymer dextran is less likely to be compartmentalized or to leak from cells than the free fluorophore. Microinjection of fura-2-dextran to final concentrations of about 20–50 µM allowed successful measurement of Ca^{2+} in germinal vesicle (GV)-intact ovarian oocytes, as well as ovulated eggs (Mehlmann and Kline 1994; Kline and Zagray 1995).

Indo-1, which like fura-2 is excited in the UV range (around 350 nm), has also been used for intracellular Ca^{2+} measurements in GV-intact oocytes, ovulated eggs, and zygotes. Indo-1 is useful where emission, rather than excitation, ratioing is to be performed. It has been loaded into eggs by incubation with indo-1-AM at a concentration of 50 µM for 20–30 minutes at 37°C in the presence of 0.02% Pluronic (Jones et al. 1995a, 1995b; Kono et al. 1995; Bos-Mikich et al. 1997). The rather high loading concentration used by this group does not appear to reflect any difficulty of loading indo-1, as they employed similarly high levels of fluo-3-AM (Jones et al. 1995a; Kono et al. 1995).

Fluo-3 has the advantage that it is excited in the visible range (490 nm), and thus is less likely to be associated with significant phototoxicity than either fura-2 or indo-1. It is not, however, ratiometric, since it exhibits only a change in intensity but not a spectral shift upon Ca^{2+} binding (Haugland 1996). Fluo-3 has been used with GV-intact oocytes and ovulated eggs (Kline and Kline 1994; Jones et al. 1995a; Kono et al. 1995; Parrington et al. 1996) and morulae (Stachecki et al. 1994). Loading protocols using either low levels of fluo-3-AM (1 µM) for 30 minutes (Kline and Kline 1994) or high levels (50 µM) for 20–30 minutes (Jones et al. 1995a; Kono et al. 1995) have been reported for GV-stage oocytes and eggs, in the presence of 0.02–0.025% Pluronic. For morulae, 5 µM fluo-3-AM for 60 min at 37°C was used, apparently without Pluronic (Stachecki et al. 1994).

Because Ca^{2+}-sensitive fluorophores are prone to becoming compartmentalized in cells, probenecid (2.5 mM), an organic anion pump inhibitor reported to block compartmentalization and export of fura-2 (DiVirgilio et al. 1990), has been included by some groups during loading and throughout experiments to inhibit compartmentalization and retain fluorophores in the egg or embryo cytoplasm (see Kline and Kline 1994). Compartmentalization can be an especially serious problem with Ca^{2+} measurements, as some subcellular compartments have a very high level of Ca^{2+}, which could affect measurements and calibration.

Calibration of Ratiometric Ca^{2+}-Sensitive Fluorophores

Most papers which report ratiometric determinations of Ca^{2+} use the calibration method of Grynkiewicz et al. (1985). Basically, this entails measurement of the fluorescence intensities and ratios in the presence of fully saturating levels of Ca^{2+} (R_{max}) and in nominally Ca^{2+}-free conditions (R_{min}), and then fitting to the equilibrium-binding equation for Ca^{2+} formulated in terms of fluorescence. The equation has the form

$$[Ca^{2+}] = K_d \beta (R - R_{min})/(R_{max} - R) \qquad (2)$$

where K_d is the dissociation constant for Ca^{2+}, and β is the intensity in Ca^{2+}-free conditions divided by the intensity in saturating Ca^{2+}, both taken at that excitation wavelength which, of the pair, least excites the Ca^{2+}-bound form. For example, in the case of fura-2, an excitation pair of 350 nm and 380 nm are often chosen, with 350 nm being near the excitation maximum of the spectrum of the Ca^{2+}-bound form, and 380 nm being near the maximum of the Ca^{2+}-free form. The value of β is then calculated from the intensities at 380 nm in Ca^{2+}-free and -saturating conditions. It should be noted that β evaluates to 1 if the isosbestic wavelength (a wavelength where the fluorophore is Ca^{2+}-insensitive) is used. Fura-2 is essentially Ca^{2+}-insensitive when excited at 360 nm, and the use of this wavelength eliminates the β term (Grynkiewicz et al. 1985). However, the ratio will not depend as steeply upon Ca^{2+} concentration if the isosbestic wavelength is used, since only one wavelength which exhibits a Ca^{2+} response, rather than two which exhibit opposite responses, produce the ratio.

The simplest way to determine R_{min}, R_{max}, and β is to use solutions of the Ca^{2+}-sensitive fluorophore in two buffer solutions, one of which is nominally Ca^{2+}-free [no added Ca^{2+} and millimolar levels of the Ca^{2+} chelator ethylene glycol-bis(β-aminoethyl ether)-N,N,N',N',-tetraacetic acid, EGTA], and the other of which contains saturating Ca^{2+} (usually above 100 µM). The solutions should mimic the intracellular ion concentrations of the cell (for embryos, see Lee 1987 and Baltz et al. 1997). Measuring the fluorescence intensities and ratios then yields the required values. This is, however, only an approximation, as the fluorescence properties of most fluorophores are very different in cytoplasm than in simple buffer solutions.

A somewhat better protocol involves the use of Ca^{2+} ionophores to determine R_{min}, R_{max}, and β in situ in the cytoplasm. In this way, these parameters are determined under the same conditions as the ratios are measured during intracellular Ca^{2+} concentration measurements. Ionomycin and A23187 are commonly used Ca^{2+} ionophores. For UV-excited fluorophores such as fura-2 and indo-1, A23187 cannot be used due to its strong UV-excited fluorescence, but its nonfluorescent analog, 4-bromo-A23187 (4-Br-A23187) can be used instead. We have successfully used 5 µM 4-Br-A23187 at 37°C to clamp intracellular Ca^{2+} concentrations in zygotes (Séguin and Baltz 1997) for measurement of R_{min}, R_{max}, and β on our system. Stachecki et al. (1994) have used 2.5 µM A23187 with morulae. Thus, micromolar amounts of these ionophores seem sufficient to equilibrate intracellular Ca^{2+} with the external solution. We routinely allow 15 minutes for equilibration before taking the first measurement after adding ionophore, and a few minutes before measurements after changing the external Ca^{2+} concentration. Since high intracellular Ca^{2+} is likely to damage the cell, we start with the lowest Ca^{2+} solution when employing ionophore, and then proceed to higher concentrations.

The use of equation (2) for calibration is convenient, but requires knowledge of the value of K_d. It is common practice to assume, rather than measure, a value for K_d. A value which is often used for fura-2 is 0.224 µM, which was determined at 37°C in a buffer solution and appears in the report of the original synthesis of fura-2 (Grynkiewicz et al. 1985). However, it is well known that the value of K_d can vary greatly depending on the environment in which the fluorophore resides, and any calibration which uses an assumed value is an approximation. Another factor which must be considered is that the fluorescence spectrum of fura-2, at least, is quite sensitive to the viscosity of its environment. Since cytoplasm has a higher viscosity than buffer, the in situ spectrum and K_d will differ from those measured in buffer, even if the ionic composition mimics that of cytoplasm. Corrections for the effects of viscosity can be applied to the equation of Grynkiewicz et al. or achieved through choice of excitation wavelength (Poenie 1990; Busa 1992; Kline and Kline 1994).

A better conversion of ratio to Ca^{2+} can be obtained if an in situ calibration of ratio is performed. This can be done by using a Ca^{2+} ionophore to set the intracellular Ca^{2+} concentration to a series of values from essentially Ca^{2+}-free to saturating Ca^{2+}. The observed dependence of ratio on Ca^{2+} concentration is approximately sigmoid, and thus the data obtained can be fitted to a convenient relationship using nonlinear regression. We have (in other cell types) used the equation: Ratio = $1/[A + Be^{-C[Ca^{2+}]}]$, where A, B, and C are arbitrary constants, which produced a good fit, but other equations which conveniently fit the data could be used. This can then be used as a calibration curve, similar to those used for pH_i calibrations, for conversion of ratio to Ca^{2+} concentration.

Alternatively, a plot of $\log[Ca^{2+}]$ vs $\log((R - R_{min})/(R_{max} - R))$, which has an x-intercept of $K_d + \beta$ (Etter et al. 1994), can be used to determine K_d. Finally, a nonlinear least-squares fit to equation (2) itself can be used to find K_d, provided that β is determined independently.

Calibration of Nonratiometric Ca^{2+}-Sensitive Fluorophores

Nonratiometric fluorophores, such as fluo-3, can be calibrated if sufficient care is taken. However, the proper precautions need to be taken to ensure that any changes in fluorescence, during both measurement and calibration, arise from changes in Ca^{2+} and not in other parameters (such as concentration, optical pathlength, etc.) that are virtually eliminated by ratiometric measurements. Calibration is carried out by the same methods outlined above. Again, because fluo-3 and other fluorophores are significantly affected by viscosity and interactions with protein (Harkins et al. 1993), in situ calibration is preferable where possible.

The fraction of fluorophore (f) in the Ca^{2+}-bound form is given by (Harkins et al. 1993) as

$$f = (F - F_{min})/(F_{max} - F_{min}) \qquad (3)$$

where F is the fluorescence emission intensity at any Ca^{2+} concentration, and F_{min} and F_{max} are the intensities in Ca^{2+}-free and saturating conditions, respectively. The fraction of bound Ca^{2+} is also given by

$$f = [Ca^{2+}]/([Ca^{2+}] + K_d) \qquad (4)$$

Therefore, the concentration of Ca^{2+} is given by the relation

$$[Ca^{2+}] = K_d(F - F_{min})/(F_{max} - F) \qquad (5)$$

This last equation can be used to convert fluorescence intensity to Ca^{2+} concentration, if K_d is known, analogous to the method of Grynkiewicz et al. (1985) for ratiometric fluorophores. Preferably, however, the response of in situ fluorophore to graded concentrations of Ca^{2+} (including Ca^{2+}-free and -saturated) would be used to construct a calibration curve. The data could be fitted to equation (5) to yield K_d and the appropriate conversion. Alternatively, a linear regression to a plot of $\log((F - F_{min})/(F_{max} - F))$ vs $\log[Ca^{2+}]$ gives $\log(K_d)$ directly from the x-intercept.

In practice, because it is often only changes in Ca^{2+} which are of interest, rather than absolute concentration, and because of the difficulty of calibrating nonratiometric fluorophores, the fluorescence intensity itself is reported without any attempt at conversion to Ca^{2+} concentration.

Manipulating the Intracellular Ca^{2+} Concentration

During investigations of Ca^{2+} signaling, transport, or homeostasis, it is often necessary to alter the concentration or dynamics of intracellular Ca^{2+}. Most investigations of cytoplasmic Ca^{2+} are concerned with changes in concentration rather than absolute concentration. In eggs and embryos, the transient increase in Ca^{2+} upon fertilization or egg activation is the subject of intense study, as are the repetitive Ca^{2+} oscillations which follow fertilization in mammals. Thus, methods for preventing such changes are useful.

Increasing the intracellular Ca^{2+} buffering capacity can greatly attenuate or eliminate increases in Ca^{2+} concentration, as well as reduce its resting concentration. The Ca^{2+} buffer, BAPTA, has a K_d for Ca^{2+} well below 1 μM (the precise value depends on the environment and the concentrations of other divalent cations such as Mg^{2+}; Haugland 1996). BAPTA can be loaded as BAPTA-AM, using techniques similar to those used for loading fluorophores. In eggs and embryos, 5–100 μM BAPTA-AM, loaded for 30 minutes at 37°C, has been used successfully (Mehlmann and Kline 1994; Kono et al. 1995; Masumoto et al. 1996; Tahara et al. 1996; Séguin and Baltz, 1997). BAPTA has also been microinjected to a final intracellular concentration of 1.6–1.8 mM (Mehlmann and Kline 1994; Xu et al. 1996). Such BAPTA loading prevented cortical granule exocytosis in eggs, greatly inhibited or eliminated the transient Ca^{2+} increases which follow fertilization, and also inhibited a swelling-correlated Ca^{2+} increase in zygotes (Kono et al. 1995; Masumoto et al. 1996; Tahara et al. 1996; Séguin and Baltz 1997). The baseline Ca^{2+} concentration was also reduced to about 50% of its value without BAPTA (Séguin and Baltz 1997). Precise control over intracellular Ca^{2+} concentrations has been achieved by injection of Ca^{2+}-BAPTA and free BAPTA in different proportions, to achieve the desired free Ca^{2+} concentrations in eggs (Xu et al. 1996).

The contents of the intracellular Ca^{2+} stores can also be perturbed by several means. Ionomycin, a Ca^{2+} ionophore, will insert into membranes throughout the cell and thus cause intracellular Ca^{2+} stores to empty. By adding ionomycin (5–10 μM) in Ca^{2+}- and Mg^{2+}-free medium (with EGTA), intracellular Ca^{2+} stores in GV-stage oocytes and ovulated eggs can be emptied (Mehlmann and Kline 1994; Jones et al. 1995a). The removal of external Ca^{2+} itself will also eventually lead to the depletion of intracellular Ca^{2+} stores, as the cell will have no means of refilling them following normal, physiological Ca^{2+} release (e.g., the repetitive Ca^{2+} transients following fertilization).

Intracellular Ca^{2+} stores are replenished by active uptake of Ca^{2+} from the cytoplasm, a mechanism that must be functional to maintain Ca^{2+} at high concentrations within the intracellular pools. Such uptake of intracellular Ca^{2+} requires activity of the endoplasmic reticulum Ca^{2+}-ATPase. Thapsigargin (at about 10 μM) inhibits this ATPase, and thus causes emptying of Ca^{2+} stores (Kline and Kline 1992; Jones et al. 1995b), which do not refill in the absence of extracellular Ca^{2+}.

Refilling of Intracellular Ca^{2+} Stores

If intracellular Ca^{2+} stores are depleted (either during normal physiological processes or by experimental intervention), the cell will attempt to refill them by importing external Ca^{2+} across the plasma membrane. This influx of Ca^{2+} can be investigated by measuring the increase in cytoplasmic Ca^{2+}, using Ca^{2+}-sensitive fluorophores, upon reintroduction of external Ca^{2+}.

More elegantly, the increased divalent cation permeability in the plasma membrane can be detected by exploiting a property of fura-2. Fura-2 fluorescence is quenched by Mn^{2+}, and thus fura-2 fluorescence decreases with increasing Mn^{2+} concentrations. By measuring fura-2 fluorescence excited at 360 nm (the Ca^{2+}-insensitive wavelength), Mn^{2+} quenching can be detected independently of any Ca^{2+} changes. Therefore, the divalent cation permeability of the plasma membrane can be quantified by measuring the decrease in 360 nm fura-2 fluorescence (Hallam et al. 1988). This technique has been used to investigate enhanced divalent cation entry into eggs and zygotes, and to demonstrate the increased divalent cation influx when intracellular Ca^{2+} stores are depleted (Cheek et al. 1993; McGuinness et al. 1996).

Intracellular Cl^-

We have recently begun using the Cl^--sensitive fluorophore, MQAE, in eggs and zygotes. Although the chemistry of MQAE has not allowed the synthesis of AM or other cleavable derivatives, we have found that sufficient levels of the fluorophore can be loaded by exposure of the egg or embryo to 20 mM MQAE for 30 minutes at 37°C. MQAE is not ratiometric (see above). Because of this, and since the loading technique results in varying amounts of the fluorophore being loaded, we have used the fluorophore only for measuring relative changes in intracellular Cl^- concentration, rather than absolute Cl^- concentrations.

The dependence of fluorescence intensity on Cl^- concentration for any of the fluorophores which detect Cl^- by a quenching mechanism is given by the Stern–Volmer equation:

$$1/F = (1 + K_{Cl}[Cl^-])/F_0 \qquad (6)$$

where F is the measured fluorescence in the presence of Cl^-, K_{Cl} is the Stern–Volmer constant (in M^{-1}), and F_0 is the fluorescence in the absence of Cl^- (Verkman 1990). The value of the Stern–Volmer constant is equal to the inverse of the Cl^- concentration which half-maximally quenches fluorescence. Note that fluorescence intensity (F) decreases asymptotically towards zero as the concentration of Cl^- increases. From this equation, it is evident that $1/F$ is proportional to the Cl^- concentration. Therefore, to plot relative Cl^- as a function of time, $1/F$ vs time should be plotted. The slope of such a plot is

$$d(1/F)/dt = (K_{Cl}/F_0) \cdot d[Cl^-]/dt \qquad (7)$$

indicating that the slope gives the relative rate of change of Cl^- concentration.

Calibration of Cl^--Sensitive Fluorophores

Although this has not been done for eggs or embryos, it is possible to attempt to calibrate the fluorescence of MQAE, or one of the other Cl^--sensitive fluorophores, by one of two methods. In the first method, the fluorescence in the absence of Cl^- (F_0) is found by incubating the cells in Cl^--free solution until all the intracellular Cl^- is depleted. For embryos, electron probe measurements have shown that all Cl^- is depleted within 30 minutes (Baltz et al. 1991b). Since these fluorophores leak out of cells, including embryos, at a significant rate, the value for F_0 will decrease with time, and therefore the value at different times in the experiment must be estimated from the rate of fluorescence decrease due to leakage (Verkman 1990). It is possible to then use this estimate of F_0, along with the published estimates of K_{Cl} for the fluorophore, to derive $[Cl^-]$ from the Stern–Volmer equation (above). This method is, at best, approximate, since the estimation of F_0 is uncertain, and the value of K_{Cl} varies for different cellular environments.

The second method involves the use of the synthetic Cl^-/OH^- exchanger, tributyltin (TBT) to equalize the intracellular Cl^- concentration with the external Cl^- concentration (Verkman 1990; Engblom and Akerman 1993), in a method analogous to that used for calibration of pH- and Ca^{2+}-sensitive fluorophores (above). Since TBT is a Cl^-/OH^- exchanger, the H^+ and Cl^- gradients will be equal at equilibrium [analogous to equation (1)], and thus any pH gradient must be eliminated. This is done by including the K^+/H^+ exchanger nigericin and high K^+, as described for calibration of pH-sensitive fluorophores (above). Thus, by exposing MQAE-loaded cells to several different concentrations of Cl^- in the presence of TBT, nigericin, and high K^+, a linear calibration curve relating $1/F$ to $[Cl^-]$ can be constructed (Krapf et al. 1988). Again, the effect of loss of fluorescence due to fluorophore leakage must be corrected.

As mentioned, we have not performed such Cl^- calibrations in embryos. However, we have utilized TBT to make zygotes permeable to Cl^- in studies of zygote cell volume regulation (Séguin and Baltz 1997). TBT at 10 µM, added from freshly prepared 1000× stocks in DMSO, was effective.

Examining Cl^- Transport

Intracellular Cl^--sensitive fluorophores have thus far been used only in eggs and zygotes (K. P. Phillips and J. M. Baltz, unpublished data), but not, to our knowledge, in later stage embryos. Loading into zygotes was as described above, by incubation with 20 mM MQAE. This resulted in adequate fluorescence levels, especially since we were examining Cl^- efflux, which results in depletion of intracellular Cl^- and hence an increase in fluorescence. Figure 4-6 shows a plot of normalized $1/F$ vs time for a zygote in normal Cl^--containing solution, and subsequently in Cl^--free medium.

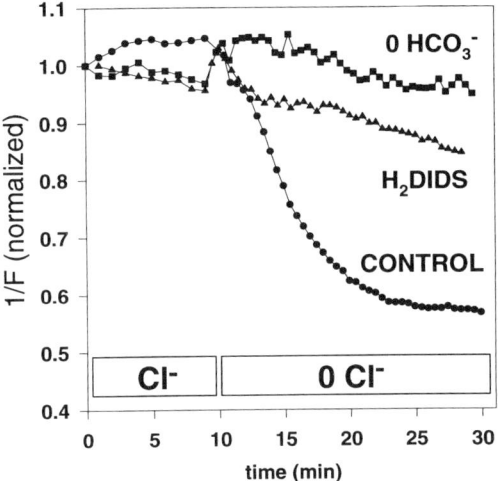

Figure 4-6 Intracellular Cl^- efflux monitored using MQAE. Zygotes were loaded with MQAE as described in the text, and the efflux of intracellular Cl^-, whose concentration is proportional to $1/F$ (1/fluorescence intensity), was detected. The Cl^- efflux is mainly via HCO_3^-/Cl^- exchange, as it is largely inhibited by lack of HCO_3^- or presence of DIDS. The Cl^- efflux mirrors the change in pH_i induced by Cl^- removal, as expected for HCO_3^-/Cl^- exchange (Figure 4-4). Note the artifactual jumps in $1/F$ upon solution changes, probably due to small changes in the position of the zygotes; this illustrates a drawback of nonratiometric fluorophores.

$1/F$ was normalized by dividing by its initial value in Cl^--containing medium. Efflux of Cl^- from the cytoplasm is thus revealed as an increase in fluorescence over time (decrease in normalized $1/F$ from its initial value of 1) following Cl^- removal. The normalization strategy used here is aimed at correcting for unequal loading of fluorophores between embryos. It is strictly valid only if the initial Cl^- concentrations are identical for all embryos. In practice, there is, of course, some variability; however, the approximation is probably acceptable since independent measurements of intracellular Cl^- concentrations in embryos using electron probe X-ray microanalysis revealed fairly uniform Cl^- concentrations among embryos (Baltz et al. 1997).

This technique is useful for following changes in Cl^- concentration. The relative initial rate of change of Cl^- upon removal of external Cl^- is calculated by determining the initial slope by regression analysis. This slope is then used to compare rates of change of Cl^- for different treatments. For example, the rate of Cl^- efflux is significantly decreased by inclusion of the anion transport inhibitor, H_2DIDS (dihydro DIDS, an analogue of DIDS which is nonfluorescent when exposed to UV excitation) or by nominally HCO_3^--free solutions at atmospheric CO_2 (Figure 4-6), which together indicate HCO_3^-/Cl^- exchanger activity. Note that this is the same Cl^--removal protocol described above for revealing HCO_3^-/Cl^- exchanger activity using pH-sensitive fluorophores, and that the similar time courses of pH_i increase and Cl^- decrease upon Cl^- removal indicate their coupling via this transporter (Figure 4-4; K. P. Phillips and J. M. Baltz, unpublished data).

Na^+- and K^+-Sensitive Fluorophores

As described above, fluorophores which are sensitive to either Na^+ or K^+ are available. We have tested SBFI, a UV-excitable Na^+-sensitive fluorophore which can be ratioed; sodium green, a nonratioable Na^+-sensitive fluorophore excited in the visible range, and PBFI, a K^+-sensitive fluorophore with properties similar to those of SBFI; in two-cell embryos. In our hands, none of these were useful for detection of the relevant ions in embryos. Although good loading of the cells and fluorescence could be obtained, we were unable to elicit significant changes in fluorescence in response to manipulations which would produce large changes in intracellular Na^+ or K^+ (Y. Zhao and J. M. Baltz, unpublished data). Therefore, the use of fluorescent indicators to detect Na^+ and K^+ in embryos awaits further refinement of the techniques, or a new generation of fluorescent indicators for these ions.

Ion Concentrations in Blastocoel Fluid

Blastocoel fluid is produced by fluid transport across the trophectoderm of the blastocyst (Biggers et al. 1988). This fluid is accumulated secondary to Na^+ and Cl^- transport into the blastocoel cavity (Manejwala et al. 1989), and differs in composition from the external fluid, including the concentrations of several ions (Borland et al. 1977). Transport of Cl^- across the trophectoderm has been measured successfully following introduction of a Cl^--sensitive fluorophore into the blastocoel fluid. Brison and Leese (1993) introduced the Cl^--sensitive fluorophore, SPQ, into blastocoel fluid by incubating rat blastocysts with 20 mM SPQ for 2 hours. They found this method to be at least as good as microinjection, culturing from the eight-cell stage in the presence of SPQ, or collapse and re-expansion of blastocysts in the presence of SPQ. There was some loading of SPQ into the trophectodermal and inner cell mass cells, but their calculations showed that this made a minimal contribution to the overall signal.

Recently, we have used a similar technique to examine Cl^- transport in the mouse blastocoel (Zhao et al. 1997a). The Cl^--sensitive fluorophore MQAE was loaded into mouse blastocoels by incubating with 20 mM MQAE in hypotonic medium (diluted 1:1 with water) for 30 minutes at 37°C, followed by 30 minutes in normal medium with 20 mM MQAE. We found that most blastocoel cavities were loaded with the fluorophore by this technique. We did not attempt to calibrate the fluorescence to absolute Cl⁻ concentrations, but rather examined only relative changes in Cl⁻ concentration.

Changes in blastocoel Cl⁻ concentrations could be detected using this technique. Upon removal of external Cl⁻, fluorescence increased linearly for at least 1 hour, indicating efflux of Cl⁻ from the blastocoel cavities (Brison and Leese 1993; Zhao et al. 1997a). In our case, we were also able to demonstrate reuptake of Cl⁻ into the blastocoel upon restoration of external Cl⁻ following depletion.

Both groups found linear changes in fluorescence over long periods of time, after Cl⁻ removal (cf. our data, Figure 4-7). It would appear difficult, at first to reconcile the simple linear dependence of fluorescence (F) on time, i.e., a constant dF/dt, with the Stern–Volmer equation [equation (6)] Differentiation of the Stern–Volmer equation instead yields a complex dependence of dF/dt on Cl⁻ concentration. However, measurement of blastocoel fluid Cl⁻ is unlike the measurement of intracellular Cl⁻, since Cl⁻ is present at very high concentrations in blastocoel fluid (Borland et al. 1977; Biggers et al. 1988). At such high Cl⁻ concentrations and for small relative changes in Cl⁻, dF/dt can be shown to be approximately equal to $-d[Cl^-]/dt$, yielding the observed dependence. Hence, in high Cl⁻, F rather than $1/F$ should be plotted (Figure 4-7).

In addition to Cl⁻-sensitive fluorophores, we have also successfully loaded the pH-sensitive fluorophore, BCECF, into blastocoel cavities using the same method with 50–100 µM of BCECF-free acid (Y. Zhao and J. M. Baltz, unpublished data). This could potentially be useful in studies of regulation of the blastocoel fluid pH by the trophectoderm. Presumably, other ion-sensitive fluorophores are also loadable into the blastocoel, and may be similarly useful.

Toxicity of Ion-Sensitive Fluorophores in Eggs and Embryos

Several of the steps in the procedures of fluorophore loading and excitation of fluorescence are potentially damaging to cells. The most common method of fluorophore loading, which makes use of endogenous intracellular esterases to cleave AM or similar ester-bonded groups from cell-permeable fluorophore precursors, produces toxic by-products. Hydrolysis of each AM group yields a molecule of acetic acid and one of formaldehyde (Haugland 1996), either of which, in sufficient amounts, would clearly cause damage. Furthermore, the fluorophore itself could be toxic to embryos. In addition, many of the Cl⁻ fluorophores are halide salts (e.g., MQAE is a bromide), and these ions could damage eggs or embryos.

Excitation of fluorescence can also negatively affect cells. Exposure to light per se usually does not harm cells, although excessive amounts of the UV illumination required to excite some fluorophores is detrimental. However, photobleaching is an unavoidable consequence of fluorescence, and the products of photobleaching can be toxic. Photobleaching occurs

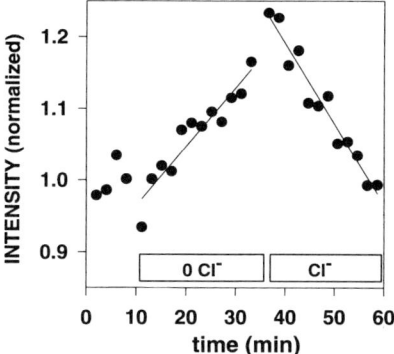

Figure 4-7 Blastocoel Cl⁻ monitored with MQAE. MQAE was used to follow changes in Cl⁻ in blastocoel fluid. Upon removal of external Cl⁻, fluorescence increased as Cl⁻ efflux caused a decrease in the Cl⁻ concentration in the blastocoel, which reduced quenching of fluorescence. Then, Cl⁻ reintroduction caused a decrease in fluorescence as Cl⁻ re-entered the blastocoel. This figure was adapted from Zhao et al. (1997a).

when the excitation light causes breakage of intramolecular bonds in the fluorophore. The products of such photolysis are highly reactive, and can damage the cell. Thus, photodamage will occur progressively during fluorescence measurements.

Therefore, it is important to minimize the exposure of cells to excitation light. This is done by shutting off the excitation light pathway between periodic measurements, which are done as quickly and infrequently as is practical. Most systems have an automatic shutter or similar device which cuts off the excitation between measurements. For example, when we perform pH_i measurements, we routinely excite fluorescence to obtain a ratio for less than 1 second, with measurements repeated only every 20–30 seconds. In addition, the intensity of the excitation light should be kept as low as is feasible for the characteristics of the particular detection system used. Even so, it is wise to ensure that the measurement is not excessively damaging to the egg or embryo.

There are two classes of damage which could occur to cells during fluorescence measurements. One is acute damage which compromises the function of the cell during the measurements. The other is damage which occurs during the measurement but which only affects the function of the cell much later. An example of the latter type would be damage to DNA. In practice, such latent damage is usually not a problem, since the eggs or embryos are discarded immediately after most experiments. This would only become a concern if fluorescence measurements are to be used at the beginning of a long-term experiment, e.g., to predict embryo viability or development based on the behavior of some intracellular ion.

Acute damage would always be problematic if it affects any parameter related to that being measured. Thus, if metabolism, membrane properties, or any of a number of cellular functions, such as signaling pathways, were affected by the fluorophore loading or fluorescence measurements, then experimental results could be rendered invalid.

If the assumption is made that any acute toxicity serious enough to affect the function of a cell during fluorescence measurements would also result in a decrease in its subsequent viability and developmental potential, then assessing the viability of embryos or eggs after dye loading and fluorescence excitation should reveal both short- and long-term damage.

There are a number of indications that eggs and embryos are not affected too adversely by such techniques. Most simply, the morphology and appearance of eggs and embryos does not usually change upon fluorophore loading or fluorescence excitation. More convincing evidence of viability is that eggs loaded even with high concentrations of Ca^{2+}-sensitive fluorophores have been repeatedly shown to exhibit the transient Ca^{2+} increase at fertilization, and exhibit repeated Ca^{2+} transients for an extended period of time thereafter, indicative of normal Ca^{2+} homeostasis and signaling pathways (Kline and Kline 1994; Jones et al. 1995b). Similarly, functional pH_i regulatory mechanisms such as the HCO_3^-/Cl^- exchanger can be detected using pH-sensitive fluorophores (Baltz et al. 1991b; Zhao et al. 1995; Zhao and Baltz, 1996). Therefore, eggs and embryos have appeared to tolerate fluorophore-based intracellular ion measurements well.

Originally, to assess the state of embryos after fluorophore loading and exposure to excitation light, we cultured mouse two-cell embryos after they had been loaded with either carboxy-SNARF-1 or BCECF by incubation with the AM forms, as described above, and then exposed them to the excitation light appropriate for each fluorophore (Baltz et al. 1990; Zhao et al. 1995), and recorded subsequent development to the four-cell stage and the blastocyst stage. Control embryos developed to the four-cell stage and to blastocysts at a rate of about 95% (Zhao and Baltz 1996, and unpublished data). BCECF, loaded at 1.2 µM BCECF-AM for 15 minutes (the highest concentration we have used), had some effect on two-cell embryo development. After fluorophore loading, 93% developed to the four-cell stage, but only about 40% developed to blastocysts. Excitation light exposure of BCECF-loaded embryos further decreased development, with 80% reaching the four-cell stage and only about 30% developing to blastocysts. Increasing either the BCECF-AM concentration or the light intensity by 10-fold completely stopped development (Baltz et al. 1990, and unpublished data). Although we have not tested this directly, we anticipate that using 0.1 µM loading concentration of BCECF-AM would reduce this toxicity. Carboxy-SNARF-1-loaded two-cell embryos (5 µM carboxy-SNARF-1-AM, 30 minutes) developed to the four-cell stage at a rate of 95–99% and to the blastocyst stage at a rate of about 80%, whether or not exposed to excitation light (Zhao and Baltz 1996 and unpublished data). Therefore, by this assessment, carboxy-SNARF-1 has less of a detrimental effect than BCECF at these concentrations. Thus, there seems to be some damage to embryos which is revealed by long-term culture. However, since most embryos are able to cleave normally for at least one cleavage after being loaded with these fluorophores and exposed to excitation light, they can probably be presumed viable throughout the duration of the experiment.

We have recently performed more rigorous experiments aimed at assessing whether eggs and embryos are viable and intact in the short-term immediately following intracellular ion concentration measurements using intracellular fluorophores (Phillips et al. 1998). Some results of these experiments are discussed below.

Testing Toxicity in Eggs using IVF

To assess the viability of eggs following fluorophore loading and the light exposure experienced during intracellular ion measurements, we have used in vitro fertilization (IVF) of cumulus-free eggs to test several fluorophores (Phillips et al. 1998). The ability of eggs to cleave to the two-cell stage by 24 hours post-IVF was assessed. To determine the maximal rate of development to the two-cell stage which could be obtained with unperturbed eggs, IVF was performed immediately after isolation of cumulus-free eggs, without fluorophore loading or any other manipulations. These developed to the two-cell stage at a rate of $67 \pm 9\%$. To then determine the effect of AM-group cleavage alone without light exposure, eggs were loaded with carboxy-SNARF-1-AM (5 µM carboxy-SNARF-1-AM for 30 minutes at 37°C), and then IVF was immediately performed. These developed to the two-cell stage at the essentially the same rate ($68 \pm 8\%$) as the control group, indicating that AM hydrolysis alone has no effect (Figure 4-8).

Experiments were then performed which were designed to mimic intracellular ion concentration measurements using the pH-sensitive fluorophores carboxy-SNARF-1 (loaded as above) and BCECF (1.2 µM BCECF-AM, 10 minutes), the Ca^{2+}-sensitive fluorophore fura-2 (2 µM fura-2-AM, 30 minutes), and the Cl^--sensitive fluorophore MQAE (20 mM MQAE, 30 minutes). Parallel control groups for each fluorophore, consisting of eggs without fluorophore subjected simultaneously to the excitation illumination, showed fertilization and cleavage rates similar to that of the control group (above) which had been subjected to minimal handling (not shown). This indicates that the extended handling and light exposure had no significant detrimental effect on eggs, even with the UV illumination used with fura-2 and MQAE.

For eggs loaded with carboxy-SNARF-1, BCECF, and fura-2, the rates of cleavage to the two-cell stage were not affected by fluorophore loading and excitation light exposure (Figure 4-8). In contrast, the rate of cleavage to the two-cell stage was significantly reduced for MQAE-loaded eggs exposed to excitation light (Figure 4-8). Thus, with MQAE, there appears

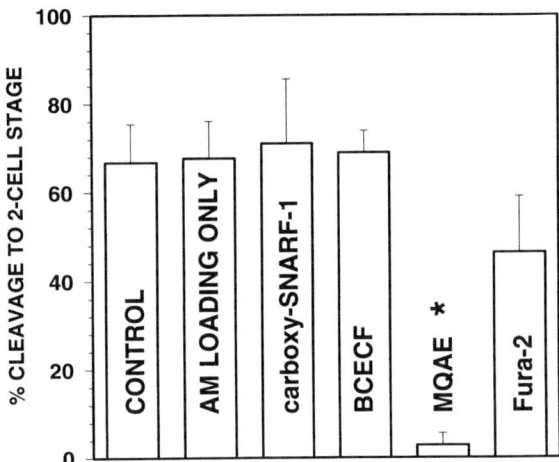

Figure 4-8 In vitro fertilization and cleavage to the two-cell stage after fluorophore loading and excitation. Eggs were loaded with carboxy-SNARF-1, BCECF, MQAE, or fura-2, exposed to excitation light, and then used for IVF. Control eggs were not loaded with fluorophore, while the group designated "AM loading only" were loaded with carboxy-SNARF-1 but not exposed to excitation light. All groups cleaved normally to the two-cell stage, except for those loaded with MQAE ($*P < .05$). See text for discussion. This figure was adapted from Phillips et al. (1998).

to be damage to the eggs which is evident by the time of the first cleavage. The MQAE-loaded eggs were, however, successfully fertilized, since 83% developed two pronuclei, which is not significantly different from control groups (although pronuclei developed more slowly).

These results indicate that, with the exception of MQAE, the techniques used for intracellular ion concentration measurements do not appear to cause any appreciable damage to eggs, as they are capable of being fertilized and developing to at least the two-cell stage. Even with MQAE, the eggs were capable of being fertilized and forming two pronuclei, indicating that acute damage during measurements is not likely to be so great as to invalidate the results.

Testing Membrane Properties with Whole-Cell Patch
Recordings after Fluorophore Measurements

The electrophysiological properties of a cell should be a sensitive indicator of the condition of the plasma membrane and the rheogenic ion transporters it contains. Indeed, it has been reported that electrophysiological recordings revealed damage to leech glial cells caused by high levels of BCECF and excitation illumination (Nett and Dietmer 1996). We have therefore used whole-cell patch-clamp recordings (Hamill et al. 1981) from mouse zygotes to determine if there is any detrimental effect of loading fluorophores into them by AM hydrolysis or of exposing fluorophore-loaded zygotes to excitation light (Phillips et al. 1998). As with the IVF experiments, a control group was used which was subjected to minimal manipulations, and the effect of fluorophore loading alone was assessed by loading with carboxy-SNARF-1-AM (5 µM for 30 minutes).

The same four fluorophores (carboxy-SNARF-1, BCECF, fura-2, and MQAE, loaded as described for eggs) which had been assessed in eggs using IVF were investigated electrophysiologically in zygotes (Phillips et al. 1998). The whole-cell conductance measured at +30 mV (during voltage ramps imposed from a holding potential of −20 mV) served as an indicator of the condition of the plasma membrane. As can be seen in Figure 4-9, fluorophore loading and AM-group hydrolysis alone had no significant affect on the electrophysiological properties of the cell. Carboxy-SNARF-1, BCECF, or fura-2 loading with excitation light illumination also did not affect membrane conductance. In contrast, MQAE loading and excitation caused a very significant increase in conductance, indicating that this fluorophore increased the membrane permeability and ion conductance of the zygote.

To obtain an example of membrane properties of zygotes which had been purposely photodamaged, we loaded zygotes with BCECF and then exposed them to continuous 450 nm illumination for 30 minutes, prior to electrophysiological recordings. Many of these zygotes were then refractory to establishing gigaohm seals, and thus recordings could not be obtained. A number of those from which recordings were successfully obtained had significantly higher conductances, indicative of membrane damage, although several others did not. The two classes of photobleached zygote responses produced a mean conductance between that of control and the high conductance of MQAE-loaded zygotes (Figure 4-9). Thus, photodamage apparently can be revealed as an increase in membrane conductance.

From these IVF and electrophysiological experiments (Phillips et al. 1998), we conclude that routine intracellular ion measurements using carboxy-SNARF-1, BCECF, or fura-2 probably do not damage the plasma membrane excessively, increase ionic permeability of the plasma membrane, or compromise egg or embryo viability. In contrast, Cl^- measurements with MQAE should be carried out with caution, as this fluorophore seems to have a marked detrimental effect on eggs and zygotes; however, our ability to measure Cl^- transport via a functional HCO_3^-/Cl^- exchanger using this method (K. P. Phillips and J. M. Baltz, unpublished data, and see Figure 4-6) indicates that the damage is not so great as to preclude its use entirely.

Possible Pitfalls in Intracellular Ion Concentration Measurements with Fluorophores

Although the premise behind using ion-sensitive fluorophores to detect intracellular ions is simple and straightforward, there are, nonetheless,

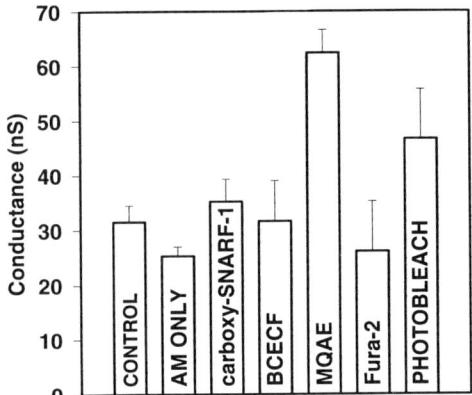

Figure 4-9 Whole-cell conductance measured by patch-clamp electrophysiological recordings after fluorophore loading and excitation in zygotes. Zygotes were loaded with carboxy-SNARF-1, BCECF, MQAE, or fura-2, exposed to excitation light, and then used for whole-cell patch-clamp recording. A control group with no fluorophore, and a group subjected only to carboxy-SNARF-1 loading without excitation ("AM only"), were also assessed. The whole-cell conductance was determined at 30 mV from a voltage ramp as described in the text. No effect on conductance was found, except for MQAE, where conductance increased sharply. A deliberately photobleached group also showed evidence of increased conductance (see text). This figure was adapted from Phillips et al. (1998).

numerous things which can conspire to give incorrect results. Many variables can affect the fluorescent properties of fluorophores. Although ratiometric measurement methods correct for many of these, there are still many possible sources of problems.

Of course, before beginning to use any fluorescence quantitation system, whether it is an imaging system or a simpler photomultiplier-type detector, it is vital to validate the operation of the system. The methods for such validation are beyond the scope of this chapter, but have been covered extensively elsewhere (Bright et al. 1987, 1989; Jericevic et al. 1989).

Fluorophore Sequestration

Cytoplasmic ion concentration measurements depend on the assumption that the majority of the fluorophore resides free in the cytoplasm. Either binding of the fluorophore to intracellular components or sequestration of fluorophore into intracellular compartments other than the cytosol can render this assumption invalid. Sequestration of fluorophores has been reported in many types of cells, including sea urchin eggs and zygotes, which have been reported to completely sequester BCECF and fura-2 (Tsien 1989; Mozingo and Chandler 1990).

The simplest test of whether a fluorophore is cytoplasmically localized is to determine how much of the fluorophore is retained if the plasma membrane is disrupted while intracellular organelles are left intact. One method of disrupting the membrane of a large cell, such as those of eggs or cleavage-stage embryos is to use a microneedle (which can be made of sharp, finely drawn glass as for a micropipette) to breach the membrane. If done gently, the contents of the cytoplasm can be left trapped in the zona pellucida. Measurement of fluorescence during disruption allows quantitation of the amount of fluorophore released compared with that retained. For BCECF, we found that >90% of fluorescence is lost with a half-time of around 12 seconds upon disruption of the membrane (Baltz et al. 1993), indicating a cytoplasmic localization.

Another method used to lyse the plasma membrane employs digitonin. Low concentrations of digitonin (e.g., 10–20 µM) preferentially disrupt the cholesterol-containing plasma membrane (Nieminen et al. 1990; House 1994). We found that BCECF-loaded mouse two-cell embryos lost all their fluorescence after exposure to 10 µM digitonin, with a residual fluorescence of less than 10%, similar to the results with mechanical disruption (Baltz et al. 1993). At this concentration, the loss of fluorescence began at varying times after digitonin introduction for each embryo, but, once commenced, fluorescence was lost rapidly with a half-time of about 10 seconds. Exposure to 100 µM digitonin is sufficient to disrupt not only the plasma membrane but organelle membranes as well. Exposure to 100 µM digitonin caused the immediate loss of fluorescence from all embryos, but the time course of fluorescence decrease was no more rapid than for 10 µM digitonin, nor was there any less residual fluorescence. Thus, for BCECF, the fluorophore seems to be localized almost entirely in the cytoplasm. House (1994) used confocal microscopy to investigate the distribution of carboxy-SNARF-1 in ovulated eggs. He found no evidence of sequestration or regions of different pH, except for a few acidic vesicles which appeared in some eggs. In addition, he used 20 µM digitonin, and found that >90% of fluorescence was lost; no further decrease was seen upon exposure to 100 µM digitonin. We have repeated this with carboxy-SNARF-1-loaded GV-intact oocytes, ovulated eggs, and zygotes, using 20 µM digitonin (Phillips and Baltz 1996). Again, almost all fluorescence was lost, with a mean of 4–6% retained after digitonin. Thus, carboxy-SNARF-1 also appears to be almost entirely localized in the cytoplasm of eggs and zygotes.

For each different fluorophore, embryo stage, and probably for each different loading protocol, there is a danger that there might be significant sequestration of the fluorophore away from the cytoplasm. If there is any doubt, cytoplasmic localization of the fluorophore should be confirmed.

Artifacts Due To Overloading

Measurement errors can arise from overloading cells with fluorophores. At high enough concentrations, processes such as absorption of emitted fluorescence by the fluorophores themselves, and nonradiative transfer of energy between fluorophores, become significant. This can lead to distortion of the fluorescence spectrum, and unpredictability of fluorophore response.

Another source of possible error due to high loading levels is the incomplete hydrolysis of AM ester groups when loading fluorophores using the AM form. For some ion-sensitive fluorophores, such as BCECF, the AM-conjugated form is nonfluorescent, and will not be problematic. However, for other fluorophores, such as fura-2, the AM form is highly fluorescent. The fluorescence spectrum of unhydrolyzed or partially hydrolyzed fura-2-AM is not affected by Ca^{2+}, and the presence of fura-2-AM in the cell will therefore lead to errors in the estimation of Ca^{2+} concentration (Zhao et al. 1997b). Thus, it is important to use the minimum intracellular fluorophore concentration that is practical, and to validate that the fluorophore is performing as expected. At the least, it should be verified that ratiometric measurements are insensitive to excitation illumination intensity, and that calibration curves are insensitive to moderate changes in the levels of fluorophore loading.

Artifacts Due to Changes in Physical Parameters

Variables other than ion concentrations will affect the fluorescence properties of ion-sensitive fluorophores. It is known that the excitation spectrum of fura-2 is significantly red-shifted in cytoplasm relative to that in buffer solutions, resulting in decreased ratios for a given Ca^{2+} concentration. At least part of this shift is due to the higher viscosity of cytoplasm (Poenie 1990; Busa 1992). Fluo-3 also exhibits a viscosity artifact (Harkins et al. 1993), as does the pH-sensitive fluorophore BCECF (Cardullo and Dandala 1997). Careful in situ calibration may help to eliminate viscosity artifacts as a source of error in intracellular ion concentration measurements, provided that the viscosity of the cytoplasm remains constant throughout the experiment. However, the constancy of cytoplasmic viscosity may not always be a valid assumption, particularly in instances where the cell swells or shrinks, and the concentrations of cytoplasmic components change accordingly. Thus, an apparent change in Ca^{2+} concentration could actually be an artifact of a viscosity change. Methods for correcting for, or avoiding, viscosity artifacts with fura-2 and BCECF have been developed (Poenie 1990; Busa 1992; Cardullo and Dandala 1997).

Artifactual apparent ion concentration changes can also be the result of changes in cell dimensions or position during the course of measurements, if the rate of change is large enough that significant changes occur on the time scale of individual fluorescence measurements or ratio pair acquisitions.

Such movement artifacts can be the result of mismatches between the two pairs of a ratio, where the cell has changed size, shape, or position between measurements at the two wavelengths. However, errors due to changes in physical parameters are especially severe for nonratiometric fluorophores. In this case, any shrinking or swelling will alter intracellular fluorophore concentration and show up as an apparent change in fluorescence, and any shift in position is also likely to affect the intensity measured at a given location in the microscopic field. Therefore, we monitor embryos for size changes when using nonratiometric fluorophores (e.g., MQAE for Cl^- measurements), and reject any measurement obtained from embryos whose dimensions changed significantly (e.g., >5%) during the experiment (Zhao et al. 1997a).

In any case, in experiments where any change in physical parameters occurs, it is important to perform appropriate controls to eliminate this variable as the source of any apparent changes in ion concentration reported by a fluorophore. This can be done by repeating the manipulation under conditions where the intracellular ion concentration is clamped at a fixed value. For example, pH_i is held constant by nigericin and high external K^+, and Ca^{2+} by the use of Ca^{2+} ionophores, as described above, or using an intracellularly loaded buffer such as BAPTA. BAPTA effectively lowers and clamps Ca^{2+} concentrations and can easily be loaded coincidentally with fura-2 using BAPTA-AM, as we did to investigate Ca^{2+} during swelling of zygotes (Séguin and Baltz 1997). If the apparent change in ion concentration reported by the ion-sensitive fluorophore is found to be abolished after ion concentration is clamped, even when the physical parameters of the embryo are changed, then it is probable that any changes in concentration reported by the the ion-sensitive fluorophore are real.

Another possible source of an artifactual change in apparent ion concentration is any change in temperature. The properties of fluorophores are often temperature dependent. In addition, intracellular components which buffer the ion of interest are also generally temperature dependent. A good example is the measurement of pH_i with BCECF. Because BCECF has a small temperature dependence, any changes in temperature might therefore be thought to have little effect. However, the endogenous buffers in the cytoplasm have a relatively steep temperature dependence, and therefore pH_i changes rather significantly with temperature (a little less than 0.02 pH units per degree Celsius; Somero 1985). Thus, it is important to ensure that temperature is held constant during such measurements, and that temperature fluctuations, e.g., during solution changes, are accounted for.

Artifacts Due to Interactions with Other Compounds

Many inhibitors and other compounds which might prove useful during experiments involving intracellular ion concentration measurements with eggs or embryos are themselves fluorescent. This is especially true for excitation in the UV range, where many molecules fluoresce strongly. Thus, it is important always to check whether any new compound fluoresces at the wavelengths to be used. In some cases, substitutions are possible to circumvent inhibitor fluorescence. For example, H_2DIDS is a nonfluorescent analog of the anion transport inhibitor DIDS, and the Cl^- channel blockers 5-nitro-2-(3-phenylpropylamino)-benzoic acid (NPPB) and indanyl-oxyacetic acid (IAA-94) can be used instead of the fluorescent anthracene-9-carboxylic acid (A9C).

In addition to directly fluorescing, it is possible that a given compound may alter the fluorescence properties of an ion-sensitive fluorophore by direct interaction. This could occur, for example, if the compound quenches fluorescence, or if it binds to the fluorophore or reacts with it chemically, thus affecting fluorescence.

Testing for problems with inhibitors or other compounds can be done in vitro, by mixing the compound in question with a solution of the fluorophore and comparing fluorescence of this mixture with that of fluorophore alone. Even in the case of significant alteration of fluorescence by a given compound, it could still be used if it is certain that it cannot permeate the egg or embryo and thus does not come into contact with the fluorophore.

Calibration Problems

A number of different calibration methods for each group of fluorophores were described above. The in vitro methods, in which the fluorescence of solutions of the fluorophore is measured at several different concentrations of the relevant ion, are simplest. However, they are approximate, at best, as discussed above, and should only be used if it is either impossible to carry out in situ calibrations or if only approximate calibrations are necessary. In such cases, however, it may be best to avoid attempting to calibrate at all, and simply to report the fluorescence intensity or ratio as the data.

In situ calibrations should be performed for most intracellular ion measurements. However, even when the calibration is carried out using an intracellularly located fluorophore, it should be recognized that there can be very significant errors in the estimated values of ion concentrations. Most in situ calibration methods rely on equilibrating the relevant intracellular ion concentration to the external value, and then varying the external concentration to generate a calibration curve. The most significant source of error in such schemes is the failure of the method to completely collapse any gradient of the ion across the membrane. A careful study with simultaneous pH-sensitive microelectrodes and BCECF in leech glial cells indicated an error of about 0.15 pH units, with BCECF reporting a higher value than the actual pH_i given by the microelectrode (Nett and Dietmer 1996). A report using phospholipid vesicles and the Ca^{2+} indicator quin-2 indicated a similar error in Ca^{2+} calibrations using ionophores (Erdahl et al. 1995). Errors of similar types would likely exist in embryos, and care should be taken in asserting that intracellular ion concentrations determined with ion-sensitive fluorophores represent the actual, absolute values of intracellular concentration.

Conclusions

Ion-sensitive fluorophores have allowed important advances in our understanding of the egg and preimplantation embryo. Methods are established for the measurement of intracellular Ca^{2+} and pH_i. In addition, intracellular Cl^- measurements and measurements of blastocoel Cl^- have been used to investigate Cl^- transport in eggs and embryos. These methods, which allow information to be obtained from single cells, are especially useful in the study of mouse and other mammalian eggs and preimplantation embryos, where only very limited numbers of eggs or embryos can be obtained. In the future, the development of fluorophore-based methods for detecting other ions, as well as other intracellular molecules of interest — such as components of signaling pathways, will surely be used to expand our knowledge of the earliest stages of development.

ACKNOWLEDGMENTS We would like to thank Ms. Kerri Dawson for excellent technical assistance and help with the manuscript, and Dr. Richard Cardullo for providing us with prepublication results. This work was supported by a Biomedical Engineering Research Grant from the Whitaker Foundation for Biomedical Engineering. Some of the data discussed arose from projects supported by an Operating Grant from the Medical Research Council of Canada, and J. M. B. held an MRC Scholarship. We also gratefully acknowledge support from the Lalor Foundation, the Loeb Research Institute, and the Department of Obstetrics/Gynecology and the Division of Reproductive Medicine of the Ottawa Hospital. Some of this work was done by one of us (J. M. B.) while in the laboratories of Dr. John D. Biggers and Dr. Claude Lechene, and we thank them.

References

Aickin C. C. (1988). Movement of acid equivalents across the mammalian smooth muscle cell membrane. Proton Passage Across Cell Membranes, Chichester, CIBA Foundation Symposium 139, pp. 3–22.

Alper, S. L. (1994). The band 3-related AE anion exchanger gene family. Cell Physiol. Biochem. 4:265–281.

Baltz, J. M. (1993). Intracellular pH regulation in the early embryo. BioEssays 15:523–530.

Baltz, J. M., Biggers, J. D., and Lechene, C. (1990). Apparent absence of Na/H antiport activity in two-cell mouse embryos. Dev. Biol. 138:421–429.

Baltz, J. M., Biggers, J. D., and Lechene, C. (1991a). Two-cell stage mouse embryos appear to lack mechanisms for alleviating intracellular acid loads. J. Biol. Chem. 266:6052–6057.

Baltz, J. M., Biggers, J. D., and Lechene, C. (1991b). Relief from alkaline load in two-cell stage mouse embryos by bicarbonate/chloride exchange. J. Biol. Chem. 266:17212–17217.

Baltz, J. M., Biggers, J. D., and Lechene, C. (1993). A novel H^+ permeability dominating intracellular pH regulation in the early mouse embryo. Development 118:1353–1361.

Baltz, J. M., Smith, S. S., Biggers, J. D., and Lechene, C. (1997). Intracellular ion concentrations and their maintenance by Na^+/K^+ ATPase in preimplantation mouse embryos. Zygote 5:1–9.

Bassnett, S., Reinisch, L., and Beebe, D. C. (1990). Intracellular pH measurement using single excitation-dual emission fluorescence ratios. Am. J. Physiol.: Cell Physiol. 258:C171–C178.

Ben-Yosef, D., Oron, Y., and Shalgi, R. (1995). Low temperature and fertilization-induced Ca^{2+} changes in rat eggs. Molec. Reprod. Dev. 42:122–129.

Biggers, J. D., Bell, J. E., and Benos, D. J. (1988). Mammalian blastocyst: transport functions in a developing epithelium. Am. J. Physiol.: Cell Physiol. 255:C419-C432.

Biwersi, J., and Verkman, A. S. (1991). Cell-permeable fluorescent indicator for cytosolic chloride. Biochemistry 30:7879–7883.

Borland, R. M., Biggers, J. D., and Lechene, C. P. (1977). Studies on the composition and formation of mouse blastocoele fluid using electron probe microanalysis. Dev. Biol. 55:1–8.

Boron, W. F. (1977). Intracellular pH transients in giant barnacle muscle fibers. Am. J. Physiol.: Cell Physiol. 233:C61–C73.

Boron, W. F. and DeWeer, P. (1976). Intracellular pH transients in squid giant axons caused by CO_2, NH_3, and metabolic inhibitors. J. Gen. Physiol. 67:91–112.

Bos-Mikich, A., Whittingham, D. G., and Jones, K. T. (1997). Meiotic and mitotic Ca^{++} oscillations affect cell composition in resulting blastocysts. Dev. Biol. 182:172–179.

Boyarsky, G., Ganz, M. B., Sterzel, R. B., and Boron, W. F. (1998a). pH regulation in single glomerular mesangial cells I. Acid extrusion in absence and presence of HCO_3^-. Am. J. Physiol.: Cell Physiol. 255:C844–C856.

Boyarsky, G., Ganz, M. B., Sterzel, R. B., and Boron, W. F. (1988b). pH regulation in single glomerular mesangial cells II. Na^+-dependent and -independent Cl^--HCO_3^- exchangers. Am. J. Physiol.: Cell Physiol. 255:C857–C869.

Bright, G. R., Fisher, G. W., Rogowska, J., and Taylor, D. L. (1987). Fluorescence ratio imaging microscopy: temporal and spatial measurements of cytoplasmic pH. J. Cell Biol. 104:1019–1033.

Bright, G. R., Fisher, G. W., Rogowska, J., and Taylor, D. L. (1987). Fluorescence ratio imaging microscopy. In *Fluorescence Microscopy of Living Cells in Culture*, Taylor, D. L., and Wang, Y.-L., eds. San Diego, Academic Press, pp. 157–192.

Brison, D. R., and Leese, J. J. (1993). Role of chloride transport in the development of the rat blastocyst. Biol. Reprod. 48:692–702.

Busa, W. B. (1992). Spectral characterization of the effect of viscosity on Fura-2 fluorescence: excitation wavelength optimization abolishes the viscosity artifact. Cell Calcium 13:313–319.

Cardullo, R. A. and Dandala, S. (1997). Correcting for viscosity induced artifacts when using the pH sensitive dye, BCECF. Anal. Biochem., in press.

Cheek, T. R., McGuinness, O. M., Vincent, C., Moreton, R. B., Berridge, M. J., and Johnson, M. H. (1993). Development 119:179–189.

Cherr, G. N., and Cross, N. L. (1986). Immobilization of mammalian eggs on solid substrates by lectins for electron microscopy. J. Microscopy 145:341–345.

DeHemptinne, A., Marrannes, R., and Vanheel, B. (1983). Influence of organic acids on intracellular pH. Am. J. Physiol.: Cell Physiol. 245:C178–C183.

DiVirgilio, F., Steinberg, T. H., and Silverstein, S. C. (1990). Inhibition of fura-2 sequestration and secretion with organic anion transport blockers. Cell Calcium 11:57–62.

Edwards, L. J., Williams, D. A., and Gardner, D. K. (1997). Stage specific effects of lactate on intracellular pH in the preimplantation mouse embryo. Biol. Reprod. 56(Suppl. 1):88.

Engblom, A. C., and Akerman, K. E. (1993). Determination of the intracellular free chloride concentration in rat brain synaptoneurosomes using a chloride-sensitive fluorescent indicator. Biochim. Biophys. Acta 1153:262–266.

Erdahl, W. L., Chapman, C. J., Taylor, R. W., and Pfeiffer, D. R. (1995). Effects of pH conditions on Ca^{++} transport catalyzed by ionophores A23187, 4-Br-A23187, and ionomycin suggest problems with common applications of these compounds in biological systems. Biophys. J. 69:2350–2363.

Etter, E. F., Kuhn, M. A., and Fay, F. S. (1994). Detection of changes in near-membrane Ca^{++} concentration using a novel membrane associated Ca^{++} indicator. J. Biol. Chem. 269:10141–10149.

Fafournoux, P., and Pouysségur, J. (1996). The vertebrate Na^+/H^+ exchangers: structure, function, and hormonal regulation. In *Transport Processes in Eukaryotic and Prokaryotic Organisms*, Konings, W. N., Kaback, H R., and Lolkema, J. S., eds. Amsterdam, Elsevier, pp. 369–380.

Fineman, I., Hart, D., and Nord, E. P., (1990). Intracellular pH regulates Na^+-independent Cl^--base exchange in JTC-12 (proximal tubule) cells. Am. J. Physiol.: Renal Fluid Electrolyte Physiol. 258:F883–F892.

Ganz, M. B., Boyarsky, G., Sterzel, R. B., and Boron, W. F. (1989). Arginine vasopressin enhances pH_i regulation in the presence of HCO_3^- by stimulating three acid–base transport systems. Nature 337:648–651

Gibb, C. A., Poronnik, P., Day, M. L., and Cook, D. I. (1995). Control of cytosolic pH in early mouse embryos. Proc. Austr. Soc. Reprod. Biol. 27:78(Abstr.).

Grinstein, S. (1988). *Na^+/H^+ Exchange*. Boca Raton, FL, CRC Press.

Groden, D. L., Guan, Z., and Stokes, B. T. (1991). Determination of fura-2 dissociation constants following adjustment of the apparent Ca-EGTA association constant for temperature and ionic strength. Cell Calcium 12:279–287.

Grynkiewicz, G., Poenie, M., and Tsien, R. Y. (1985). A new generation of Ca^{++} indicators with greatly improved fluorescence properties. J. Biol. Chem. 260:3440–3450.

Hallam, T. J., Jacob, R., and Merritt, J. E. (1988). Evidence that agonists stimulate bivalent cation influx into human endothelial cells. Biochem. J. 255:179–184.

Hamill, O. P., Marty, A., Neher, E., Sakmann, B., and Sigworth, F. J. (1981). Improved patch-clamp techniques for high-resolution current recording from cells and cell-free membrane patches. Pflügers Arch. 391:85–100.

Harkins, A. B., Kurebayashi, N., and Baylor, S. M. (1993). Resting myoplasmic free calcium in frog skeletal muscle fibers estimated with fluo-3. Biophys. J. 65:865–881.

Haugland, R. P. (1996). *Handbook of Fluorescent Probes and Research Chemicals*, 6th ed. Eugene, OR, Molecular Probes, Inc..

House, C. R. (1994). Confocal ratio-imaging of intracellular pH in unfertilized mouse oocytes. Zygote 2:37–45.

Jericevic, Z., Wiese, B., Bryan, J., and Smith, L. C. (1989). Validation of an imaging system: steps to evaluate and validate a microscope imaging system for quantitative studies. In *Fluorescence Microscopy of Living Cells in Culture*, Taylor, D. L., and Wang, Y.-L., eds. San Diego, Academic Press, pp. 47–83.

Jezek, P., Mahdi, F., and Garlid, K. D. (1990). Reconstitution of the beef heart and rat liver mitochondrial K^+/H^+ (Na^+/H^+) antiporter: quantitation of K^+ transport with the novel fluorescent probe, PBFI. J. Biol. Chem. 265:10522–10526.

Jiang, L., Stuart-Tilley, A., Parkash, J., and Alper, S. L. (1994). pH_i and serum regulate AE2-mediated Cl^-/HCO_3^- exchange in CHOP cells of defined transient transfection status. Am. J. Physiol.: Cell Physiol. 267:C845–C856.

Jones, K. T., Carroll, J., Merriman, J. A., Whittingham, D. G., and Kono, T. (1995a). Repetitive sperm-induced Ca^{2+} transients in mouse oocytes are cell cycle dependent. Development 121:3259–3266.

Jones, K. T., Carroll, J., and Whittingham, D. G. (1995b). Ionomycin, thapsigargin, ryanodine, and sperm induced Ca^{++} release increase during meiotic maturation of mouse oocytes. J. Biol. Chem. 270:6671–6677.

Kim, J. H., Demaurex, N., and Grinstein, S. (1996). Intracellular pH: measurement, manipulation, and physiological regulation. In *Transport Processes in Eukaryotic and Prokaryotic Organisms*, Konings, W. N., Kaback, H. R., and Lolkema, J. S., eds. Amsterdam, Elsevier, pp. 447–472.

Kline, D., and Kline, J. T. (1992). Thapsigargin activates a calcium influx pathway in the unfertilized mouse egg and suppresses repetitive calcium transients in the fertilized egg. J. Biol. Chem. 267:17624–17630.

Kline, J. T., and Kline, D. (1994). Regulation of intracellular calcium in the mouse egg: evidence for inositol triphosphate-induced calcium release, but not calcium-induced calcium release. Biol. Reprod. 50:193–203.

Kline, D., and Zagray, J. A. (1995). Absence of an intracellular pH change following fertilization of the mouse egg. Zygote 3:305–311.

Kono, T., Carroll, J., Swann, K., and Whittingham, D. G. (1995). Nuclei from fertilized mouse embryos have calcium-releasing activity. Development 121:1123–1128.

Kopito, R. R. (1990). Molecular biology of the anion exchanger gene family. In *International Review of Cytology*, Vol. 123, Jeon, K. W., and Friedlander, M., eds. San Diego, Academic Press, pp. 177–199.

Krapf, R., Berry, C. A., and Verkman, A. S. (1988). Estimation of intracellular chloride activity in isolated perfused rabbit proximal convoluted tubules using a fluorescent indicator. Biophys. J. 53:955–962.

LeClerc, C., Becker, D., Buehr, M., and Warner, A. (1994). Low intracellular pH is involved in the early embryonic death of DDK mouse eggs fertilized by alien sperm. Dev. Dynamics 200:257–267.

Lee, S. (1987). Membrane properties in preimplantation mouse embryos. J. in Vitro Fertil. Embryo Transf. 4:331–333.

Manejwala, F. M., Cragoe, E. J., and Schultz, R. M. (1989). Blastocoel expansion in the preimplantation mouse embryo: role of extracellular sodium and chloride and possible apical routes of their entry. Dev. Biol. 133:210–220.

Masumoto, N., Sasaki, T., Tahara, M., Mammoto, A., Ikebuchi, Y., Tasaka, K., Tokunaga, M., Takai, Y., and Miyake, A. (1996). Involvement of rabphilin-3A in cortical granules exocytosis in mouse eggs. J. Cell Biol. 135:1741–1747.

Mehlmann, L. M., and Kline, D. (1994). Regulation of intracellular calcium in the mouse egg: calcium release in response to sperm or inositol triphosphate is enhanced after meiotic maturation. Biol. Reprod. 51:1088–1098.

McGuinness, O. M., Moreton, R. B., Johnson, M. H., and Berridge, M. J. (1996). A direct measurement of increased divalent cation influx in fertilised mouse oocytes. Development 122:2199–2206.

Mozingo, N. M., and Chandler, D. E. (1990). The fluorescent probe BCECF has a heterogeneous distribution in sea urchin eggs. Cell Biol. Int. Rep. 14:689–699.

Murer, H., Hopfer, U., and Kinne, R. (1976). Sodium/Proton antiport in brush-border-membrane vesicles isolated from rat small intestine and kidney. Biochem. J. 154:597–604.

Nett, W., and Deitmer, J. W. (1996). Simultaneous measurements of intracellular pH in the leech giant glial cell using $2'7'$-bis-(2-carboxyethyl)-5,6-carboxyfluorescein and ion-sensitive microelectrodes. Biophys. J. 71:394–402.

Nieminen, A. L., Gores, G. J., Dawson, T. L., Herman, B., and Lemasters, J. J. (1990). J. Biol. Chem. 265:2399–2408.

Nord, E. P., Brown, S. E. S., and Crandall, E. D. (1988). Cl^-/HCO_3^- exchange modulates intracellular pH in rat type II alveolar epithelial cells. J. Biol. Chem. 263:5599–5606.

Olsnes, S., Tønnessen, T. I., and Sandvig, K. (1986). pH-regulated anion antiport in nucleated mammalian cells. J. Cell Biol. 102:967–971.

Olsnes, S., Tønnessen, T. I., Ludt, J., and Sandvig, K. (1987). Effect of intracellular pH on the rate of chloride uptake and efflux in different mammalian cell lines. Biochemistry 26:2778–2785.

Parrington, J., Swann, K., Shevchenko, V. I., Sesay, A. K., and Lai, F. A. (1996). Calcium oscillations in mammalian eggs triggered by a soluble sperm protein. Nature 379:364–368.

Phillips, K. P., and Baltz, J. M. (1996). Intracellular pH change does not accompany egg activation in the mouse. Molec. Reprod. Devel. 45:52–60.

Phillips, K. P., Zhov, W.-L., and Baltz, J. M. (1998). Are intracellularly-loaded ion-sensitive fluorophores toxic to mouse eggs and zygotes? Zygote (in press).

Poenie, M. (1990). Alteration of intracellular fura-2 fluorescence by viscosity: a simple correction. Cell Calcium 11:85–91.

Rink, T. J., Tsien, R. Y., and Pozzan, T. (1982). Cytoplasmic pH and free Mg^{++} in lymphocytes. J. Cell Biol. 95:189–196.

Roos, A., and Boron, W. F. (1981). Intracellular pH. Physiol. Rev. 61:296–434.

Séguin, D. G., and Baltz, J. M. (1997). Cell volume regulation by the mouse zygote: mechanism of recovery from a volume increase. Am. J. Physiol.: Cell Physiol. 41:C1854–C1861.

Sekler, I., Kobayashi, S., and Kopito, R. R. (1996). A cluster of cytoplasmic histidine residues specifies pH dependence of the AE2 plasma membrane anion exchanger. Cell 86:929–935.

Somero, G. N. (1985). Intracellular pH, buffering substances and proteins: imidazole protonation and the conservation of protein structure and function. In *Transport Processes, Iono- and Osmoregulation*, Gillies, R., and Gilles-Baillien, M., eds. Heidelberg, Springer-Verlag, pp. 454–468.

Stachecki, J. J., Yelian, F. D., Schultz, J. F., Leach, R. E., and Armant, D. R. (1994). Blastocyst cavitation is accelerated by ethanol- or ionophore-induced elevation of intracellular calcium. Biol. Reprod. 50:1–9.

Szatkowski, M. S. (1989). The effect of extracellular weak acids and bases on the intracellular buffering power of snail neurones. J. Physiol. 409:103–120.

Szatkowski, M. S., and Thomas, R. C. (1989). The intrinsic intracellular H^+ buffering power of snail neurones. J. Physiol. 409:89–101.

Tahara, M., Tasaka, K., Masumoto, N., Mammoto, A., Ikebuchi, Y., and Miyake, A. (1996). Dynamics of cortical granule exocytosis at fertilization in living mouse eggs. Am. J. Physiol.:Cell Physiol. 270:C1354–C1361.

Thomas, J. A., Buchsbaum, R. N., Zimniak, A., and Racker, E. (1979). Intracellular pH measurements in Ehrlich ascites tumor cells utilizing spectroscopic probes generated in situ. Biochemistry 18:2210–2218.

Tsien, R. Y. (1989). Fluorescent indicators of ion concentrations. In *Fluorescence Microscopy of Living Cells in Culture*, Taylor, D. L., and Wang, Y.-L., eds. San Diego, Academic Press, pp. 127–156.

Vaughan-Jones, R. D. (1988). Regulation of intracellular pH in cardiac muscles. Proton Passage Across Cell Membranes, Chichester, CIBA Foundation Symposium 139, pp. 23–46.

Verkman, A. S. (1990). Development and biological applications of chloride-sensitive fluorescent indicators. Am. J. Physiol. 259:C375–C388.

Wong, M. M., and Foskett, J. K. (1991). Oscillations of cytosolic sodium during calcium oscillations in exocrine acinar cells. Science 254:1014–1016.

Xu, Z., LeFevre, L., Ducibella, T., Schultz, R. M., and Kopf, G. (1996). Effects of calcium-BAPTA buffers and the calmodulin antagonist W-7 on mouse egg activation. Dev. Biol. 180:594–604.

Zhang Y., Chernova, M. N., Stuart-Tilley, A. K., Jiang, L., and Alper, S. L. (1996). The cytoplasmic and transmembrane domains of AE2 both contribute to regulation of anion exchange by pH. J. Biol. Chem. 271:5741–5749.

Zhao, Y., and Baltz, J. M. (1996). Characterization of bicarbonate/chloride exchange during preimplantation mouse embryo development. Am. J. Physiol.: Cell Physiol. 40:C1512–C1520.

Zhao Y., Chauvet, P. J.-P., Alper, S. L., and Baltz, J. M. (1995). Expression and function of bicarbonate/chloride exchangers in the preimplantation mouse embryo. J. Biol. Chem. 270:24428–24434.

Zhao, Y., Doroshenko, P. A., Alper, S. L., and Baltz, J. M. (1997a). Routes of Cl^- transport across the trophectoderm of the mouse blastocyst. Dev. Biol. 189:148–160.

Zhao, M., Hollingworth, S., and Baylor, S. M. (1997b). AM-loading of fluorescent Ca^{2+} indicators into intact single fibers of frog muscle. Biophys. J. 72:2736–2747.

5

Approaches to Study Translational Control in Mouse Oocytes

André Stutz
Joachim Huarte
Béatrice Conne
Jean-Dominique Vassalli

The control of mRNA translation plays a decisive role in modulating gene expression in a number of specific instances. In oocytes, a subset of mRNAs is transcribed during the growth phase of these cells, to be stored, translationally dormant, in the cytoplasm. These mRNAs are then specifically recruited for translation at later stages, during meiotic maturation and early development, when the genome is transcriptionally silent (Richter 1995; Vassalli and Stutz 1995; Wickens et al. 1995). One emerging concept, developed from studies performed in a wide range of biological systems, is that the 3' untranslated region (3'UTR) of dormant cytoplasmic mRNAs is critical for the regulation of their translation.

The study of translational control in mammalian oocytes is particularly challenging. The limited availability of these cells and their relatively small size places a number of constraints on the types of experiments that can be performed. However, over the last 10 years, specific approaches have been designed that have allowed significant progress. A mouse oocyte dormant mRNA has been identified that serves as a useful paradigm to explore translational silencing and activation in mammalian oocytes (Huarte et al. 1985, 1987, 1992; Vassalli et al. 1989). Microinjections of mouse oocytes have proven to be a remarkably efficient tool to probe, in the living cell, the structures and molecular events that modulate the rate of translation of specific mRNAs, either endogenous to the oocyte or transcribed in vitro.

Furthermore, improvements in the sensitivity of detection procedures have allowed experimental work to be accomplished with relatively scarce material. The results from studies in mouse oocytes have contributed to our general understanding of translational control, in that they have revealed concepts applicable to oocytes from other species, and perhaps also to cells other than oocytes.

This chapter focuses on the technical approaches that we have developed to study the *cis*- and *trans*-acting determinants involved in the translational control of an endogenous mRNA in the intact mouse oocyte. We describe a highly sensitive zymographic assay that can be used to reveal the translation of an endogenous oocyte mRNA, which encodes the serine protease tissue-type plasminogen activator (tPA), as well as that of a reporter mRNA encoding an enzyme of similar catalytic specificity, urokinase-type plasminogen activator (uPA). We also detail several techniques that allow structural studies to be performed on endogenous mRNAs, following injection of antisense oligodeoxynucleotides, or on microinjected transcripts. Finally, in vivo UV-induced cross-linking experiments of microinjected radiolabeled transcripts, which can help identify *trans*-acting factors interacting with specific mRNA sequences, will be described.

Preparation, Injection and Culture of Mouse Oocytes
Materials
1. Swiss albino (BRL, Basel, Switzerland) 3-week-old female mice (see Note 1).
2. MEM*: minimal essential medium (MEM), supplemented with 5% fetal calf serum (FCS), 0.1 mg/ml sodium pyruvate, 2.5 mg/ml polyvinylpyrrolidone (PVP, see Note 2).
3. Light paraffin oil (Sigma).
4. Cell culture incubator (5% CO_2).
5. Maturation inhibitor: dibutyryl cAMP (db-cAMP), from aliquots of a 50× stock solution prepared in deionized sterile water and stored at −20°C.

Procedure
1. Primary oocytes from excised ovaries are collected in MEM* culture medium containing 50 µg/ml db-cAMP, a concentration that prevents, in a reversible fashion, meiotic maturation.
2. Injections are performed in the same medium as before. Surviving oocytes are then transferred in paraffin oil-covered drops of MEM* medium without PVP and without db-cAMP to allow meiotic maturation (see Note 3). If resumption of meiosis has to be prevented, db-cAMP is added at 100 µg/ml.
3. Oocytes are incubated in a cell culture incubator at 37°C in a humidified atmosphere of 5% CO_2 in air.
4. Oocytes are scored for their stage of maturation (primary oocytes: presence of the germinal vesicle; maturing oocytes: no germinal vesicle; secondary oocytes: presence of the first polar body) and collected for analysis.

Detailed descriptions of the handling of oocytes, the preparation of pipettes, and the equipment necessary for microinjection can be found elsewhere (DePamphilis et al. 1988; Sallés et al. 1993).

Notes
1. 40–80 oocytes that have reached near full size (~70 μm in diameter, 180 pl in volume) can be obtained from the ovaries of a 3-week-old mouse; 80% of these fully grown primary oocytes will spontaneously resume meiosis in vitro when released from their follicular environment and cultured in the absence of db-cAMP. Germinal vesicle break down (GVBD) takes place after 1–3 hours of culture and meiotic maturation progresses to metaphase 2, with emission of the first polar body ~10–12 hours after GVBD.
2. PVP serves as a bulking agent that decreases "bleeding" of the oocytes after microinjection and does not inhibit meiotic maturation.
3. After injections are completed, the oocytes are sequentially washed through three drops of culture medium and transferred to clean drops for further incubation. The volume of medium transferred between drops should be minimized so that db-cAMP is removed for optimal maturation.

Translation

The mRNA for tPA accumulates in growing mouse oocytes and is maintained untranslated in the cytoplasm of meiotically arrested primary oocytes (Huarte et al. 1985). Following resumption of meiosis, translational activation of tPA mRNA is accompanied by elongation of its $3'$-poly(A) tail (Huarte et al. 1987, 1992; Vassalli et al. 1989). The tPA can be detected starting 4 hours after GVBD. Plasminogen activators (PAs) are serine proteases that convert plasminogen into plasmin, a trypsin-like protease of broad specificity; PAs can be detected by the zymographic method of Granelli Piperno (Granelli Piperno and Reich 1978), as modified by Vassalli et al. (1984). To study the translation of injected mRNAs, the coding region of mouse uPA mRNA provides a useful reporter sequence, since uPA can be revealed by the same zymographic procedure used to assay for tPA. The different molecular weights of the two enzymes allow them to be visualized simultaneously in the same sample. The high sensitivity of the zymographic assay allows detection of both enzymes in samples from individual oocytes.

SDS–PAGE and Zymography
Materials
1. Oocyte washing medium: MEM supplemented with 1 mg/ml crystallized bovine serum albumin (BSA; Miles).
2. Oocyte lysis buffer (TB) containing:
    0.25% Triton X-100 (Merck),
    2 mg/ml BSA, in water.
3. 2× electrophoresis sample buffer (2 × SB) containing:
    1 M Tris-HCl, pH 6.8,
    2% SDS,
    20% glycerol,
    0.005% bromophenol blue, in water.
4. SDS–PAGE slab gels: 7 × 15 cm.

5. Indicator gels prepared with:
   a. 8% commercial nonfat dry milk in phosphate-buffered saline (PBS),
   b. PBS,
   c. 10% azide (NaN$_3$),
   d. Plasminogen,
   e. 2.5% agar in water.

Procedure
1. Before analysis, the oocytes are sequentially washed in washing medium to eliminate proteases and inhibitors present in the FCS.
2. Groups of three oocytes are solubilized in 10 µl of TB, and rapidly mixed with an equal volume of 2× SB. Samples can be stored for several days at −20°C or directly subjected to 10% SDS–PAGE using the buffer system of Laemmli (1970), under nonreducing conditions.
3. Gels are run at a constant current of 15–20 mA for the stacking gel and 30–40 mA for the running gel (see Note 1).
4. The gel is washed twice for 10 minutes in 2.5% Triton X-100 and twice for 10 minutes in PBS (see Note 2), on a rocking platform.
5. The electrophoretic gel is layered on an indicator gel prepared as follows:
   a. A glass plate of approximately the same dimensions as the electrophoretic gel is extensively washed with ethanol and prewarmed at 50°C on a hotplate.
   b. The solution of plasminogen (Note 3) is centrifuged at 25,000g for 15 minutes at 4°C to remove any precipitate, and then kept on ice.
   c. Agar (2.5%) is melted for ∼15 minutes at 90°C.
   d. A mixture (e.g., total volume of 6 ml for a gel of the suggested dimensions) is prepared by adding reagents, in the following order to the indicated concentration:
   PBS
   nonfat dry milk   1.3%
   Azide   0.2%
   (This solution is prewarmed at 50C in a water bath.)
   Plasminogen   40 µg/ml
   agar   0.8%
   e. The mixture is poured on the prewarmed glass plate, and allowed to spread over the entire surface.
   f. When the indicator gel has solidified, the electrophoretic gel is carefully layered on top of it.
6. Zymograms are allowed to develop in a humidified atmosphere at 37°C for 12–48 hours (see Note 4).
7. Photographs of the zymograms are taken using dark-ground illumination.

Notes
1. SDS can solubilize the proteolytic substrate, generating zones of "lysis" indistinguishable from those indicative of the presence of PAs and due to the activity of plasmin; the bromophenol blue/SDS-enriched electrophoretic front should therefore either be allowed to migrate out or be removed by cutting the gel.

2. PAs are reversibly denatured by SDS. To allow their reactivation after SDS–PAGE, the SDS must be removed by extensive washing of the gel in Triton X-100; further washes in PBS remove the Triton X-100.
3. The plasminogen is purified from human plasma by lysine-Sepharose (Pharmacia) affinity chromatography (Deutsch and Mertz 1970) or can be purchased (American Diagnostica GmbH).
4. PAs present in the electrophoretic gel diffuse into the indicator gel, where they catalyze the conversion of plasminogen to plasmin, which, in turn, digests the casein of the milk. This results in the development of transparent zones in the translucent indicator gel, localized at sites corresponding to the migration of PAs. Quantitative assessment of the relative PA content of different samples requires comparison of the rates of development and areas of lytic zones with those obtained with dilutions of a standard preparation of enzyme.

Biosynthetic Labeling of Oocytes

To evaluate overall rates and patterns of protein synthesis — for instance, to control for possible nonspecific effects when exploring specific modulations of translation of endogenous or injected mRNAs — the oocytes can be cultured in the presence of ^{35}S-labeled amino acids and the newly-synthesized proteins revealed by SDS-PAGE and autoradiography.

Materials
1. Biggers' culture medium (Biggers et al. 1971) or another amino acid-free medium.
2. ^{35}S methionine (1370 Ci/mmol, Amersham).
3. TB oocyte lysis buffer (see above).
4. Staining solution:
   30% methanol,
   7.5% acetic acid,
   0.2% Coomassie brillant blue, in water.
5. Destaining solution:
   30% methanol,
   7.5% acetic acid, in water.
6. Amplify fluorographic solution (Amersham).
7. Kodak SB X-ray film.

Procedure
1. The oocytes are cultured in Biggers' medium, which does not contain amino acids.
2. 5–10 oocytes are metabolically labeled in the presence of 200 µCi/ml ^{35}S methionine, in 20 µl drops of culture medium covered with paraffin oil, for 4–6 hours.
3. Lysis is performed in TB, and the samples are subjected to SDS–PAGE as described above (see "SDS–PAGE and Zymography").
4. The gel is incubated in the staining solution for 30 minutes and destained overnight.
5. The gel is incubated for 30 minutes in Amplify, dried and exposed for fluorography at −80°C between intensifying screens for at least 2 days.

In Vitro Modification of RNA for Microinjection

Polyadenylation

In fully grown primary oocytes, the length of the poly(A) tail on dormant mRNAs appears to be unusually short, i.e. it consists of only 30–50 A residues (Sallés and Strickland 1995; Stutz et al. 1997). In the case of tPA mRNA, this is due to a process of sequence-specific cytoplasmic deadenylation, which removes most of the As that are added at the time of transcription (Huarte et al. 1992). During meiotic maturation, tPA mRNA undergoes cytoplasmic readenylation, and the presence of a poly(A) tail is required for translation (Vassalli et al. 1989). Injection of in vitro polyadenylated synthetic transcripts allows the study of the metabolism and the role of the poly(A) tail.

Materials
Remarks: particular precautions are required when working with RNA, since it is critical to eliminate RNase contamination. Gloves should be worn and frequently changed. Sterile plasticware should be used and glassware should be incubated at 180°C in a dried baking oven for several hours. Stock solutions should be filtered (to avoid microbial contamination), treated with diethylpyrocarbonate (DEPC, 0.1%) for at least 12 hours at 37°C, and autoclaved (121°C, 30 minutes).
Warning: DEPC reacts with reagents containing amines; Tris-HCl should be prepared with DEPC-treated water. Alternatively, double-distilled water can be used for the preparation of stock solutions, which should be filtered and autoclaved.

1. In vitro-synthesized capped RNA transcripts (Note 1) resuspended in DEPC-treated water.
2. ATP (Pharmacia), 100 mM stock solution.
3. Ribonuclease inhibitor, 40 U/μl (RNAguard, Pharmacia).
4. *Escherichia coli* poly(A) polymerase (PAP) 1 U/μl (BRL).
5. Dithiothreitol (DTT), 40 mM (aliquots are stored at −20°C, and used only once).
6. Bovine serum albumin (BSA), 2 mg/ml.
7. NaCl, 5 M.
8. Tris-HCl, 1 M, pH 8.1.
9. $MgCl_2$ 0.2 M.
10. $MnCl_2$ 50 mM.
11. 10× R-Stop solution:
    1% SDS,
    10 mM EDTA, in water.
12. Phenol/$CHCl_3$/isoamylalcohol (25:24:1).
13. TS buffer:
    10 mM Tris-HCl, pH 8.1, 1% SDS, in water.
14. Sodium acetate, 2 M, pH 5.4.
15. Ethanol, 100% and 70% (prepared with DEPC-treated water).

Procedure
1. All reagents should be thawed on ice and subjected to a rapid spin.
2. The polyadenylation mixture (total volume 40 μl) is prepared by adding, in the following order and to the indicated concentration:
   a. Water (as required)
   b. RNA   0.1–0.2 μg/ml
   c. 2 μl NaCl   250 mM
   d. 2 μl Tris-HCl pH 8.1   50 mM
   e. 2 μl MgCl$_2$ 10 mM
   f. 2 μl BSA   100 μg/ml
   g. 2 μl DTT   2 mM
   h. 1 μl RNAguard   1 U/μl
   i. 4 μl ATP 1 mM   100 μM
   j. 2 μl MnCl$_2$ 2.5 mM
   k. 1 μl PAP   0.025 U/μl
3. Incubate for 30 minutes at 37°C.
4. Stop the reaction by adding 10 μl of 10× R-stop, complete to 100 μl with DEPC-treated water.
5. Proceed with two extractions with phenol/CHCl$_3$/isoamyl alcohol, in which the organic phases are back-extracted with 100 μl of TS buffer.
6. The RNAs from the combined aqueous phases are precipitated with 1/10 sodium acetate and three volumes of 100% ethanol and kept at −20°C for 30 minutes. Before precipitation, the size of the transcript may be verified by electrophoresis of an aliquot of the solution in a polyacrylamide/urea gel.
7. Centrifuge at 25,000g for 15 minutes at 4°C. Discard the supernatant and wash the pellet with 300 μl of 70% ethanol, centrifuging for 2 minutes and discarding the supernatant.
8. Resuspend the pellet of RNA in the adequate buffer for injection.

Note
1. The cap structure contributes to the stability of mRNAs by protecting their 5′ end from phosphatases and nucleases. In addition, caps enhance the translation of mRNAs by eukaryotic protein-synthesizing systems.

Inhibition of Poly(A) Tail Elongation

Poly(A) tail elongation accompanies the translational activation of maternal mRNAs during meiotic maturation; this polyadenylation requires the presence of *cis*-acting sequences (CPEs and AAUAAA hexanucleotide) in the 3′UTR (Richter 1995; Vassalli and Stutz 1995). To determine whether polyadenylation is necessary for translational activation, reporter mRNAs can be modified before injection by incorporation of a 3′-terminal cordycepin residue (3′-deoxyadenosine) to prevent 3′ elongation during meiotic maturation. To account for the incomplete efficiency of cordycepin incorporation, modified transcripts should be subjected to a polyadenylation reaction followed by purification on an affinity oligo(dT) resin to eliminate the transcript molecules that have not incorporated cordycepin.

Materials
1. Same reagents as before, except that ATP is replaced with cordycepin 5'-triphosphate (3'dATP, Sigma).
2. TEN-8:
   10 mM Tris-HCl, pH 8.1,
   1 mM EDTA,
   100 mM NaCl, in water.
3. Disposable column (QS-Q, Isolab).
4. G-50 Sephadex (Pharmacia). The spun column is prepared by filling a disposable column with a sterile 50% slurry of G-50 Sephadex in TEN-8 (10 ml) followed by centrifugation at 200g for 5 minutes at 4°C.
5. Oligotex-dT poly(A^+) in vitro transcript purification kit (Qiagen).

Procedure
1. Addition of cordycepin to in vitro capped transcripts is performed according to the polyadenylation protocol described above, except that ATP is replaced by cordycepin 5'-triphosphate.
2. After phenol/$CHCl_3$/isoamyl alcohol extraction (see above, step 5), the cordycepin-modified transcripts are separated from unincorporated cordycepin by spun-column centrifugation:
   a. The samples are deposited on freshly prepared, spin-drained resin and the columns are placed in sterile conical centrifuge tubes.
   b. After centrifugation at 200g for 5 minutes at 4°C, 200 µl of TEN-8 are deposited on top of the resin and the columns are recentrifuged.
   c. The transcripts are collected in the flow through, and ethanol-precipitated as above.
3. To remove molecules that may not have incorporated cordycepin, the transcripts are subjected to polyadenylation (as above), purified, dissolved in water, and adsorbed onto oligo(dT) spun columns.
4. The flow through, containing the cordycepin-poly(A^-) transcripts, is collected and ethanol-precipitated. The poly(A^+) transcript fraction is retained in the resin.
5. The integrity and size of the transcripts is verified by electrophoresis in polyacrylamide/urea gels.

Analysis of mRNA Poly(A) Tails

This section deals with the analysis of changes in the poly(A) tail length of endogenous oocyte mRNAs or injected transcripts. While changes in the electrophoretic migration of RNA molecules (as detected, for instance, by Northern blot hybridization) are indicative of changes in poly(A) tail length, this needs to be verified; to this effect, an RNase H-based approach has proven very useful. To compare the rates and extent of polyadenylation of different injected transcripts, RNase T1-mediated cleavage provides a valuable tool. These two approaches will be described here. A very sensitive method, termed RT–PCR–PAT, which allows determination of the poly(A) tail length of a specific endogenous mRNA using only very small amounts of material (less than one oocyte), has recently been developed (Sallés and Strickland 1995).

RNase H-Mediated Cleavage of RNA

In vitro treatment of RNA with RNase H (which degrades RNA in DNA–RNA hybrids), in the presence of oligo(dT), will selectively remove the poly(A) tail (Schibler et al. 1980) and thus verifies that the cause of a change in electrophoretic migration of a given RNA is due to a change in the length of its poly(A) tail. To refine this analysis, particularly in the case of relatively long transcripts, it may be necessary to cleave internally the mRNA of interest to generate a sufficiently small 3' fragment so that the poly(A) tail represents a significant fraction of the molecule. This can be achieved by RNase H digestion in the presence of an antisense oligodeoxynucleotide (as-ODN) complementary to an internal segment of the target transcript (Huarte et al. 1987; Sheets et al. 1994). After electrophoresis of the digestion products and Northern blot transfer, the filters are hybridized with a specific probe corresponding to the 3' region of the target transcript. If oocytes are injected with radiolabeled transcripts, samples can be analyzed directly after RNase H treatment on polyacrylamide/urea gels.

Materials
1. Total RNA prepared from injected or noninjected oocytes.
2. oligo(dT) (Boehringer) and/or an as-ODN complementary to an internal region of the mRNA of interest.
3. Carrier yeast tRNA (20 µg/µl).
4. 2× RNase H buffer containing:
    20 mM Tris-HCl, pH 7.4,
    260 mM amonium chloride,
    20 mM magnesium acetate,
    10% sucrose, in water.
5. *Escherichia coli* RNase H (1.5 U/ml) (Promega).
6. 10× R-stop solution:
    1% SDS,
    10 mM EDTA, in water.
7. Phenol/CHCl$_3$/isoamyl alcohol (25:24:1).
8. Sodium acetate, 2 M, pH 5.4.
9. Ethanol.
10. Equipment necessary for Northern blots.

Procedure
1. Synthesize a specific as-ODN complementary to a region of the target transcript ~400 nt upstream of its 3' end (see Note 1).
2. Extract total RNA from 10–20 oocytes (Huarte et al. 1987). Nucleic acid precipitation must be performed after addition of 20 µg of carrier yeast tRNA.
3. RNA resuspended in 5 µl DEPC-treated water is mixed with:
    a. Water as required (total volume of 15 µl),

b. 1 µl oligo(dT) (0.5 µg/µl), with or without 1 µl of the specific as-ODN (0.5 µg/µl).

After 2 minutes at 100°C, the samples are chilled on ice and supplemented with:

c. 7.5 µl of 2× RNase H buffer,

d. 1 µl RNase H, and incubated at 37°C for 20 minutes.

4. Stop the reaction by adding 5 µl of 10× R-stop solution and 30 µl DEPC-treated water.
5. After two extractions with phenol/CHCl$_3$/isoamylalcohol, the nucleic acids are ethanol-precipitated.
6. RNAs resuspended in the appropriate sample buffer are electrophoresed either in a 1.2–1.6% agarose gel for Northern blot or in a 5–6% polyacrylamide/urea gel if the injected transcript was radiolabeled.

Note
1. Larger or smaller fragments will either be more difficult to resolve or will give a diffuse signal in an agarose gel.

RNase T1-Mediated Cleavage of Injected Transcripts

We have observed that, depending on structural features of radiolabeled injected transcripts, such as their length or the number of CPE-like sequences they contain, the extension of their poly(A) tail occurs with different kinetics and/or varies in its extent (Huarte et al. 1992). To reveal such subtle differences, the length of the poly(A) tail added to transcripts of different initial sizes must be compared with sufficient precision. This can be best achieved by cleaving the RNA recovered from injected oocytes, so that the fragments analyzed (which have received poly(A) tails of different lengths) have heteropolymeric regions of the same size. Total RNAs recovered from injected oocytes, containing radiolabeled transcripts that have been polyadenylated in vivo (for instance, during meiotic maturation), are allowed to hybridize in vitro with oligodeoxynucleotides complementary to approximately the 3′-terminal 30 nt of the radiolabeled transcripts. In the presence of RNase T1 (which cleaves single-stranded RNA after G residues), the heteropolymeric regions are degraded, except their protected 3′-terminal 30 nt (plus a few 5′ nt up to the nearest G residue); this yields 3′ fragments of the different transcripts that differ, with respect to their size, only in the number of A residues that have been added by the oocyte's polyadenylation machinery.

Materials
1. Total RNA prepared from injected oocytes.
2. as-ODN complementary to the 3′ 30 nt of the radiolabeled injected transcript and containing a poly(dT)$_{10}$ stretch at its 5′ end.
3. Carrier yeast tRNA (20 µg/µl).
4. 2× RNase T1 buffer containing:
   2 M NaCl,
   0.33 M HEPES, pH 7.3,
   0.66 mM EDTA, in water.

5. RNase T1 (5 U/µl) (Reactolab).
6. 10× R-stop solution.
7. Sodium acetate 2 M, pH 5.4.
8. Ethanol.

Procedure
1. Extract total RNA from 10–20 oocytes (as for RNase H treatment).
2. RNAs resuspended in 5 µl of DEPC-treated water are mixed with:
   a. Water as required (total volume of 10 µl),
   b. 1 µl of as-ODN (0.5 µg/µl),
   c. 5 µl of 2× RNase H buffer.
   After 5 minutes at 75°C, the samples are placed at 45°C for 2 hours, then supplemented with:
   d. 1 µl RNase T1, and incubated at room temperature for 60 minutes.
3. Stop the reaction by adding 5 µl of 10× R-stop solution and 35 µl of water.
4. The nucleic acids are directly ethanol-precipitated, resuspended in the appropriate sample buffer, and electrophoresed on an 8% polyacrylamide/urea gel.

In Vivo Antisense Oligodeoxynucleotide Mapping

Antisense strategies have proven very useful to suppress a gene's activity and thus to deduce its function by examining the resulting phenotype (Melton 1985; Strickland et al. 1988; Gebauer et al. 1994; Sheets et al. 1995). Hybridization of as-ODNs should also provide a tool to explore, in the physiological context of the intact cell, the structural features (secondary structure or interaction with other macromolecules) of mRNAs that may be involved in the control of their fate (localization, translation, degradation). Specifically, it should be possible to use as-ODNs to discriminate regions of an endogenous mRNA that are accessible to, or protected from, hybridization with the ODN. To explore this possibility, we have taken advantage of the antisense-directed mRNA cleavage that occurs following cytoplasmic microinjection of antisense RNAs (Strickland et al. 1988) or as-ODNs in primary oocytes (Stutz et al. 1997): injection of antisense sequences complementary to portions of oocyte mRNAs results in the degradation of the target mRNA in the duplex region by RNase H-like (Woolf et al. 1990) or RNase III-like activities endogenous to the oocyte (Melton 1985; Strickland et al. 1988). Depending on where the cleaved region lies, this prevents the translation of a targeted maternal mRNA (Strickland et al. 1988; Gebauer et al. 1994; Sheets et al. 1995), because the 3'UTR sequences involved in its translational activation during meiotic maturation have been amputated. Even when translation of the mRNA of interest cannot be easily tested, analysis of its structure will reveal whether it has undergone as-ODN-directed cleavage. One very appealing aspect of this strategy is that it can be used to explore structural features of native, endogenous mRNAs in intact cells. In oocytes and early embryos, the absence of transcription, the slow turnover of maternal messages and the relative ease with which as-

ODNs can be introduced in the cytoplasm could make this strategy particularly powerful.

Designing as-ODNs for In Vivo Mapping

The design of ODNs should be such that they are complementary only to the desired target region. Secondary structure should be avoided to prevent intramolecular hybridization (Sallés et al. 1993; Woolf et al. 1990). Low melting temperatures (T_m values, which correspond to the computed DNA–DNA duplex melting temperature of the ODN) will not allow efficient hybridization in the cytoplasm of injected oocytes at 37°C. ODNs of 25 nt should be used to increase both specificity and RNase recognition.

Synthesis and Purification of ODNs Used For Microinjection

Unmodified ODNs are synthesized using the phosphoramidite method. They are purified by extraction with *n*-butanol (Sawadogo and Van Dyke 1991), extracted once with phenol/chloroform, and precipitated with ethanol. ODNs can be used without HPLC purification and need not be filtered.

Assaying for the Effect of as-ODNs

In the case of tPA mRNA, two types of tests can be performed on injected oocytes to evaluate the effect of different as-ODNs: the translation of the targeted mRNA can be monitored by zymography (see "Translation"), an enzymatic assay sensitive enough to be performed on individual oocytes, and the structure of the mRNA can be investigated by RNase protection using only 10–15 oocytes. Cleavage of accessible, unprotected regions of an mRNA occurs within 1 hour after injection of the complementary as-ODN (Stutz et al. 1997).

Controlling as-ODNs for their Capacity to Direct Specific mRNA Cleavage

While as-ODNs can be used to discriminate between accessible and protected regions of an endogenous maternal mRNA, these differences in accessibility are, of course, not known a priori. Several considerations should thus be kept in mind when evaluating ODNs for antisense-directed cleavage. In particular, as-ODNs have to be tested in vitro for their intrinsic capacity to induce RNase H-mediated antisense-directed cleavage of a targeted radiolabeled synthetic transcript or of the targeted sequence using purified oocyte mRNA and RNase protection analysis. To confirm that the inaccessibility of a given sequence in an endogenous targeted mRNA is due to specific structural features, a corresponding synthetic transcript can be mixed with the as-ODN at the time of injection, so that complementary sequences can hybridize before structure develops in vivo, and assessed for its translation or cleavage after introduction in the oocytes (Stutz et al. 1997).

Effects of Concentration, Melting Temperature and Length of Injected ODNs

We summarize here experiments designed to evaluate the use of as-ODNs complementary to accessible regions of tPA mRNA. The purified ODNs were dissolved in 150 mM KCl at 1 µg/µl (1 A_{260} = 33 µg/ml). At this concentration, the injected volume (~10 pl) contains ~10^9 molecules of a 25-nt-long ODN. The effects of two 23-nt as-ODNs [as-ODN-1 (5′-GTCCC AAGAG TTGAG GAGTG TGG) and as-ODN-2 (5′-GTTTA TAAAG AAAAA GACAT TTA], differing in their melting temperatures (72°C and 54°C, respectively), were determined. When injected at 1 µg/µl, both ODNs completely inhibited tPA production. Decreasing by 10-fold the amount of ODNs injected abolished the inhibitory effect of as-ODN-2 but not that of as-ODN-1, presumably because of the difference in their predicted T_m values. as-ODNs consisting of the 18, 13, and 10 5′-most nt of as-ODN-1 were tested: when injected at a concentration of 1 µg/µl, both the 18-mer (T_m = 54°C) and the 13-mer (T_m = 40°C) completely inhibited tPA production, whereas the 10-mer (T_m = 32°C) was only partly effective. Twelve different as-ODNs with predicted T_m values of 40°C or above, complementary to accessible regions of tPA mRNA around the translation initiation site, in the coding region, and in the 3′UTR all inhibited tPA production by more than 90% when injected at a concentration of 1 µg/µl.

Stability of Injected ODNs

The half-life of unmodified 5′-radiolabeled ODNs injected in mouse primary oocytes is less than 30 minutes, a result consistent with those of studies using end-labeled or internally labeled ODNs injected into *Xenopus* oocytes (Woolf et al. 1990). This relatively short half-life allows an exploration of changes in the accessibility of target sequences during resumption of meiosis: for instance, an as-ODN complementary to a tPA mRNA sequence that is masked in primary oocytes does not direct cleavage of its target even though this sequence is unmasked when meiosis resumes, while the same as-ODN injected after GVBD effectively hybridizes and prevents tPA mRNA translational activation by directing its amputation (Stutz et al. 1997).

Nonspecific Effects of ODNs

Injection of ODNs could have "nonspecific" effects on oocyte protein synthesis or the capacity to resume meiosis. Thus, the proportion of as-ODN-injected oocytes undergoing resumption of meiosis and first meiotic division should be monitored, keeping in mind that inhibiting the translation of a specific maternal mRNA should not prevent GVBD (which does not depend on protein synthesis in the case of mouse oocytes), but could affect further progression through meiosis. The extent and overall pattern of protein synthesis can be analyzed by ^{35}S-amino acid labeling; the translation of an irrelevant reporter mRNA can also be verified. Inhibition of tPA synthesis by as-ODNs complementary to accessible regions of the mRNA, and used at a concentration of 1 μg/μl, appeared specific and did not reflect a generalized decrease in protein synthesis or an arrest of meiotic maturation (Stutz et al. 1997).

Interpretation of Results

When a given region of an mRNA molecule is engaged in a structural interaction with another macromolecule, or is part of a secondary structure, it should be less accessible than the rest of the transcript to hybridization with a complementary nucleic acid. as-ODN mapping, which can be achieved in intact oocytes and can probe the structure of endogenous mRNAs, is thus a unique tool to identify regions of an mRNA that take part in intra- or intermolecular interactions. In view of the relative simplicity of the approach, particularly when assay of an easily detectable translation product such as tPA is used as the end-point, it is possible to "scan" rapidly along an mRNA molecule of interest for protected regions; using overlapping sets of as-ODNs, information can also be obtained as to the size of a given protected region. Because of the time scale of this type of experiment, only quite stable interactions should be revealed: transient interactions — as occur, for instance, in the 5'UTR and the coding region during translation — are probably not detected, while regions involved in mRNA localization, stabilization, masking, translational arrest, polyadenylation, and translational activation appear as particularly good candidates to be identified by as-ODN mapping.

In Vivo UV-Induced RNA-Protein Cross-Linking

The identification of proteins that interact with specific regions of mouse oocyte mRNAs is an obligatory step to understand better their translational control. To purify such proteins, classical in vitro approaches that rely on the use of tissue or cell extracts cannot be easily adapted, because of the very limited amounts of material available. As an alternate approach, we have attempted to reveal such proteins by in vivo UV-induced RNA-protein cross-linking of injected radiolabeled transcripts. Oocyte lysates are then digested with RNase T1, to remove regions of the cross-linked RNA that are not bound to oocyte proteins, and the spectrum of proteins that have become isotopically tagged by label transfer is revealed by SDS–PAGE and autoradiography.

Materials
1. Terasaki tissue culture plates (flat-bottom 10-μ wells) (Falcon).
2. Stratalinker device (UV Stratalinker™ 2400, Stratagen).
3. Cross-linking lysis buffer containing:
    10 mM Tris-HCl, pH 7.4,
    1 mM EDTA,
    1 mM DTT,
    0.25% Triton X-100,
    1 μg/μl yeast tRNA,
    1 U/μl ribonuclease inhibitor (RNAguard, Pharmacia), in water.
4. RNase T1 (5 U/μl) (Reactolab).
5. 2× SB$^+$: 2× electrophoresis reducing sample buffer containing:
    1 M Tris-HCl, pH 6.8,
    2% SDS,
    20% glycerol,
    20% mercaptoethanol,
    0.005% bromophenol blue, in water.

Procedure
1. 20–30 oocytes are injected with 5–10 × 10^6 cpm/μl of a radiolabeled capped transcript.
2. The oocytes are cultured, in standard conditions, for 1–3 hours to allow binding of oocyte proteins to the injected transcript.
3. The oocytes are transferred into wells of Terasaki tissue culture plates and exposed for 7 minutes to UV radiation in a Stratalinker device (1 μJ/cm^2).
4. The oocytes are lysed in 20 μl of cross-linking lysis buffer; 2 μl of RNase T1 is then added and the samples are digested for 30 minutes at 37°C.
5. An equal volume of 2× SB$^+$ is added, and the samples are subjected to SDS–PAGE (Laemmli 1970). The gel is analyzed by autoradiography at −80°C betwen intensifying screens for 4–10 days.

To evaluate whether radiolabeled RNA-protein complexes that are revealed by this approach are due to specific interactions of the injected transcript with oocyte proteins, results obtained following injection of a panel of radiolabeled RNA molecules must be compared. Competition experiments, using different unlabeled transcripts, can also be informative.

In vivo UV-induced RNA-protein cross-linking should be a useful approach to reveal candidate regulatory proteins. Because there is no need to prepare an extract of oocyte proteins before addition of radiolabeled RNA, the potentially interacting species are kept at a high concentraion and, in the case of the proteins, a physiological concentration. Also, both the interaction and the cross-linking occur in the live oocyte, a condition that should minimize the influence of potential artifacts.

Conclusions

The approaches and procedures that have been described in this chapter have allowed considerable progress to be made in our understanding of the translational control of maternal mRNAs in mammalian oocytes. They have identified a critical role for the 3'UTR of tPA mRNA, in both its translational silencing in growing primary oocytes and its translational activation during meiotic maturation. They have shown that a discrete portion of the 3'UTR, termed the adenylation control element (ACE), is involved in silencing as well as in activation, and they have demonstrated that the presence of a poly(A) tail is necessary for translation. In vivo as-ODN mapping has confirmed the importance of the ACE, by demonstrating that it is a protected region of tPA mRNA in both primary and maturing oocytes, and therefore that it is a candidate regulatory element engaged in interactions with oocyte proteins.

A number of technical issues will have to be addressed in order to achieve further progress. For instance, it would be interesting to elucidate the role of the specific deadenylation of tPA mRNA that occurs shortly after its transcription in inducing its dormancy. To this effect, a procedure to prevent deadenylation of a reporter transcript would be valuable. Also, in view of a recent report implicating cap methylation in the translational recruitment of dormant *Xenopus* oocyte mRNAs (Kuge and Richter 1995), sufficiently sensitive methods to study this modification in mouse oocytes will need to be developed.

Clearly, the major challenge will be the identification of the mouse oocyte proteins involved in translational regulation. Because it seems difficult to prepare enough mouse oocytes to proceed according to classical biochemical methods, alternative approaches will be required. Similar sequences are responsible for the translational regulation of mouse and *Xenopus* oocyte mRNAs, proteins purified from amphibians may thus provide a practical approach. The identification of the murine equivalent of such a *Xenopus* protein is an important step in this direction (Gebauer and Richter 1996). The in vivo UV-induced cross-linking of mouse oocyte proteins should help in this essential part of future studies.

References

Biggers, J. D., Whitten, W. J., and Whittingham, D. G. (1971). The culture of mouse embryos *in vitro*. In *Methods in Mammalian Embryology*, Daniel, G. C., ed. Academic Press, New York, pp. 86–116.

DePamphilis, M. L., Herman, S. A., Martinez-Salas, E., Chalifour, L. E., Wirak, D. O., Cupo, D. Y., and Miranda, M. (1988). Microinjecting DNA into mouse ova to study DNA replication and gene expression and to produce transgenic animals. BioTechniques 6(7):662–680.

Deutsch, D. G., and Mertz, E. T. (1970). Plasminogen: purification from human plasma by affinity chromatography. Science 170:1095–1096.

Gebauer, F., and Richter, J. D. (1996). Mouse cytoplasmic polyadenylation element binding protein: an evolutionarily conserved protein that interacts with the cytoplasmic polyadenylation elements of c-mos mRNA. Proc. Natl. Acad. Sci. USA 93:14602–14607.

Gebauer, F., Xu, W., Cooper, G. M., and Richter, J. D. (1994). Translational control by cytoplasmic polyadenylation of c-mos mRNA is necessary for oocyte maturation in the mouse. EMBO J 13:5712.

Granelli Piperno, A., and Reich, E. (1978). A study of proteases and protease-inhibitor complexes in biological fluids. J. Exp. Med. 148:223–234.

Huarte, J., Belin, D., and Vassalli, J.-D. (1985). Plasminogen activator in mouse and rat oocytes: induction during meiotic maturation. Cell 43:551–558.

Huarte, J., Belin, D., Vassalli, A., Strickland, S., and Vassalli, J.-D. (1987). Meiotic maturation of mouse oocytes triggers the translation and polyadenylation of dormant tissue-type plasminogen activator mRNA. Genes Dev. 1:1201–1211.

Huarte, J., Stutz, A., O'Connell, M. L., Gubler, P., Belin, D., Darrow, A. L., Strickland, S., and Vassalli, J.-D. (1992). Transient translational silencing by reversible mRNA deadenylation. Cell 69:1021–1030.

Kuge, H., and Richter, J. D. (1995). Cytoplasmic 3′ poly(A) addition induces 5′ cap ribose methylation: implications for translational control of maternal mRNA. EMBO J 14:6301–6310.

Laemmli, U. K. (1970). Cleavage of structural proteins during the assembly of the head of bacteriophage T4. Nature 227:680–685.

Melton, D. A. (1985). Injected anti-sense RNAs specifically block messenger RNA translation *in vivo*. Proc. Natl. Acad. Sci. USA 82:144–148.

Richter, J. D. (1995). Dynamics of poly(A) addition and removal during development. In *Translational Control*, Hershey, J. W. B., Mathews, M. B., and Sonenberg, N., eds. Cold Spring Harbor, NY, Cold Spring Harbor Laboratory Press, pp. 481–503.

Sallés, F. J., and Strickland, S. (1995). Rapid and sensitive analysis of mRNA polyadenylation states by PCR. PCR Methods Appl. 4:317–321.

Sallés, F. J., Richards, W. G., Huarte, J., Vassalli, J.-D., and Strickland, S. (1993). Microinjecting antisense sequences into oocytes. Methods Enzymol. 225:351–361.

Sawadogo, M., and Van Dyke, M. W. (1991). A rapid method for the purification of deprotected oligodeoxynucleotides. Nucleic Acids Res. 19:674.

Schibler, U., Tosi, M., Pittet, A. C., Fabiani, L., and Wellauer, P. K. (1980). Tissue-specific expression of mouse alpha-amylase genes. J. Mol. Biol. 142:93–116.

Sheets, M. D., Fox, C. A., Hunt, T., Vande Woude, G., and Wickens, M. (1994). The 3′-untranslated regions of c-mos and cyclin mRNAs stimulate translation by regulating cytoplasmic polyadenylation. Genes Dev. 8:926–938.

Sheets, M. D., Wu, M., and Wickens, M. (1995). Polyadenylation of c-mos mRNA as a control point in Xenopus meiotic maturation. Nature 374:511–516.

Strickland, S., Huarte, J., Belin, D., Vassalli, A., Rickles, R. J., and Vassalli, J.-D. (1988). Antisense RNA directed against the 3′ noncoding region prevents dormant mRNA activation in mouse oocytes. Science 241:680–684.

Stutz, A., Huarte, J., Gubler, P., Conne, B., Belin, D., and Vassalli, J. D. (1997). *In vivo* antisense oligodeoxynucleotide mapping reveals masked regulatory elements in an mRNA dormant in mouse oocytes. Mol. Cell. Biol. 17:1759–1767.

Vassalli, J. D., and Stutz, A. (1995). Translational control. Awakening dormant mRNAs. Curr. Biol. 5:476–479.

Vassalli, J.-D., Dayer, J. M., Wohlwend, A., and Belin, D. (1984). Concomitant secretion of prourokinase and of a plasminogen activator-specific inhibitor by cultured human monocytes-macrophages. J. Exp. Med. 159:1653–1668.

Vassalli, J.-D., Huarte, J., Belin, D., Gubler, P., Vassalli, A., O'Connell, M. L., Parton, L. A., Rickles, R. J., and Strickland, S. (1989). Regulated polyadenylation controls mRNA translation during meiotic maturation of mouse oocytes. Genes Dev. 3:2163–2171.

Wickens, M., Kimble, J., and Strickland, S. (1995). Translational control of developmental decisions. In *Translational Control*, Hershey, J. W. B., Mathews, M. B., and Sonenberg, N., eds. Cold Spring Harbor, NY, Cold Spring Harbor Laboratory Press, pp. 411–450.

Woolf, T. M., Jennings, C. G., Rebagliati, M., and Melton, D. A. (1990). The stability, toxicity and effectiveness of unmodified and phosphorothioate antisense oligodeoxynucleotides in Xenopus oocytes and embryos. Nucleic Acids Res. 18:1763–1769.

6

Identification of Genes and Processes Guiding the Transition between the Mammalian Gamete and Embryo

Bermseok Oh
Sue-Yun Hwang
Wilhelmine N. De Vries
Davor Solter
Barbara B. Knowles

Identification of Maternal Transcripts

In all metazoan species, the mature female gamete is arrested in meiotic prophase until the cell cycle resumes upon fertilization or parthenogenetic activation. Transcription is inhibited in the oocyte and is not immediately initiated in the newly formed embryonic genome upon fertilization. Rather, it is activated at different times in the various species, following a strict "zygotic clock," perhaps so-named because the events that occur during preimplantation embryogenesis have been accurately timed but not molecularly understood. Accordingly, the developmental processes active during this transcriptionally inert period are controlled by stored maternal messages that were previously transcribed in the ovarian oocyte.

The timing of the switch from maternal to embryonic control of development varies from species to species. It occurs between the two- and eight-cell stage in most mammals, whereas in other well-studied vertebrates such as *Xenopus*, and invertebrates such as *Drosophila*, many more cell divisions take place before new transcription initiates from the embryonic genome (Thompson et al. 1997). Furthermore, the length of time required for the first 2–3 cell divisions in mammals is measured in days, whereas the 11 cleavage divisions which precede *Xenopus* genome activation, for example, are accomplished in but 7 hours (Gilbert 1997). In fact, when activation of the embryonic genome occurs in the late two-cell stage in the mouse, between one-tenth and one twentieth of the time required for the entire embryonic development of the mouse has elapsed.

The molecular processes which occur during the time when the male and female gametes fuse and the zygotic genome becomes transcriptionally active have been particularly intractable to investigation in mammals. The primary obstacle has been the difficulty of collecting a sufficient quantity of mammalian eggs/embryos for meaningful molecular exploration. To solve this problem, we prepared large and representative cDNA libraries from the mRNAs present in the mouse unfertilized egg and various preimplantation-stage mouse embryos (Rothstein et al. 1992, 1993). These primary libraries were used to prepare a partial subtraction library, enriched in stored maternal messages that are abundant in the egg and still present in the two-cell embryo, but not expressed in the eight-cell stage. Because the two-cell embryo is transcriptionally dormant, it is those maternal transcripts with this differential expression pattern that are most likely to control the major processes taking place during the earliest phases of preimplantation development, when the first mitotic cell cycle is being traversed and the embryonic genome is becoming activated. It has been our working principle that by exploring these transcribed genes and the biochemical pathways in which they participate, the key molecular mechanisms controlling the transition from gamete to embryo can be identified. Indeed, following this "expression-screening" protocol, novel genes were found which function in this time period that would not have been discovered by the usual homology search strategies (Hwang et al. 1996, 1997; Oh et al. 1997). We have also applied differential display analysis (Liang and Pardee 1992) using these cDNA libraries (Heyer et al. 1997) or embryo RNAs (Struwe, M., Hwang, S. and Solter, D., unpublished data) to uncover genes with a differential expression pattern in the egg and embryo. These latter approaches have not yet been as effective as the subtraction library strategy at uncovering novel genes with obvious potential for controlling key processes in the newly formed embryo.

In this chapter, we will illustrate the molecular complexities controlling the newly formed embryo using examples from the novel and known genes we have studied. The routine methods we used or devised to investigate translation of maternal transcripts, and to detect the post-translational modifications which likely control embryonic genome activation will be described in detail. Also, an innovative strategy we devised to specifically delete maternally encoded protein products to understand how they normally contribute to embryo development will be discussed.

Controlled Translation of Maternal mRNAs

Studies of maternal transcripts in *Xenopus* revealed that those messages that are translated into protein during the transcriptionally inert period are marked with unique elements in their 3' untranslated region (3'UTR; see Richter 1996 for a review). When this UA-rich consensus sequence lies proximal to the nuclear polyadenylation signal, it serves as a binding site for a protein complex that leads to deadenylation of these transcripts after their synthesis. Later, when the proper *trans*-factors become available in the egg or embryo, they bind to the cytoplasmic polyadenylation element (CPE), enabling regeneration of the poly(A) tail and translation. Cytoplasmic polyadenylation of mRNAs appears to occur at different times in mammals, and may involve different 3'UTR *cis*-elements and/or *trans*-factors, in addition to the CPE and its binding proteins. Controlled translation may engage the same basic mechanisms in all metazoans, but the detailed features may vary with the mode of development of the species.

Sequence Analysis of 3'UTRs — Screening for Cytoplasmic Polyadenylation Elements

In the mouse, transcription in the oocyte ceases upon meiotic maturation, when the chromosomes of the oocyte are completely condensed (Knowland and Graham 1972; Moore 1975), and does not reinitiate until the embryonic genome is activated in the two-cell-stage embryo (Schultz 1993). Thus, meiotic maturation of oocytes as well as progression through the first cleavage divisions, depends on the timely translation of stored maternal mRNAs. Those mRNAs that contain the CPE, ([A]UUUU[U]A[A]U) in the 3'UTR, are deadenylated upon transport into the oocyte cytoplasm, where they are stored until their readenylation and translation. Cytoplasmic polyadenylation requires the presence of two proximal *cis*-acting elements in the 3'UTR: AAUAAA, a nuclear polyadenylation signal, and the U-rich CPE (Fox et al. 1989; McGrew and Richter 1990; Huarte et al. 1992; Salles et al. 1992; reviewed in Bachvarova 1992 and Richter 1996). Interestingly, activation of these stored mRNAs can be initiated by resumption of the meiotic cell cycle as well as by fertilization, suggesting that these processes are controlled by signal transduction cascades (Fox et al. 1989; McGrew and Richter 1990; Salles et al. 1992; Simon et al. 1992; Simon and Richter 1994; Oh et al. 1997). This also implies that timely synthesis of proteins from stored mRNAs is required for both the progress of the meiotic cell cycle and the mitotic divisions following fertilization (Simon et al. 1992; Gebauer et al. 1994; Sheets et al. 1994).

Detection of Cytoplasmic Polyadenylation in the Egg and Preimplantation Embryo — Determining Changes in Transcript Size

In *Xenopus* and *Drosophila*, cytoplasmic polyadenylation has been shown to be necessary for translation of stored maternal mRNAs in a number of genes. Cytoplasmic polyadenylation of the tissue-type plasminogen activator (tPA) maternal transcript triggers its translation in the mouse embryo

(Huarte et al. 1987). However, it is not known whether this process is common to the translation of all maternal transcripts in mammals. Accordingly, we began an extensive investigation of maternal transcripts in the mouse egg and embryo. Northern blot analysis (Huarte et al. 1987), PCR-based amplification of 3'UTR sequences (Huarte et al. 1992; Salles et al. 1992), and RNase protection assays (Gebauer et al. 1994) have each been used to detect cytoplasmic polyadenylation of mRNAs in embryos. Northern blot analysis is the conventional way to detect size changes of specific mRNAs during the transition from oocyte to embryo; yet it presents difficulties if the size of the mRNA is large relative to the size increase resulting from polyadenylation. The PCR-based polyadenylation assay utilizes a powerful amplification method so that a size change can be detected using very little RNA. However, one problem of this PCR-based method is that the size differences between mRNAs from eggs and embryos at different stages in development are difficult to compare since the PCR product appears as a smear rather than a distinct band. In the RNase protection assay, in vitro synthesized non-polyadenylated RNA is microinjected into the oocyte, the RNA from the microinjected oocytes is isolated and annealed to radiolabeled antisense RNA containing a poly(U) tail, and the size of the annealed fragment is examined after RNase A treatment (Gebauer et al. 1994). Because the results are clear and only a few cells are needed, this method is useful when testing mutated constructs for mapping the *cis*-elements essential for polyadenylation. The disadvantages of this method are that it requires additional experimental steps (subcloning, in vitro transcription, and microinjection), and the results only show whether the RNA is polyadenylated or not, but do not indicate whether the length of the poly(A) tail varies in the various samples. Used in combination, these three methods are of value for determining the molecular basis of translation of specific transcripts at particular developmental time points.

We have found conventional northern blot analysis to be particularly useful. Transcript size changes are readily discernible, the method is relatively artifact-free, and, once prepared, the polyadenylation pattern of different genes can be tested by repeatedly hybridizing the same blot with different probes. An example is shown in Figure 6-1A, where RNAs from different embryonic stages are blotted and hybridized to an α-catenin probe. Here, the upward size shift in the mRNA isolated from the ovulated egg and the one-cell embryo, when compared with the mRNA from the full-grown oocyte, is clearly discernible. Deadenylation and degradation of this mRNA in the two-cell-stage embryo and de novo transcription from the embryonic genome in the four- to eight-cell-stage embryo and blastocyst can also be readily distinguished. To prepare RNA samples for northern blots, embryos were either collected at defined times after mating, designated as midnight of the day on which the male and female were caged together, or "delayed mating" (see below) was utilized to collect exactly synchronized early embryos. By combining delayed mating and northern analysis, we have

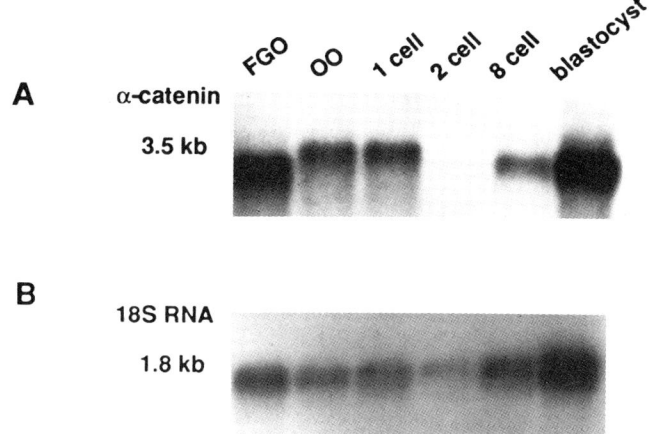

Figure 6-1 Northern analysis of oocytes and early embryos. (A) RNA, from 500 oocytes and embryos in each lane, was hybridized with an α-catenin open reading frame probe. The α-catenin transcript is approximately 3.5 kb. FGO, full-grown oocyte; OO, ovulated oocyte; 1-cell embryos, embryonic day (E) 0.5; 2-cell embryos, E1.5; 8-cell embryos, E2; blastocyst, E4. (B) A small portion of the RNAs (10 oocytes and embryos in each lane) was hybridized with 18S RNA probe to test the integrity and amount of sample per lane.

comparatively examined transcript stability and cytoplasmic polyadenylation patterns of a number of maternal transcripts throughout preimplantation embryo development and found differences in their profiles (B. Oh et al. unpublished data).

Collection of Oocytes and Embryos Synchronized in vivo Full-grown, germinal vesicle (GV)-intact oocytes are collected in HEPES-buffered Eagle's minimal essential medium (MEM) supplemented with 3 mg/ml bovine serum albumin (BSA) from 22–24-day-old female mice (B6D2 F1 or B6A F1, The Jackson Laboratory) by puncturing ovarian follicles with a 30G needle. Normally, mice were not stimulated with pregnant mare's serum gonadotropin (PMSG), for ease in obtaining oocytes without the surrounding granulosa cells. Oocytes are denuded by pipetting them several times in a drop of medium with a P200 micropipette until the surrounding cells are detached from the oocytes. To collect metaphase II-arrested ovulated oocytes, mice are superovulated by injecting 5 units of human chorionic gonadotropin (hCG) 46–48 hours after administering 5 units of PMSG, and are sacrificed 13 hours later. Cumulus cells are removed from the oocyte–cumulus complexes by incubating them in collection medium containing 500 μg/ml hyaluronidase, at 37°C for a few minutes, and then washing the oocytes several times in fresh medium. To collect synchronized embryos, superovulated females, 13 hours after hCG injection, are caged with male mice for 1–2 hours and then checked for vaginal plugs. Embryos are collected at 12-hour intervals: mid one-cell-stage embryos are obtained 12 hours after mating, early two-cell-stage embryos are obtained 24 hours after mating, late two-cell-stage embryos are obtained 36 hours after mating; at 48 hours post coitus, more than half the embryos contain eight cells, the remainder contain four cells. Morulae, collected 3 days after fertilization, are cultured in vitro to blastocysts in KSOM (Lawitts and Biggers 1993). All embryos are collected by flushing the oviducts with medium as described (Hogan et al. 1986).

Extraction of RNA and Northern Hybridization The RNAs are prepared by the method of Huarte et al. (1987). Oocytes and embryos are directly placed, without removing the zona pellucida, into lysis buffer containing 0.1 M Tris-HCl, pH 7.4, 1 M β-mercaptoethanol, and 4 M guanidium thiocyanate. Usually, 200 oocytes or embryos are lysed in 100 ml buffer. To precipitate RNA, 20 µg tRNA (RNase-free, Boehringer Mannheim), 8 µl of 1 M acetic acid, 5 µl of 2 M KCl, and 60 µl of absolute ethanol are added to the lysate. RNA from 500 oocytes or embryos is required for each gel lane, therefore, the ethanol-precipitated RNAs are kept at $-70°C$ until all the samples are collected. The RNAs are pelleted by centrifugation, rinsed with 80% ethanol, dried briefly, and resuspended in 20 µl of resuspension buffer containing 10 mM Tris-HCl, pH 8.0, 5 mM EDTA, and 0.1% SDS. The resuspended samples are pooled so that each tube contains RNA from 500 oocytes or embryos. The lysates are then extracted twice with phenol/chloroform and once with chloroform. The extracted RNAs are ethanol-precipitated by adding $\frac{1}{10}$ volume of 3 M sodium acetate and 2 volumes of ethanol. After 2-minute centrifugation at 14,000g, the pellets are resuspended in 50 µl of dH_2O. Then 1 µl, containing the equivalent of RNA extracted from 10 oocytes or 10 embryos, is removed for testing the quantity and the quality of the RNA by northern blot analysis, and the remainder is stored in ethanol at $-70°C$. For gel electrophoresis, the RNA is mixed with 3.5 µl of dH_2O, 2 µl of 5× RNA gel running buffer, 3.5 µl of formaldehyde and 10 µl of formamide, and is then denatured by incubating at 65°C for 5 minutes. The denatured RNA is kept on ice until it is loaded onto the gel. The RNA marker (9.4–0.24 kb, Gibco BRL) is visualized by adding ethidium bromide to the treated marker RNA at a concentration of 160 µg/ml, and then incubating at 65°C for 5 minutes before loading onto the gel. The RNAs are separated by migration on a 0.9% formaldehyde agarose gel at 65 V, transferred to a nylon membrane (Micro Separation, Inc.), and cross-linked by UV-irradiation (Sambrook et al. 1989). To determine the relative amount of RNAs in each sample, the RNA test blot is hybridized to an 18S rRNA probe at 1.5×10^6 cpm/ml hybridization buffer containing 50% formamide, 5× Denhardt's solution, 5× SSC, 0.2% SDS, 100 µg/ml denatured salmon sperm DNA, and 10% dextran sulfate at 42°C (Figure 6-1B, Hwang et al. 1997). Once it has been ascertained that the RNAs in each lane are of high quality and quantity, the remainder is ethanol-precipitated and the RNA is dissolved in 4.5 µl of dH_2O and prepared for the experimental RNA blot

(Figure 6-1A). This same membrane can be hybridized with a succession of different probes. The hybridization of each probe is visualized by autoradiography, then the probe is stripped from the membrane by incubating it for 2 minutes in boiling buffer containing 0.25% SDS and 0.05% 20× SSC.

Estimating the Extent of Polyadenylation — RNase H Treatment The length of the poly(A) tail of a given transcript can be determined by annealing a test RNA to oligo(dT) and then treating the mixture with RNase H (Mercer and Wake, 1985, diagrammed in Figure 6-2A). The RNA is denatured in 20 μl of 1 mM EDTA by heating at 95°C for 5 minutes and cooling on ice. Then, 1 μl of oligo (dT) (0.5 μg) is added to the denatured RNA and allowed to anneal for 10 minutes at 25°C. The concentration of KCl is adjusted to 50 μM by adding 1.5 μl of 1 M KCl and 7.5 μl dH$_2$O and incubating for 10 minutes at 25°C. Then, 10 μl of 4× RNase H buffer (100 mM KCl, 112 mM MgCl$_2$, and 80 mM Tris-HCl, pH 8.0) and 1 unit of RNase H (Boehringer Mannheim) is added and the reaction mixture is incubated for 30 minutes at 37°C. Following two phenol/chloroform and one chloroform extraction, the RNA is ethanol-precipitated, resuspended in 50 μl dH$_2$O and a 1-μl aliquot is separated on a formaldehyde agarose gel and blotted. To confirm the quality of the RNase H-treated RNA, the blot is probed with an 18 S rRNA probe. The length of the poly(A) tail of a particular transcript in the oocyte, egg, and embryo is estimated by hybridization with a corresponding probe. In the example shown, it is difficult to detect any size differences among the various stages from a large transcript such as β-catenin (Figure 6-2B). However, when the 3.7 kb β-catenin transcript is foreshortened to a 1.2 kb fragment by hybridization with the complementary antisense oligonucleotide and then treated with RNase H, the change in size can be readily detected. Here, the size of the RNA in the ovulated oocyte is distinctly different from that in the full-grown oocyte and zygote samples. The difference between the ovulated oocyte samples treated, or not, with oligo(dT) before RNase H digestion, indicates that the size increase is the result of polyadenylation.

Assessing Controlled Translation in the Oocyte, Zygote and Two-Cell-Stage Embryo — The Reporter Gene Approach

To determine whether the polyadenylation of a maternal mRNA in the cytoplasm is directly related to its translational activation, the presence or absence of the encoded polypeptide must be determined. The relationship between polyadenylation and the activation of translation was clearly demonstrated by making use of the enzymatic activity of tPA to demonstrate its synthesis (Huarte et al. 1987). In other cases, more indirect approaches to documenting translation were employed (Paynton et al., 1988). The time of translation from a particular maternal transcript during the oocyte–gamete transition can be documented if antibody to the protein under putative translational control is available. The relationship between

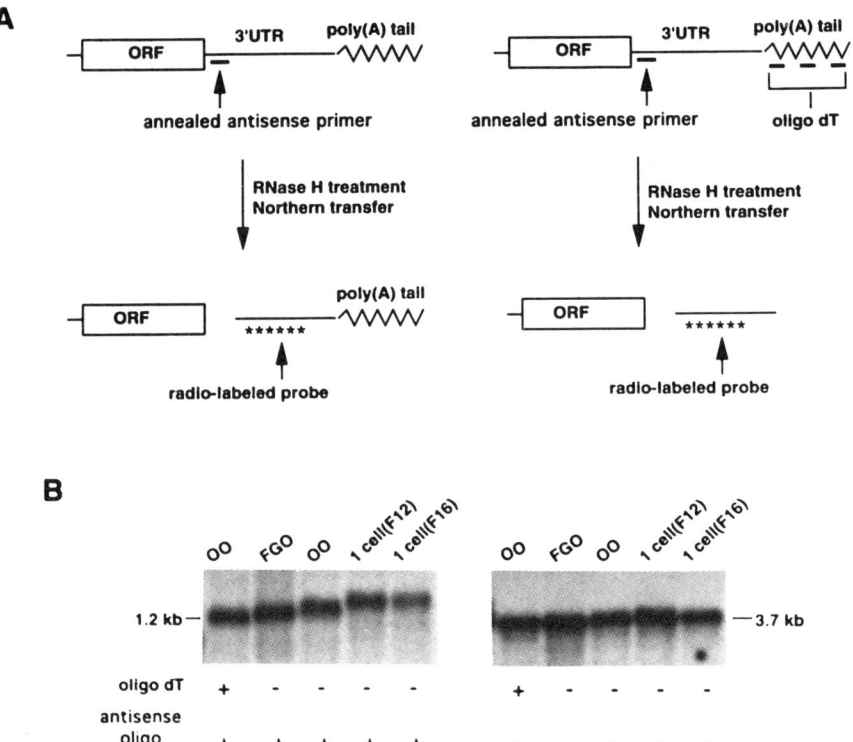

Figure 6-2 RNase H treatment of oocytes' and embryos' RNAs. (A) Schematic diagram of an RNase H experiment. (B) RNA from 500 full-grown or ovulated oocytes was annealed with 500 ng of an antisense oligonucleotide specific to the 3' end of β-catenin mRNA (5'-GGAACTGGTCAGCTCGACTG-3') and then treated with RNase H to digest the regions containing DNA/RNA hybrids. RNA in the first lane was also annealed with 500 ng of oligo (dT)$_{15}$ (Boehringer Mannheim, Indianapolis, IN) to remove the poly(A) tail. Hybridization was performed with a 3'UTR β-catenin probe. Full-length β-catenin is 3.7 kb; the expected size of the fragment resulting from annealing with oligo (dT) and the specific oligo (above) after RNase H treatment is 1.2 kb. F12, 12 hours after fertilization; F16, 16 hours after fertilization. OO, FGO — see above.

transcript polyadenylation and translation can be investigated by pulse-labeling the egg or embryo, immunoprecipitating the polypeptide from the cell extract, and autoradiographically determining whether it was synthesized during the labeling period (see below). Recently we described a novel gene, *Spin*, which has three maternal transcripts (one of 0.8 kb, one of 1.7 kb, and one of 4.1 kb), each with an identical open reading frame (ORF) but each with a different 3'UTR (Oh et al. 1997). Translation of SPIN is continuous from the time the oocyte is first formed until the end of the two-cell-stage. The unique combination of different CPE sequences in these transcripts may influence their ability to be translated. Thus, the *Spin* gene provided an experimental system to evaluate whether each of these 3′UTR sequences controlled translation at a specific time in the oocyte/embryo. To exploit this opportunity, we devised a reporter gene assay in which the coding sequence of β-galactosidase is fused to the 3′UTR of each SPIN transcript (see Figure 6-3).

The Reporter Gene To prepare the constructs for analysis, we used a reporter gene cassette, pBlue-β-gal, containing the coding sequence of β-galactosidase (HindIII 414-DraI 3819) from pSV-β-gal (Promega), subcloned into the HindIII and SmaI sites of pBluescript SK(+). In this construct, the T7 promoter is available for in vitro transcription and the multiple cloning sites are available for further subcloning. The effect of a particular 3′UTR sequence on translational activation is assessed by inserting it downstream of the β-galactosidase-coding region and microinjecting the fusion RNA transcribed from the T7 promoter into appropriately staged eggs and embryos. The amount of β-galactosidase translated is assessed by incubating extract from the injected eggs and embryos with a luminescent *lacZ* substrate and measuring the emitted light.

To place each of the *Spin* 3′UTRs downstream of the β-galactosidase coding region, they were PCR-amplified from the relevant *Spin* cDNA using a BamHI-anchored 5′ primer and a XbaI-anchored 3′ primer. After digesting the PCR product with BamHI and XbaI, it was ligated into the BamHI/XbaI site of pBlue-β-gal, downstream of the coding sequence. Capped RNA was transcribed in vitro from the linearized reporter construct using T7 RNA polymerase (Ambion, mMESSAGE mMACHINE). After in vitro transcription, the RNA was digested with DNase to remove template DNA and extracted twice with phenol/chloroform. Quantitation of the in vitro transcribed RNAs was performed by running an aliquot on a gel and comparing it with a RNA marker of known concentration (see Figure 6-4A). Normally, 10 μg of capped RNA is made from 1 μg template DNA. The RNAs are aliquoted as 2 μg in 5 μl of dH$_2$O and kept at −70°C. Just before injection, each aliquot was vacuum-dried, resuspended in 10 μl of 0.15 M KCl, and microfuged for 5 minutes to remove any particulates which might clog the microinjection needle.

As a control, the RNAs can be in vitro polyadenylated before microinjection (Vassalli et al. 1989). Briefly, a polyadenylation reaction containing 0.25 M NaCl, 0.05 M Tris-HCl, pH 8.0, 0.01 M MgCl$_2$, 0.1 mg/ml BSA, 2 mM DTT, 0.25 mM ATP, 2.5 mM MnCl$_2$, 1 unit/ml RNase inhibitor, and 0.1 unit/ml poly(A) polymerase (Gibco BRL) is incubated for 30 minutes at 37°C, at 5–10 μg of mRNA in a 40 μl reaction volume. It is important to point out that in vitro polyadenylated RNA does not separate well on an agarose gel and looks degraded because it appears as a long smear. However, relatively distinct bands can be seen on a RNA gel after two or more phenol/chloroform extractions (Figure 6-4B, compare lanes 3 and 4).

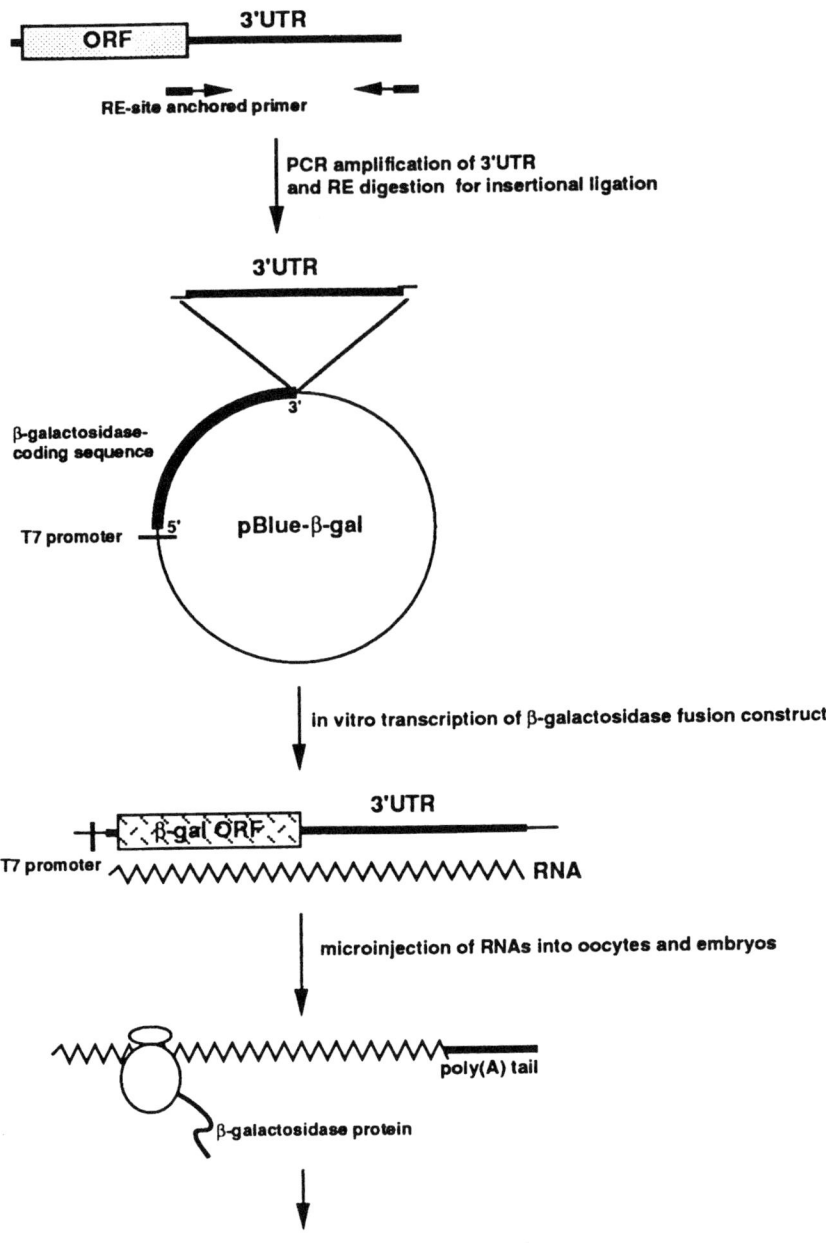

Figure 6-3 Diagrammatic representation of the reporter gene experiment used to determine the effect of various 3'UTRs on translation in the egg and embryo.

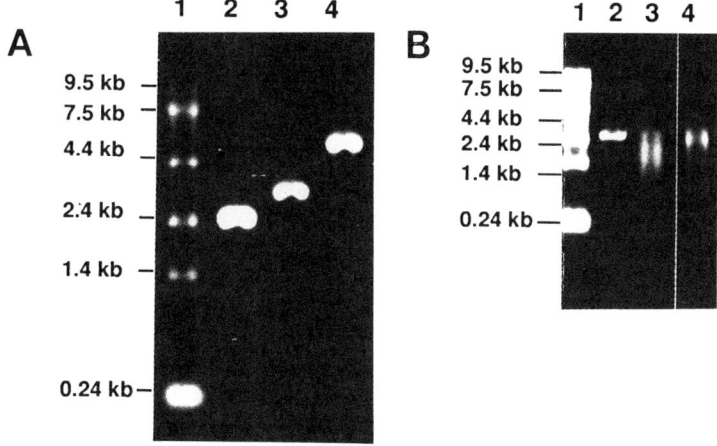

Figure 6-4 In vitro transcription and polyadenylation of RNAs for microinjection. (A) In vitro transcription was of the β-galactosidase constructs from the three different 3′UTRs of *Spin*. Template DNAs were linearized at the end of the 3′UTR and transcribed using T7 polymerase (Ambion). Lanes: 1, RNA molecular marker (Gibco BRL); 2, β-galactosidase RNA fused with the 3′UTR of the 0.8-kb *Spin* transcript; 3, β-galactosidase RNA fused with the 3′UTR of the 1.7-kb *Spin* transcript; 4, β-galactosidase RNA fused with the 3′UTR of the 4.1-kb *Spin* transcript. (B) In vitro polyadenylation was performed and the RNAs were run on a formaldehyde 0.9% agarose gel. Lane 1, RNA marker; 2, untreated β-galactosidase RNA fused with the 3′UTR of the 1.7-kb *Spin* transcript; 3, In vitro polyadenylated β-galactosidase RNA fused with the 3′UTR of the 1.7-kb *Spin* transcript. 4, In vitro polyadenylated 1.7-kb construct RNA after two phenol/chloroform extractions.

Microinjection of the Reporter Gene Constructs into Oocytes and Zygotes
Before analyzing the effect of the different 3′UTRs on the translation of β-galactosidase, the optimum incubation time and RNA concentration to be injected was determined in maturing oocytes using the 3′UTR of the 1.7 kb transcript. The amount of β-galactosidase synthesized after microinjection of 0.2 mg/ml non-polyadenylated fusion transcript in 0.15 M KCl solution was monitored over an 18-hour time course by measurement of conversion of a chemiluminescent substrate with a luminometer (Figure 6-5A). At this concentration, β-galactosidase activity continuously increased throughout the time course. To determine the effect of the RNA concentration, serially diluted RNA, from 25 to 400 ng/µl, was injected into oocytes that had undergone germinal vesicle breakdown (GVB). Within the test range, β-galactosidase activities increased almost proportional to the amount of microinjected RNA (Figure 6-5B). We chose 200 ng/µl for further experimentation. It might be possible to increase the sensitivity of these single-cell measurements by injecting higher RNA concentrations, but practical considerations, such as the difficulty of producing large amounts of mRNA, have dampened our enthusiasm for exploring the use of higher concentrations.

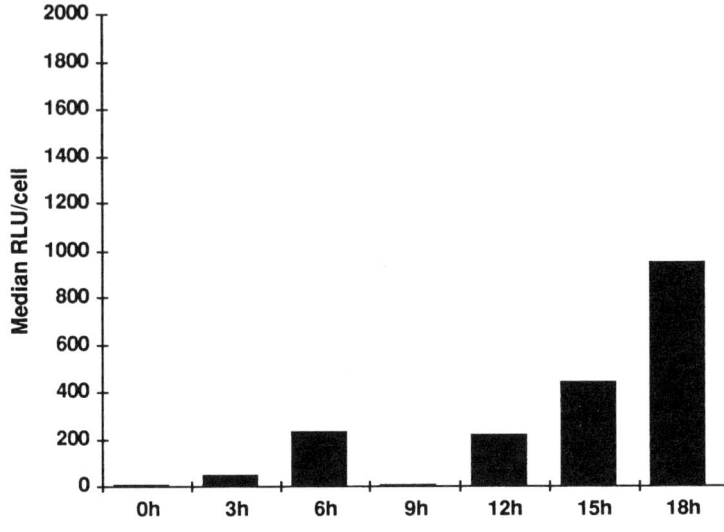
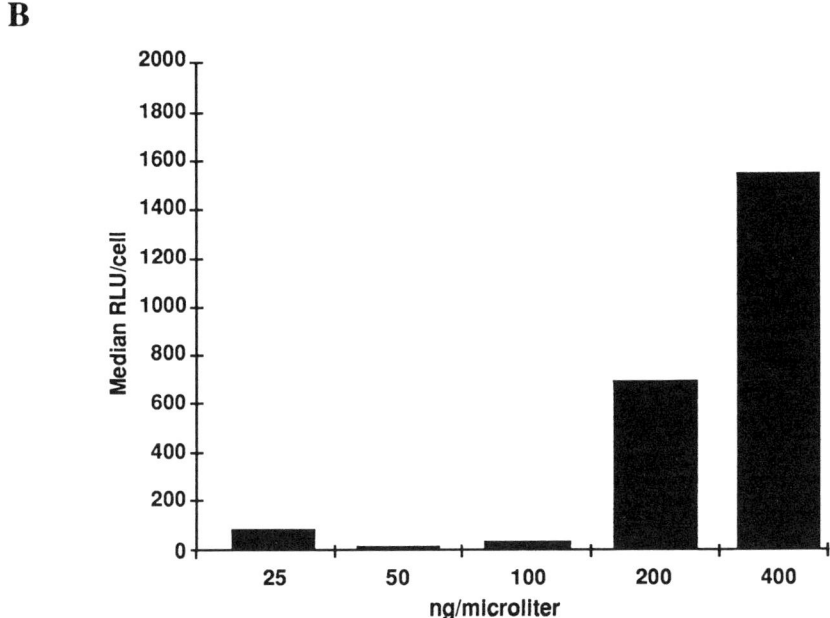

Figure 6-5 Standardization of the β-galactosidase reporter gene assay. (A) Time course of the translation of the non-polyadenylated β-galactosidase 1.7 kb *Spin* 3'UTR RNA. In this experiment, three or four injected oocytes were pooled together incubated with Galacton-Star (Clontech) and the relative light units (RLU) of the chemiluminescent enzyme product was measured fluorometrically. (B) The effect of the amount of RNA injected on β-galactosidase translation.

To relate the amount of protein synthesized to the changes in transcript polyadenylation in the various stages, β-galactosidase activity was measured in GV-oocytes, GVB-oocytes, and zygotes. Full-grown GV-oocytes were injected and cultured for 16–18 hours in modified MEM supplemented with 100 mM IBMX to inhibit meiotic maturation. To test GVB-oocytes, full-grown GV-oocytes were injected, cultured for 16–18 hours in modified MEM, and those that had undergone germinal vesicle breakdown were selected for assay. Zygotes were collected from superovulated females which had been caged with males overnight. The cell membrane of the zygote is extremely elastic, making cytoplasmic injection extremely difficult. To overcome this problem, the microinjection needle is pushed towards a pronucleus until it just touches and as the needle is drawn back the RNA is released into the cytoplasm. Zygotes were injected as soon as their pronuclei became prominent, approximately 10 hours after mating. Injected zygotes were incubated for 16–18 hours in modified MEM, and only those developing to the two-cell stage were selected for assay with a luminescent β-galactosidase substrate (Genetic Reporter System, Clontech). Briefly, individual oocytes or zygotes were lysed in 5 µl of 0.1 M potassium phosphate buffer (pH 7.8) by three freeze–thaw cycles, mixed with 50 µl of reaction buffer, and β-galactosidase activity was measured in an Auto Lumat LB953 (Berthold, Germany) or TD-20/20 Luminometer (Turner Design). The relative light units (RLU) measured using either machine were proportional throughout a broad range (Figure 6-6). The RLU values obtained from individual oocytes varied (Figure 6-7); therefore, the median of RLU obtained from individual assays is taken and the results from a series of assays are compared and statistically evaluated. The results presented clearly indicate the effect of each 3′UTR on translation in the GVB-oocyte; the 1.7-kb 3′UTR strongly enables translation in the ovulated oocyte (Figure 6-7 and B. Oh et al. unpublished data).

Post-translational Modification of Proteins in the Egg and Embryo

Signal transduction, perpetuated by sequential phosphorylation/dephosphorylation of specific substrates, most likely controls timely translation of specific mRNAs in the egg and the preimplantation embryo. In the most direct case, post-translational modification of a given *trans*-factor might be required for its binding to a *cis*-factor. The overall pattern of protein biosynthesis changes during mouse oocyte maturation (Globus and Stein 1976; Schultz and Wassarman 1977) and after fertilization (Levinson et al. 1978; Howe and Solter 1979; Cullen et al., 1980; Van Blerkom 1981; Cascio and Wassarman 1982; Pratt et al. 1983; Howlett and Bolton 1985, Latham et al. 1991). These pattern shifts may reflect neotranslation of stored maternal transcripts or post-translational modification of pre-existing proteins. Many of the gel mobility/isoelectric point shifts detected at this

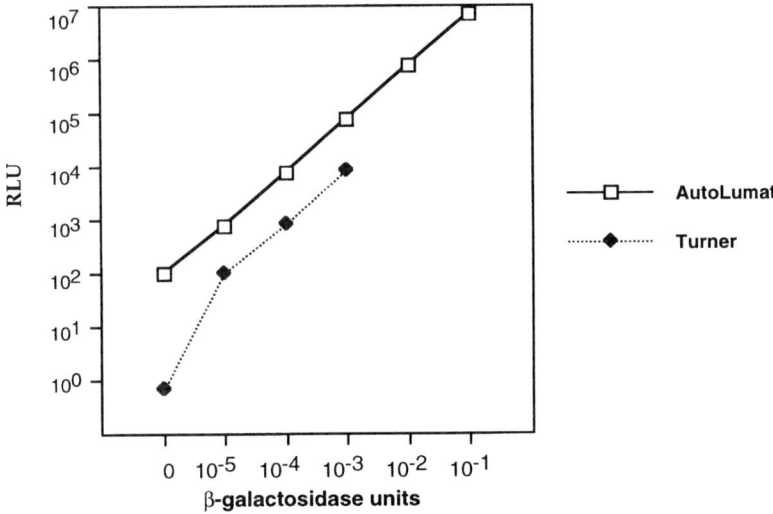

Figure 6-6 Standard curve of relative light units (RLUs) from a β-galactosidase assay as determined with a Turner Design 20/20 and with a Barthold luminometer.

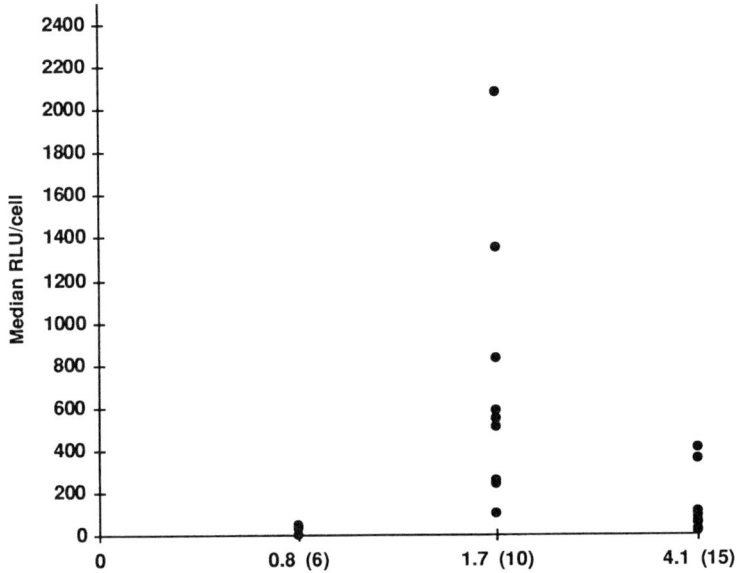

Figure 6-7 Scatter plot from individual assays using β-galactosidase constructs containing either the 0.8-kb, 1.7-kb or 4.1-kb *Spin* 3'UTR. Each dot represents an assay from an individual GVB-oocyte. Numbers in parentheses on the abscissa are the total number of embryos assayed.

developmental time have been attributed to phosphorylation and/or glycosylation (Van Blerkom 1981; Pratt et al. 1983; Howlett 1986; Oh et al. 1997). Moreover, several molecules that regulate controlled translation of maternal transcripts during this transition period, such as the CPE-binding protein (CPEB), are post-translationally modified (Roy et al. 1990; Gautier et al. 1991; Hake and Richter, 1994).

Antibody, A Tool for Investigating Post-translational Modification

Antibodies are ideal reagents for investigating protein synthesis and post-translational modification. By varying the method of analysis, the time of synthesis, the subtle shifts in size and isoelectric point, and the changes in subcellular localization that connote antigen modification are all readily detected. Antibodies to specific peptides or proteins can be obtained following any number of protocols (Harlow and Lane 1988). Antigenic SPIN polypeptides were prepared from bacterial cultures transfected with SPIN cDNA under the control of an inducible promoter. Excellent vectors are commercially available that make various epitope-tagged fusion polypeptides (e.g. pTrcHis, Invitrogen; pPET, Novagen; pGST, Pharmacia). For example, nickel chelation chromatography was used to isolate histidine-tagged specific SPIN polypeptide from bacterial cell lysates for injection into animals (Oh et al. 1997). Most antibodies are useful for a variety of techniques, i.e., immunolocalization of antigen in cells or sections, immunoprecipitation, or immunodetection of polypeptides containing detectable antigenic determinants on western blots. However, some antibodies do not recognize antigens that have been exposed to specific fixatives or have been permanently denatured. Therefore, it is important to evaluate each antibody preparation using the assay planned for analysis.

To produce polyclonal antiserum, a rabbit was injected three times with 100 mg of purified SPIN protein (the first boost followed the initial immunization after 2 weeks; the second boost followed a week after the first boost, and the third boost was administered 4 weeks after the second, by CoCalico Inc., Reamstown, PA). The rabbit was test bled 1 week following each immunization. To determine whether each immunization was successful, in vitro translated (Promega rabbit reticulocyte lysate) [^{35}S]methionine-labeled polypeptide was used as the antigen in immunoprecipitation reactions (Figure 6-8). As the number of immunizations increased, increasing amounts of SPIN polypeptides were precipitated by a constant volume of serum as detected by SDS–polyacrylamide gel electrophoresis (SDS–PAGE).

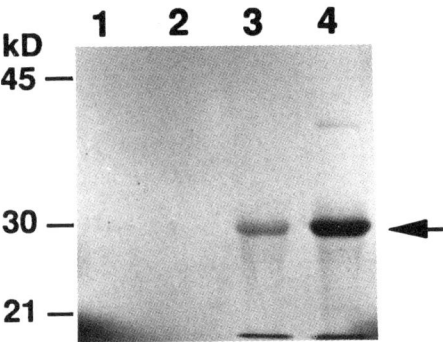

Figure 6-8 Immunoprecipitation of in vitro translated SPIN polypeptide using a constant volume (10 µl) of rabbit test sera. Lanes: 1, preimmune serum; 2, first test serum after a single immunization; 3, second test serum 1 week after the second immunization; 4, antiserum obtained 1 month after the third immunization.

Immunoprecipitation Reveals Cell-Cycle-Dependent Phosphorylation

Immunoprecipitation of SPIN from [^{35}S]methionine-labeled full-grown and ovulated oocytes appears as a single band on SDS–PAGE, but when SPIN is immunoprecipitated from labeled ovulated eggs, two or three closely spaced bands are seen. When extracts of inorganic [^{32}P] phosphate-labeled full-grown oocytes are similarly analyzed, no band is detectable by autoradiography, whereas SPIN is detected in [^{32}P]-labeled ovulated oocytes (Oh et al., 1997). Full-grown oocytes have not yet entered meiotic maturation whereas ovulated oocytes are in meiotic metaphase. Therefore, these oocytes are labeled at different times in the cell cycle. To investigate whether SPIN is differentially phosphorylated during the cell cycle, ovulated eggs were pulse-labeled and embryos were collected by the delayed mating protocol so that the zygotic clock can be accurately read in regards to the first mitotic cell cycles. The slower migrating form of SPIN (upper band) is phosphorylated while the faster migrating form is not (Figure 6-9; Howlett 1986; Oh et al. 1997, and unpublished data). Both forms are clearly identified in meiotic metaphase II-arrested ovulated eggs and in the one cell embryo as it enters M phase of the first mitotic cell cycle (Figure 6-9 — 0 h, 16 h). Both forms are also abundant in the G1 phases of the first and second mitotic cell cycles (Figure 6-9 — 5 h, 18 h), but the phosphorylated form of SPIN becomes much less discernible in embryos as they progress into the first and second mitotic S phases (Figure 6-9 — 11 h, 14 h, 20 h). Thus, by pulse-labeling throughout the first mitotic cell cycles, immunoprecipitation, and SDS–PAGE analysis, we have demonstrated that post-translational processing of SPIN coincides with progression through the cell cycle.

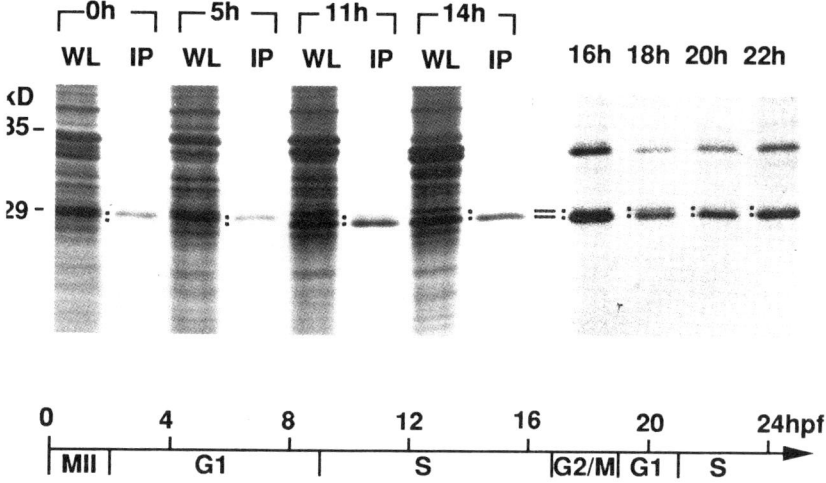

Figure 6-9 Immunoprecipitation of newly synthesized SPIN during the first mitotic cell cycle. Autoradiogram of SDS–PAGE of whole lysate (WL) and SPIN immunoprecipitated (IP) from ^{35}S-methionine labeled metaphase II-arrested ovulated oocytes (Oh) and embryos at 5, 11, 14, 16, 18, 20, and 22 hours after fertilization. Dots indicate the two sizes of SPIN, the result of differential phosphorylation (Oh et al. 1997). The diagram below shows the phases of the first mitotic cell at various hours post-fertilization (hpf).

To obtain these results, oocytes and synchronized embryos, collected from females in the delayed mating scheme described above, were labeled by incubating them in modified MEM containing 1 mCi/ml [^{35}S]methionine (specific activity > 1000 Ci/mmol; Amersham) for 3 hours or in phosphate-free modified MEM containing 0.5 mCi/ml [^{32}P]orthophosphate (New England Nuclear) for 6 hours. Normally, 100 oocytes or fertilized eggs are labeled in 50-µl drops of medium, under mineral oil in a Billups-Rothenberg chamber, at 37°C in a 5% O_2, 5% CO_2 and N_2 atmosphere. The medium for [^{35}S]methionine labeling does not contain any nonlabeled amino acids other than glutamine, as they greatly reduce the efficiency of incorporation of [^{35}S]methionine. Normally, the efficiency of labeling is 5×10^3 cpm/oocyte or embryo, as measured by TCA precipitation (Sambrook et al. 1989). Labeled oocytes and fertilized eggs are transferred through several drops of PBS supplemented with 4 mg/ml polyvinylpyrrolidone (PVP) and lysed in 100 µl of immunoprecipitation (IP) buffer (50 mM Tris-HCl, pH 7.4, 0.25 mM NaCl, 0.1% Nonidet-P 40, 5 mM EDTA, 50 mM sodium fluoride, 0.1 mM sodium orthovanadate, Hampl and Eppig 1995). Just before lysis, proteinase inhibitors, including 1 mM phenylmethylsulfonyl fluoride (PMSF), 0.1 mM dithiothreitol (DTT), 1 µg/ml leupeptin, 1 µg/ml aprotinin, and 10 mg/ml N-tosyl-L-phenylalanine chloromethyl ketone (TPCK), or proteinase inhibitor Complete™ (Boehringer Mannheim), are added to the IP buffer.

Lysates are precleared of polypeptides that nonspecifically bind to the beads by two 1-hour incubations at 4°C with 50 µl of a 50% (v/v) slurry of protein A-Sepharose CL-4B (Sigma) preincubated with 25 µg of normal rabbit immunoglobulin (Ig). The beads are removed by centrifugation and the precleared supernatant is then incubated for 1 hour at 4°C with 50 µl of protein A-Sepharose CL-4B preincubated with immune Ig. The beads are then removed by centrifugation and the antigen–antibody complexes are prepared for SDS–PAGE by boiling the beads in SDS buffer (Sambrook et al. 1989), or for two-dimensional gel electrophoresis by incubating the beads in urea sample buffer for 20 minutes at 42°C (BioRad). Either Rainbow Markers (Amersham) or MW-SDS-70L (Sigma) are used as protein molecular-weight standards on the gel. For autoradiography, the gel is dried and exposed to Kodak Biomax film.

Intracellular Compartmentalization

The subcellular localization of a gene product in the oocyte and embryo may change as the cause or effect of its post-translational processing. This localization can be immunodetected in the growing and full-grown oocyte in the ovary, or in the ovulated egg or embryo in the oviduct, after fixation and sectioning. SPIN is localized to the cytoplasm of the full-grown oocyte within the ovarian follicle but it is not detectable in the nucleus (cf. Figure 6-10A with control Figure 6-10B). However, in ovulated oocytes, which are arrested in MII, the spindle and spindle poles are also SPIN-positive (Figure 6-10C, Oh et al. 1997). We were interested in determining whether decoration of the spindle by SPIN antibody results from the coincident appearance of an intracellular structure to which SPIN binds, or is the result of the phosphorylation of SPIN previously demonstrated in the M and G1 portion of the first mitotic cell cycles (Figure 6-9). To do this, oocytes were isolated from *Mos* null mutant mice deficient in MAP kinase, one of the controlling kinases in the signal transduction cascades active at this time in development (Sagata et al. 1989; Yew et al. 1992; Colledge et al. 1994). SPIN phosphorylation is defective in oocytes from *Mos* null mutant mice and SPIN detection in the spindle is haphazard, suggesting that post-translational phosphorylation indeed affects its subcellular localization (Oh et al. 1998).

To prepare samples for immunolocalization, mouse ovaries are dissected from females primed with PMSG for 9 hours to stimulate a proportion of oocytes to undergo meiotic maturation. Oviducts containing fertilized eggs or embryos are obtained from mice which have been superovulated and mated. Ovaries or oviducts are fixed by overnight incubation in modified Bouin's solution (85 parts saturated picric acid, 10 parts formaldehyde solution and 5 parts glacial acetic acid), or Tellyesniczky/

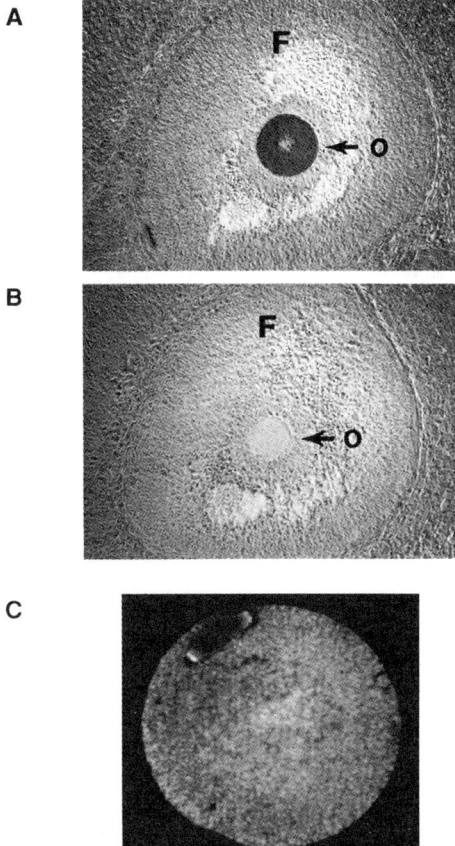

Figure 6-10 Immunodetection of proteins in the mouse oocyte. (A) An ovary section from a 3-week-old mouse was incubated with rabbit antibody to SPIN, and then, consecutively with goat antirabbit immunoglobulin coupled to peroxidase conjugated and diaminobenzidine. F, follicle; O, oocyte. (B) Control section incubated with preimmune serum and then treated as above. (C) Immunodetection of β-tubulin in an ovulated oocyte using mouse monoclonal antibody (Sigma). Ovulated oocytes were fixed and permeabilized before reacting with mouse antibody to tubulin. Goat antimouse immunoglobulin conjugated with FITC was used for visualization.

Fekete (Telly's — 100 µl of 70% ethanol, 5 ml of glacial acetic acid, and 10 ml of formaldehyde solution) or 4% paraformaldehyde in phosphate-buffered saline (PBS), then embedded in paraffin, sectioned, and mounted on slides. The sections are deparaffinized by heating the slides at 55°C for 15 minutes, incubating them in two changes of xylene and one change of xylene/ethanol (1:1) each for 5 minutes. When peroxidase is to be used for visualization, the deparaffinized sections are incubated in 3% H_2O_2/methanol for 15 minutes to reduce endogenous peroxidase activity. To hydrate the sections, the slides are serially passed through graded ethanol/dH_2O solutions (two changes, 5 minutes each, of absolute ethanol, 95% ethanol, 80% ethanol, 70% ethanol), and then rinsed in tap water. Before starting an immunoreaction, the area of the slide containing the sections is encircled with paraffin (PAP pen, Kiyota). They are equilibrated in three changes of PBS, each 5 minutes, and blocked by incubating the area within the paraffin in a drop of 3% BSA/PBS for 30 minutes in a humidified chamber (several layers of wet 3MM paper placed in a Nunc plate). The slides are rinsed in 0.25% BSA/PBS and the sections are incubated with purified antibody diluted in 1% BSA/PBS for 60 minutes. Dilution factors should be determined experimentally, but usually fall in the range of 100- to 1000-fold in the case of serum. The slides are washed three times for 10 minutes in an excess volume of PBS, and the sections are incubated for 40 minutes with biotinylated second antibody (goat antirabbit), washed in two changes of PBS, each 5 minutes, and the sections are then incubated in peroxidase-conjugated avidin diluted 20-fold in 1% BSA/PBS (Sigma ExtrAvidin kit) and then rinsed three times with PBS. For visualization of the bound enzyme-tagged antibody, the slides are incubated with 3,3'-diaminobenzidine (Sigma Fast DAB), the reaction is monitored for optimum color development, and then stopped by rinsing the slides in water. Hematoxylin is frequently used as a counterstain.

Whole-mount immunostaining can be performed by several well-established methods. Depending on the cellular localization of the protein in question, the oocyte or embryo is processed by either first permeabilizing or first fixing the sample (see Messinger and Albertini 1991; Simerly and Schatten 1993). To localize SPIN, ovulated oocytes obtained from superovulated mice were washed through several drops of 0.4% PVP/PBS, transferred to a drop of Bouin's fixative, and incubated for 30 minutes at room temperature (Ohsugi et al. 1996; Oh et al. 1997). The incubations were performed in an organ culture dish (Falcon #3037) in which the outer ring was filled with dH_2O to prevent evaporation. The fixed oocytes were washed twice by transferring them through 0.4% PVP/PBS drops, permeabilized by incubating in 0.5% Triton X-100/PBS for 10 minutes, and incubated in 1% fetal calf serum/PBS for 1 hour at 4°C and then in a 50-µl drop containing anti-SPIN immunoglobulin diluted in 1% BSA/PBS for 2 hours at 37°C. After incubation in two changes of an excess volume of 1% BSA/PBS for 15 minutes at room temperature with constant shaking, the eggs were transferred to Cy3-conjugated goat antirabbit antibody. After 1 hour at 37°C, oocytes were washed, equilibrated in a 50% glycerol/50% PBS mounting solution containing 100 mg/ml 1,4-diazabicyclooctane as an antibleaching agent (Sigma), mounted onto slides, and examined by confocal microscopy.

Functional Analysis of Genes Expressed in the Egg and Preimplantation Embryo

Clues to the function of individual genes have recently been found by positional- or tag-cloning of the genes responsible for spontaneous mutations (e.g. Ackerman et al. 1997). The phenotype of the mutant hints at the developmental and/or biochemical pathway in which the gene functions. However, there are very few spontaneous mutations known to affect the ovulated oocyte, zygote, and two-cell embryo, and none have yielded the function of particular genes at this time of development. On the other hand, the mutant phenotype of any cloned gene can be found by null mutagenesis. Traditionally, the targeted gene is disrupted by homologous recombination in ES cells that are injected into blastocysts to form chimeras. Mutant stocks, established from the targeted ES cells that colonized the germ line, are then characterized. Using this approach, it was found that oocytes from targeted *Mos* null mutants are prone to parthenogenetic activation (Colledge et al. 1994). However, conventional null mutagenesis is not suitable to elucidate the role of all genes involved in the control of preimplantation development. Most of these genes are also expressed in adult tissues; therefore, the homozygous null mutant females, required to produce oocytes lacking the targeted maternal transcript, might never be born or may not live to reproduce. Moreover, the stored maternal messages that are translated after fertilization are stable and the protein synthesized appears to be able to carry the null embryo for quite a way into embryogenesis (Haegel et al. 1995).

To overcome these difficulties and to understand the function of individual gene products in this transitional period, null mutation of these genes should be initiated in, and restricted to, the developing egg. Using conditional mutagenesis, mature eggs devoid of the expressed gene product can be produced and investigated.

To eliminate the presence of targeted genes in the ovulated egg and preimplantation embryo we are testing a modified Cre/*lox*P-facultative mutagenesis system. Two mouse stocks are required: a transgenic lineage containing the P1 bacteriophage recombinase protein Cre under the control of an oocyte-specific promoter, and a line in which the gene of interest is flanked by two *lox*P sites, the 34-bp DNA site at which the Cre recombinase acts (Figure 6-11; Barinaga 1994; Hoess and Abremski 1985). A transgenic line containing Cre under control of the *Zona pellucida 3 (Zp3)* promoter has been described that appears to be uniquely suited to activate Cre in the oocyte and zygote (Lewandowski et al. 1997). *Zp3* is expressed prior to the completion of the first meiotic division, therefore, *lox*P-flanked sequences will be excised at this time. In a test case of this line, recombination was found exclusively in the oocytes (Lewandowski et al. 1997). ES cells containing maternally expressed and translationally controlled genes flanked by *lox*P sites can be prepared and injected into blastocysts. Highly chimeric

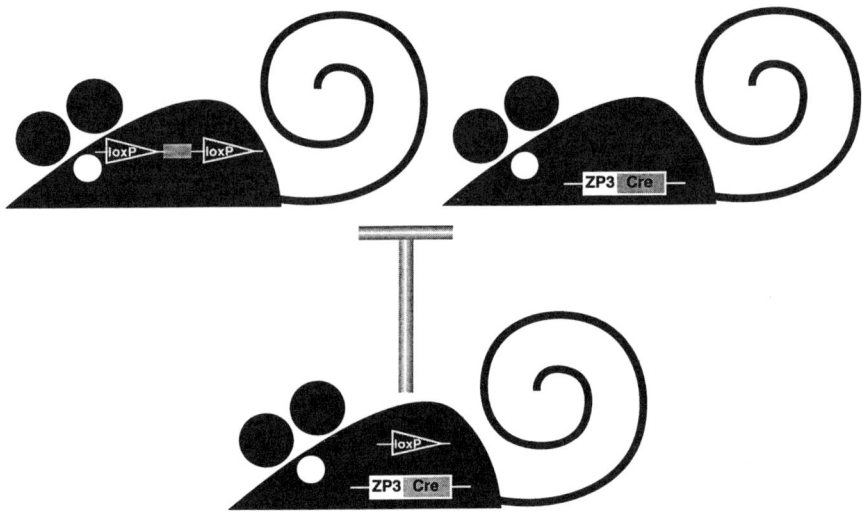

Figure 6-11 Diagram of oocyte Cre/*lox* conditional mutagenesis schema.

mice are then test-mated to obtain those which transmit the *lox*P targeted gene through the germ line. From their offspring females homozygous for the *lox*P-targeted gene is bred and mated to a *Zp3*-Cre transgenic male (Figure 6-11). Accordingly, the targeted gene can be eliminated, no stored message will be available in the egg and preimplantation embryo, and the phenotypic effect of this deficiency can be analyzed. This type of analysis tests the hypothesis that the gene identified is critical to the initiation of embryogenesis.

In this chapter, we have described the experimental approaches that enabled us to identify the genes transcribed in the oocyte that are likely to control egg and embryo development prior to activation of the embryonic genome. We have used these genes as examples to point up the process affecting timely translation of proteins in this unique cell type. We also have shown how post-translational modifications can change the intracellular distribution of a protein, a distribution which may hint at the function of the gene product. Finally, we have described a conditional mutagenesis schema to reveal the role of each of these maternally encoded genes at the origin of the life of a new individual.

References

Ackerman, S. L., Kozak, L. P., Przyborski, S. A., Rund, L. A., Boyer, B. B., and Knowles, B. B. (1997). The mouse rostral cerebellar malformation gene encodes an UNC-5-like protein. Nature 386:838–842.

Bachvarova, R. F. (1992). A maternal tail of poly(A): the long and the short of it. Cell 69:895–897.
Barinaga, M. (1994). Knockout mice: round two. Science 265:26–28.
Cascio, S. M., and Wassarman, P. M. (1982). Program of early development in the mammal: post-transcriptional control of a class of proteins synthesized by mouse oocytes and early embryos. Dev. Biol. 89:397–408.
Colledge, W. H., Carlton, M. B. L., Udy, G. B., and Evans, M. J. (1994). Disruption of c-*mos* causes parthenogenetic development of unfertilized mouse eggs. Nature 370:65–68.
Cullen, B. R., Emigholz, K., and Monahan, J. J. (1980). Protein patterns of early mouse embryos during development. Differentiation 17:151–160.
Fox, C. A., Sheets, M. D., and Wickens, M. P. (1989). Poly(A) addition during maturation of frog oocytes: distinct nuclear and cytoplasmic activities and regulation by the sequence UUUUUAU. Genes Dev. 3:2151–2162.
Gautier, J., Solomon, M. J., Booher, R. N., Bazan, J. F., and Kirschner, M. W. (1991). cdc25 is a specific tyrosine phosphatase that directly activates p34^{cdc2}. Cell 67:197–211.
Gebauer, F., Xu, W., Cooper, G. M., and Richter, J. D. (1994). Translational control by cytoplasmic polyadenylation of c-mos mRNA is necessary for oocyte maturation in the mouse. EMBO J. 13:5712–5720.
Gilbert, S. F. (1997). *Developmental Biology*, 5th ed. Sunderland, MA, Sinauer.
Globus, M. S., and Stein, M. P. (1976). Qualitative patterns of protein synthesis in the mouse oocyte. J. Exp. Zool. 198:337–342.
Haegel, H., Larue, L., Ohsugi, M., Fedorov, L., Herrenknecht, K., and Kemler, R. (1995). Lack of β-catenin affects mouse development at gastrulation. Development 121:3529–3537.
Hake, L. E., and Richter, J. D. (1994). CPEB is a specificity factor that mediates cytoplasmic polyadenylation during Xenopus oocyte maturation. Cell 79:617–627.
Hampl, A., and Eppig, J. (1995). Analysis of the mechanism(s) of metaphase I arrest in maturing mouse oocytes. Development 121:925–933.
Harlow, E., and Lane, D. (1988). *Antibodies: A Laboratory Manual*. Cold Spring Harbor, NY, Cold Spring Harbor Laboratory Press.
Heyer, B. S., Warsowe, J., Solter, D., Knowles, B. B., and Ackerman, S. L. (1997). New member of the Snf1/AMPK kinase family, *Melk*, is expressed in the mouse egg and preimplantation embryo. Mol. Reprod. Dev. 47:148–156.
Hoess, R. H., and Abremski, K. (1985). Mechanism of strand cleavage and exchange in the Cre-lox site-specific recombination system. J. Mol. Biol. 181:351–362.
Hogan, B., Costantini, F., and Lacy, E. (1986). *Manipulating the Mouse Embryo*. Cold Spring Harbor, NY, Cold Spring Harbor Laboratory Press.
Howe, C. C., and Solter, D. (1979). Cytoplasmic and nuclear protein synthesis in preimplantation mouse embryo. J. Embryol. Exp. Morphol. 52:209–225.
Howlett, S. K. (1986). A set of proteins showing cell cycle dependent modification in the early mouse embryo. Cell 45:387–396.
Howlett, S., and Bolton, V. (1985). Sequence and regulation of morphological and molecular events during the first cell cycle of mouse embryogenesis. J. Embryol. Exp. Morphol. 87:175–206.

Huarte, J., Belin, D., Vassalli, A., Strickland, S., and Vassalli, J.-D. (1987). Meiotic maturation of mouse oocytes triggers the translation and polyadenylation of dormant tissue-type plasminogen activator mRNA. Genes Dev. 1:1201–1211.

Huarte, J., Stutz, A., O'Connell, M. L., Gubler, P., Belin, D., Darrow, A. L., Strickland, S., and Vassalli, J.-D. (1992). Transient translational silencing by reversible mRNA deadenylation. Cell 69:1021–1030.

Hwang, S.-Y., Benjamin, L. E., Oh, B., Rothstein, J. L., Ackerman, S. L., Beddington, R. S. P., Solter, D., and Knowles, B. B. (1996). Genetic mapping and embryonic expression of a novel, maternally transcribed gene *Mem3*. Mamm. Genome 7:586–590.

Hwang, S.-Y., Oh, B., Fuchtbauer, A., Fuchtbauer, E.-M., Solter, D., and Knowles, B. B. (1997). *Maid*: a maternally transcribed novel gene encoding a potential negative regulator of bHLH proteins in the mouse egg and zygote. Dev. Dyn. 209:217–226.

Knowland, J., and Graham, C. (1972). RNA synthesis at the two-cell stage of mouse development. J. Embryol. Exp. Morphol. 27:167–176.

Latham, K. E., Garrels, J. I., Chang, C., and Solter, D. (1991). Quantitative analysis of protein synthesis in mouse embryos. I. Extensive reprogramming at the one- and two-cell stages. Development 112:921–932.

Lawitts, J. A., and Biggers, J. D. (1993). Culture of preimplantation embryos. Methods Enzymol. 225:153–164.

Levinson, J., Goodfellow, P., Vadeboncoeur, M., and McDevitt, H. (1978). Identification of stage-specific polypeptides synthesized during murine preimplantation development. Proc. Natl. Acad. Sci. USA 75:3332–3336.

Lewandowski, M., Wassarman, K. M., and Martin, G. R. (1997) *Zp3-cre*, a transgenic mouse line for the activation of *lox*P-flanked target genes specifically in the female germ line. Curr. Biol. 7:148–151.

Liang, P., and Pardee, A. B. (1992). Differential display of eukaryotic messenger RNA by means of the polymerase chain reaction. Science 257:967–971.

McGrew, L. L., and Richter, J. D. (1990). Translational control by cytoplasmic polyadenylation during *Xenopus* oocyte maturation: characterization of cis and trans elements and regulation by cyclin/MPF. EMBO J. 9(11):3743–3751.

Mercer, J. F., and Wake, S. A. (1985). An analysis of the rate of metallothionein mRNA poly(A)-shortening using RNA blot hybridization. Nucleic Acids Res. 13:7929–7943.

Messinger, S. M., and Albertini, D. F. (1991). Centrosome and microtubule dynamics during meiotic progression in the mouse oocyte. J. Cell Sci. 100:289–298.

Moore, G. P. M. (1975). The RNA polymerase activity of the preimplantation mouse embryo. J. Embryol. Exp. Morphol. 34:291–298.

Oh, B., Hwang, S.-Y., Solter, D., and Knowles, B. B. (1997). Spindlin, a major maternal transcript expressed in the mouse during the transition from oocyte to embryo. Development 124:493–503.

Oh, B., Hampl, A., Eppig, J., Solter, D. and Knowles, B. B. (1998). SPIN, a substrate in the MAP kinase pathway in mouse oocytes. Mol. Rep. Dev. 50:240–249.

Ohsugi, M., Hwang, S.-Y., Butz, S., Knowles, B. B., Solter, D., and Kemler, R. (1996). Expression and cell membrane localization of catenins during mouse preimplantation development. Dev. Dyn. 206:391–402.

Paynton, B. S., Rempel, R., and Bachvarova, R. (1988). Changes in the state of adenylation and time course of degradation of maternal RNAs during oocyte maturation and early embryonic development in the mouse. Dev. Biol. 129:304–314.
Pratt, H. P. M., Bolton, V. N., and Gudgeon, K. A. (1983). The legacy from the oocyte and its role in controlling early development of the mouse embryo. In *Molecular Biology of Egg Maturation.* London, Pitman, pp. 197–227.
Richter, J. D. (1996). Dynamics of poly(A) addition and removal during development. In *Translation Control.* Cold Spring Harbor, NY, Cold Spring Harbor Laboratory Press, pp. 481–503.
Rothstein, J. L., Johnson, D., DeLoia, J. A., Skowronski, J., Solter, D., and Knowles, B. (1992). Gene expression during preimplantation mouse development. Genes Dev. 6:1190–1201.
Rothstein, J. L., Johnson, D., Jesse, J., Skowronski, J., DeLoia, J., Solter, D., and Knowles, B. B. (1993). Construction of primary and subtracted cDNA libraries from early embryos. Methods Enzymol. 225:587–610.
Roy, L. M., Singh, B., Gautier, J., Arlinghous, R. B., Nordeen, S. K., and Maller, J. M. (1990). The cyclin B2 component of MPF is a substrate for the c-mos^{XE} proto-oncogene product. Cell 61:825–831.
Sagata, N., Daar, I., Oskarsson, M., Showwalter, S. D., and Vande Woude, G. F. (1989). The product of the c-mos proto-oncogene as a candidate initiator for oocyte maturation. Science 245:643–645.
Salles, F. J., Darrow, A. L., O'Connell, M. L., and Strickland, S. (1992). Isolation of novel murine maternal mRNAs regulated by cytoplasmic polyadenylation. Genes Dev. 6:1202–1212.
Sambrook, J., Fritsch, E. F., amd Maniatis, T. (1989). *Molecular Cloning.* Cold Spring Harbor, NY, Cold Spring Harbor Laboratory Press.
Schultz, R. M. (1993). Regulation of zygotic gene activation in the mouse. BioEssays 15:531–538.
Schultz, R. M., and Wassarman, P. M. (1977). Biochemical studies of mammalian oogenesis: protein synthesis during oocyte growth and meiotic maturation in the mouse. J. Cell Sci. 24:167–194.
Sheets, M., Fox, C., Hunt, T., Vande Woude, G. F., and Wickens, M. (1994). The 3'-untranslated region of c-*mos* and *cyclin* mRNAs stimulate translation by regulating cytoplasmic polyadenylation. Genes Dev. 8:926–938.
Simerly, C., and Schatten, G. (1993). Techniques for localization of specific molecules in oocytes and embryos. Methods Enzymol. 225:516–553.
Simon, R., and Richter, J. D. (1994). Further analysis of cytoplasmic polyadenylation in *Xenopus* embryos and identification of embryonic cytoplasmic polyadenylation element-binding proteins. Mol. Cell. Biol. 14:7867–7875.
Simon, R., Tassan, J.-P., and Richter, J. (1992). Translational control by poly(A) elongation during *Xenopus* development: differential repression and enhancement by a novel cytoplasmic polyadenylation element. Genes Dev. 6:2580–2591.
Thompson, E. M., Legouy, E., and Renard, J.-P. (1998). Mouse embryos do not wait for the MBT: chromatin and RNA polymerase remodeling in genome activation at the outset of development. Dev. Genet. 22:31–42.
Van Blerkom, J. (1981). Structural relationship and posttranslational modification of stage-specific proteins synthesized during early preimplantation development of the mouse. Proc. Natl. Acad. Sci. USA 78:7629–7633.

Vassalli, J. D., Huarte, J., Belin, D., Gubler, P., Vassalli, A., O'Connell, M. L., Parton, L. A., Rickles, R. J., and Strickland, S. (1989). Regulated polyadenylation controls mRNA translation during meiotic maturation of mouse oocytes. Genes Dev. 3:2163–2171.

Yew, N., Mellini, M. L., and Vande Woude, G. F. (1992). Meiotic initiation by the mos protein in *Xenopus*. Nature 355:649–652.

7

Approaches for the Analysis of Genomic Imprinting in Oocytes and Early Mouse Embryos

Angela J. Villar

Genomic imprinting is an epigenetic phenomenon unique to mammalian development, whereby gene expression is dependent on parent of origin (for reviews see Pedersen et al. 1993; Efstratiadis 1994). The consequences of genomic imprinting are evidenced by the failure of parthenogenetic and androgenetic development in mammals (Mann and Lovell-Badge 1984; McGrath and Solter 1984), human ovarian teratomas and gestational trophoblastic neoplasia (Bagshawe and Lawler 1982; Ariel et al. 1994), and parental-specific effects in the etiology of a number of human diseases (Reik 1989; Hall 1990; Langois 1994; Nicholls 1994).

Evidence that the maternal and paternal genomes are not functionally equivalent arose from the study of chromosomally balanced mice that inherited maternal duplication/paternal deficiency or paternal duplication/maternal deficiency for certain chromosomal regions (Cattanach and Kirk 1985; Searle and Beechey 1985; Beechey and Searle 1987). Although a substantial portion of the genome can be complemented, i.e., uniparental disomy is compatible with normal development, certain chromosomes contain regions displaying a parental effect. This phenomenon was recognized to result from the differential expression of the maternal and paternal chromosomes; the implication being that there are specific genes preferentially expressed when inherited from one parent but not the other (Lyon and Glenister 1977). Table 7-1 summarizes the data on some of the chromosomal regions

Table 7-1 Consequences of uniparental disomy during early mouse development

Chromosome region	Origin	Effect	Imprinted loci within region
Distal 7	Maternal	Mid-fetal lethality	*Igf2* (DeChiara et al. 1991) *Ins 2* (Giddings et al. 1994) *H19* (Bartolomei et al. 1991; Tremblay et al. 1995) *Mash2* (Guillemot et al. 1995)
	Paternal	Early embryo lethality	*p57kip2* (Hatada and Mukai 1995)
Central 7	Maternal	Postnatal lethality	*ZNF127* (Efstratiadis 1994) *Snrpn* (Cattanach et al. 1992; Leff et al. 1992) *Ipw* (Wevrick and Francke 1997)
	Paternal	Postnatal growth/behavior	
Proximal 7	Maternal	Neonatal lethality	*Peg3* (Kaneko-Ishino et al. 1995)
Distal 2	Maternal	Neonatal lethality/hypokinetic	*Gnas* (Williamson et al. 1996)
	Paternal	Neonatal lethality/hyperkinetic	
Proximal 6	Maternal	Early embryo lethality	*Peg1/Mest* (Kaneko-Ishino et al. 1995)
Proximal 11	Maternal	Reduced prenatal growth	*U2 afbp-rs* (Hatada et al. 1993; Hayashizaki et al. 1994)
	Paternal	Enhanced prenatal growth	
Proximal 17	Paternal	Neonatal lethality (Tme)	*Igf2r* (Barlow et al. 1991) *Mas* (Villar and Pedersen 1994; Walther et al. 1998)

shown to have defective complementation depending on parent of origin in the mouse (Searle et al. 1989; Cattanach and Beechey 1990; Efstratiadis 1994) and lists the known imprinted genes within these regions.

Experimental evidence for genomic imprinting has generally been restricted to the mouse, but has also been observed in humans (for reviews see Solter 1992; Lalande 1996; Nakao and Sasaki 1996) and sheep (Cockett et al. 1996; Hagemann et al. 1998). It is generally presumed that the specialization of the parental genomes takes place during gametogenesis when the paternal and maternal germ lines are separated (Swain et al. 1987; Surani et al. 1988). Thus, one function of gametogenesis is to switch the parent-of-origin imprints from the previous generation such that males transmit only the paternal imprint and females transmit only the maternal imprint to the next generation (Figure 7-1) (Hadchouel et al. 1987; Reik et al. 1987; Sapienza et al. 1987; Swain et al. 1987). Although it is not clear whether

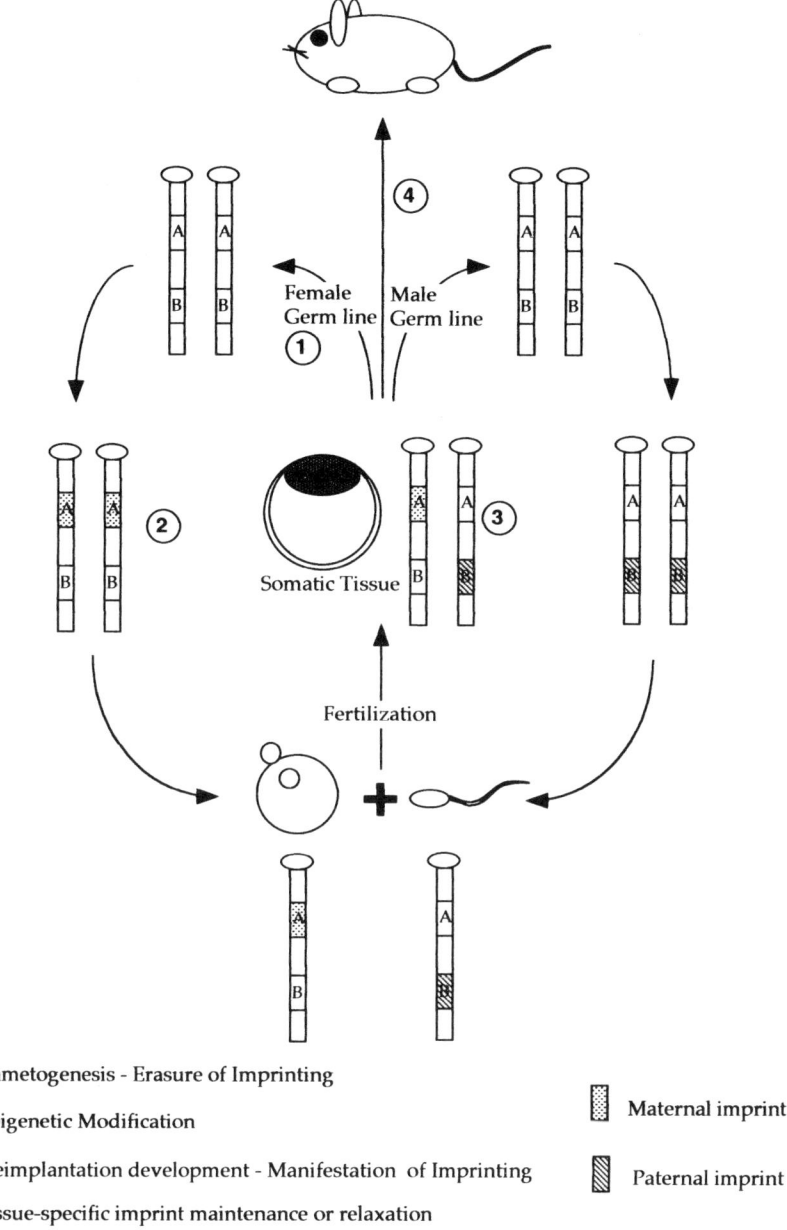

① Gametogenesis - Erasure of Imprinting
② Epigenetic Modification
③ Preimplantation development - Manifestation of Imprinting
④ Tissue-specific imprint maintenance or relaxation

▒ Maternal imprint
▨ Paternal imprint

Figure 7-1 A diagrammatic model representing the developmental regulation of genomic imprinting. The parental imprints are propagated in the embryonic and extraembryonic tissues and erased in the germ line. The developmental potentials of the male and female gamete are limited due to epigenetic modifications acquired during transmission through the germ line. As a result, expression of an undefined number of genes is hemizygous and dependent on the parent of origin. The changes in epigenetic modification during development are indicated.

the acquisition of the imprint is a discrete event or a gradual process during gametogenesis, it is clear that epigenetic modifications during oocyte growth have a profound effect on subsequent embryonic development, as evidenced by the extended development of parthenogenones containing chromosomes from nongrowing primary oocytes (Kono et al. 1996; Obata et al. 1998).

The consequences of genomic imprinting in the germ line and its wide ranging implications for mammalian embryogenesis and human disease emphasize the importance of identifying the developmental stage(s) involved in the "erasure" and subsequent acquisition of the parent-specific imprint.

Imprinted Genes

Imprint switching has been observed for several transgenes, as well as a number of endogenous genes (reviewed by Reik 1992; Gold and Pedersen 1994), four of which have been extensively studied in the mouse and human. Briefly, insulin-like growth factor 2 (*Igf2*) is a single-copy gene mapping to the distal end of mouse chromosome 7, encoding a 67 amino acid polypeptide with a mitogenic function mediated through the *Igf1r* gene (Roth 1988). It is highly expressed in both embryonic and extraembryonic tissue during gestation and exhibits paternal-specific expression (DeChiara et al. 1991) as early as the two-cell stage (Rappolee et al. 1992). *H19* is tightly linked to *Igf2* on the distal portion of mouse chromosome 7 (Zemel et al. 1992). Despite their proximity, however, *H19* is oppositely imprinted with the paternally inherited allele being transcriptionally silent. In mice, *H19* is expressed as early as 4.5 days of gestation (dg) in extraembryonic tissue and in the embryo proper by 8.5 dg (Poirier et al. 1991). Deletion of the T-associated maternal effect (*Tme*) locus is a naturally occurring mutation in the mouse spanning 800–1100 kb on chromosome 17 (Johnson 1975; Artzt et al. 1991). Embryos inheriting the deletion from their mother die at 15 dg but survive if the deletion is paternally inherited. Within this region, the insulin-like growth factor 2 receptor (*Igf2r*) gene is exclusively expressed from the maternal allele; therefore, when the *Tme* locus deletion is inherited from the female, Igf-2r protein is absent (Barlow et al. 1991). Finally, the *Snrpn* gene encodes a nuclear protein associated with the small nuclear ribonucleoprotein particles (snRNPs) and maps to the central portion of mouse chromosome 7 (Leff et al. 1992). Like *Igf2*, the mouse *Snrpn* gene is expressed exclusively by the paternal allele, but primarily in the brain and heart (McAllister et al. 1988).

For these imprinted genes and others, the key to the analysis of imprinting is the ability to distinguish between the parental genomes. Several approaches are applicable depending on whether parental imprinting is to be assessed by differential methylation or differential expression (for review see Kelsey and Reik 1998). Of these approaches—which include single-strand conformational polymorphism (SSCP) (Giddings et al. 1994), RNase protection (Bartolomei et al. 1991), and denaturing gradient gel electrophoresis (DGGE) (Nakao et al. 1996) — reverse transcription–polymerase chain reaction (RT–PCR) in combination with either mRNA phenotyping (Villar et al. 1995) or a SNuPE assay (Szabo and Mann 1995a,b) has been specifically applied to the analysis of imprinting during gametogenesis and early embryo development.

Analysis of Allele-Specific Expression

It is not possible to observe the changes in the epigenetic modification of the germ line because the mechanism of imprinting is not known. It is possible, however, to detect allele-specific expression. Assuming changes in imprinted expression reflect alterations in the physical imprint, it is possible to follow the developmental regulation of genomic imprinting by assessing allele-specific transcription in germ cells as they progress through gametogenesis.

To analyze allele-specific expression during gametogenesis and early mouse development, two approaches have been developed: mRNA phenotyping (Villar et al. 1995) and RT–PCR SNuPE (Szabo and Mann 1995a,b). Both strategies use RT–PCR to detect small quantities of RNA and exploit the genetic diversity between mouse species, subspecies, or strains to assess allelic variants in RNA from hybrid progeny (Figure 7-2).

mRNA phenotyping assesses allele-specific expression using RT–PCR to amplify the cDNA of interest and restriction fragment length polymorphisms (RFLPs) unique to the species. Thus, the "mRNA phenotype" exhibited by the progeny from an interspecies cross reflects the parent of origin (Figure 7-3). This approach does not use radioactivity or an elaborate electrophoresis protocol, permitting efficient low-cost screening of imprinted genes. One disadvantage is that quantification of allele-specific expression is relative and differential levels of expression may not be detected. Another consideration is that sequence polymorphisms may not be informative because they do not translate into endonuclease restriction site differences.

The second approach combines RT–PCR with a single nucleotide primer extension (SNuPE) (Singer-Sam et al. 1992) assay, and is based on defined nucleotide differences between allelic RNAs of hybrid progeny that permit sensitive and quantitative analysis of allele-specific expression (Figure 7-4). This approach uses radioactivity and two gel electrophoresis protocols: one to purify the amplified PCR product and a second to measure the incorporation of each radioactive allele-specific nucleotide. The most significant advantage of this assay is that SNuPE allows sensitive quantitative measurement of the relative amounts of allelic transcripts (Singer-Sam 1994) and absolute quantification when an internal standard is used (Buzine et al. 1994).

To determine the development regulation of genomic imprinting, i.e., the erasure and subsequent establishment, in the oocyte and early mouse embryo, these approaches have been applied to the analysis of four imprinted genes, *Igf2*, *H19*, *Snrpn*, and *Igf2r* (described above). Table 7-2

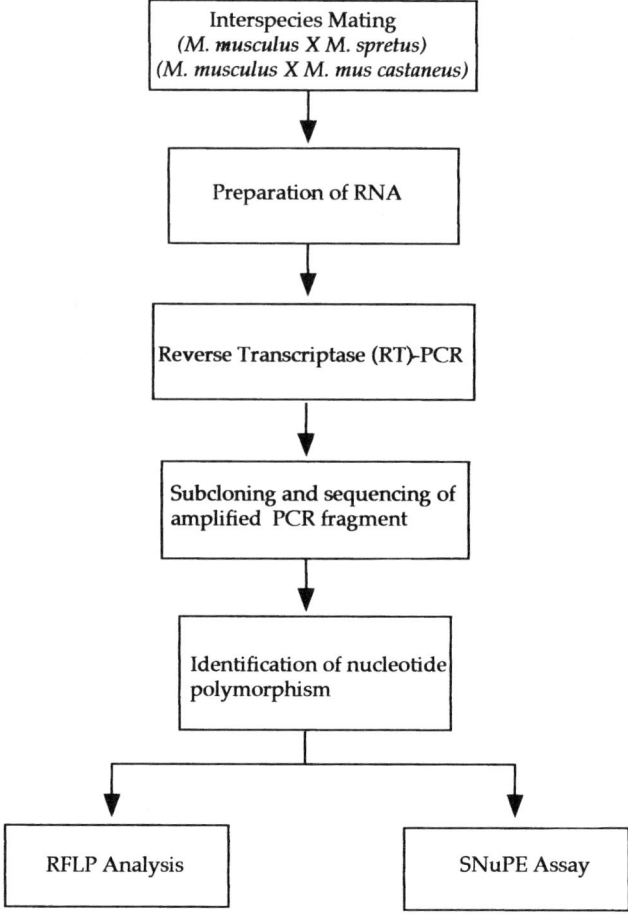

Figure 7-2 Interspecies mouse hybrid approaches for the detection of allele-specific expression applicable to the analysis of imprinted genes at any stage of development.

presents some of the cumulative data from several experiments using the following materials and methods. The developmental aspects of oogenesis in mouse are summarized in Figure 7-5.

Genetic Crosses

Natural interspecies matings are performed between *Mus musculus* (C57BL/6J, Jackson Laboratories, Bar Harbor, ME) females and *Mus spretus* (Jackson Laboratories) males to produce F1 hybrids. Because reciprocal cross-matings between *M. musculus* males and *M. spretus* females frequently fail to produce progeny, it may be necessary to perform backcross matings

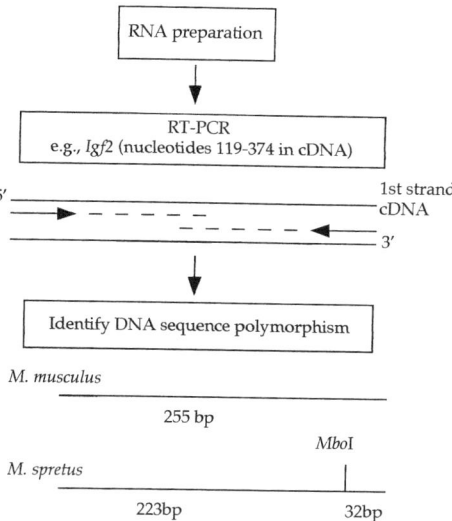

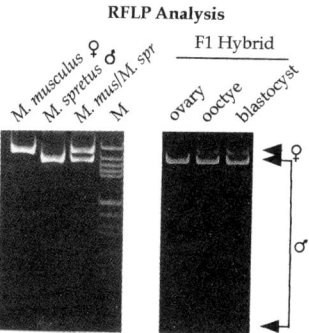

Figure 7-3 mRNA phenotyping analysis scheme. Total RNA is isolated from *M. musculus*, *M. spretus*, and F1 hybrid progeny produced by crossing *M. musculus* and *M. spretus*. Each RNA preparation is reverse transcribed into cDNA and amplified by PCR with the use of primer pairs designed for specific detection of cDNA target sequences. The *M. musculus* and *M. spretus* PCR products are sequenced to identify base-pair differences that generate species-specific restriction enzyme sites. The PCR products generated from F1 hybrids are then digested with restriction enzymes specific to each species to determine their parental origin. Restriction fragments are resolved electrophoretically on polyacrylamide or agarose gels and visualized with ethidium bromide. The mRNA phenotype of the F1 hybrid is compared with the *M. musculus* and *M. spretus* parental mRNA phenotypes to determine whether expression is biallelic or allele-specific. In this example, *Igf2* PCR products are digested with *Mbo*I. Lane 1, tongue RNA from *M. musculus*; lane 2, tongue RNA from *M. spretus*; lane 3 equal amounts (0.5 µg) of tongue RNA from *M. musculus* and *M. spretus*; lane 5, neonate ovary; lane 6, ovulated oocyte; lane 7, blastocyst. The molecular size marker (lane 4) is DNA molecular weight marker pBR322 DNA-*Hae*III (Boehringer Mannheim, Indianapolis, IN).

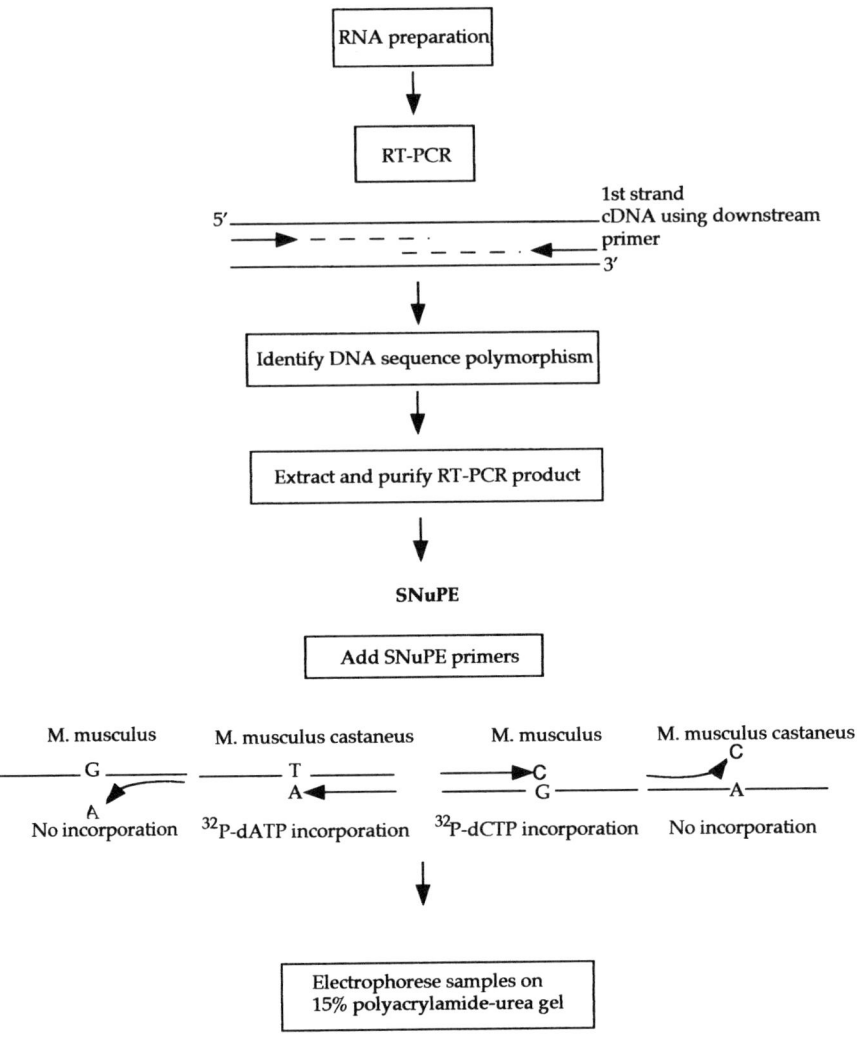

Figure 7-4 Outline of the single nucelotide primer extension assay (SNuPE). The assay consists of the enzymatic extension of a SNuPE primer positioned 5′ to the polymorphism. The primer is extended by one base in the presence of a [^{32}P]dNTP specific for either the *M. musculus* or *M. musculus castaneus* sequence. The 5′ to 3′ orientation of the primers is indicated by the direction of the arrows. The ratio of radioactivity in the primer products of these reactions is determined after gel electrophoresis.

Table 7-2 RT–PCR primer and polymorphism information for the analysis of allele-specific RNA

Loci	Primer pairs	Nucleotide/restriction site polymorphism	Assay	Ref.
Igf2	5'-GACGTGTCTACCTCTCAGGCCGTACTT-3' 5'-GGGTGTCAATTGGGTTGTTTAGAGCCA-3'	C/T	SNuPE	Rotwein and Hall 1990
	5'-GGCCCCGGAGAGACTCTGTGC-3' 5'-GCCCACGGGGTATCTGGGGAA-3'	AluI/MboI	RFLP	Dull et al. 1984
Snrpm	5'-TGCTGCTGTTGCTGCTACTG-3' 5'-GCAGTAAGAGGGGTCAAAAGC-3'	G/A	SNuPE	Gerelli et al. 1991
H19	5'-CCACTACACTACCTGCCTCAGAATCTGC-3' 5'-GGTGGGTACTGGGGCAGCATTG-3'	A/C	SNuPE	Pachnis et al. 1988
	5'-GAATTCAAACAGGGCAAGATGGGGTCA-3' 5'-GAATTCGGCGCCACATGGTGTTCAAGAAG-3'	BsrI	RFLP	Villar et al. 1995
Igf2r	5'-CTGGAGGTGATGAGTGTAGCTCTGGC=3' 5'-GAGTGACGAGCCAACACAGACAGGTC-3'	A/G	SNuPE	Szebenyi and Rotwein 1994
	5'-ATGATGACAGCGACGAAGACC-3' 5'-AAACCTAGGCACTCAGGGACC-3'	ScrFI	RFLP	Villar et al. 1995

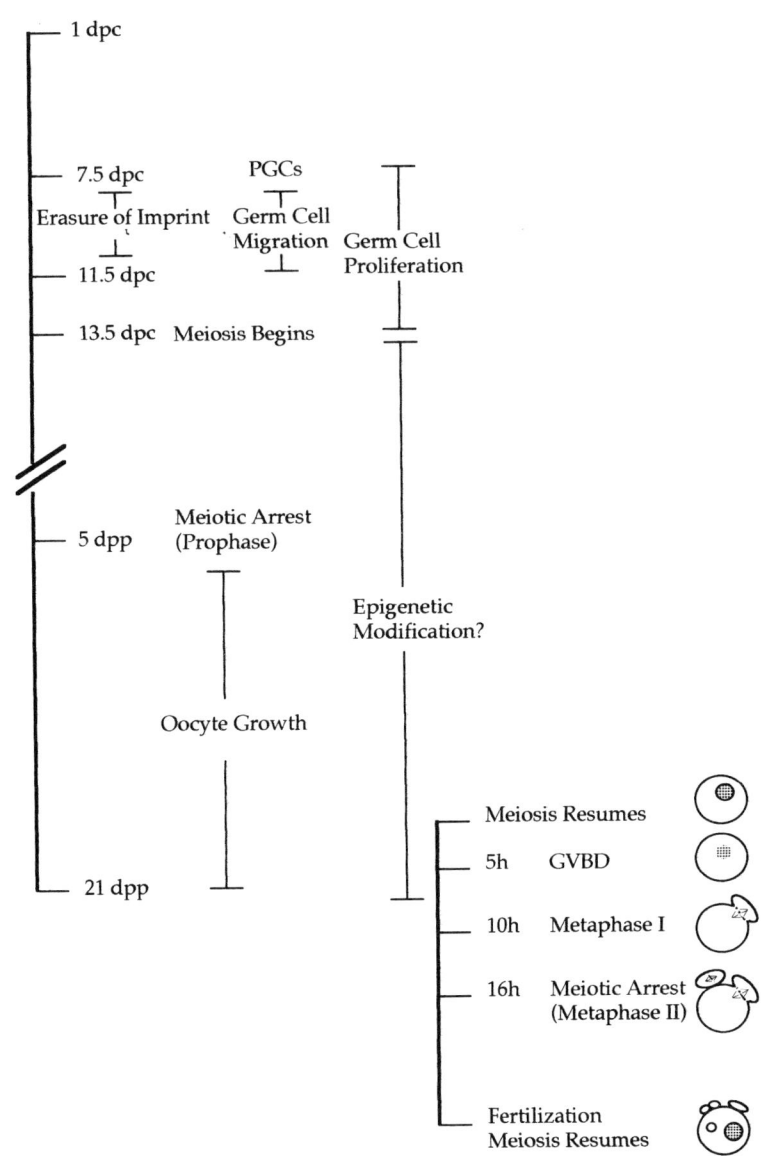

Figure 7-5 A schematic summary of oogenesis in the mouse. Oogenesis begins with the formation of the primordial germ cells (PGCs) in the 7.5-day-old embryo (McLaren 1984). By 11.5 days post coitus (dpc), most of the PGCs have migrated from the hindgut to colonize the presumptive gonad. The oogonia undergo a last round of DNA synthesis prior to entering meiotic prophase at 13.5 dpc. The primary oocytes enter the dictyate stage at 5 days post partum (dpp) and remain arrested in the first meiotic prophase until just prior to ovulation (Eddy et al. 1981; Ginsgurg et al. 1990). Biallelic expression of imprinted genes is detected in postmigratory PGCs by 11.5 dpc (Szabo and Mann 1995a). Onset and duration of epigenetic modifications are unknown. GVBD, germinal vesicle breakdown.

beteen F1 females (derived from matings of *M. musculus* females and *M. spretus* males) and *M. musculus* males to verify parental imprinting. By demonstrating that the mRNA phenotype expressed by backcross progeny is dependent on the parent of origin rather than the species of origin precludes the possibility that the allele-specific expression is a result of a dominant species effect or selective PCR amplification.

Hybrid progeny derived from matings between *M. musculus* and *M. musculus castaneus* (Jackson Laboratories) also provide informative polymorphisms (Szabo and Mann 1995a,b) for *Igf2*, *H19*, *Snrpn*, and *Igf2r* (See Table 7-3).

Embryo and Oocyte Recovery

Female C57BL/6 (*M. musculus*) mice (Jackson Laboratories) are superovulated by injection of 5 IU pregnant mare's serum (PMS) (Equitech-Bio, West Ingram, TX) followed by 5 IU human chorionic gonadotropin (hCG) (Serono, Randolph, MA) 46–48 h later. Immediately after injection of hCG, cage females individualy with *M. spretus* males (Jackson Laboratories). The following morning, check for copulatory plugs. Females with plugs are considered to be 0.5 days pregnant. Flush embryos from the oviduct on the second day (two to four-cell stage), on the third day (eight-cell stage), and from the uterus on the fourth day (blastocyst stage) with flusing medium II (FMII) (Cutherbertson 1983). Remove zona pellucida by treating the embryos with 0.5% Pronase in phosphate-buffered saline for 2–3 minutes at 37°C. After 3–4 washes in FMII, pool and transfer zona-free embryos to a 0.5-ml microfuge tube.

Isolation of primordial germ cells (PGCs) from F1 hybrid embryos requires the dissection of PGC-containing tissue. By 7.0 days post coitus (dpc), PGCs are identified as alkaline phosphatase-positive staining cells at the posterior end of the primitive streak (Ginsburg et al. 1990). By 11.5 dpc, the mitotic PGCs have migrated to the genital ridge where they continue to divide, increasing in number to approximately 25,000 (Tam and Snow 1981). Perform embryo dissections according to Buehr and McLaren (1993) and release PGCs from the genital ridges by disrupting tissue with fine hypodermic needles. Collect and rinse cells in flushing medium I (FMI) (Spindle 1990).

To release oocytes from ovaries, press ovaries through nylon mesh (mesh size 20–30 μm) by using a rounded glass rod in a Petri dish containing FMI. Pool the oocytes free from contaminating cells and rinse in media.

Recover the ovulated oocytes from the oviducts of F1 hybrid females injected with superovulatory hormones, as described above, using the flushing method for cleavage stages (Spindle 1990). Collect and treat the oocytes with hyaluronidase (Sigma) to remove adhering cumulus cells (Gordon and Talansky 1986) and wash thoroughly in FMI medium.

Table 7-3 Summary of imprinted gene expression in oocytes and early mouse development

Loci	Imprint origin	Methylated allele	Fetal ovary[a]	Mitotic PGCs[b]	Ovulated oocyte	4-cell	8-cell	Blastocyst
Igf2	Maternal	Paternal	P	M/P	M/P	P	P	M/P
Snrpn	Maternal	Paternal	M/P	M/P	M/P	P	P	P
H19	Paternal	Paternal	M	M/P	n.d.	n.d.	n.d.	n.d.
Igf2r	Paternal	Maternal	M/P	M/P	M/P	M/P	M/P	M/P

n.d., not detected; M, maternal; P, paternal.
[a]Postmigratory PGCs purified from 11.5-days-post-coitus (dpc) genital ridge.
[b]Ovary at 12.5–15.5 dpc.

RNA Isolation

Prepare poly(A)$^+$ mRNA using the polyATtract system 1000 from Promega (Madison, WI) as modified for the isolation of mRNA from mandible tissue of chicken embryo. Centrifuge to pellet cells and carefully discard medium. Resuspend cells (25–50 oocytes, ≥10 blastocysts, or a single ovary) in 100 µl of extraction solution [100 µl of GTC extraction buffer and 4.1 µl of 48.7% β-mercaptoethanol (β-ME) stock]. Vortex briefly and freeze in a dry ice/ethanol bath for at least 5 minutes. Thaw in a heating block at 37°C, vortex and repeat the freeze–thaw procedure. Dilute at 1:10 the biotinylated oligo(dT) in prewarmed dilution buffer (65°C). After the second thaw, add 200 µl of the dilution buffer and 4 µl of 48.7% β-ME. Mix briefly and add 1 µl of diluted biotinylated oligo(dT). Mix again and centrifuge (15 K) for 10 minutes at room temperature. Resuspend the Streptavidin Magnesphere® Paramagnetic Particles (PMP) stock. Aliquot 200 µl into a 0.5-ml microfuge tube. Place in a PolyATtract® System 1000 magnetic separation stand and allow at least 30 seconds separation time. Discard the supernatant and resuspend the PMP pellet in 200 µl 0.5× SSC solution (75 mM sodium chloride/7.5 mM tri-sodium citrate). Place in the stand to separate again. Wash the PMP pellet three times with a final resuspension in 100 µl of 0.5× SSC. After the final thaw, centrifuge to pellet cell debris and then carefully pipette the supernatant into a tube containing washed PMPs. Gently mix by tapping and let this stand at room temperature for 5 minutes. Place in the separation stand and discard supernatant. Wash the pellet in 100 µl of 0.5× SSC. Repeat the magnetic separation and wash steps four times. After the last wash, discard the supernatant and elute the mRNA in 10 µl of nuclease-free water. Incubate at 65°C for 2 minutes and then place in the stand. Allow the PMPs to separate completely and then transfer the supernatant to an RNase-free 0.5 ml microfuge tube. To further concentrate the mRNA, dry under vacuum.

Reverse Transcription

For RFLP analysis, reverse transcribe mRNA into cDNA as described by Rappolee et al. (1988a,b). Briefly, preincubate mRNAs at 42°C for 20 minutes with 0.2 µg of random hexamer (Pharmacia, Gaithersburg, MD) in a 10-µl mixture [10 mM Tris-Cl, pH 8.3; 50 mM KCl; 5 mM MgCl$_2$; 1 mM dNTPs; RNasin (Promega) at 2 units (U)/ml]. Add 40 units of Moloney murine leukenia vrius (MMLV) reverse transcriptase (Gibco BRL, Gaithersburg, MD) and incubate at 37°C for 1 hour. Heat to 99°C for 5 minutes and then dilute the total reaction mixture with 20 µl of RT reaction buffer.

For SNuPE analysis, reverse transcribe RNA in 20 µl of RT mixture containing 2.5 U/µl MMLV reverse transcriptase (Gibco BRL) and 1 µM of downstream primer. Incubate at 42°C for 15 minutes and 99°C for 5 minutes.

Polymerase Chain Reaction

For RFLP analysis, amplify 3–10 µl of the RT reaction mixture with 2.5 units of AmpliTaq polymerase (Perkin-Elmer Cetus, Norwalk, CT) in a final volume of 50 µl containing 10 mM Tris-HCl buffer (pH 8.3), 1.5 mM $MgCl_2$ (*Ifg2r*), 3.0 mM $MgCl_2$ (*H19* and *Igf2*), 50 mM KCl, 1 µg acetylated bovine serum albumin, 0.25 µg of each sequence-specific primer, and 2.5 mM each dNTP (Pharmacia). Amplify the cDNA sequences by PCR for 35–40 cycles in a DNA thermal cycler (Perkin-Elmer Cetus) programmed for a 94°C denaturation, 57.4°C (*Igf2r*), 61.1°C (*H19*), 62°C (*Igf2*) annealing, and 72°C primer extension step for each cycle. Primer pairs for *Igf2r*, *H19*, and *Igf2* are generalized by using PCR Mate (Applied Biosystems, Foster City, CA) and are listed in Table 7-2. Also amplify each sample using *Gapdh* primers to verify mRNA yield and production of cDNA. The *Gapdh* primer sequences are 5'-TGATGACATCAAGAAGGTGGTGAAG-3' and 5'-ATGGCCTACATGGCCTCCAAGGA-3'. For SNuPE analysis, equilibrate RT reactions to 65°C. Add RT reaction samples to a final volume of 100 µl containing 10 mM Tris-Cl (pH 8.3), 50 mM KCl, 1 mM $MgCl_2$, 200 µM dNTPs, 0.5 µM of lower and upper primer (see Table 7-2), and 2.5 U of AmpliTaq polymerase (Perkin-Elmer Cetus). Amplify cDNA sequences by PCR for 40 cycles programmed for a 95°C denaturation, 42°C annealing, and 72°C primer extension step for each cycle. Gel purify the PCR products using the Prep-A-Gene DNA Purification Kit (BioRad).

To control for DNA contamination, design primer pairs to span an intron and reverse transcribe in the absence of reverse transcriptase. Presence of a PCR product from RNA that has not been reverse transcribed indicates contaminating genomic sequences.

Subclone PCR products into the pCRII vector with the TA Cloning Kit (Invitrogen, San Diego, CA) and sequence by the dideoxy-chain termination method (Sanger et al. 1977). Verify the identity of the PCR products by comparison with published sequences.

Species-Specific Restriction Enzyme Digestion of PCR Products

Digest PCR products with restriction enzymes that produce restriction fragment length polymorphisms (RFLPs) specific to each species to determine their parental origin. Visualize restriction fragments electrophoretically on 8% polyacrylamide gels.

To determine the relative level of *Igf2*, *Igf2r*, and *H19* RNAs transcribed from the relaxed parental allele, perform RT–PCR for each gene with known ratios of *M. musculus* to *M. spretus* RNAs in a total of 1 µg as described by Villar and Pedersen (1994). Electrophorese the digested PCR products amplified from known ratios of input RNA on 8% polyacrylamide gels, stain with ethidium bromide, photograph with Polaroid type 55 positive–negative film and scan by laser densitometry. Quantify the band intensities by Molecular Dynamics ImageQuant software (Molecular Dynamics, Mountain View, CA) to determine the percentage of species-specific fragments relative to the total PCR product. Compare the percentage of PCR product generated by the *M. musculus*-specific fragments (for *Igf2*) and *M. spretus*-specific fragments (for *Igf2r* and *H19*) with the species-specific fragments from known ratios of RNA to estimate the relative level of maternal and paternal transcripts, respectively, in the embryo, ovary, and gamete at each stage of development.

Single Nucleotide Primer Extension Assay

Perform the assay as described elsewhere (Singer-Sam et al. 1992; Singer-Sam 1994; Szabo and Mann 1995a,b). Briefly, in a total volume of 10 µl, add ~10 ng DNA to a reaction mixture containing 2 mM $MgCl_2$, 10 mM Tris-Cl (pH 8.3), 50 mM KCl, 0.75 U of AmpliTaq polymerse (Perkin-Elmer Cetus), and 1 µM of the appropriate SNuPE primer [5'-TGCGGGGCCATC-3' (*Igf2r*; Szebenyi and Rotwein 1994); 5'-ACATTCATACGGAGAGACTC-3' (*H19*; Pachnis et al. 1988); 5'-TCAAATTTGGTTTTTTAGAA-3' (*Igf2*; Rotwein and Hall 1990); 5'-CAATTTCACAAGCATT-3' (*Snrpn*; Gerrelli et al. 1991). Incubate the samples for one round of denaturation, annealing, and extension with 2 µM of the "incorrect" and appropriate $[^{32}P]dNTP$ to control for background and maximum incorporation, respectively. Mix 5 µl of each reaction with the DNA loading buffer, denature at 90°C for 1 minute and electrophorese on a 15% polyacrylamide-urea gel. Visualize by autoradiography and quantify with a PhosphorImager (Molecular Dynamics, Sunnyvale, CA). Measure the radioactivity in each band and subtract the background to determine the relative amount of each allelic transcript in the embryos, ovary, and gamete.

Conclusions

Current evidence suggests that specialization of the parental genomes is established during gametogenesis when the imprint pattern inherited from the parent is switched to reflect the sex of the progeny (for review see Surani 1998). To examine the epigenetic modification of specific imprinted genes and subsequent allele-specific expression, an assay is required that can distinguish between the maternal and paternal alleles and their respective transcripts. mRNA phenotyping and the RT–PCR SNuPE assay are powerful and sensitive approaches for the detection of allele-specific expression during gametogenesis and early mouse development because they permit the analysis of low-abundance transcripts. Based on the analysis of four imprinted genes — *Igf2*, *H19*, *Igf2r*, and *Snrpn* — these approaches have provided evidence for:

1. Erasure of imprinting during gametogenesis (Szabo and Mann 1995a; Villar et al. 1995).
2. Imprinted gene expression in the early embryo (Szabo and Mann 1995b).
3. Functional independence of the parental genomes in establishing allele-specific expression (Szabo and Mann, 1996).

Future application of these approaches may permit prenatal diagnosis of chromosome imbalance and imprinting mutations based on the inappropriate expression, i.e., biallelic or null, of genes involved in imprinted diseases; the diagnosis and intervention of ovarian teratomas and gestational trophoblastic neoplasia; and may be also effective in assessing the consequences, if any, of using immature gametes for in vitro fertilization.

Note Added in Proof

Since submission of this manuscript additional imprinted genes (Impact, Htr2a, Meg1/Grb10, Rasgrf1, Ip1, Impt1, Ube3a, Igf2as, Tapa1/Cd81, DN34, Ndn, Nnat/Peg5, kvlqt1) have been identified.

ACKNOWLEDGMENTS My thanks to R. Pedersen, F. Shamanski, and W. Reik for helpful comments and suggestions. Original work from the laboratory of the author referred to in this review was supported by NIH Program Project grant HD26732 and by the Office of Health and Environmental Research, U.S. Department of Energy contract DE-AC03-76-SF01012. A.J.V. is a Hitchings-Elion Fellow.

References

Ariel, I., Lustig, O., Oyer, C. E., Elkin, M., Gonik, B., Rachmilewitz, J., Biran, H., Goshen, R., deGroot, N., and Hochberg, A. (1994). Relaxation of imprinting in trophoblast disease. Gynecol. Oncol. 53:212–219.

Artzt, K., Barlow, D., Dove, W. F., Fischer-Lindahl, K., Klein, J., Lyon, M. F., and Silver, L. M. (1991). Mouse chromosome 17. Mamm. Genome 1:S280–S300.

Bagshawe, K. D., and Lawler, S. D. (1982). Unmasking moles. Brit. J. Obst. Gyn. 89:255–257.

Barlow, D. P. (1995). Gametic imprinting in mammals. Science 270:1610–1613.

Barlow, D. P., Stoger, R., Herrmann, B. G., Saito, K., and Schweifer, N. (1991). The mouse insulin-like growth factor type-2 receptor is imprinted and closely linked to the Tme locus. Nature 349:84–87.

Bartolomei, M. S., Zemel, S., and Tilghman, S. M. (1991). Parental imprinting of the mouse H19 gene. Nature 351:153–155.

Beechey, C. V., and Searle, A. G. (1987). Chromosome 7 and genetic imprinting. Mouse News Letter 77:126–127.

Buehr, M., and McLaren, A. (1993). Isolation and culture of mouse primordial germ cells. Guide to techniques in mouse development. Methods Enzymol. 225:58–87.

Buzin, C. H., Mann, J. R. and Singer-Sam, J. (1994). Quantitative RT–PCR assays show Xist RNA levels are low in mouse female adult tissue, embryos and embryoid bodies. Development 120(12):3529–3536.

Cattanach, B. M., and Beechey, C. V. (1990). Autosomal and X-chromosome imprinting. Development (Suppl.): 63–72.

Cattanach, B. M., and Kirk, M. (1985). Differential activity of maternally and paternally derived chromosome regions. Nature (London) 315:496–498.

Cattanach, B. M., Barr, J. A., Evans, E. P., Burtenshaw, M., Beechey, C. V., Leff, S. E., Brannan, C. I., Copeland, N. G., Jenkins, N. A., and Jones, J. (1992). A candidate mouse model for Prader-Willi syndrome which shows an absence of Snrpn expression. Nat. Genet. 2:270–274.

Cockett, N. E., Jackson, S. P., Shay, T. L., Farnir, F., Berghmans, S., Snowder, G. D., Nielsen, D. M., and Georges, M. (1996). Polar overdominance at the ovine callpyge locus. Science 273:236–238.

Cuthbertson, K. S. R. (1983). Parthenogenetic activation of mouse oocytes *in vitro* with ethanol and benzyl alcohol. J. Exp. Zool. 226:311–314.

DeChiara, T. M., Robertson, E. J., and Efstratiadis, A. (1991). Parental imprinting of the mouse insulin-like growth factor II gene. Cell 64:849–859.

Dull, T. J., Gray, A., Hayflick, J. S., and Ullrich, A. (1984). Insulin-like growth factor II precursor gene organization in relation to insulin gene family. Nature 310:777–781.

Eddy, E. M., Clark, J. M., Gong, D., and Fenderson, B. A. (1981). Origin and migration of primordial germ cells in mammals. Gamete Res. 4:333–362.

Efstratiadis, A. (1994). Parental imprinting of autosomal mammalian genes. Curr. Opin. Genet. Dev. 4:265–280.

Gerrelli, D., Sharpe, N. G., and Latchman, D. S. (1991). Cloning and sequencing of a mouse embryonal carcinomas cell mRNA encoding the tissue specific RNA splicing protein SmN. Nucleic Acids Res. 19:6642.

Giddings, S. J., King, C. D., Harman, K. W., Flood, J. F., and Carnaghi, L. R. (1994). Allele specific inactivation of insulin 1 and 2, in the mouse yolk sac, indicates imprinting. Nat. Genet. 6:310–313.

Ginsburg, M., Snow, M. H., and McLaren, A. (1990). Primordial germ cells in the mouse embryo during gastrulation. Development 110:521–528.

Gold, J. D., and Pedersen, R. A. (1994). Mechanisms of genomic imprinting in mammals. Curr. Topics Dev. Biol. 29:227–280.

Gordon, J. W., and Talansky, B. E. (1986). Assisted fertilization by zona drilling: a mouse model for correction of oligospermia. J. Exp. Zool. 239:347–354.

Guillemot, F., Caspary, T., Tilghman, S. M., Copeland, N. G., Gilbert, D. J., Jenkins, N. A., Anderson, D. J., Joyner, A. L., Rossant, J., and Nagy, A. (1995). Genomic imprinting of Mash2, a mouse gene required for trophoblast development. Nat. Genet. 9:235–241.

Hadchouel, M., Farza, H., Simon, D., Tiollias, P., and Pourcel, C. (1987). Maternal inhibition of hepatitis B surface antigen gene expression in transgenic mice correlates with de novo methylation. Nature (London) 329:454–456.

Hagemann, L. J., Peterson, A. J., Weilert, L. L., Lee, R. S., and Tervit, H. R. (1998). In vitro and early in vivo development of sheep gynogenones and putative androgenomes. Mol. Reprod. Dev. 50(2):154–162.

Hall, J. (1990). Genomic imprinting: review and relevance to human disease. Am. J. Hum. Genet. 46:857–873.

Hatada, I., and Mukai, T. (1995). Genomic imprinting of p57kip2, a cyclin-dependent kinase inhibitor, in mouse. Nat. Genet. 11:204–206.

Hatada, I., Kitagawa, K., Yamaoka, T., Wang, X., Arai, Y., Hashido, K., Ohishi, S., Masuda, J., Ogata, J., and Mukai, T. (1993). Allele-specific methylation and expression of an imprinted U2af1-rs1 (SP2) gene. Nucleic Acids. Res. 21:5577–5582.

Hayashizaki, Y., Shibata, H., Hirotsune, S., Sugino, H., Okazaki, Y., Sasaki, N., Hirose, K., Imoto, H., Okuizumi, H., Muramatsu, M., et al. (1994). Identification of an imprinted U2af binding protein related sequence on mouse chromosome 11 using the RLGS method. Nat. Genet. 6:33–40.

Johnson, D. R. (1975). Further observations on the hairpintail (Thp) mutation in the mouse. Genet. Res. 24:207–213.

Kaneko-Ishino, T., Kuroiwa, Y., Miyoshi, N., Kohda, T., Suzuki, R., Yokoyama, M., Viville, S., Barton, S. C., Ishino, F., and Surani, M. A. (1995). Peg1/Mest imprinted gene on chromosome 6 identified by cDNA subtraction hybridization. Nat. Genet. 11:52–59.

Kelsey, G., and Reik. W. (1998). Analysis and identification of imprinted genes. Methods 14(2):211–234.

Kono, T., Obata., Y., Yoshimzu, T., Nakahara, T., and Carroll, J. (1996). Epigenetic modifications during oocyte growth correlates with extended parthenogenetic development in the mouse. Nat. Genet. 13:91–94.

Lalande, M. (1996). Parental imprinting and human disease. Annu. Rev. Genet. 30:173–195.

Langois, S. (1994). Genomic imprinting: a new mechanism for disease. Pediatr. Pathol. 14:161–165.

Leff, S. E., Brannan, C. I., Reed, M. L., Ozcelik, T., Francke, U., Copeland, N. G., and Jenkins, N. A. (1992). Maternal imprinting of the mouse Snrpn gene and conserved linkage homology with the human Prader-Willi syndrome region. Nat. Genet. 2:259–264.

Lyon, M. F., and Glenister, P. (1977). Factors affecting the observed number of young resulting from adjacent-2 disjunction in mice carrying a translocation. Genet. Res. 29:83–92.

Mann, J. R., and Lovell-Badge, R. H. (1984). Inviablity of parthenogenones is determined by pronuclei, not egg cytoplasm. Nature (London) 310:66–67.

McAllister, G., Amara, S. G., and Lerner, M. R. (1988). Tissue-specific expression and cDNA cloning of small nuclear ribonucleoprotein-associated polypeptide. N. Proc. Natl. Acad. Sci. USA 85:5296–5300.

McLaren, A. (1984). Meiosis and differentiation of mouse germ cells. Symp. Soc. Exp. Biol. 38:7–23.

McGrath, J., and Solter, D. (1984). Completion of mouse embryogenesis requires both the maternal and paternal genomes. Cell 37:179–183.

Nakao, M., and Sasaki, H. (1996). Genomic imprinting: significance in development and diseases and the molecular mechanisms. Biochemistry (Tokyo) 120:467–473.

Nakao, M., Sutcliffe, J. S., and Beaudet, A. L. (1996). Advantages of RT-PCR and denaturing gradient gel electrophoresis for analysis of genomic imprinting:

detection of new mouse and human expressed polymorphisms. Hum. Mutat. 7:144–148.
Nicholls, R. D. (1994). New insights reveal complex mechanisms involved in genomic imprinting. Am. J. Hum. Genet. 54:733–740.
Obata, Y., Kaneko-Ishino, T., Koide, T., Takai, Y., Ueda, T., Domeki, I., Shiroishi, T., Ishino, F. and Kono, T. (1998). Disruption of primary imprinting during oocyte growth leads to the modified expression of imprinted genes during embryogenesis. Development 125(8):1553–1560.
Pachnis, V., Brannan, C. I., and Tilghman, S. M. (1988). The structure and expression of a novel gene activated in early mouse embryogenesis. EMBO J. 7:673–688.
Pedersen, R. A., Sturm, K. S., Rappolee, D. A., and Werb, Z. (1993). Effects of imprinting on early development of mouse embryos. In *Preimplantation Embryo Development*, Bavister, B. D., ed. New York, Springer-Verlag, pp. 212–226.
Poirier, F., Chan, C. T. J., Timmons, P. M., Robertson, E. J., Evans, M. J., and Rigby, P. W. J. (1991). The murine H19 gene is activated during embryonic stem cell differentiation in vitro and at the time of implantation in the developing embryo. Development 113:1105–1114.
Rappolee, D. A., Brenner, C. A., Schultz, R., Mark, D., and Werb, Z. (1988a). Developmental expression of PDGF, TGF-alpha, and TGF-beta genes in preimplantation mouse embryos. Science 241:1823–1825.
Rappolee, D. A., Mark, D., Banda, M. J., and Werb, Z. (1988b). Wound macrophages express TGF-alpha and other growth factors in vivo: analysis by mRNA phenotyping. Science 241:708–712.
Rappolee, D. A., Sturm, K. S., Behrendtsen, O., Schultz, G. A., Pedersen, R. A., and Werb, Z. (1992). Insulin-like growth factor II acts through an endogenous growth pathway regulated by imprinting in early mouse embryos. Genes Dev. 6:939–952.
Reik, W. (1989). Genomic imprinting and genetic disorders in man. Trends Genet. 5:331–336.
Reik, W. (1992). Genome imprinting. In *Transgenic Animals*, New York, Academic Press, pp. 99–125.
Reik, W., Collick, A., Norris, M. L., Barton, S. C., and Surani, M. A. H. (1987). Genomic imprinting determines methylation and parental alleles in transgenic mice. Nature (London) 328:248–251.
Roth, R. A. (1988). Structure of the receptor for the insulin-like growth factor II. The puzzle amplified. Science 238:1269–1271.
Rotwein, P., and Hall, L. J. (1990). Evolution of insulin-like growth factor II: characterization of the mouse IGF-II gene and identification of two pseudo-exons. DNA Cell Biol. 9:725–735.
Sanger, F., Nicklen, S., and Coulson, A. R. (1977). DNA sequencing with chain-terminating inhibitors. Proc. Natl. Acad. Sci. USA 74:5463–5467.
Sapienza, C., Peterson, A. C., Rossant, J., and Balling, R. (1987). Degree of methylation of transgenes is dependent on gamete of origin. Nature (London) 328:251–254.
Searle, A. G., and Beechey, C. V. (1985). Noncomplementation phenomena and their bearing on nondisjunctional effects. Basic Life Sci. 36:363–376.

Searle, A. G., Peters, J., Lyon, M. F., Hall, J. G., Evans, E. P., Edwards, J. H., and Buckle, V. J. (1989). Chromosome maps of man and mouse. IV. Ann. Hum. Genet. 53:89–140.

Singer-Sam, J. (1994). Quantitation of specific transcripts by RT-PCR SNuPE assay. PCR Methods Appl. 3:S48–S50.

Singer-Sam, J., LeBon, J. M., Dai, A., and Riggs, A. D. (1992). A sensitive, quantitative assay for measurement of allele-specific transcripts differing by a single nucleotide. PCR Methods Appl. 1:160–163.

Solter, D. (1992). Relevance of genomic imprinting to human diseases. Curr. Opin. Diotechnol. 3:632–636.

Spindle, A. (1990). In vitro development of one-cell embryos from outbred mice: influence of culture medium composition. In Vitro Cell Devel. Biol. 26:151–156.

Surani, M. A. (1998). Imprinting and the initiation of gene silencing in the germ line. Cell 93(3):309–312.

Surani, M. A. H., Reik, W., and Allen, N. D. (1988). Transgenes as molecular probes for genomic imprinting. Trends Genet. 4:59–62.

Swain, J. L., Stewart, T. A., and Leder, P. (1987). Parental legacy determines methylation and expression of an autosomal transgene: a molecular mechanism for parental imprinting. Cell 50:719–727.

Szabo, P. E., and Mann, J. R. (1995a). Biallelic expression of imprinted genes in the mouse germ line: implications for erasure, establishment, and mechanisms of genomic imprinting. Genes Dev. 9:1857–1868.

Szabo, P. E., and Mann, J. R. (1995b). Allele-specific expression and total expression levels of imprinted genes during early mouse development: implications for imprinting mechanisms. Gene Dev. 9:3097–3108.

Szabo, P. E., and Mann, J. R. (1996). Maternal and paternal genomes function independently in mouse ova in establishing expression of the imprinted genes Snrpn and Igf2r: no evidence for allelic trans-sensing and counting mechanisms. EMBO J. 15:6018–6025.

Szebenyi, G., and Rotwein, P. (1994). The mouse insulin-like growth factor II/cation-independent mannose 6-phosphate (IGF-II/MPR) receptor gene: molecular cloning and genomic organization. Genomics 19:120–129.

Tam, P. P., and Snow, M. H. (1981). Proliferation and migration of primordial germ cells during compensatory growth in mouse embryos. J. Embryol. Exp. Morphol. 64:133–147.

Tremblay, K. D., Saam, J. R., Ingrams, R. S., Tilghman, S. M., and Bartolomei, M. S. (1995). A paternal-specific methylation imprint marks the alleles of the mouse H19 gene. Nat. Genet. 9:407–413.

Villar, A. J., and Pedersen, R. A. (1994). Parental imprinting of the Mas protooncogene in mouse. Nat. Genet. 8:373–379.

Villar, A. J., Eddy, E. M., and Pedersen, R. A. (1995). Developmental regulation of genomic imprinting during gametogenesis. Dev. Biol. 172:264–271.

Walther, T., Balschun, D., Voigt, J. P., Fink, H., Zuschratter, W., Birchmeier, C., Ganten, D., and Bader, M. (1998). Sustained long term potentiation and anxiety in mice lacking the Mas protooncogene. J. Biol. Chem. 273(19):11867–11873.

Wevrick, R., and Francke, U. (1997). An imprinted mouse transcript homologous to the human imprinted in Prader-Willi syndrome (IPW) gene. Hum. Mol. Genet. 6:325–332.

Williamson, C. M., Schofield, J., Dutton, E. R., Seymour, A., Beechey, C. V., Edwards, Y. H., and Peters, J. (1996). Glomerular-specific imprinting of the mouse gsalpha gene: how does this relate to hormone resistance in Albright hereditary osteodystrophy? Genomics 36:280–287.

Zemel, S., Bartolomei, M. S., and Tilghman, S. M. (1992). Physical linkage of two mammalian imprinted genes, H19 and insulin-like growth factor 2. Nat. Genet. 2:61–65.

8

Gene Expression in Mouse Embryos: Use of mRNA Differential Display

Richard M. Schultz

A fundamental problem in early mammalian development is the transformation of the highly differentiated oocyte, which is the only cell in the female that can undergo meiosis and also expresses oocyte-specific genes (e.g., the *zona pellucida* genes), to totipotent blastomeres by the two-cell stage. This remarkable transformation probably entails, in addition to the requirement for both the maternal and paternal genomes (McGrath and Solter 1984), reprogramming of the pattern of gene expression. In fact, a dramatic change in gene expression occurs during the two-cell stage (Latham et al. 1991), and failure to execute successfully this reprogramming likely accounts for early development arrest (Latham et al. 1994). The molecular basis for how this reprogramming of gene expression occurs and the identity of the genes whose expression pattern is reprogrammed is very poorly understood.

Changes in the spatial pattern of gene expression that are accompanied by changes in cell potency also occur during preimplantation development. Blastomeres of the eight-cell embryo are totipotent (Pedersen 1986). Following compaction, cell division allocates blastomeres to either the inside or outside of the morula. Associated with this allocation are changes in cell potency in the late morula; whereas the potency of the outer cells, which give rise to the trophectoderm (TE), becomes restricted (i.e., they lose the ability to give rise to inner cells), the inner cells, which give rise to the inner cell mass (ICM), retain their ability to give rise to TE cells (Pedersen 1986). By the early blastocyst stage, the TE cells are clearly differentiated and unable to generate inner cells, while the ICM cells can still give rise to TE cells. Accompanying these changes in cell specification and fate (i.e., inner and outer cells of the morula give rise to the ICM and TE, respectively, of the blastocyst) are changes in the spatial pattern of gene expression that were first detected by two-dimensional gels (Van Blerkom et al. 1976) and may initiate in the morula (Handyside and Johnson 1978). These spatial changes in gene expression are likely critical for the establishment and differentiation of the two cell lineages in the blastocyst, i.e., they are intimately coupled to cell specification and the maintenance or loss of totipotency. The identity of these genes, however, is largely unknown.

What has previously severely hampered progress in identifying genes whose expression is reprogrammed during the maternal-to-zygotic transition, as well as during the differentiation of cells in the blastocyst, is the paucity of biological material. For example, the two-cell embryo contains only about 10 pg of poly(A)$^+$ RNA and only about 20–30 embryos can be obtained from a superovulated mouse. While cDNA libraries to preimplantation embryos at different stages of development have been generated, these libraries were made only to the egg, two-cell, eight-cell, and blastocyst stages, and only for one time during each stage, e.g., late two-cell stage (Rothstein et al. 1992).

The mRNA differential display method (Liang and Pardee 1992), as adapted for preimplantation mouse embryos (Zimmermann and Schultz 1994), provides an ideal approach to analyze changes in the temporal and spatial pattern of gene expression in the preimplantation embryo. The major advantages of the mRNA differential display method to analyze gene expression, as opposed to the use of the conventional cDNA libraries generated to preimplantation embryos at different stages of development, are that the mRNA differential display method requires readily obtainable numbers of embryos and permits analysis of gene expression during very narrow developmental windows, e.g., embryos that are only a few hours apart in developmental time. Because the mRNA differential display method generates amplified cDNAs that tend to be small and at the 3′ untranslated region, these cDNAs frequently do not contain informative sequence information. The marriage of mRNA differential display with either the cDNA libraries generated from preimplantation embryos or 5′-RACE (Rapid Amplification of cDNA Ends) methods circumvents this problem and permits the isolation and characterization of longer or full-length cDNAs.

Presented below are two methods for analyzing gene expression in the preimplantation mouse embryo by mRNA differential display. In the first method, total RNA is isolated from the embryos and then subjected to reverse transcription. In the second method, the embryos are directly lysed in a tube and subjected to reverse transcription with no prior isolation of the RNA. In both methods, an aliquot is removed for PCR, and the products (amplicons) are resolved on a sequencing gel.

Materials and Methods

Embryo Collection and Culture

Embryos are collected from superovulated mice by standard methods. The collection medium is bicarbonate-free minimal essential medium (Earl's salts), supplemented with pyruvate (100 µg/ml), gentamycin (10 µg/ml), polyvinylpyrrolidone (3 mg/ml), and 25 mM HEPES (pH 7.2) Embryo cultures is conducted in KSOM medium (Ho et al. 1995): 95 mM NaCl, 2.5 mM KCl, 0.35 mM KH_2PO_4, 0.2 mM $MgSO_4$, 10 mM Na lactate, 0.2 mM glucose, 0.2 mM Na pyruvate, 25 mM $NaHCO_3$, 1.71 mM $CaCl_2$, 1 mM glutamine, 0.01 mM EDTA, BSA at 1 mg/ml, and gentamycin at 10 µg/ml. The gas phase is 5% CO_2/5% O_2/90% N_2.

Isolation of RNA

This is the method as originally described by Zimmermann and Schultz (1994). Its main advantage is that it allows a single, proven RNA preparation to serve as the basis for a large number of differential displays. Its main disadvantage is that the efficiency of the RNA recovery can be variable; recoveries range from 20% to 60%.

Prepare a 0.5-ml polypropylene tube with 100 µ of lysis buffer [4 M guanidine thiocyanate, 1 M 2-mercaptoethanol, 0.1 M Tris-HCl (pH 7.4)] and 20 µg of poly(C) [2.5 µl of the stock solution; poly(C) (Boehringer Mannheim) at 8 µg per µl in DEPC-H_2O] and mix well by vortexing; keep this tube on ice. The embryos (typically 50–100), in as small a volume of medium as possible (preferably <1 µl, but up to 5 µl should not be a problem), are then transferred to the tube with a mouth-operated micropipette and the tube is mixed by vortexing. Add 8 µl of 1 M acetic acid, 5 µl of 2 M potassium acetate, and 300 µl of 100% ethanol to the tube and mix well by vortexing. The nucleic acids are precipitated overnight at −20°C and then harvested by centrifugation at 10,000g for 15 minutes at 4°C. The pellet is washed with 300 µl of ice-cold 75% ethanol and the supernatant is then carefully removed such that approximately 0.5 µl of fluid remains over the pellet. Note that it is critical not to allow the pellet to dry. Poly(C) pellets become virtually impossible to resuspend if allowed to dry out. The pellet is then resuspended in 20 µl of resuspension buffer [40 mM Tris-HCl (pH 7.9), 10 mM NaCl, 6 mM $MgCl_2$ in DEPC-H_2O].

Contaminating DNA is degraded by adding 1 unit of RQ1 DNase (Promega) and incubating at 37°C for 1 hour. The reaction is terminated by the addition of 50 µl of DEPC-H_2O and 80 µl of resuspension buffer-saturated phenol. The sample is then mixed by vortexing and centrifuged for

8 minutes at 10,000g. The aqueous phase is then transferred to a new 0.5-ml polypropylene tube and the RNA is precipitated by adding 8 µl of 3 M potassium acetate (pH 5.2) and 300 µl of 100% ethanol. The sample is then mixed by vortexing and the RNA is precipitated overnight at $-20°C$. The RNA is collected by centrifugation at 10,000g for 15 minutes at 4°C and the pellet is then washed with 300 µl of ice-cold 75% ethanol. After carefully removing the supernatant, the pellet is resuspended in 10 µl of DEPC-H_2O; the RNA can be stored at $-20°C$. The efficiency of the recovery should be calculated by determining the A_{260} of an aliquot and comparing this value to that of 20 µg of poly(C). The RNA can now be reverse transcribed with the 3′ primer of your choice.

Reverse Transcription

Mix 2 µl of the 3′-primer stock [25 µM 3′ primer (a 13-mer oligonucleotide: either $T_{11}MG$, $T_{11}MC$, $T_{11}MA$, or $T_{11}MT$, where "M" signifies any base but T) in DEPC-H_2O] in a thin-walled PCR tube with an appropriate amount of the RNA preparation. Note that this amount will vary according to your stage of interest and your purposes; 10 blastocyst-equivalents worth of RNA works well. Add a sufficient amount of DEPC-H_2O such that the total volume is brought to 9.5 µl. The sample is incubated at 70°C for 10 minutes in a model 9600 thermocycler (Perkin-Elmer Cetus). The temperature is then reduced to 45°C at which time 9 µl of reverse transcription master-mix is added. To make the reverse transcription master-mix, add 4 parts 250 mM Tris-HCl, pH 8.3, containing 375 mM KCl, 15 mM $MgCl_2$, 2 parts of a solution containing 100 mM dithiothreitol, 1 part of a solution containing RNasin (at 40 units/µl; Promega), and 2 parts of a solution containing 200 µM dATP, 200 µM dTTP, 200 µM dCTP, and 200 µM dGTP. Agitate the tube with your fingertip and allow the sample to equilibrate to 45°C for approximately 1 minute. Reverse transcription is initiated by adding 1.5 µl (300 units) of Superscript II reverse transcriptase (Gibco BRL). The sample is incubated for 1 hour at 45°C and the reaction is then terminated by incubating the sample at 95°C for 5 minutes. The tube is then chilled on ice, centrifuged briefly, and vortexed lightly. The cDNA is stored at $-20°C$.

It is also possible to conduct the reverse transcription reaction following direct lysis. The main advantage of this method is that it allows the investigator to conduct the reverse transcription and PCR reactions quickly and simply, with low background, from a specific, selected, small group of embryos. Its main disadvantage is the danger of genomic DNA contamination of the cDNA preparation.

Transfer a small number of embryos in a minimal volume of medium (< 0.5 µl) to 9 µl of direct-lysis reverse transcription master-mix in a thin-walled PCR tube. The composition of the direct-lysis reverse transcription master-mix is 75 mM Tris-HCl, pH 8.3, containing 112.5 mM KCl, 4.5 mM

MgCl$_2$, 10 mM dithiothreitol, 30 μM dATP, 30 μM dTTP, 30 μM dCTP, 30 μM dGTP, 2.5 μM 3′ primer, RNasin to a final concentration of 2 units/μl, and 0.5% Nonidet P-40. (Note that using three blastocysts is clearly superior to using five blastocysts; too much cellular debris possibly begins to inhibit the reverse transcription.) The sample is mixed by agitation with the fingertip and then incubated at 70°C for 1 minute in a model 9600 thermocycler. The tubes are then removed from the thermocycler and incubated for 3 minutes at room temperature on the bench top. Reverse transcription is initiated by the addition of 1 μl (200 units) of Superscript II reverse transcriptase and the samples are then returned to the thermocycler where they are incubated at 37°C for 10 minutes and then for 20 minutes at 45°C. The reaction is terminated by incubating the sample at 95°C for 5 minutes. The tube is then chilled on ice, centrifuged briefly, and vortexed lightly. The cDNA is then stored at −20°C.

mRNA Differential Display

cDNA prepared by either of the two methods described above is used as a template for PCR amplification. A PCR master-mix is prepared, and 18 μl aliquots are distributed to thin-walled PCR tubes on ice. PCR master-mix has the following composition per 18 μl aliquot: 2 μl of a 100 mM Tris-HCl, pH 8.3, solution containing 500 mM KCl; 2 μl of 25 mM MgCl$_2$; 2 μl of 5 μM 5′ primer; 2 μl of 25 μM 3′ primer; 0.2 μl of a solution containing 200 μM dATP, 200 μM dTTP, 200 μM dCTP, and 200 μM dGTP; 8.3 μl of DEPC-H$_2$O; 1 μl [10 μCi] of [^{35}S]dATP (specific activity > 1000 Ci/mmol; Amersham); and 0.5 μl (2.5 units) of AmpliTaq DNA polymerase (Perkin-Elmer Cetus). The KCl, Tris-HCl, and MgCl$_2$ may be contributed by the buffer(s) supplied with Perkin-Elmer's AmpliTaq. The primers may be suspended in DEPC-H$_2$O; note that under some conditions diethyl pyrocarbonate may inhibit PCR amplification (although we have not experienced this problem). The cDNA preparation (2 μl) is added to the tube, and is mixed by vortexing. The sample is next overlaid with 50 μl of ice-cold mineral oil and subjected to 40 cycles of PCR amplication using the following three-step cycle parameters: 94°C for 30 seconds, 42°C for 60 seconds, and then 72°C for 30 seconds. The last round of amplification is followed by a 5-minute extension at 72°C. The samples are stored at −20°C.

Visualization and Isolation of Amplified cDNAs

The PCR sample (5 μl) is resolved on a standard 6% DNA sequencing gel. It is recommended that three separate amplifications are run for each stage of interest. Run stages to be compared in adjacent lanes; shark's-tooth combs are not recommended. Following electrophoresis, the gel is dried onto filter paper without fixation; phosphorescent ink is applied to specific positions on the gel to aid in the location of the desired band to be eluted and amplified. The bands are visualized following autoradiography using Kodak XAR-5 X-ray film at −80°C for 1–5 days. The resulting autoradiogram serves as a template to align the gel with the aid of the spots generated by the phosphorescent ink. A band of interest is located, excised with a razor blade, and the gel slice is then transferred to a 0.5-ml polypropylene tube containing 100 μl of DEPC-H_2O and incubated at room temperature for 10 minutes and then in boiling water for 15 minutes. Debris is removed by centrifugation at 10,000g for 2 minutes, and the supernatant is transferred to a fresh tube. The eluted DNA is precipitated by adding 2.5 μl of glycogen (20 mg/ml), 8 μl of 3 M potassium acetate, and then 300 μl of 100% ethanol. After an overnight incubation at −20°C, the DNA and glycogen are harvested by centrifugation at 10,000g for 15 minutes at 4°C. The pellet is washed with 200 μl of ice-cold 85% ethanol, and then resuspended in 10 μl of DEPC-H_2O. To reamplify the eluted cDNA, remove 5 μl and subject to PCR as described above, with the following modifications: use a 40 μl reaction volume, use the dNTPs at 20 μM, and do not include radionucleotide. Visualize the amplified products on an agarose minigel. This procedure will almost invariably result in clearly reamplified bands. These amplicons can then be cloned (the Invitrogen TA cloning kit is recommended) and sequenced using standard techniques.

Comments

We have used the poly(C) carrier method to analyze gene expression in preimplantation embryos (Zimmermann and Schultz 1994), as well to identify genes whose expression is modulated by growth factors in the blastocyst (Babalola and Schultz 1995). In addition, we have used this method to identify genes (e.g., translation factor eIF-4C) that are transiently expressed during the two-cell stage by comparing the amplicon profiles for one-cell, mide two-cell and late two-cell/early four-cell embryos (Davis et al., 1996). We have also used the method to identify amplicons that are differentially expressed in the ICM and trophectoderm (Figure 8-1). In this approach, we subject equal numbers of ICM cells and total blastocyst cells to the RT–PCR reactions that are then analyzed by mRNA differential display; this corresponds to about three ICMs for each intact blastocyst. A gene that is preferentially expressed in the ICM should, in principle, give rise to an amplicon whose intensity is greater in the ICM than in the TE. Reciprocally, a gene that is preferentially expressed in the TE should give rise to an amplicon whose intensity is greater in the intact blastocyst than in the ICM. Genes that are expressed in both cell types to essentially similar extents will give rise to amplicons of similar intensity in both samples. (For further details regarding the rationale for the experimental design, see Brison and Schultz 1996.)

To make valid comparisons of the differences in amplicon intensity for embryos at different stages of preimplantation development, it is necessary

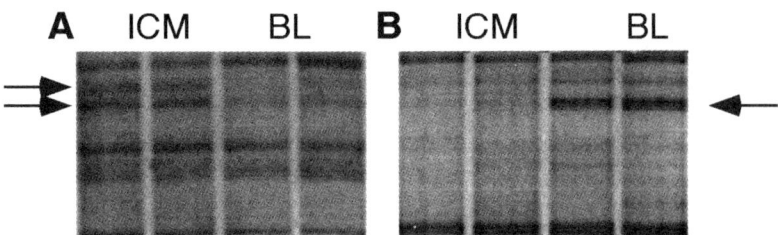

Figure 8-1 Regions of an autoradiogram showing amplicons that are preferentially expressed in either the ICM or TE. (A) The arrows points to two amplicons that are preferentially expressed in the ICM. (B) The arrow points to an amplicon that is preferentially expressed in the TE. Note that the intensity of the other amplicons is essentially the same. Duplicate samples are shown. ICM, inner cell mass; BL, intact blastocyst.

to subject equivalent amounts of poly(A)$^+$ RNA to reverse transcription. Thus, we correct for the developmental changes in the amount of poly(A)$^+$ RNA in the embryos. For example, the amount of poly(A)$^+$ in the blastocyst is about 2.5 and 5 times that present in the eight-cell and two-cell embryos, respectively. Thus, for every blastocyst equivalent, 2.5 eight-cell embryo equivalents or 5 two-cell embryo equivalents are subjected to the reverse transcription reaction. In addition, is critical to run triplicate samples to ensure reproducibility with respect to both the presence and intensity of the amplicon bands.

We have had good success with the more direct alternative method in which the embryos are directly subjected to the reverse transcription reaction without any isolation of RNA. To ensure that DNA contamination is not a problem, initial experiments should be conducted to demonstrate that, in the absence of the reverse transcription reaction, essentially no amplicons are detected following the PCR, and that the pattern of those that are detected differs markedly from the pattern obtained when the reverse transcription reaction is conducted. To date, we have not observed that DNA contamination is a problem.

We have also noted that the alternative method gives higher reproducibility of the amplicon band pattern and, moreover, it is probably a more fair representation of the reproducibility of the method, since each sample is subjected to reverse transcription and then each sample is subjected to PCR. In contrast, when poly(C) is used as the carrier, a single sample is subjected to reverse transcription and then this sample is subjected to multiple PCR.

A new primer strategy that improves the reproducibility and reduces the number of false positives can also be entertained (Ayala et al. 1995). In this variation, the primers are modified such they are 10 bases longer than those previously used. This permits PCR to be conducted at a higher annealing temperature; an EcoRI restriction site is included to facilitate subcloning of the PCR product. Following reverse transcription of equal numbers of embryos in each group, PCR should be conducted for one low-stringency cycle (40°C for annealing) followed by 35 high-stringency cycles at 60°C for annealing.

Sequence analysis of a number of amplicons reveals that the 3' primer primes internal to the junction between the end of the 3' untranslated region and the poly(A) tail at a frequency of about 10–15%. In these cases, the 3' primer appears to prime as a hexamer and useful sequence information is obtained, e.g., one can identify the gene if the sequence is already in the databases. Last, it must be emphasized that an independent method must be employed to confirm the expression pattern obtained with the mRNA differential display method. We use the DNA sequence obtained to generate gene-specific primers and a reverse transcription/PCR assay that permits determination of relative changes in the abundance of the mRNA under investigation (Temeles et al. 1994).

References

Ayala, M., Balint, R. F., Fernández-de-Cossío, M. E., Canaán-Haden, L., Lark, J. W., and Gavilondo, J. V. (1995). New primer strategy improves precision of differential display. BioTechniques 18:842–850.

Babalola, G. O., and Schultz, R. M. (1995). Effect of TGF-α and TGF-β on gene expression in the preimplantation mouse embryo. Mol. Reprod. Dev. 41:133–139.

Brison, D. A., and Schultz, R. M. (1996). An RT-PCR-based method to localize the spatial expression of genes in the mouse blastocyst. Mol. Reprod. Dev. 44:171–178.

Davis, W., Jr., De Sousa, P. D., and Schultz, R. M. (1996). Transient expression of translation initiation factor eIF-4C during the 2-cell stage of the preimplantation mouse embryo: identification by mRNA differential display and the role of DNA replication. Dev. Biol. 174:190–201.

Handyside, A. H., and Johnson, M. H. (1978). Temporal and spatial patterns of synthesis of tissue-specific polypeptides in the preimplantation mouse embryo. J. Embryol. Exp. Morph. 44:191–199.

Ho, Y., Wigglesworth, K., Eppig, J. J., and Schultz, R. M. (1995). Preimplantation development of mouse embryos in KSOM: augmentation by amino acids and analysis of gene expression. Mol. Reprod. Dev. 41:232–238.

Latham, K. E., Garrels, J. I., Chang, C., and Solter, D. (1991). Quantitative analysis of protein synthesis in mouse embryos. I. Extensive reprogramming at the one- and two-cell stages. Development 112:921–932.

Latham, K. E., Garrels, J. I., and Solter, D. (1994). Alterations in protein synthesis following transplantation of mouse 8-cell stage nuclei to enucleated 1-cell embryos. Dev. Biol. 163:341–350.

Liang, P., and Pardee, A. B. (1992). Differential display of eukaryotic messenger RNA by means of the polymerase chain reaction. Science 257:967–971.

McGrath, J., and Solter, D. (1984). Completion of mouse embryogenesis requires both maternal and paternal genomes. Cell 37:179–183.

Pedersen, R. A. (1986). Potency, lineage, and allocation in preimplantation mouse embryos. In *Experimental Approaches to Mammalian Development*, Rossant J., and Pedersen, R. A., eds. Cambridge, Cambride University Press, pp. 3–33.

Rothstein, J., Johnson, D., DeLoia, J., Skowronski, J., Solter, D., Knowles, B. (1992). Gene expression during preimplantation mouse development. Genes Dev. 6:1190–1201.

Temeles, G. L., Ram, P. T., Rothstein, J. L., and Schultz, R. M. (1994). Expression patterns of novel genes during mouse preimplantation embryogenesis. Mol. Reprod. Dev. 37:121–129.

Van Blerkom, J., Barton, S. S. C., and Johnson, M. H. (1976). Molecular differentiation of the preimplantation mouse embryo. Nature 259:319–321.

Zimmermann, J. W., and Schultz, R. M. (1994). Analysis of gene expression in the preimplantation mouse embryo: use of mRNA differential display. Proc. Natl. Acad. Sci. USA 91:5456–5460.

9

Use of Differential Display to Identify Gene Expression Differences in Male and Female Mouse Genital Ridges

Jay Brewster
Jill O'Moore
William Crain

Differential display (DD) is a powerful technique for distinguishing specific gene expression differences among tissues, cells, embryonic stages, etc., which is particularly useful when only relatively small amounts of material are available. The procedure is designed to reproducibly display subsets of the mRNAs (50–100 at a time) from total mRNA preparations as distinct bands on polyacrylamide gels so that the mRNAs from different tissues can be compared. While the procedure can, in principle, be used to compare many samples at once, in practice it is most often employed to look for differences in mRNAs between two samples. As an illustration of this we will describe how we have applied DD to find gene expression differences between the genital ridges of male and female mouse fetuses, just as they are beginning their divergent pathways of differentiation. This approach makes it possible to identify genes whose expression is crucial during the early differentiation of testes and ovaries from an already somewhat differentiated tissue, the genital ridge, and thus ultimately will allow us to understand the molecular details of these alternate differentiation pathways. Furthermore, using DD comparisons of the genital ridges at progressively earlier times, relative to overt gonadal differentiation, should make it possible to identify those messages which encode factors that determine the fate of this group of cells, and thus the sex of the animal. A combination of genetic analysis of sex-reversed mice and humans and molecular biology has demonstrated that a gene on the Y chromosome, known as *Sry*, encodes the testis-determining factor that serves as the switch to activate the testis pathway of differentiation (reviewed by McLaren 1990, 1991 and Goodfellow and Lovell-Badge 1993). It is thus predicted that an extensive analysis of gene expression differences between male and female genital ridges by differential display will reveal *Sry* mRNA and messages that are activated or repressed by the *Sry* gene product.

The original DD procedure was published in 1992 by Liang and Pardee. However, since then there has been some controversy about its reliability (Liang et al. 1993) and few reports of its successful use to identify differentially expressed mRNAs, until recently (Linskens et al. 1995; Zhao et al. 1995). Because of this, we use and will describe a modification that increases substantially the reliability of the original procedure (referred to as enhanced differential display, EDD).

In DD, mRNAs are copied into cDNA with reverse transcriptase using an oligo (dT) primer that contains one or two nucleotides on its 3' end that serve to anchor the primer to a specific subset of the poly(A)-containing RNAs. We use single nucleotide anchors, $5'T_{12}N$ (N is A, G, or C), so that each primer will promote reverse transcription of approximately $\frac{1}{3}$ of the mRNAs. Subsequently, another oligonucleotide primer, of arbitrarily selected sequence (AP), is used in combination with the same $T_{12}N$ to amplify by PCR portions of sequences from a further reduced subset of the available mRNAs. A crucial aspect of this step is that the selected AP anneals to between 50 and 100 different mRNAs [within several hundred nucleotides of the 3'-poly(A) tract]; this is accomplished only if the stringency of hybridization is low enough to allow up to two or three mismatches within a 10 nucleotide primer sequence. The PCR is carried out in the presence of radioactive precursors so that the amplified bands can be visualized autoradiographically after separation on a polyacrylamide gel. The problem that has arisen frequently is that the bands produced show a high level of nonreproducibility, which results in apparent differences between samples that often turn out to be false (Liang et al. 1993, and many personal communications). It has been suggested that the reason for the high level of DD false positives in the original protocol is that every cycle in the PCR is carried out at low hybridization stringency (40°C in standard PCR buffer), allowing additional mismatched duplexes to form at each step, which will introduce new fragments at each cycle, which will be amplified further (Linskens et al. 1995; Zhao et al. 1995). This hypothesis has been tested by introducing two modifications to the original procedure. First, an additional 10 nt linker sequence was added to the 5' end of the $T_{12}N$ and the AP primers (now referred to as $LT_{12}N$ and LAP). Second, only the first (or first few) PCR cycle was carried out at low stringency (40°C), with the remainder of the cycles proceeding at much higher stringency (60°C). The prediction was that the low-temperature cycle (or cycles) would still allow the crucial mismatching to occur so that enough different messages would be targeted, and that then raising the temperature to 60°C would allow the longer primers to anneal during the PCR only if they matched perfectly with the set of DNAs copied during the first step. Several labs have reported much greater success with this protocol (Zhao et al. 1995 were able to verify six out of eight of their originally observed differences using only one low-stringency cycle; Linskens et al. 1995 confirmed 23 out of 42 differences using four low-stringency cycles). We describe the successful use of this modification for comparing male and female mouse genital ridge RNAs.

Collection of Fetuses: Staging and Sexing

In mice, differentiation of the genital ridges along the male or female pathways is first detectable at about 11.5 days post coitus (dpc) and continues for several days. In the analysis described here, we examine genital ridges from 12.5 dpc fetuses. This stage was chosen initially because differentiation should have advanced to the point where there are a small, but detectable, number of gene expression differences.

To obtain high numbers of embryos, superovulation of female mice was carried out according to established protocols (Hogan et al. 1986). Superovulated females were mated with young males (all mice were C57BL/6J, Jackson Labs), and checked for vaginal plugs early the next morning. Pregnant females were sacrificed at 12.5 dpc, and the uterus dissected out into a sterile tissue culture dish containing prewarmed Brinster's BMOC media (Gibco BRL). Each fetus was dissected away from any maternal tissue and transferred to a dish of fresh prewarmed media. The handling and dissection of each fetus was done as quickly as possible to reduce RNA degradation within the genital ridge.

When mouse fetuses are collected, some asynchrony in development is often seen, both among fetuses within the same mother and between those from different mothers. Because of this, and because we are looking for changes in gene expression which could occur within relatively short periods, it is important that the embryos are staged accurately. The stage of each fetus was determined according to the criteria described by Hacker et al. (1995) by counting the number of tail somites posterior to the hind limb bud: 12.5 dpc fetuses contain 30 somites, 11.5 dpc, 18 somites, and 10.5 dpc, 8 somites. While the anatomical stage of most embryos agreed with the time since fertilization, this was not always true, and inclusion of variants will complicate the gene expression comparisons.

At 12.5 dpc it is difficult to distinguish male and female genital ridges during dissection, yet it is absolutely critical to know which are males and females for the differential display comparisons. To determine the sex of each fetus, DNA was extracted from the head and used in PCR analysis to determine whether Y chromosome-specific genes were present. Heads were removed by cutting just above the anterior limb buds and were placed in a microfuge tube for freezing in a dry ice/ethanol bath. It is very important to note that cross-contamination of small amounts of tissue among these samples is a real possibility that we have encountered. It is critical to prevent this because PCR detection of Y chromosome genes is very sensitive and low levels of male DNA contamination can cause female fetuses to be misidentified as male. Therefore, all surgical tools should be cleaned and flamed between the handling of each fetus, and reagents should be carefully managed to avoid contamination. To isolate genomic DNA from heads, a standard proteinase K digestion is used. The frozen heads are digested for 4 hours at 55°C in 200–300 µl of a buffer containing 20 mM Tris, pH 8.0, 0.5% NP40, 0.5% Tween 20, 1 mM $MgCl_2$, and 1 mg/ml proteinase K. The soft tissue should digest quickly and completely. After inactivating the proteinase K by heating at 95°C for 15 minutes, the samples are extracted with phenol/chloroform (1:1), and chloroform, and then ethanol-precipitated. The pellet is dissolved in 200 µl TE buffer (10 mM Tris, pH 8.0, 1 mM EDTA) and 1 µl of this solution is used in PCR reactions. If this results in smearing of the DNA on the agarose gel analysis, then the DNA solution can be diluted 1/10 before using 1 µl in PCR.

To determine the sex of the fetuses, we used PCR to detect *Zfy-1*, a gene that is located on the Y chromosome and therefore is male-specific (Page et al. 1987; Cao et al. 1995). An example is shown in Figure 9-1A. Because the identification of a female is a negative result (i.e., no band), it is important to test for integrity of each DNA sample with a second set of primers that will detect a gene that is not located on the Y chromosome. Figure 9-1B shows controls for the integrity of the DNAs used in Figure 9-1A by detection of the *Thy-1* gene sequence (Seki et al. 1985). The reactions were carried out in a solution of 10 mM Tris, pH 8.3, 50 mM KCl, 1.5 mM $MgCl_2$, 0.01% gelatin, 200 µM dNTPs, 1 µM *Thy-1* or *Zfy-1* primers (see Table 9-1 for sequences), and 1 Unit AmpliTaq DNA polymerase (Perkin-Elmer). Both primer pairs amplified DNA fragments of the predicted sizes using 30 cycles (1 minute/step), with denaturation/annealing/elongation at 95°C/63°C/72°C, respectively. Amplified DNA fragments were visualized after electrophoresis on an agarose gel.

Genital Ridge Dissection and RNA Purification

Dissection of genital ridges is not trivial, but can be accomplished quickly and efficiently with some practice. Instruments and plastics should be as clean as possible to reduce the likelihood of RNase contamination within the samples. Also, as noted above for head samples, considerable care should be taken to prevent cross-contamination of the samples. We autoclaved all instruments before use. Although the same instruments are used for each fetus they are dipped into ddH_2O, wiped with a clean paper towel, dipped in 100% EtOH, and flamed after each fetus is removed from the extraembryonic membrane. After the genital ridges are isolated and frozen, the tools should be washed and sterilized again. To expose the ridges, the

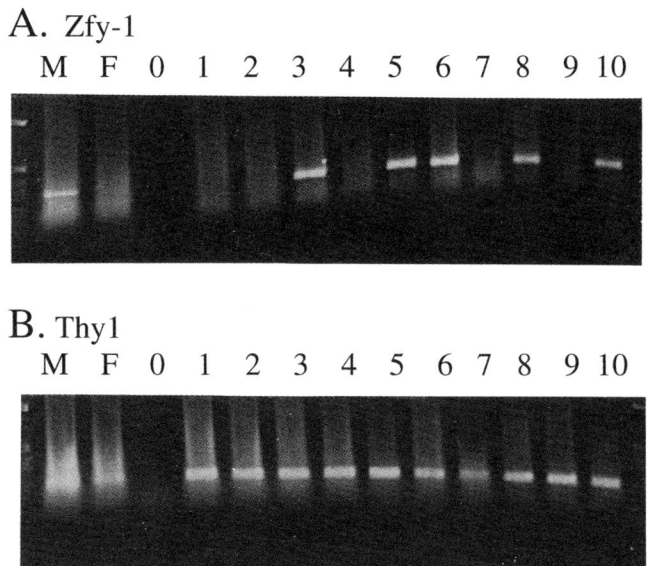

Figure 9-1 Determination of the sex of 12.5 dpc fetuses by PCR. Panel A shows the results of the amplification of a 310-nucleotide fragment from the *Zfy-1* gene using 10 different head DNA samples as template (1–10). Samples 3, 5, 6, 8, and 10 are identified as male by the presence of the *Zfy-1* band. Panel B shows the testing of the same DNA preparations for a 188-nucleotide fragment amplified from the *Thy-1* gene. M, male control DNA, F, female control DNA; 0, PCR control carried out with no-DNA template.

visible organs (heart, lungs, kidneys, etc.) are grasped firmly with surgical tweezers and pulled downward exposing the genital ridges. The genital ridges are located just anterior to the posterior limb buds, and are attached to the inside of the dorsal abdominal wall. Using a good dissecting scope, the ridges can be seen as two distinct and symmetrical structures, aligned anterior to posterior. At 12.5 dpc the structure resembles a hot dog sitting on one side of a hot dog bun with the hot dog being the developing gonad and the bun being the mesonephros. The mesonephros lies toward the outer edge of the fetus with the gonad lying toward the midline. Using fine-curved surgical scissors, the intact ridge can be isolated quickly from the fetus by cutting under the ridge and teasing the tissue free with a pair of fine tweezers. Both ridges from a single animal are placed in a microfuge tube and are frozen in a dry ice/ethanol bath.

RNA was isolated from genital ridges of sexed embryos according to a modified protocol, based on the method of Sargent et al. (1986). Pairs of frozen male or female ridges were suspended in 250 µl of lysis buffer (4.2 M guanidinium thiocyanate, 0.5% sarkosyl, 25 mM Tris, pH 8.0, and 0.7% 2-

mercaptoethanol) at room temperature and pipetted until they dissolved. Lysis of the ridge should be fairly rapid and not require extensive pipetting. The homogenate is then transferred to another tube containing a pair of same-sex ridges and the lysis is repeated. As more samples are added, the homogenate will become increasingly viscous due to the presence of genomic DNA. In a second set of tubes, 250 μl of phenol is overlaid with 250 μl of PEB (1% SDS, 100 mM Tris, pH 8.0, 10 mM EDTA). The homogenate is added directly to the aqueous phase and mixed vigorously. This is followed by the addition of 250 μl of chloroform with mixing, and centrifugation for 5 minutes at 14,000 rpm. Extractions of the aqueous phase should be repeated with phenol/chloroform (1:1) until no interface is observed, and then followed by one extraction with chloroform. Glycogen (20 μg) is added as carrier, and the nucleic acids are precipitated by the addition of one volume of isopropanol (15 minutes at room temperature, then centrifuge). The pellet is washed with 70% ethanol, resuspended in 50 μl of TE, and the RNA precipitated by adding 17 μl of 10 M LiCl (on ice, 1 hour). The centrifugation and wash are repeated, and the pellet is resuspended in 40 μl of DNase digestion mix (50 mM Tris, pH 7.5, 1 mM $MgCl_2$, 1 unit DNase I). After incubating the samples at 37°C for 30 minutes, 10 μl of 5× PEB is added, the samples are extracted once with phenol/chloroform, once with chloroform, and precipitated with one volume of isopropanol. The pellets are resuspended in 20–50 μl of water and the yield is determined by A_{260}/A_{280} analysis. Our preparations have yielded an average of 2.8 μg/genital ridge with no detectable differences between males and females.

Enhanced Differential Display of Expressed Genes

We have now carried out the EDD procedure with 10 primer pairs on RNA from 12.5 dpc genital ridges of XX and XY mice and verified three differentially expressed bands (in the range of 0.5–1.0% differences in mRNAs). Figure 9-2 shows the banding pattern for the first primer pair set that yielded a difference. Two things are important about these results. First, most of the displayed bands are not different among the four samples. This agrees with the expectation that there would not be many gene expression differences this early in the differentiation of the genital ridges. Second, one of the approximately 50 bands appears only in the female (XX) genital ridge RNA profiles (referred to as fs1 for female-specific 1). A DNA database search with the sequence of this fragment revealed that it matches with the *Xist* RNA which plays a critical role in inactivation of one of the two X chromosomes and is expressed only in females (Brockdorff et al. 1992; Brown et al. 1992). Thus, in this case the sequence of the fragment serves as a verification of the validity and reliability of the EDD procedure. We have also found two mRNAs that are present in males and not females, ms1 (Figure 9-3) and ms2 (Figure 9-4), and a second putative female-specific RNA, fs2 (Figure 9-5).

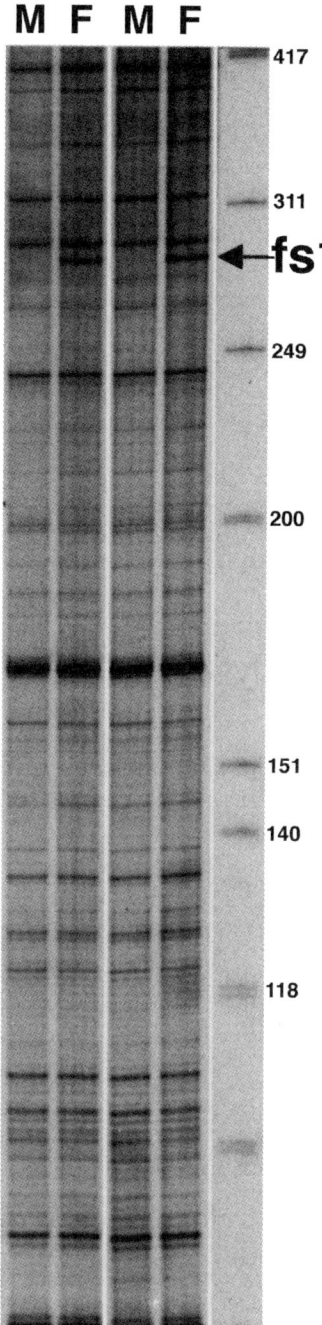

Figure 9-2 EDD comparing RNAs from 12.5 dpc genital ridges of XX and XY fetuses using primers $LT_{12}A$ and LAP6. M, male, F, female. The two sets of male and female samples represent two different EDD experiments that were carried out with the same RNA preparations. The arrow labeled fs1 points to a female-specific band of approximately 300 nucleotides.

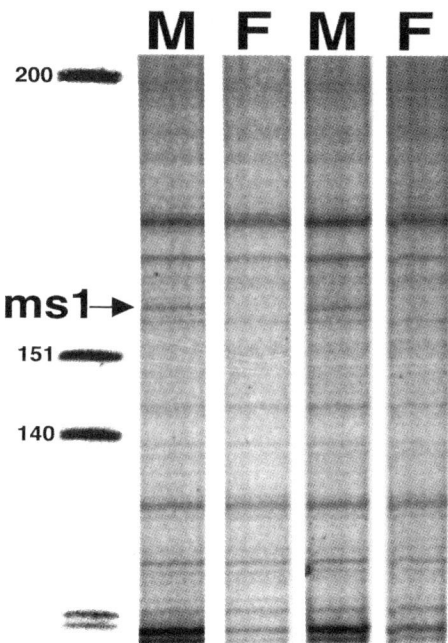

Figure 9-3 EDD comparing RNAs from 12.5 dpc genital ridges of XX and XY fetuses using primers $LT_{12}A$ and LAP7. The two sets of male and female samples represent two different EDD experiments that were carried out with the same RNA preparations. The arrow labeled ms1 points to a band of approximately 160 nucleotides that is present in XY but not in XX fetuses.

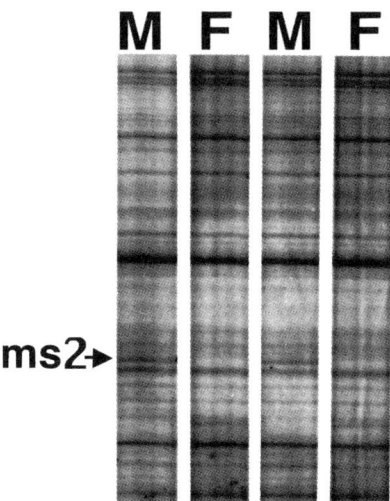

Figure 9-4 EDD comparing RNAs from 12.5 dpc genital ridges of XX and XY fetuses using primers $LT_{12}A$ and LAP2. The two sets of male and female samples represent two different EDD experiments that were carried out with the same RNA preparations. The arrow labeled ms2 points to a fragment of approximately 350 nucleotides that is present in the genital ridges of males but not females.

Gene Expression Differences in Mouse Genital Ridges 165

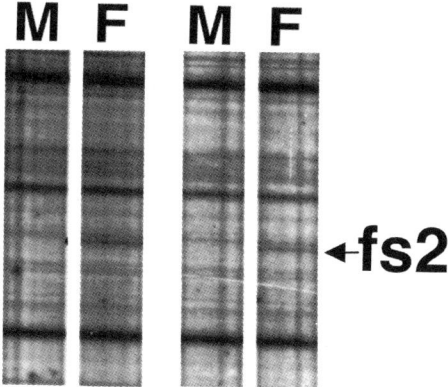

Figure 9-5 EDD comparing RNAs from 12.5 dpc genital ridges of XX and XY fetuses using primers $LT_{12}A$ and LAP1. The two sets of male and female samples represent two different EDD experiments that were carried out with the same RNA preparations. The arrow labeled fs2 points to a band of approximately 325 nucleotides that is present in the genital ridges of XX but not XY fetuses.

Total RNA from genital ridges is reverse transcribed using standard protocols. The reactions are set up on ice using RNase-free reagents. Each reaction contains 2 μg of male or female total RNA (or water for a no-RNA control) in 50 mM Tris, pH 8.3, 75 mM KCl, 3 mM $MgCl_2$, 10 mM DTT, 0.5 μM dNTPs, and 1 μM LT_{12}G, C, or A (Table 9-1) in a total volume of 28 μl. Samples are heated to 70°C for 5 minutes, cooled to 37°C, and 1 μl RNasin (Promega) is added to inhibit any surviving RNase activity. Before addition of the reverse transcriptase, a 1 μl-aliquot of the mixture is removed as a "no-RT" control for the EDD reactions. To synthesize cDNA, add 1 μl of Superscript II MMLV reverse transcriptase (Life Technologies), and incubate at 37°C for 2 hours. The reverse transcriptase is inactivated by incubation at 95°C for 5 minutes. Aliquots of 1 μl of this mix can be added directly into PCR reactions for EDD analysis, while any remainder is stored at −70°C for use with additional primer pairs later.

The EDD PCR mix contains 10 mM Tris, pH 8.3, 50 mM KCl, 1.5 mM $MgCl_2$, 0.01% gelatin, 0.4 μM LAP and $LT_{12}N$ primers, 2.5 μM dNTPs, 1 μCi ^{33}P-dATP (2000 Ci/mmol, NEN), 1 μl of the RT mix, and 0.4 U AmpliTaq DNA polymerase (Perkin-Elmer) in a final volume of 18 μl. After testing ^{35}S-dATP, ^{32}P-dATP, and ^{33}P-dATP in this reaction, we prefer ^{33}P-dATP. The original DDRT/PCR procedures recommend ^{35}S-dATP because of the resolving power of the low-energy emission, but we found that reactions with ^{35}S-labeled nucleotides were inconsistent. Use of ^{32}P-dATP resulted in reliable amplification of DNA bands but also produced broader bands in autoradiography, making the identification and isolation

Table 9-1 Primers used for EDD and sexing embryos

Differential displayprimers	
$LT_{12}C$	5'-CGGAATTCGGTTTTTTTTTTTC-3'
$LT_{12}G$	5'-CGGAATTCGGTTTTTTTTTTTG-3'
$LT_{12}A$	5'-CGGAATTCGGTTTTTTTTTTTA-3'
LAP1	5'-CGTGAATTCGAGCCAGCGAA-3'
LAP2	5'-CGTGAATTCGGACCGCTTGT-3'
LAP3	5'-CGTGAATTCGAGGTGACCGT-3'
LAP4	5'-CGTGAATTCGGGTACTCCAC-3'
LAP5	5'-CGTGAATTCGGTTGCGATCC-3'
LAP6	5'-CGTGAATTCGCTGATACTTG-3'
LAP7	5'-CGTGAATTCGAAGTATTCCC-3'
Zfy-1 primers	
	5'-GAGCCACAAGCTAACCATTAAGAC-3'
	5'-ACTGACATTCATAGGGCTTCTCTCC-3'
Thy-1 primers	
	5'-CTCTGGGAACCAGTCTCAC-3'
	5'-GATACCCGAGCACACGTAC-3'

of single bands more difficult. ^{33}P-dATP has the resolving power of ^{35}S, yet generates highly reproducible results in our experience and was therefore chosen for these analyses. The samples are cycled once at low stringency (94°C for 1 minute, 40°C for 4 minutes, 72°C for 1 minute) and 35× at high stringency (94°C for 45 seconds, 60°C for 2 minutes, 72°C for 1 minute). The EDD products are analyzed by combining 3.5 µl of the reaction mix with 3.5 µl of denaturing loading buffer (98% formamide, 10 mM EDTA, 0.25% xylene cyanol FF, and 0.25% bromophenol blue), heating to 90–95°C for 3 minutes, and analyzing on a sequencing size, 6% polyacrylamide, 8 M urea gel. Banding is revealed by autoradiography, and male or female specificity is indicated by the presence of a band in one lane and the absence of a corresponding band in the lane of comparison. Generally, we repeat the EDD three times to verify an observed difference before attempting to isolate the band of interest. Using these conditions, we find high reproducibility and an excellent density of signal (50–100 bands/lane).

Bands of amplified DNA that displayed differential expression three times were cloned after carefully cutting the band of interest from the dried gel. This step is critical. After the band is removed from the gel, a second autoradiogram should be carried out to verify that the correct band was isolated. The filter paper containing the band is boiled for 15 minutes in sterile water in a microfuge tube to release the DNA. After removing debris by centrifugation, an aliquot of the supernatant is used in PCR to amplify the fragment for cloning. This PCR should be carried out under the conditions described above for sexing the embryos using the EDD primers that yielded the band initially. We use the TA cloning system from Invitrogen.

The pCRII vector DNA contains a TT overhang which ligates efficiently with the AA overhang left on the PCR product by AmpliTaq polymerase. An aliquot of the PCR product is analyzed by gel electrophoresis and if a single band of the expected molecular weight is observed approximately 10 ng of PCR product (1–3 µl) is used in a 10-µl ligation reaction with 25–50 ng of vector DNA. This is somewhat risky because other bands may be present that are not seen with ethidium bromide staining. However, we have sometimes found it difficult to clone the gel-purified fragments, and speculate that this is due to digestion of the AA overhang left on the PCR product by the Amplitaq polymerase, thus preventing ligation. When cloning fragments from a DD gel, it is fairly common to clone more than one sequence because of the density of bands on the gel. Thus, it is a good idea to grow up 3–5 bacterial clones initially and determine the sequence of the cloned fragment in each. If all are the same, you probably have the fragment of interest. If there is more than one sequence, each should be tested for the expected differential expression.

Verification of Gene Expression Differences

After cloning of the candidate fragments from DD, it is essential that their differential expression be verified independently in the tissues that were originally compared — in this case, male and female genital ridges. If sufficient amounts of RNA are available from the tissues and the mRNA is fairly abundant, this can be accomplished using Northern blot analysis. However, because of limited amounts of genital ridge RNA, we chose RT–PCR to verify our candidate sequences, with detection of the amplified DNA fragments by hybridization with radioactive probes after Southern transfers (Gaudette and Crain 1991). An upstream primer internal to the EDD primer was synthesized based on the sequence of the cloned fragment, and RT–PCR analysis was performed with this primer and the original $LT_{12}N$ primer, as described above. The four candidate sequences identified in Figures 9-2 to 9-5 were tested in this way. As seen in Figure 9-6, fs1 (*Xist*), ms1, and ms2 were confirmed as having the sex-specific expression patterns originally detected in EDD. The fs2 sequence, however, could not be confirmed as showing female-specific expression. There are two possible explanations for this. First, the originally observed difference may have been an unexplained artifact of differential display. As discussed previously, this certainly occurs, although it seems to be much less frequent with EDD. The other possibility is that we have not yet cloned the original differentially expressed band. It can be seen that the band of interest (Figure 9-5) is not particularly prominent and it lies quite near at least one other strong band, which increases the chances of removing additional sequences from the gel. Among the clones of this band that have been recovered, we found and tested three different sequences, none of which displayed female-specific expression. For the other three sequences, which were confirmed, all of the clones tested (3–4

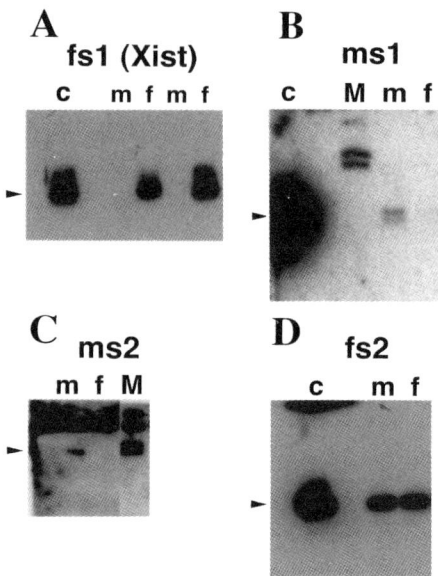

Figure 9-6 Detection of four different candidate sequences in genital ridge RNAs from XX and XY fetuses. In each case, RNA prepared from pools of dissected 12.5 dpc genital ridges (six each) was reverse transcribed and a sequence-specific primer and the original $LT_{12}A$ primer were used to amplify the sequence of interest by PCR. After agarose gel electrophoresis and blotting, the specific sequences were detected by hybridization with ^{32}P-labeled riboprobes synthesized from the appropriate cloned sequence. Panel A: fs1; panel B: ms1; panel C: ms2; panel D: fs2. m, male (XY); f, female (XX); c, positive control, using the cloned sequence; M, DNA size markers. In every case, the size of the amplified fragment matched the positive control (not shown in panel C) and agreed with the prediction based on sequence. The arrows point to the appropriate band in each panel.

clones for each) contained the same sequence, which displayed the expected pattern of expression.

ACKNOWLEDGMENT   This work was supported by NIH grant HD34755 to W. C.

References

Brockdorff, N., Ashworth, A., Kay, G. F., McCabe, V. M., Norris, D. P. Cooper, P. J., Swift, S., and Rastan, S. (1992). The product of the mouse *Xist* gene is a 15 kb inactive X-specific transcript containing no conserved ORF and located in the nucleus. Cell 71:515–526.

Brown, C. J., Hendrich, B. D., Rupert, J. L., Lafreniere, R. G., Xing, Y., Lawrence, J., and Willard, H. F. (1992). The human *XIST* gene: analysis of a 17 kb inactive X-specific RNA that contains conserved repeats and is highly localized within the nucleus. Cell 71:527–542.

Cao, Q. P., Gaudette, M. F., Robinson, D. H., and Crain, W. R. (1995). Expression of the mouse testis-determining gene *Sry* in male preimplantation embryos. Mol. Reprod. Dev. 40:196–204.

Gaudette, M. F., and Crain, W. R. (1991). A simple method for quantifying specific mRNAs in small numbers of early mouse embryos. Nucleic Acids Res. 19:1879–1884.

Goodfellow, P. N., and Lovell-Badge, R. (1993). *SRY* and sex determination in mammals. Annu. Rev. Genet. 27:71–92.

Hacker, A., Capel, B., Goodfellow, P., and Lovell-Badge, R. (1995). Expression of *Sry*, the mouse sex determining gene. Development 121:1603–1614.

Hogan, B., Costantini, F., and Lacy E. (1986). *Manipulating the Mouse Embryo; A Laboratory Manual.* Cold Spring Harbor, NY, Cold Spring Harbor Laboratory Press.

Liang, P., and Pardee, A. (1992). Differential display of eukaryotic messenger RNA by means of polymerase chain reaction. Science 257:967–971.

Liang, P., Averboukh, L., and Pardee, A. (1993). Distribution and cloning of eukaryotic mRNAs by means of differential display: refinements and optimization. Nucleic Acids Res. 21:3269–3275.

Linskens, M. H. K., Feng, J., Andrews, W. H., Enlow, B. E., Saati, S. M., Tonkin, L. A., Funk, W. D., and Villeponteau, B. (1995). Cataloging altered gene expression in young and senescent cells using enhanced differential display. Nucleic. Acids Res. 23:3244–3251.

McLaren, A. (1990). What makes a man a man? Nature 346:216–217.

McLaren, A. (1991). The making of male mice. Nature 351:96.

Page, D. C., Mosher, R., Simpson, E. M., Fisher, E. M. C., Mardon, G., Pollack, J., McGillivray, B., de la Chapelle, A., and Brown, L. G. (1987). The sex-determining region of the human Y chromosome encodes a finger protein. Cell 51:1091–1104.

Sargent, T. D., Jamrich, M., and Dawid, I. (1986). Cell interactions and the control of gene activity during early development of *Xenopus laevis*. Dev. Biol., 114:238–246.

Seki, T., Chang, H., Moriuchi, T., Denome, R., Ploegh, H., and Silver, J. (1985). A hydrophobic transmembrane segment at the carboxyl terminus of *Thy-1*. Science 227:649–651.

Zhao, S., Ooi, S. L., and Pardee, A. B. (1995). New primer strategy improves precision of differential display. Biotechniques 18:842–850.

Part II

METHODS FOR STUDYING *XENOPUS*
OOCYTES AND EMBRYOS

10

DNA Recombination and Repair in *Xenopus* Oocytes and Eggs: Substrate Design, Direct Microinjection, and Extract Preparation

Dana Carroll

The popularity of the South African clawed frog, *Xenopus laevis*, as an experimental animal (Kay and Peng 1991) can be attributed to a number of different considerations. One is the ease and economy of maintaining this species in the laboratory. Another is the rapid development of its embryos as free-living forms that are very amenable to observation and manipulation. A third is the large size and corresponding functional capacity of its oocytes, which allow probing of cellular processes by experimental injection of designed substrates or inhibitors. In fact, a full-grown oocyte is more than 1 mm in diameter, has an internal volume of about 1 µl, and contains quantities of many enzymes and organelles equivalent to about 10^5 typical vertebrate somatic cells (Dumont 1972, Gurdon and Wakefield 1986, Hausen and Riebesell 1991).

Why are *Xenopus* oocytes so large? The answer reflects the strategy that amphibians have adopted in response to the need to proceed through even the earliest stages of development fully exposed to environmental hazards. The main chromosomal events of meiosis are essentially completed at a very early stage in small oocytes (Hausen and Riebesell 1991). Subsequent enlargement is due to the accumulation of cellular components and nutrients that will be used after fertilization during the very rapid cleavage stages of embryogenesis (Gerhart 1980). Because cell divisions occur every 30 minutes during this phase, there is insufficient time to duplicate the cellular contents.

Instead, many of the constituents of the mid-blastula stage are stockpiled in the oocyte and partitioned to the blastomeres by cell division. Although the full-grown oocyte is technically in meiosis, most of its contents are destined for somatic tissues, and the activities that can be probed by injection are largely those of mitotic, not meiotic, cells.

Because of their large size, direct injection of oocytes is quite easy, and they have the capacity to process relatively large amounts of introduced substrates. This has been demonstrated for translation of exogenous messenger RNAs, transcription of genes injected into the oocyte nucleus, processing of RNA precursors, and targeting and functional assembly of various proteins (Kay and Peng 1991).

Xenopus Eggs

Full-grown oocytes are arrested at prophase of the first meiotic division. The transition to fertilizable eggs requires hormone stimulation, in vivo or in vitro, and leads to progression to metaphase of the second meiotic cycle. In addition to these nuclear and other internal changes, the exterior of the egg is prepared to interact with sperm and endowed a jelly coat. The final transition to fertilized egg can be achieved with live sperm, or mimicked by activation with calcium influx. In practice, some care must be taken if one wants to avoid activation, since penetration with an injection needle in calcium-containing medium is sufficient to induce it.

During the maturation from oocyte to egg, the molecular contents of the cell are not altered much by biosynthesis, but the properties of eggs differ in many significant respects due to the activation and inactivation of specific functions. These changes reflect differences in the metabolic requirements of these cell types. For example, DNA synthesis is quiescent in oocytes, and injected double-stranded DNAs are stable but not replicated (Wyllie et al. 1978, Gurdon and Melton 1981). Eggs embark, very shortly after fertilization, into a phase of rapid DNA synthesis, and injected DNAs are readily replicated in eggs (Harland and Laskey 1980; Mechali and Kearsey 1984). Oocytes maintain their stockpile of cellular components with a modest level of replacement synthesis, and injected templates are transcribed continuously for many days (Gurdon and Wickens 1983; Gurdon and Wakefield 1986). RNA synthesis is largely shut down in cleavage-stage embryos, and the transcription of templates injected into fertilized eggs proceeds only briefly before being turned off (Newport and Kirschner, 1982).

The functions we will be concerned with in this chapter are processes of DNA repair. Oocytes and eggs from *Xenopus*, and extracts prepared from these cells, have the capability to repair many types of DNA damage (Carroll 1998).

Recombination

In meiosis, genetic recombination is orchestrated to ensure the proper segregation of homologous chromosomes (Hawley 1988). This is reflected in the number and locations of crossovers and the timing of chiasma formation and resolution. Mutations that disable meiotic recombination in yeast or flies lead to rampant chromosome nondisjunction and elevated inviability.

In contrast, recombination in somatic cells functions largely as a mechanism for the repair of potentially lethal double-strand breaks (DSBs) in DNA. In principle, there are several ways that DSBs can be remedied, and we will focus on three of them: one that depends on extensive sequence homology and two that do not. These account for the majority of DSB repair events that have been characterized in *Xenopus* and in cells of most higher organisms.

Homologous recombination proceeds largely by the nonconservative mechanism called single-strand annealing (SSA), illustrated in Figure 10-1A (Lin et al. 1984; Carroll 1996). This process depends on exonuclease resection to expose complementary sequences, formation of annealed junctions by simple base pairing, and resolution by nuclease, polymerase, and ligase activities. In *Xenopus*, and probably in other systems as well, the exonuclease responsible for generating the recombinagenic tails has $5' \to 3'$ polarity, so the tails have free $3'$ ends.

In the category of nonhomologous, or illegitimate, recombination, some direct ligation may occur, even when the ends are neither blunt nor well matched (Pfeiffer and Vielmetter 1988; Pfeiffer et al. 1994; Lehman et al. 1994b). Other joining events are accompanied by deletion of sequences from one or both molecular ends. Close inspection of the joints reveals microhomologies that appear to support recombination (Lehman et al. 1994b). The proposed mechanism (Figure 10-1B) (Roth et al. 1985; Lehman et al. 1994b; Merrihew et al. 1996) involves exonuclease action to expose free 3' ends, transient base pairing of a very few nucleotides (usually no more than five), and extension by DNA polymerase to stabilize the joint.

All of these processes have been observed and rather well characterized using *Xenopus* oocytes, eggs, and/or extracts from them. The data supporting SSA in oocytes are quite compelling, and it is the only mode of recombination that those cells can support (Carroll 1996). The fact that a single oocyte can process in excess of 10^9 injected molecules from substrate to recombination product in a few hours has made it possible to characterize reaction intermediates. Both biochemical data (Maryon and Carroll 1991) and electron microscopic visualization (Pont-Kingdon et al. 1993) are consistent with SSA. Manipulation of substrate structure has shown that oocytes do not support a mode of conservative recombination that is prominent in yeast (Jeong-Yu and Carroll 1992b). Nuclear extracts, but not whole cell extracts, of oocytes also catalyze SSA (Lehman and Carroll 1991, 1993). Homologous recombination capability appears in mid-oogenesis

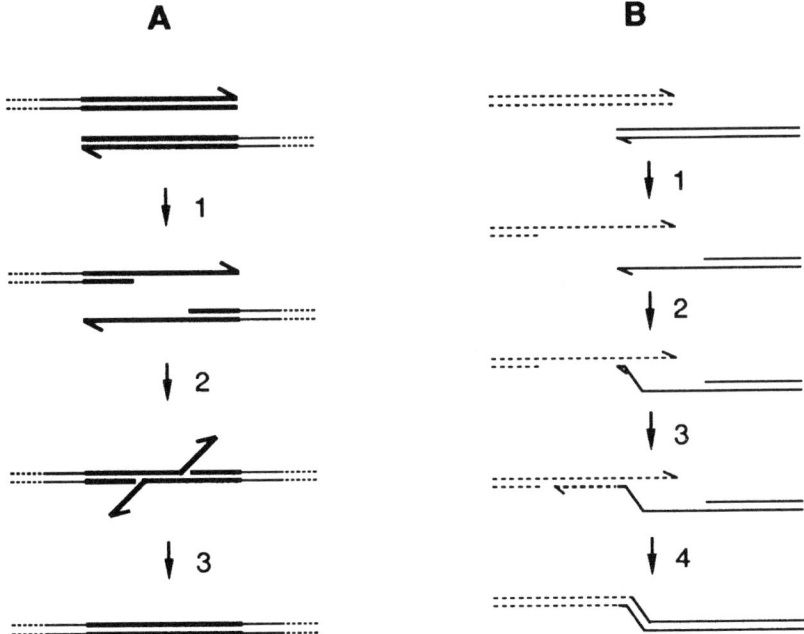

Figure 10-1 Recombination mechanisms. (A) Homologous recombination by single-strand annealing (SSA). Heavy lines denote homologous sequences at two molecular ends. The ends are resected by a $5' \rightarrow 3'$ exonuclease (step 1), exposing complementary sequences. These complements associate, forming an annealed junction (2), and the junction is resolved by further exonuclease action, DNA polymerase, and DNA ligase (3). (B) Illegitimate end joining by primed synthesis. The dashed and solid lines represent nonhomologous sequences at two molecular ends. Resection by a $5' \rightarrow 3'$ exonuclease (1) is followed by transient base pairing of a few nucleotides from each single-stranded region (2). At low frequency, such an unstable junction is extended by DNA polymerase (3) and resolved to a finished joint by nuclease, DNA polymerase, and DNA ligase activities (4). In both diagrams, half arrowheads indicate $3'$ ends.

and persists into activated eggs, where it competes for substrate ends with illegitimate joining (Lehman et al. 1993).

Neither ligation nor microhomology-based illegitimate recombination is observed in full-grown oocytes, but both occur in eggs and egg extracts. It is not clear what natural functions are reflected in this difference, since oocytes have DNA ligase activity and the capacity for repair-type DNA synthesis. The nature of the joints formed is very similar to that in other systems, particularly those seen in mammalian cells (Roth et al. 1985, Roth and Wilson 1988; Meuth 1989).

Repair of Local DNA Lesions

DSBs can be particularly devastating, but cells also have to remedy quite a variety of other insults to their genomes. Most of these other modes of DNA repair have been demonstrated in *Xenopus* oocytes and/or eggs, at varying levels of sophistication. Base excision repair (BER) is quite effective in oocyte extracts (Matsumoto and Bogenhagen 1989, 1991), and fractionation of the responsible activities has identified two distinct pathways, as in mammalian systems (Matsumoto et al. 1994, Matsumoto and Kim 1995). A typical DNA polymerase β has been identified in *Xenopus* (Reichenberger and Pfeiffer 1998). Nucleotide excision repair (NER) has not been extensively studied, but oocytes have a large capacity for correcting UV-induced lesions (Legerski et al. 1987; Hays et al. 1990; Saxena et al. 1990). The incisions made around a specific thymine dimer are identical to those made in human cells (Svodboda et al. 1993), at least one enzyme of NER is highly homologous to its counterpart in mammals (Scherly et al. 1993), and repair in egg extracts is dependent on PCNA (Shivji et al. 1994).

Mismatch repair has also been observed in injected oocytes and in egg extracts (Brooks et al. 1989, Varlet et al. 1990, Lehman et al. 1994a). In the latter system, all possible mismatches are repaired with comparable efficiencies (Varlet et al. 1990). An intriguing observation is that repair is initiated on covalently closed DNA substrates. Mismatch correction in mammalian cell extracts is dependent on pre-existing nicks (Holmes et al. 1990, Thomas et al. 1991); thus, it may be possible to elucidate the initiation step using *Xenopus* materials.

Substrate Design

The Importance of Substrate Design

I have a friend who says that, in genetics, you always get what you ask for; the problem is knowing what you are asking for. The same applies to many types of investigation: you can only see outcomes that are allowed by your experimental design. In many situations, the goal is to limit the range of outcomes so they will report definitively on the process that you are studying. The better defined the substrate, the more information you can derive from the products. In other situations a specific outcome cannot be anticipated, and the substrate must allow detection of a wider range of possibilities.

Substrates for Recombination

For assaying homologous recombination in *Xenopus* oocytes or eggs, we have found substrates of the form illustrated in Figure 10-2A to be very useful (Carroll 1996). The prototype is a linear molecule with terminal direct repeats that can support intra- or intermolecular SSA events. In our most

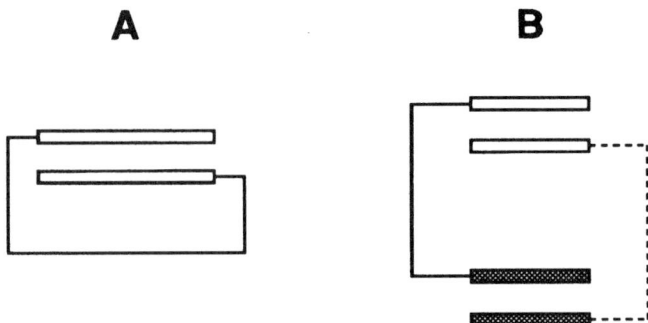

Figure 10-2 Substrates for homologous recombination. (A) Intramolecular: a linear DNA molecule with terminal direct repeats (open boxes). (B) Intermolecular: two linear DNA molecules with two different terminal homologies (open and shaded boxes). Solid and dashed lines indicate sequences not shared by the substrates.

common substrate, the length of the homologous overlap is 1250 bp, but this can range from several thousand base pairs down to 200 bp or smaller without affecting the overall yield substantially (Carroll et al. 1986; Segal and Carroll 1995). Because the rate of recombination appears to be determined by the initial exonuclease digestion, it takes longer to reach completion with longer overlaps. It is important not to have nonhomologous sequences longer that a few tens of base pairs at the ends of the substrate, because they are removed only rather slowly and can effectively block the reaction if they are too extensive (Jeong-Yu and Carroll 1992a,b).

As suggested above, recombination can proceed intermolecularly in oocytes, and if the only homologies presented are on separate molecules, only intermolecular events will occur (Carroll et al. 1986; Sweigert and Carroll 1990; Segal and Carroll 1995). A simple substrate design is shown in Figure 10-2B. We used this approach to demonstrate that only linear DNAs can participate in oocyte recombination (Segal and Carroll 1995): one molecule was injected in linear form, the other either in linear or circular. Only when both partners were linear did measurable recombination proceed. This reflects the need for access by the exonuclease to the homologous sequences.

Substrates for illegitimate recombination are linear DNAs, usually without substantial repeats. The nature of the molecular ends determines, in part, what products are formed. In injected eggs and in egg extracts, compatible ends generated by restriction enzyme digestion will often simply be ligated, restoring the original site (Pfeiffer and Vielmetter 1988; Lehman et al. 1993). Pfeiffer and colleagues have focused on joints formed between unmatched ends by cutting a single plasmid molecule with two enzymes, then isolating the circular products of intramolecular joining (Pfeiffer and Vielmetter 1988). They have also recently designed oligonucleotide substrates that permit greater control over the structure of the ends (Beyert et al. 1994). In injected eggs, a substantial fraction of all illegitimate joints results from microhomology priming, irrespective of the structure of the original ends (Lehman et al. 1993, 1994b).

Substrates for Repair

For investigation of the repair of internal lesions in DNA, two classes of substrates have been employed. In both, the damage is introduced into a DNA molecule prior to injection or incubation in extracts. To determine whether a particular repair process operates at all, it is sufficient to produce some level of random lesions and to employ an assay for their removal (Legerski et al. 1987; Hays et al. 1990; Sweigert and Carroll 1990). For example, the efficacy of NER can be determined with UV-irradiated DNA and either a physical or a biological assay.

More information can be obtained with substrates that have unique, targeted lesions. Matsumoto and Bogenhagen (1989) placed a single, model abasic site at one location in a circular plasmid DNA, while Sancar and colleagues (Svodboda et al. 1993) examined a molecule that had both thymine dimers and ^{32}P labels at unique positions.

Three procedures have been used to introduce these unique lesions into DNA substrates. An oligonucleotide can be synthesized with the desired alteration, if the appropriate precursor is available, or the damage can be created by treatment with an appropriate reagent. This oligo can be paired with a complementary oligo, and ligated into a larger molecule (Matsumoto and Bogenhagen 1989). As an alternative, the modified oligo can be hybridized to a single-stranded circular DNA and used as a primer for DNA synthesis, thereby completing the duplex (Svodboda et al. 1993). It is important in these cases to purify fully double-stranded, covalently closed molecules from the reaction mixture, since any nicks, breaks, or gaps that remain may bias the processing of the substrate.

A third approach has been to use triplex-forming oligos to deliver damage to a unique site in an intact duplex (Wang et al. 1995; Faruqi et al. 1996; Segal et al. 1997). This has potential limitations, since only particular families of sequences form triplexes efficiently, and the presence of the third-strand oligo may affect processing of the lesion, although there are ways to avoid this complication (Wang and Glazer 1995). Site-specific psoralen cross-links have been introduced successfully and efficiently by this method.

Substrates with specific mismatches are commonly prepared by annealing parental strands that differ only at the desired site(s). Varlet et al. (1990) assembled a set of M13 variants with the four different base pairs at one position. To prepare all possible mismatches, they annealed a single-stranded circular form of one variant with strands from a linear duplex form of a second variant. After DNA ligase treatment, covalently closed double-stranded circular DNA was isolated. We used an alternative procedure that involves cleavage of parental duplex DNAs into circularly permuted linears, which are denatured, annealed, and ligated prior to injection (Lehman et al. 1994a).

One might expect that simple oligonucleotide substrates would be preferable to large plasmid DNAs, but these have not been used extensively with *Xenopus* oocytes and eggs. One concern is that linear oligos would be degraded rather rapidly; another is that ends provide an entry site for exonucleases, helicases, and potentially other activities, so that repair might not be dependent on damage recognition per se. Pfeiffer's group (Beyert et al. 1994; Reichenberger et al. 1996) has devised an effective approach for studying illegitimate recombination by blocking one end of the oligo with a hairpin. It is possible that an oligo with hairpins at both ends would allow normal recognition of internal lesions.

Injection of Oocytes and Eggs

Injection Apparatus

Oocyte injection is easy. The apparatus can be relatively inexpensive, and even inexperienced students routinely learn to inject successfully. We still use a rather primitive setup that employs house air and vacuum to deliver the injection solution (Stephens et al. 1981). The only substantial costs are for a micromanipulator (Narishige Instruments, $400) to which the injection pipette is attached, a custom-made right-angle telescope ($1000) to monitor the meniscus in the pipette, and a boom-mounted dissecting microscope ($1500) to observe the injection. A number of commercial injection stations are now available (Kay and Peng 1991), many of which use oil-filled rather than air-driven lines. The cost of these instruments ranges from several hundred to a few thousand dollars.

Xenopus oocytes are hardy cells, but one must still respect their integrity in preparing the injection pipettes. The starting material consists of standard-bore glass capillary tubes (Drummond Scientific; 0.63 mm o.d., 0.20 mm i.d.). A capillary is drawn out in a pipette puller (Industrial Science Associates or Narishige Instruments), then broken at a point where the outside diameter is close to the desired size of 20 mm. This end is sharpened either by polishing with fine-grained emery paper on a rotating platform (Narishige Instruments), or using a microforge (Narishige Instruments). Making a clean incision with a sharp tip is less traumatic to the oocyte than punching out a disk of membrane with a blunt end. A pipette that is too thick can also cause significant damage.

Preparation of Oocytes

Ovary segments are removed surgically from a mature adult female frog that has been anesthetized. Although the ovary is vascularized, generally little bleeding results from this procedure, and no special precautions are taken to prevent it. Some care is taken to keep the surgical field sterile, but extreme measures are not required. We can thank Michael Zasloff for discovering that the remarkable power of *Xenopus* to resist infection is due to potent antimicrobial peptides in their skin (Zasloff 1987). In 15 years, we have never lost a frog to a postsurgical infection.

Protocol 1: Surgical Removal of Ovary Segments
1. Prepare 500 ml of 0.3% tricaine methanesulfonate in distilled water.
2. Place a mature female frog in this solution — e.g., in an ice bucket.
3. Test periodically by turning the frog on her back. When she no longer reacts, she is adequately anesthetized. This usually takes 15–30 minutes.
4. Place the frog on her back on a clean, well-lit surface. We sometimes pack ice around the animal to reduce the chance of awakening during surgery.
5. Using sharp surgical scissors, make a small (0.5–1 cm) longitudinal incision in the belly, a little to one side of the midline. Cut through both the skin and body wall.
6. Using forceps, pull a few lobes of ovary through the incision; then cut them off with scissors. Push the remaining segments back into the abdominal cavity.
7. Suture the body wall and skin separately, each with two stitches, using 3-0 or 4-0 surgical thread.
8. Place the frog in a few hundred milliliters of filtered water and monitor her until she is swimming before returning her to a normal tank.

To separate individual oocytes from ovarian tissue and to remove the layer of follicle cells that surrounds them, the ovary segments are usually treated with collagenase. We typically isolate oocytes the day before we plan to inject and let them sit in medium overnight at 19°C in order to make sure they are healthy. Deterioration is easily detected as marbling of the pigment in the usually well-defined hemispheres. If a substantial fraction of the oocytes are marbled, we usually discard the whole batch and start with a fresh frog, anticipating that even those that appear healthy may expire before the experiment is complete.

When studying processes of DNA metabolism, it makes sense to inject substrate DNA directly into the oocyte nucleus (germinal vesicle, GV). The GV lies just below the animal (pigmented) pole and is the least dense organelle in the oocyte. It is easily visualized after gentle centrifugation with the animal hemisphere placed "up" with regard to the centrifugal field (Kressmann and Birnstiel 1980). The GV pushes pigment granules out of the way and effectively draws a ring around itself (Figure 10-3). To keep the oocytes in place during centrifugation, we use a holding dish that consists of the lid from a 60-mm Petri dish with a layer of agarose in the bottom, in which small wells of about the same diameter as the largest oocytes have

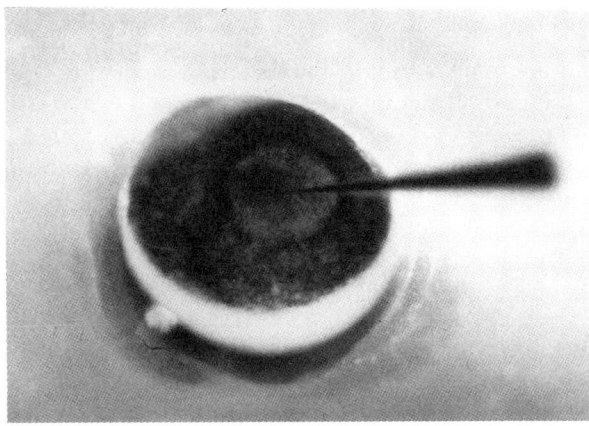

Figure 10-3 Nuclear injection. The oocyte shown has been subjected to centrifugation as described in the text. The location of the GV is evident as a light circle in the dark animal hemisphere. The injection needle has been loaded with India ink to make the mock injection more visible. The sides of the agarose well in which the oocyte sits can be partly seen.

been poked with a Pasteur pipette. Care must be taken not to spin the oocytes too hard or too long.

The two most popular media for maintaining oocytes are Modified Barth's Solution (MBS) and OR-2, which seem to give equivalent results (Kay and Peng 1991). Our lab routinely uses OR-2. A Ca^{2+}-free version of the standard solution is called for in some protocols. Oocytes are transferred between containers using a Pasteur pipette that has been cut back to produce a larger orifice and fire-polished to prevent damage to the oocytes.

OR-2
82.5 mM NaCl, 2.5 mM KCl, 1 mM $MgCl_2$, 1 mM $CaCl_2$, 1 mM Na_2HPO_4, 5 mM HEPES, pH 7.8.

Protocol 2: Preparation of Oocytes for Nuclear Injection
1. Place about 1 ml of ovary segments in 12 ml of collagenase solution (2 mg/ml in OR-2) and incubate with constant gentle inversion for about 2 hours, until essentially all the oocytes are freed from connective tissue.
2. Rinse the oocytes repeatedly with fresh OR-2.
3. Disperse the oocytes into a 100-mm Petri dish in OR-2 and incubate overnight at 19°C.
4. Place individual healthy oocytes in wells of the holding dish.
5. Spin in a refrigerated, low-speed centrifuge for about 15 minutes at 600g (e.g., 1500 rpm in a Beckman J6 centrifuge) at 15°C. Different batches of oocytes have different requirements. If the GVs are very prominent or if marbling is observed, reduce the speed or time of centrifugation. If the GVs are not readily visible, increase these parameters.

Preparation of Eggs

A mature female frog will lay hundreds, sometimes thousands, of eggs ready for fertilization in response to administration of gonadotrophic hormone by injection into the dorsal lymph sac. Both pregnant mare serum gonadotropin (PMSG) and human chorionic gonadotropin (hCG) are commonly used, and a number of different protocols have proven to be effective. Typically, a priming injection is given 2–3 days in advance, then a larger, initiating injection is given. Laying begins within about 12 hours and may continue for several hours. Once it has begun, additional eggs can be stripped by squeezing the abdomen. Ovulation is not guaranteed for any individual frog, even following hormone injection, so it is wise to begin with more than one. Alternative procedures for preparing eggs and egg extracts can be found in Kay and Peng (1991).

Protocol 3: Preparation of Eggs for Injection
1. Inject each frog with 50–100 units of hCG; leave for 2–3 days.
2. In the late afternoon, inject each frog with 500 units of hCG and place them singly in MMR solution (100 mM NaCl, 2 mM KCl, 1 mM $MgCl_2$, 2 mM $CaCl_2$, 0.1 mM EDTA, 5 mM HEPES, pH 7.8) to collect the eggs.
3. In the morning, strip additional eggs from each frog that has successfully ovulated by squeezing the abdomen toward the cloaca.
4. Dejelly the eggs by incubation in 2% cysteine for 10 minutes at room temperature. Rinse thoroughly in 0.2× MMR.
5a. Activate the eggs with electric shock, using the apparatus described by Murray (1991). Give two 1-second pulses at 12 V AC.
5b. Alternatively, the eggs can be treated with the calcium ionophore A23187, 0.5 µg/ml, for 5 minutes in 0.2× MMR. Rinse with 0.2× MMR.

The eggs should be used very shortly after activation, as they will begin to divide within about 90 minutes. Eggs prefer a medium of lower ionic strength than that used for incubating oocytes, and they are more fragile. For some experiments, it may be desirable to activate the eggs after injection. Wangh (1989) has devised a protocol for hardening the eggs that makes this possible, and we have used it with success (Lehman et al. 1993).

Injection and Recovery of DNA

Best results are obtained when the amount of DNA injected is in the range of 1–10 ng per oocyte. With larger or smaller amounts, there appear to be problems with assembly of chromatin (Gargiulo and Worcel 1983). If less than this amount of the specific substrate DNA is being injected, it can be supplemented with an irrelevant circular DNA to bring the total concentration into the desired range. Such a DNA must be chosen carefully so as not to interfere with the reaction being studied or with the subsequent analysis. We often add a circular DNA to our samples to serve as a recovery control.

Since 20 nl of injection solution is typically delivered to each GV, the DNA concentration should be approximately 50–500 µg/ml. Delivering smaller volumes does not hurt, but may be more difficult to reproduce, and some investigators inject up to 50 nl into a single GV. Concentrated DNA solutions are rather viscous, which can cause some difficulty in loading the injection pipettes. To remove particulate matter that may clog the pipette, the DNA solution is spun at top speed in a microfuge for 1 minute just prior to loading.

In eggs, the injection of large amounts of DNA can be injurious to subsequent development (Rusconi and Schaffner 1981), and some investigators make a point of keeping the quantity delivered below 1 ng per egg. For short-term incubations in which development is not an issue, larger amounts of substrate can still be processed appropriately in many cases.

An injection pipette can be front-loaded by placing the tip in a drop of injection solution and applying suction. This draws the solution through a narrow orifice, which creates a shear force sufficient to fragment long DNAs, but not the typical small, circular plasmid molecules. An alternative is to invert the pipette and back-load it.

The loaded pipette is positioned in the micromanipulator, with the appropriate pneumatic or hydraulic drive behind the column of injection solution. Our apparatus positions the pipette vertically over the dish of centrifuged oocytes, and this may be optimal for nuclear injections. As the pipette tip is lowered over the center of the nucleus, the surface of the oocyte buckles. With continued, gentle vertical pressure, the beveled point penetrates the investing layers and is seen to pop through to the interior. The pipette should then be retracted slightly to ensure that the tip is still in the nucleus and has not pushed completely through it into the cytoplasm. In teaching students to perform nuclear injections, we use a test solution of blue dextran. If the injection has been successful, a blue GV will be seen upon dissection.

Delivery of the injection solution is controlled by different means with different setups. Manual control and monitoring with a right-angle telescope and a reticulated eyepiece allows a measured volume to be expelled regardless of the speed of flow (Stephens et al. 1981). Some systems operate with continuous delivery at a constant flow rate, so only the time the needle is in each oocyte needs to be controlled. For various reasons, we limit the flow rate during injection so that it takes 5–10 seconds to deliver 20 nl into a single oocyte. Several cells can be injected per minute.

Protocol 4 — DNA Injection
1. Prepare and characterize the substrate DNA or DNA mixture.
2. Precipitate the DNA from 0.3 M sodium acetate with 2.5 volumes of ethanol; spin for 5 minutes in a microfuge; remove the supernatant. Wash the pellet with cold 70% ethanol; spin for 1 minute; remove the supernatant. Dry the pellet in a vacuum.

3. Dissolve the DNA in 17 mM Tris, pH 7.5, for about 1 hour, then add 1/9 volume 0.88 M NaCl. The final DNA concentration should be 200–400 µg/ml.
4. Spin the DNA solution for 1 minute at top speed in a microfuge just prior to loading the pipette.
5. Remove a 5-µl aliquot and place it in a drop on clean Parafilm under the injection pipette. Lower the pipette tip into the droplet and apply vacuum. Monitor filling of the pipette visually. Be careful not to allow the pipette tip to leave the drop while suction is being applied.
6. Place the dish of oocytes under the filled pipette. Manually guide the dish so that one oocyte sits just under the pipette. Lower the pipette into the oocyte nucleus, withdrawing it slightly after penetration has occurred.
7. Deliver 20 nl of injection solution into the oocyte by applying air pressure.
8. Withdraw the pipette, position another oocyte, and repeat steps 6–8 for as many oocytes as desired.
9. Move the oocytes to fresh OR-2 in a Petri dish and incubate at 19°C.

With eggs, it is not possible to see or inject a nucleus, so DNA is introduced at the animal pole. Although no centrifugation is necessary, the eggs should still be placed in wells, so they stay in place while being penetrated by the injection pipette. As noted earlier, eggs are handled and incubated in 0.2 MMR.

One convenient feature of the oocytes and eggs is that they can be harvested at any time after injection, from a minute or less to several days. If the goal is to see whether reaction products are generated from a particular substrate, an overnight incubation is often preferred. Most repair processes go to completion in this time period, although processing of psoralen crosslinks is slower (Segal et al. 1997). To monitor reaction rates or to isolate intermediates, shorter incubation times can be chosen. When using longer incubations, it is important to anticipate the fate of the reaction products. Circular DNAs are assembled into chromatin and are stable for days in the oocyte nucleus (Wyllie et al. 1978). Linear DNAs, however, are degraded over a period of several hours (Carroll et al. 1986, Maryon and Carroll 1989). Therefore, one would usually design a substrate that yields circular products that can be accumulated stably.

Depending on the analysis to be used, DNA is recovered from injected oocytes either by homogenization of the whole cells, or after manual dissection of the GV. Although laborious, the latter procedure effectively removes cytoplasmic components that may interfere with analysis of products. Particularly troublesome are the very abundant, highly phosphorylated yolk proteins that are protease resistant and fractionate with nucleic acids in most standard extraction and precipitation procedures. Yolk components inhibit some restriction enzymes, they often interfere with bacterial transformation, and they obscure nucleic acids in the electron microscope. A simple control in which an extract of uninjected oocytes is added to a standard analytical reaction will determine whether this is problematic in any particular case.

186 Methods for Studying *Xenopus* Oocytes and Embryos

Protocol 5: DNA Recovery by Homogenization
1. Place up to 30 oocytes or eggs in a 1.5-ml microfuge tube. Withdraw excess medium, and add 0.4 ml of homogenization buffer (30 mM Tris, 20 mM EDTA, pH 7.6, 1% SDS). Homogenize with a plastic pestle.
2. Add 2.5 μl of pronase solution (20 mg/ml) and incubate for 30 minutes at 37°C.
3. Add 0.4 μl of phenol/chloroform/isoamyl alcohol (25:24:1) and mix vigorously. Spin briefly in a microfuge to separate phases and carefully remove the upper, aqueous phase to a fresh tube.
4. Repeat step 3.
5. Add 0.5 ml of chloroform/isoamyl alcohol (24:1) and mix vigorously. Separate the phases as above.
6. To the final aqueous phase, add 1/9 volume 3 M sodium acetate and 2.5 volumes of ethanol. Chill for 10 minutes or more at −20°C, then spin for 5 minutes in a microfuge. Remove the supernatant and wipe the walls of the tube, being careful not to disturb the pellet.
7. Dissolve the pellet in 0.43 ml of TE. Add 22 μl of 2 M NaCl, 0.5 M Tris, 0.2 M EDTA, pH 7.9.
8. Add 5 μl of RNase A solution (10 mg/ml); incubate for 30 minutes at room temperature.
9. Add 50 μl of 10% SDS, 1 μl of pronase solution (20 mg/ml); incubate for 30 minutes at 37°C.
10. Extract once with phenol/chloroform/isoamyl alcohol, as in step 3; once with chloroform/isoamyl alcohol, as in step 5; and once with ether.
11. To the final aqueous phase, add 1/9 volume of 3 M sodium acetate and precipitate the DNA with 0.7 volume of isopropanol. Spin for 5 minutes in a microfuge. Remove the supernatant.
12. Dissolve the pellet in 0.36 ml of TE, add 40 μl of 3 M sodium acetate, and precipitate with 1.0 ml of ethanol. Spin for 5 minutes in a microfuge. Remove the supernatant.
13. Rinse the pellet with 500 μl of cold 70% ethanol; spin for 1 minute in a microfuge. Remove the supernatant and wipe the walls of the tube.
14. Dissolve the pellet in TE at a volume of 0.5–1 μl for each oocyte or egg in the original sample.

Protocol 6: DNA Recovery by GV Dissection
1. Collect a batch of (up to 30) oocytes in a 60 mm Petri dish in OR-2 medium.
2a. With two pairs of sharp forceps (e.g., Dumont INOX #5), manually dissect the GVs. Use one pair of forceps to steady the oocyte and the other to tear the membrane just behind the GV, taking care not to disrupt the GV itself. Tease the GV free of yolk.
  b. Use forceps to steady the oocyte and a sharp scalpel to make an incision in the surface just above the GV. Squeeze the oocyte with the forceps, pushing out the GV.
3. Either singly or after a few GVs have been dissected, pick them up in a micropipette in a minimum volume of medium, and transfer them to a microfuge tube on ice. Try to keep the volume per GV to about 1 μl. These samples can be frozen, if necessary, before extracting the DNA.

4. To a batch of up to 50 GVs, add 0.46 ml of 0.1 M glycine, 0.2 M NaCl, 1 mM EDTA, pH 9.6. Add 40 µl of 10% SDS and leave for 30 minutes at room temperature.
5. Extract twice with 0.5 ml of phenol/chloroform isoamyl alcohol, as in Protocol 5, step 3.
6. Extract the second aqueous phase with 1 ml ether. Spin briefly in a microfuge to separate the phases. Remove the upper ether layer completely.
7. Add 1/9 volume of 3 M sodium acetate and 2.5 volumes of ethanol. Chill for 10 minutes or more at $-20°C$, then spin for 5 minutes in a microfuge.
8. Rinse the pellet with 500 µl of cold 70% ethanol; spin for 1 minute in a microfuge. Remove the supernatant and wipe the walls of the tube.
9. Dissolve the pellet in TE at a volume of 0.5–1 µl for each GV in the original sample.

Preparation of Active Extracts

Manual GV Extracts

For small amounts of extract that is active in various facets of DNA metabolism, GVs can be dissected one at a time from oocytes. An experienced worker can isolate as many as 100 GVs per hour. The composition of the buffer solution into which the GVs are dissected is very important, and the one we have found that maintains most of the relevant activities is called IGE buffer (Lehman and Carroll 1993).

IGE Buffer
20 mM Tris, pH 7.5, 75 mM KCl, 7 mM $MgCl_2$, 2 mM dithiothreitol (DTT), 0.1 mM EDTA, 1mM EGTA, 0.2 mM PMSF, 0.5 µg/ml leupeptin, 0.7 µg/ml pepstatin A, 10% glycerol.

Protocol 7: GV Extracts by Manual Dissection
1. Place about 10 oocytes, prepared as in Protocol 2 through step 2 or 3, in IGE buffer in a small Petri dish. The quality of the yolk changes with time in this medium, and the GVs become more difficult to dissect, so it is preferable to move a relatively small number of oocytes at a time.
2. Dissect the GV, as in Protocol 6, step 2a or b.
3. Collect the GV in 0.5–1 µl using a micropipette and transfer to a microfuge tube on ice.
4. When several hundred GVs have been collected, they are disrupted by pipetting repeatedly through a 200-µl pipette tip. Debris is removed by centrifugation at $500g$ for 5 minutes at 4°C.
5. The extract can be frozen in aliquots at $-70°C$ until used.

Bulk GV Extracts

Larger quantities of active nuclear extract can be obtained by bulk isolation of GVs. The following procedure was adapted for making extracts competent for recombination and other oocyte nuclear processes (Lehman and

188 Methods for Studying *Xenopus* Oocytes and Embryos

Carroll 1993). It requires one special piece of apparatus that must be custom made: a lysis/flotation chamber that is essentially a large syringe with a threaded end to drive the plunger and a collection vessel connected to the main chamber at the narrow end via a stopcock (Figure 10-4) (Scalenghe et al. 1978; Lehman and Carroll 1993). The isolation is based on two sedimentation steps at $1g$. In the first, advantage is taken of the fact that the GV is the least dense organelle, and when the oocytes are lysed with minimum dilution, the GVs float. In the second step, the GV suspension is layered over a more standard buffer, through which the GVs settle and collect as a pellet.

Protocol 8: GV Extracts by Bulk Isolation

1. Surgically remove the whole ovary from several mature female frogs. The goal is to start with 50–80 ml of ovary.
2. Transfer to 2 volumes of OR-2 containing 1.5 mg/ml collagenase and swirl on a rotating platform overnight at room temperature.
3. Wash the oocytes 10 times with Ca^{2+}-free OR-2 by adding, swirling, and decanting the solution. This also removes most of the smaller oocytes.
4. Wash the oocytes five times in P buffer (5 mM Tris, pH 7.8, 10 mM $MgCl_2$).
5. Transfer the oocytes to a graduated cylinder and measure their volume, which will be reduced about 25% compared with the original ovary volume. Assume one oocyte per microliter of packed volume.
6. Redisperse the oocytes in a flask and incubate for 30 minutes with gentle agitation in P buffer. The oocytes swell during this treatment.
7. Rinse once with P buffer.
8. Add 1 volume (relative to the oocyte volume in step 5) of P buffer containing 1 mg/ml Pronase. Agitate gently for 20 minutes. It is important to let the oocytes flatten during this time, but not to overdigest them.
9. Wash once in 250 ml of P buffer, then three times with 80 ml of P buffer, plus 20 mg/ml BSA, 0.2 mM PMSF. The oocytes are fragile at this stage, so swirling must be quite gentle.
10. Place the flask with the oocytes on ice and wash twice with 50 ml of cold IE buffer (IGE without glycerol), plus 10 mg/ml BSA.
11. Rinse once briefly with 50 ml of cold lysis buffer (IE, plus 0.3% Nonidet P-40).
12. Add 1 volume of cold lysis buffer and agitate gently for 5 minutes at 4°C.
13. Transfer the suspension to the lysis/flotation chamber. In the first 5 minutes, occasionally rotate the chamber around its axis to provide very gentle agitation. For another 5 minutes, simply let the chamber sit in a vertical position to allow the GVs to float. You will see them as small, clear spheres in the gray sludge of the lysed oocytes.
14. Open the stopcock and gently push the top layer of fluid, with the GVs, into the collection chamber, by rotating the threaded collar.
15. Remove the GV suspension from the collection chamber carefully with a large-bore pipette. Keep on ice.

DNA Recombination and Repair in *Xenopus* Oocytes and Eggs 189

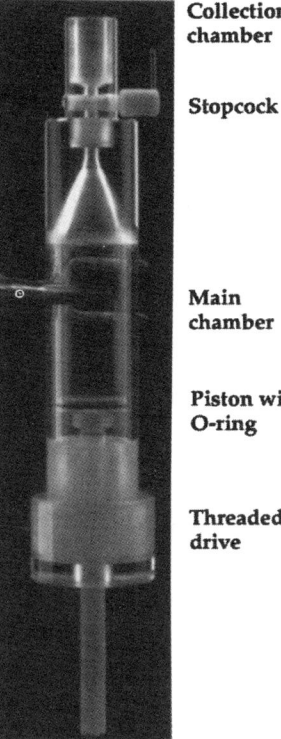

Figure 10-4 Lysis/flotation chamber. The dimensions of this device are described by Scalenghe et al. (1978) and the modifications by Lehman and Carroll (1993). Lysis occurs in the main chamber. Once the GVs have floated into the conical neck, the stopcock is opened, and they are pushed gently into the smaller collection chamber at the top.

16. Close the stopcock and mix the contents of the large chamber by inverting it a few times. Let the GVs float, and collect them as before (steps 13–15). The total volume of GV suspension should be about 10–15 ml.
17. Several 50-ml conical-bottom polypropylene centrifuge tubes are prepared with 40 ml of IGS (IGE, plus 0.1 M sucrose), overlaid with 4 ml of C buffer (IE, plus 5% glycerol), and kept on ice.
18. Layer 2 ml of GV suspension on each tube and allow the GVs to settle for 30–45 minutes. Monitor visually, noting when most of the GVs have reached the bottom of the tube, but most debris is still floating.
19. Carefully remove the overlying liquid with a pipette, leaving the GVs in the bottom of the tube in 0.5–2 ml.
20. Pool the GVs from all tubes. Gently suspend them; take two 25-µl aliquots and count the GVs in each under a dissecting microscope. Expect 1.5–4 GVs per microliter.

21. Transfer the GV suspension to several 1.5-ml microfuge tubes and spin at 16,000g (top speed in a microfuge) for 10 minutes at 4°C to break the GVs. Aliquots of the supernatant can be stored at −70°C until use.

Whole-Cell Extracts from Oocytes and Eggs

It is a simple matter to break oocytes and eggs, but the goal in preparing active extracts is to keep cellular components as concentrated as possible. Thus, the cells are usually disrupted by centrifugation in the absence of added buffer. GVs are also broken by this procedure, so nuclear activities are present in the soluble fraction.

A critical distinction is made between egg supernatants prepared by low-speed and high-speed centrifugation (Leno and Laskey 1991). In the former, yolk platelets, pigment granules, and large debris have been removed, but membrane vesicles required for forming nuclear envelopes around introduced DNA or nuclei are retained. High-speed extracts lack these vesicles, but retain many soluble activities. It is important to start with healthy eggs, removing any that show marbling or other pigment irregularities. Oocyte extracts are prepared in essentially the same manner as described here for eggs, and additional variations are presented in Kay and Peng (1991).

Protocol 9: Preparation of Egg Extracts
1. Collect dejellied, activated eggs, as described in Protocol 3.
2. Place the eggs in a 1.5-ml microfuge tube, and remove excess buffer from the surface after a gentle spin (100–200g) for 30 seconds at 4°C.
3. Break the eggs by spinning for 10 minutes at top speed in a microfuge at 4°C.
4. A thick pellet (yolk platelets, pigment granules, and other debris) and yellow pellicle (lipids) will surround a clear, amber intermediate layer. Carefully remove the intermediate layer to a fresh microfuge tube.
5. Spin again for 5 minutes in a microfuge at 4°C. The resulting supernatant is the low-speed extract. It can be stored at −70°C after addition of glycerol to 5%.
6. To prepare a high-speed extract, transfer 1.5 ml of the low-speed supernatant to a 5-ml centrifuge tube, overlay with chilled paraffin oil, and spin for 1 hour at 100,000 g in a Beckman SW50.1 rotor at 4°C.
7. Remove the clear supernatant; this is the high-speed extract. In some cases, it may still be contaminated with membranous vesicles, which can be removed by repeating the high-speed centrifugation. It can be stored at −70°C after addition of glycerol to 7%.

Recovery of DNA

After a DNA substrate has been added to an extract and incubated under the desired conditions, it can be reisolated using the protocols described for injected oocytes (see "Injection and Recovery of DNA" above). Recovery from whole-cell extracts would follow Protocol 5. Because GV extracts are inherently cleaner, the simpler procedure for dissected GVs, Protocol 6, can be used.

Analysis of Recovered DNA

Clearly, the analysis of the DNA retrieved from oocytes, eggs, or extracts depends on the process being investigated, the design of the substrate, and other aspects of the nature of the information being sought. Typically, the products are evaluated by agarose or polyacrylamide gel electrophoresis, with or without prior digestion with restriction enzymes that produce diagnostic fragments. In some cases, sufficient DNA is recovered to be visualized by staining with ethidium bromide, but Southern blot hybridization can also be performed with specific probes. It is possible to incorporate a radioactive label into a substrate prior to injection, but account must be taken of how the cell might potentially recycle the label, particularly in extended incubations. Labeled precursors can also be coinjected with the substrate.

If the substrate and/or products are plasmid DNAs that can be propagated in bacteria, they can be transformed into a competent host for analysis of individual cloned species. In the case of recombination reactions, intermolecular events can generate products (long linear molecules or multimer circles) that have reduced transformation efficiency compared with small circular DNAs. One solution is to cut the product DNA and recycle it at low-DNA concentration (Grzesiuk and Carroll 1987; Carroll et al. 1994). DNA recovered from GVs or GV extracts is suitable for visualization in the electron microscope (Pont-Kingdon et al. 1993), but further purification is required for samples isolated from whole cells or cell extracts.

Concluding Remarks

Xenopus oocytes and eggs have abundant activities to perform various processes of DNA metabolism. Because oocytes contain components that are accumulated for use during embryonic development, these activities are characteristic of vertebrate somatic cells. The ready accessibility of the *Xenopus* system makes it possible to investigate these mechanisms in biochemical detail through injection of carefully designed substrates, by isolation of reaction intermediates, and by utilization of competent cell-free extracts.

ACKNOWLEDGMENTS I am grateful to people in my laboratory, present and past, who contributed to the studies cited here and to the long-term pursuit of *Xenopus* oocyte functions. Chris Lehman, Ed Maryon, and Jon Trautman have made particularly important contributions in technical areas. I also appreciate other investigators of DNA repair in frogs for their generosity and interest. These include Petra Pfeiffer, Peter Brooks, John Hays, and Yoshihiro Matsumoto. Research in my laboratory has been supported by grants from the National Science Foundation and the National Institutes of Health.

References

Beyert, N., Reichenberger, S., Peters, M., Hartung, M., Göttlich, B., Goedecke, W., Vielmetter W., and Pfeiffer, P. (1994). Nonhomologous DNA end joining of synthetic hairpin substrates in Xenopus laevis egg extracts. Nucleic Acids Res. 22:1643–1650.

Brooks, P., Dohet, C., Almouzni, G., Méchali, M., and Radman, M. (1989). Mismatch repair involving localized DNA synthesis in extracts of *Xenopus* eggs. Proc. Natl. Acad. Sci. USA 86:4425–4429.

Carroll, D. (1996). Homologous genetic recombination in *Xenopus*: mechanism and implications for gene manipulation. Prog. Nucleic Acid Res. Mol. Biol. 54:101–125.

Carroll, D. (1998). Modes of DNA repair in *Xenopus* oocytes, eggs, and extracts. In *DNA Damage and Repair: Biochemistry, Genetics, and Cell Biology*, Vol. 1., Nickoloff, J. A., Hoekstra, M. F., eds. Totowa, NJ, Humana Press, pp. 597–616.

Carroll, D., Wright, S. H., Wolff, R. K., Grzesiuk, E., and Maryon, E. B. (1986). Efficient homologous recombination of linear DNA substrates after injection into *Xenopus laevis* oocytes. Mol. Cell. Biol. 6:2053–2061.

Carroll, D., Lehman, C. W., Jeong-Yu, S., Dohrmann, P., Dawson, R. J., and Trautman, J. K. (1994). Distribution of exchanges upon homologous recombination of exogenous DNA in *Xenopus laevis* oocytes. Genetics 138:445–457.

Dumont, J. N. (1972). Oogenesis in *Xenopus laevis* (Daudin). 1. Stages of oocyte development in laboratory maintained animals. J. Morphol. 136:153–180.

Faruqi, A. F., Seidman, M. M., Segal, D. J., Carroll, D., and Glazer, P. M. (1996). Recombination induced by triple helix-targeted DNA damage in mammalian cells. Mol. Cell. Biol. 16:6820–6828.

Gargiulo, G. and Worcel, A. (1983). Analysis of the chromatin assembled in germinal vesicles of *Xenopus* oocytes. J. Mol. Biol. 170:699–722.

Gerhart, J. C. (1980). Mechanisms regulating pattern formation in amphibian egg and early embryo. In *Molecular Organization and Cell Function*, Biological Regulation and Development, Vol. 2, Goldberger, R. F., ed. New York, Plenum Press, pp. 133–316.

Grzesiuk, E., and Carroll, D. (1987). Recombination of DNAs in *Xenopus* oocytes based on short homologous overlaps. Nucleic Acids Res. 15:971–985.

Gurdon, J. B., and Melton, D. A. (1981). Gene transfer in amphibian eggs and oocytes. Annu. Rev. Genet. 15:189–218.

Gurdon, J. B., and Wakefield, L. (1986). Microinjection of amphibian oocytes and eggs for the analysis of transcription. In *Microinjection and Organelle Transplantation Techniques*, Celis, J. G., Graessmann, A., and Loyter, A., eds. London, Academic Press, pp. 269–299.

Gurdon, J. B., and Wickens, M. P. (1983). The use of *Xenopus* oocytes for the expression of cloned genes. Methods Enzymol. 101:370–386.

Harland, R. M., and Laskey, R. A. (1980). Regulated replication of DNA microinjected into eggs of *X. laevis*. Cell 21:761–771.

Hausen, P., and Riebesell, M. (1991). *The Early Development of Xenopus Laevis. An Atlas of the Histology*. Berlin, Springer-Verlag.

Hawley, R. S. (1988). Exchange and chromosomal segregation in eucaryotes. In *Genetic Recombination*, Kucherlapati, R., and Smith, G. R., eds. Washington, DC, American Society for Microbiology, pp. 497–527.

Hays, J. B., Ackerman, E. J., and Pang, Q. (1990). Rapid and apparently error-prone excision repair of nonreplicating UV-irradiated plasmids in *Xenopus laevis* oocytes. Mol. Cell. Biol. 10:3505–3511.

Holmes, J. J., Clark, S., and Modrich, P. (1990). Strand-specific mismatch correction in nuclear extracts of human and *Drosophila melanogaster* cell lines. Proc. Natl. Acad. Sci. USA 87:5837–5841.

Jeong-Yu, S., and Carroll, D. (1992a). Effect of terminal nonhomologies on homologous recombination in *Xenopus laevis* oocytes. Mol. Cell. Biol. 12:5426–5437.

Jeong-Yu, S., and Carroll, D. (1992b). Test of the double-strand-break repair model of recombination in *Xenopus laevis* oocytes. Mol. Cell. Biol. 12:112–119.

Kay, B. K., and Peng, H. B. (1991). *Xenopus laevis: Practical Uses in Cell and Molecular Biology*, Methods in Cell Biology, Vol. 36. San Diego, Academic Press.

Kressmann, A., and Birnstiel, M. L. (1980). Surrogate genetics in the frog oocyte. In *Transfer of Cell Constituents into Eukaryotic Cells*, Celis, J. E., Graessmann, A., and Loyter, A., eds. New York, Plenum Press, pp. 383–407.

Legerski, R. J., Penkala, J. E., Peterson, C. A., and Wright, D. A. (1987). Repair of UV-induced lesions in *Xenopus laevis* oocytes. Mol. Cell. Biol. 7:4317–4323.

Lehman, C. W., and Carroll, D. (1991). Homologous recombination catalyzed by a nuclear extract from *Xenopus* oocytes. Proc. Natl. Acad. Sci. USA 88:10840–10844.

Lehman, C. W. and Carroll, D. (1993) Isolation of large quantities of functional, cytoplasm-free *Xenopus laevis* oocyte nuclei. Anal. Biochem. 211: 311-319.

Lehman, C. W., Clemens, M., Worthylake, D., Trautman, J. K., and Carroll, D. (1993). Homologous and illegitimate recombination pathways in developing *Xenopus* oocytes and eggs. Mol. Cell. Biol. 13:6897–6906.

Lehman, C. W., Jeong-Yu, S., Trautman, J. K., and Carroll, D (1994a). Repair of heteroduplex DNA in *Xenopus laevis* oocytes. Genetics 138:459–470.

Lehman, C. W., Trautman, J. K., and Carroll, D. (1994b). Illegitimate recombination in *Xenopus*: characterization of end-joined junctions. Nucleic Acids Res. 22:434–442.

Leno, G. H., and Laskey, R. A. (1991). DNA replication in cell-free extracts from *Xenopus laevis*. In *Xenopus laevis: Practical Uses in Cell and Molecular Biology*, Methods in Cell Biology, Vol. 36, Kay, B. K., and Peng, H. B., eds. San Diego, Academic Press, pp. 561–579.

Lin, F.-L., Sperle, K., and Sternberg, N. (1984). Model for homologous recombination during transfer of DNA into mouse L cells: role for the ends in the recombination process. Mol. Cell. Biol. 4:1020–1034.

Maryon, E., and Carroll, D. (1989). Degradation of linear DNA by a strand-specific exonuclease activity in *Xenopus laevis* oocytes. Mol. Cell. Biol. 9:4862–4871.

Maryon, E., and Carroll, D. (1991). Characterization of recombination intermediates from DNA injected into *Xenopus laevis* oocytes: evidence for a nonconservative mechanism of homologous recombination. Mol. Cell. Biol. 11:3278–3287.

Matsumoto, Y., and Bogenhagen, D. F. (1989). Repair of a synthetic abasic site in DNA in a *Xenopus laevis* oocyte extract. Mol. Cell. Biol. 9:3750–3757.

Matsumoto, Y., and Bogenhagen, D. F. (1991). Repair of a synthetic abasic site involves concerted reactions of DNA synthesis followed by excision and ligation. Mol. Cell. Biol. 11:4441–4447.

Matsumoto, Y., and Kim, K. (1995). Excision of deoxyribose phosphate residues by DNA polymerase β during DNA repair. Science 269:699–702.

Matsumoto, Y., Kim, K., and Bogenhagen, D. F. (1994). Proliferating cell nuclear antigen-dependent abasic site repair in *Xenopus laevis* oocytes: an alternative pathway of base excision DNA repair. Mol. Cell. Biol. 14:6187–6197.

Méchali, M., and Kearsey, S., (1984). Lack of specific sequence requirements for DNA replication in *Xenopus* eggs compared with high sequence specificity in yeast. Cell 38:55-64.

Merrihew, R. V., Marburger, K., Pennington, S. L., Roth, D. B., and Wilson, J. H. (1996). High-frequency illegitimate integration of transfected DNA at preintegrated target sites in a mammalian genome. Mol. Cell. Biol. 16:10–18.

Meuth, M. (1989). Illegitimate recombination in mammalian cells. In *Mobile DNA*. Berg, D. E., and Howe, M. M., eds. Washington, DC, American Society for Microbiology, pp. 833–860.

Murray, A. W. (1991). Cell cycle extracts. In *Xenopus laevis: Practical Uses in Cell and Molecular Biology*, Methods in Cell Biology, Vol. 36, Kay, B. K., and Peng, H. B., eds. San Diego, Academic Press, pp. 581–605.

Newport, J., and Kirschner, M. (1982). A major developmental transition in early *Xenopus* embryos: II. Control of the onset of transcription. Cell 30:687–696.

Pfeiffer, P., and Vielmetter, W. (1988). Joining of nonhomologous DNA double strand breaks in vitro. Nucleic Acids Res. 16:907–924.

Pfeiffer, P., Thode, S., Hancke, J., and Vielmetter, W. (1994). Mechanisms of overlap formation in nonhomologous DNA end joining. Mol. Cell. Biol. 14:888–895.

Pont-Kingdon, G., Dawson, R. J., and Carroll, D. (1993). Intermediates in extrachromosomal recombination in *Xenopus laevis* oocytes: characterization by electron microscopy. EMBO J. 12:23–34.

Reichenberger, S., and Pfeiffer, P., (1998). Cloning, purification and characterization of DNA polymerase β from *Xenopus laevis*. Studies on its potential role in DNA-end joining. Eur. J. Biochem. 251:81–90.

Reichenberger, S., Brüll, N., Feldmann, E., Göttlich, B., Vielmetter, W., and Pfeiffer, P. (1996). A novel nuclease activity from *Xenopus laevis* releases short oligomers from 5′ ends of double- and single-stranded DNA. Genes to Cells 1:355-367.

Roth, D., and Wilson, J. (1988). Illegitimate recombination in mammalian cells. In *Genetic Recombination*, Kucherlapati, R., and Smith, G. R., eds. Washington, DC, American Society for Microbiology, pp. 621–653.

Roth, D. B., Porter, T. N., and Wilson, J. H. (1985). Mechanisms of nonhomologous recombination in mammalian cells. Mol. Cell. Biol. 5:2599–2607.

Rusconi, S., and Schaffner, W. (1981). Transformation of frog embryos with a rabbit β-globin gene. Proc. Natl. Acad. Sci. USA 78:5051–5055.

Saxena, J. K., Hays, J. B., and Ackerman, E. J. (1990). Excision repair of UV-damaged plasmid DNA in *Xenopus* oocytes is mediated by DNA polymerase α (and/or d). Nucleic Acids Res. 18:7425–7432.

Scalenghe, F., Buscaglia, M., Steinheil, C., and Crippa, M. (1978). Large scale isolation of nuclei and nucleoli from vitellogenic oocytes of *Xenopus laevis*. Chromosoma 66:299–308.

Scherly, D., Nouspikel, T., Corlet, J., Ucla, C., Bairoch, A., and Clarkson, S. G. (1993). Complementation of the DNA repair defect in xeroderma pigmentosum group G cells by a human cDNA related to yeast *RAD2*. Nature 363:182–185.

Segal, D. J. and Carroll, D. (1995). Endonuclease-induced, targeted homologous extrachromosomal recombination in *Xenopus* oocytes. Proc. Natl. Acad. Sci. USA 92:806–810.

Segal, D. J., Faruqi, A. F., Glazer, P. M., and Carroll, D. (1997). Repair and mutagenesis of targeted psoralen crosslinks in *Xenopus* oocytes. Mol. Cell. Bibl. 17:6645–6652.

Shivji, M. K. K., Grey, S. J., Strausfeld, U. P., Wood, R. D., and Blow, J. J. (1994). Cip1 inhibits DNA replication but not PCNA-dependent nucleotide excision-repair. Curr. Biol. 4:1062–1068.

Stephens, D. L., Miller, T. J., Silver, L., Zipser, D., and Mertz, J. E. (1981). Easy-to-use equipment for the accurate microinjection of nanoliter volumes into the nuclei of amphibian oocytes. Anal. Biochem. 114:299–309.

Svodboda, D. L., Taylor, J.-S., Hearst, J. E., and Sancar, A. (1993). DNA repair by eukaryotic nucleotide excision nuclease. Removal of thymine dimer and psoralen monoadduct by HeLa cell-free extract and of thymine dimer by *Xenopus laevis* oocytes. J. Biol. Chem. 268:1931–1936.

Sweigert, S. E., and Carroll, D. (1990). Repair and recombination of X-irradiated plasmids in *Xenopus laevis* oocytes. Mol. Cell. Biol. 10:5849–5856.

Thomas, D. C., Roberts, J. D., and Kunkel, T. A. (1991). Heteroduplex repair in extracts of human HeLa cells. J. Biol. Chem. 266:3744–3751.

Varlet, I., Radman, M., and Brooks, P. (1990). DNA mismatch repair in *Xenopus* egg extracts: repair efficiency and DNA repair synthesis for all single base-pair mismatches. Proc. Natl. Acad. Sci. USA 87:7883–7887.

Wang, G., and Glazer, P. M. (1995). Altered repair of targeted psoralen photoadducts in the context of an oligonucleotide-mediated triple helix. J. Biol. Chem. 270:22595–22601.

Wang, G., Levy, D. D., Seidman, M. M., and Glazer, P. M. (1995). Targeted mutagenesis in mammalian cells mediated by intracellular triple helix formation. Mol. Cell. Biol. 15:1759–1768.

Wangh, L. J. (1989). Injection of *Xenopus* eggs before activation, achieved by control of extracellular factors, improves plasmid DNA replication after activation. J. Cell Sci. 93:1–8.

Wyllie, A. H., Laskey, R. A., Finch, J., and Gurdon, J. (1978). Selective DNA conservation and chromatin assembly after injection of SV40 DNA into *Xenopus* oocytes. Dev. Biol. 64:178–188.

Zasloff, M. (1987). Magainins, a class of antimicrobial peptides from *Xenopus* skin: isolation, characterization of two active forms, and partial cDNA sequence of a precursor. Proc. Natl. Acad. Sci. USA 84:5449–5453.

11

DNA Replication and Chromatin Assembly Using *Xenopus* Eggs or Embryos

Sophie Menut
Jean-Marc Lemaitre
Alan Hair
Marcel Méchali

The investigation of the cellular and biochemical processes that are involved in eukaryotic chromosome replication has for many years lagged behind that of other biosynthetic processes such as transcription or translation. Except in yeast, small DNA molecules transfected in eukaryotic cells replicate very poorly and a genetic approach to DNA replication is difficult. The development of cell-free systems that can sustain the replication of viral DNA molecules was the first significant step towards the dissection of DNA replication at the molecular level (Li and Kelly 1984). A fundamental stage of DNA replication, such as initiation, could not be satisfactorily analyzed, however, because viral DNA replication is entirely dependent on a virus-encoded protein. The development of cell-free systems from *Xenopus* eggs has provided the first nonviral cell-free systems capable of faithfully reproducing nuclear synthesis events, from the replication of the DNA to the assembly of chromatin and reconstitution of a functional nucleus. The large size of the egg (1.2 mm) permits the analysis of chromosomal replication in an in vivo context and relatively easy microinjection.

Certain aspects of oogenesis in *Xenopus* explain why *Xenopus* eggs are a suitable system for the molecular analysis of processes such as DNA replication and chromatin assembly. The most voluminous organ in *Xenopus* is the ovary, which harbors several thousands of oocytes. Oocytes are cells that are blocked at the prophase stage of the first meiotic division. During a period which lasts 1–2 years, they grow without dividing and accumulate a huge stockpile of components that will be used during the early developmental stages that follow fertilization. The maturation of the oocyte into an egg is induced by gonadotropin, and eggs are laid unfertilized outside of the body of the animal. During this complex process, the oocyte's nuclear envelope breaks down and the cell cycle is arrested at the metaphase of the second division. This is the stage that is usually the source of cell-free extracts, because the egg contains all the maternal components necessary for the next 12 divisions that occur at an accelerated rate (every 30 minutes) after fertilization.

Different kinds of extracts can be obtained, depending on the mechanism to be analyzed or the question to be addressed. Figure 11-1 and Table 11-1 give a summary of the main characteristics of the different *Xenopus* egg extracts. The replication of DNA molecules microinjected into *Xenopus* eggs was first reported by Gurdon et al. (1969), and demonstrated by Laskey and Gurdon (1973), and Ford and Woodland (1975) to be a semi-

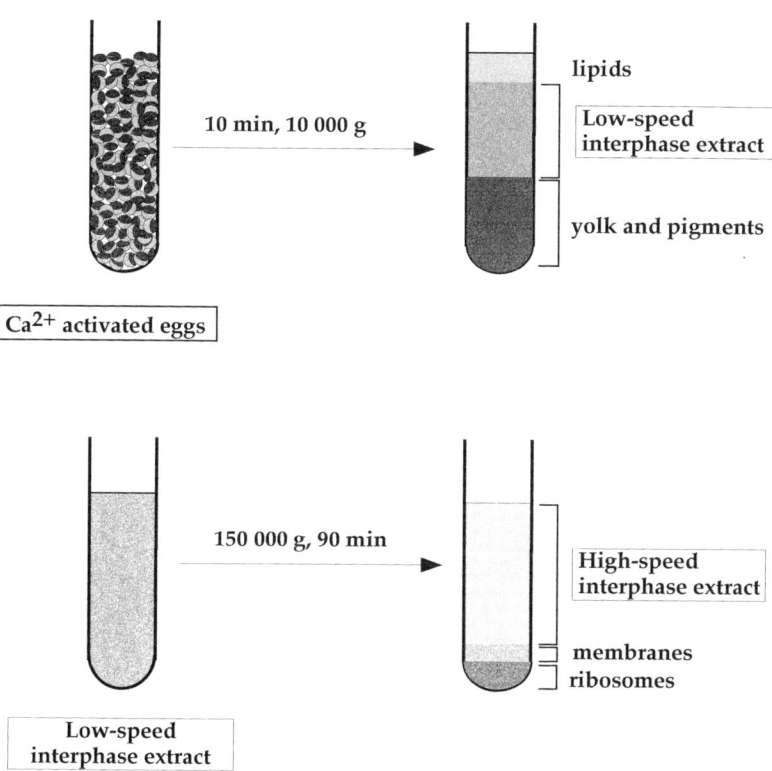

Figure 11-1 Preparation of *Xenopus* egg extracts. Low-speed and high-speed extracts are prepared from unfertilized *Xenopus* eggs. See text for details.

Table 11-1 Different types of egg extracts: properties and their use for cell-cycle studies

Extract	Obtention	Properties	Drugs currently used
Low speed Interphasic (LSE)	From activated eggs centrifuged at 10,000g	• Assembles naked DNA into chromatin • Reproduces nuclear assembly • Reproduces DNA replication and its regulation	• Cycloheximide: inhibits translation and blocks entry in mitosis • Aphidicolin: blocks elongation of DNA synthesis • Camptothecin: blocks DNA supercoiling • 6 DMAP: inhibits kinase activity and prevents chromatin licensing
Mitotic (LSE–CSF)	From nonactivated eggs centrifuged at 10,000g in presence of EGTA	• Does not replicate DNA • Condenses chromosomal DNA • Forms mitotic spindle • Can be induced in interphase and replicates after addition of Ca^{2+}	• Nocodazol or colchicin: inhibits spindle formation • Taxol: stabilizes the mitotic spindle • ICRF193, VM26: inhibit topoisomerase II activity and prevent chromosome condensation and decondensation
High speed (HSE)	From LSE centrifuged at 100,000g	• Does not replicate double-stranded DNA but assembles prereplication foci • Converts a single-stranded DNA to a double-strand minichromosome (complementary DNA strand synthesis) • Assembles naked DNA in chromatin	• Aphidicolin: blocks DNA synthesis • Camptothecin: blocks topoisomerase I activity

conservative mechanism, like that in vivo. DNA replication occurs with no preferred initiation site and under strict cellular control (Harland and Laskey 1980; Méchali and Kearsey 1984; Hyrien and Méchali 1992; Mahbubani et al. 1992). High-speed egg extract was then found to perform complementary strand DNA synthesis, in a reaction which reproduces events occurring at the replicating fork in vivo (Méchali and Harland 1982). These extracts convert a single-stranded circular DNA molecule into a complete double-stranded DNA molecule, but cannot replicate double-stranded DNA. The production of extracts for genomic DNA replication was reported by Lohka and Masui (1983), and then further demonstrated and characterized by Blow and Laskey (1986), Hutchison et al. (1987), and Murray and Kirschner (1989). Several laboratories are now analyzing DNA replication using these extracts.

These concentrated extracts are also active in transcription in vitro but as large stores of histones are present in the extract, a dynamic competition between histone assembly and transcription complex assembly occurs, leading to the repression of transcription. This phenomenon can be alleviated in several ways, including addition of transcription factors, or titration of the histone pool by DNA, or dilution of the extract (Wolffe 1989, 1994; Almouzni et al. 1990; Toyoda and Wolffe 1992; Prioleau et al. 1994 Wolffe).

We will describe in the next sections the preparation of cell-free extracts from eggs or early embryos and their use in chromatin assembly and DNA replication in vitro. We will also consider the use of microinjection for the study of this process in vivo in eggs or early embryos.

Collection of Eggs

Females are injected subcutaneously with human chorionic gonadotropin (hCG, chorulon). Usually, three females should be injected in their dorsal lymph sacs with 200 units of chorulon, then 6 hours later with 600 units. After the second injection, frogs should be left in individual tanks containing high-salt Barth's solution (HSB). The laying occurs between 8 hours and 16 hours after the second injection. It is convenient to perform the injections the day before egg collection and to leave the animals overnight in HSB at 20–22°C. If the temperature of the room falls below 18°C, significant delays in the laying can be registered. Frogs also respond to a single injection of 700–800 units of chorulon, 12–14 hours before laying, but less reproducibly.

If difficulties are reproducibly observed in obtaining eggs, the females can be primed first with an injection of 100 units of pregnant mare serum gonadotropin (PMSG), at least 3 days, and at most 10 days, before hCG injection. A single injection of 300–600 units of chorulon (hCG) 14–16 hours before egg collection is then sufficient to obtain large amounts of eggs. Although several laboratories, including our own, use this method with success, we have found that the first method gives eggs of a better quality.

Eggs are normally laid spontaneously but they can be also squeezed from frogs by gentle pressure on the abdomen immediately before the preparation of the extract. This method has the advantage of providing fresh eggs, but disadvantages such as an increase in spontaneous activation and necrosis. Moreover, frogs squeezed in this way often recover with difficulty from this manipulation and the quality of their laying decreases with time. We do not recommend this method for fertilization and production of embryos.

Freshly laid eggs (within the last 4 hours) usually give the best results, but eggs laid overnight can be also used if there is no visible sign of necrosis.

Low-Speed Egg Extract (LSE)

The two main features of this protocol are the disruption of the cells by a simple centrifugation, and the preparation and use of the extract in conditions in which no dilution of the egg content occurs (Figure 11-1).

The preparation of the extract under conditions that avoid the need for buffer dilution was first successfully achieved by Laskey et al. (1977), who developed a chromatin assembly assay. Centrifugation of the eggs to produce extracts was first described by Lohka and Masui (1983) and permits the release of the soluble egg content with minimum damage of the intracellular vesicles and cellular organelles. In these conditions, the release of nucleases and proteases is minimized.

Low-speed egg extracts replicate sperm nuclei efficiently, and DNA plasmid molecules with a lower efficiency (Blow and Laskey 1986). They form a nuclear envelope around DNA (Blow and Laskey 1986; Newport 1987) and reproduce a number of events linked to the cell cycle (Hutchison et al. 1987; Blow and Laskey 1988; Murray and Kirschner 1989). When freshly prepared egg extracts are used, several cell cycles are obtained in vitro, but this ability is easily lost after freezing, probably because a large decrease ($>90\%$) in the efficiency of translation of the extract occurs (unpublished observations). Nevertheless, frozen extracts efficiently reproduce one complete cell cycle in vitro, especially DNA replication when sperm nuclei are used.

Addition of cycloheximide to the extract blocks protein synthesis. The accumulation of cyclin B and activation of cdc2-cyclin B is thus prevented and the cell cycle is arrested in vitro in a G2-like state after DNA replication. We prefer to prepare the extract in the absence of cycloheximide, but its addition may help to prevent any progress in the cell cycle during the preparation of the extract itself.

1. Set the centrifuge at 1°C. Cool all tubes, adaptors, and syringes to 4°C before starting the preparation of the extract.
2. Transfer the eggs to a glass beaker and rinse with HSB. We do not recommend the pooling of eggs from different frogs.

3. Add distilled water and leave the external jelly coat to swell for 5 minutes at room temperature.
4. Add HSB 0.3×, cysteine 2%, pH 7.9 (the solution should be used within 6 hours of preparation), and dejelly by gentle swirling at intervals. This takes 5–10 minutes and complete removal of the jelly is obtained when the eggs can be tightly packed together, slightly deformed. It is important to obtain a complete dejellification. At this stage, success depends on both the rapidity with which the preparation is carried out and the strict observation of the cold temperature conditions after step 10.
5. Rinse immediately at least five times with 100–200 ml of HSB per milliliter of eggs. If, at this point, necrosis is visible in more than 20% of the eggs, do not proceed further. Transfer to a 50-ml glass beaker.
6. Rinse twice with 0.2× MMR.
7. Add 0.2× MMR + 0.3 µg/ml calcium ionophore to activate the eggs. Leave for 3–5 minutes at room temperature. This step can be avoided since crushing the egg in a buffer containing Ca^{2+}, as in step 8, may be sufficient to activate the extract in interphase. We found, however, that previous activation with Ca^{2+} ionophore gives more consistent results.
8. Rinse twice in 0.2× MMR and transfer the eggs to a large glass Petri dish for observation under a microscope. Activation can be monitored by the transient contraction of the animal cortex which occurs 11 minutes after addition of calcium ionophore. Using a Pasteur pipette, rapidly remove damaged or abnormal eggs. If more than 20% of eggs need to be removed, do not proceed further.
9. Transfer to 50-ml Corning tubes. Rinse twice with XB buffer to which 10 µg/ml protease inhibitors have been added.
10. Transfer the eggs in a Beckman Ultra-clear tube or equivalent. Add XB containing 10 µg/ml protease inhibitors + 100 µg/ml cytochalasin. Use 1 ml for 3 ml of eggs.
11. At 12 minutes postactivation, transfer the tube containing the eggs to ice.

All subsequent steps should be performed at 4°C, with precooled buffer solutions and tubes.

12. Leave for 10 minutes in ice to chill the eggs before extraction.
13. Remove the excess buffer and pack the eggs by centrifugation at 150g for 45 seconds at 1°C, in a Sorvall swinging rotor or equivalent.
14. Rapidly remove the excess buffer and centrifuge at 17,000g (Sorvall HB4 swinging rotor, 10K), for 10 minutes at 1°C. The centrifugation crushes the eggs and the soluble content is thus exuded.
15. Withdraw the extract by puncturing the side of the tube with a 20G needle inserted into a 1-ml to 5-ml syringe, depending on the amount of soluble extract. Insert the needle just above the black pigment layer and collect the cytoplasmic layer, avoiding the yellow lipid top layer (Figure 11.1). Transfer to a cold Ultra-clear tube. Add 10 µg/ml protease inhibitors, 10 µg/ml cytochalasin B, 1/20 volume Energy Mix 20×, and 5% glycerol. Mix gently.
16. Centrifuge again in the same conditions.
17. Collect the supernatant in a cold tube. If necessary, add 200 µg cycloheximide to prevent protein synthesis. Store at −80°C in 100- or 200-µl aliquots previously frozen in liquid nitrogen. Protein concentration in low-speed extracts

is around 50 mg/ml and RNA concentration, mainly in ribosomes, is 5–10 mg/ml. Aliquots should be used only once and should not be frozen again after thawing.

Mitotic Extract (CSF Extract)

Mitotic extracts (also referred to as CSF extracts) are useful to analyze the entry into mitosis. Mitotic extracts contain a high level of cdc2-cyclin B stabilized by CSF factor, and addition of G2 nuclei is followed by nuclear envelope breakdown, lamin depolymerization, and chromosome condensation. The cell cycle is blocked at this stage, but further addition of Ca^{2+} permits entry into the cell cycle and, more especially, DNA replication (Table 11-1).

Mitotic extracts are prepared through a procedure similar to that used for interphasic extract. Eggs should not be activated, however, and EGTA should be added to buffers to chelate traces of calcium, either present in solutions or released from intracellular stores.

Steps 1–5: Proceed exactly as for interphasic extracts, with HSB-CSF instead of HSB buffer.
6. Transfer the eggs in a large glass Petri dish for observation under a microscope. Eggs should not show any signs of spontaneous activation.
7. Transfer to a cold Ultra-clear tube. Rinse with cold XB-CSF containing 10 µg/ml protease inhibitors and 100 µg/ml cytochalasin B. Use 1 ml for 3 ml of eggs.
8. Leave the tube in ice for 5–10 minutes to chill the eggs.
9. Then proceed exactly as for interphasic extracts, steps 13–17.

Preparation of High-Speed Extract (HSE)

High-speed egg extracts (Figure 11-1) are convenient for chromatin assembly (Laskey et al. 1977; Almouzni and Méchali 1988a) and complementary DNA strand synthesis (single-stranded DNA replication; Méchali and Harland 1982). They also assemble chromatin in a process coupled with DNA synthesis (Almouzni and Méchali 1988b). These extracts are inactive in translation, nuclear assembly and double-stranded DNA replication (Table 11-1).

1. Set the centrifuge to 1°C. Cool all tubes, adaptors, and syringes to 4C before starting the preparation of the extract.
2. Transfer the eggs to a glass beaker and rinse with HSB. We do not recommend pooling the eggs from different frogs.
3. Add distilled water and leave the external jelly coat to swell for 5 minutes at room temperature.
4. Add HSB 0.3×, cysteine 2%, pH 7.9 (the solution should be used within 6 hours of preparation). Eggs are dejellied by gentle swirling at intervals. This takes 5–10 minutes and complete removal of the jelly is obtained when the

eggs can be tightly packed together, and are slightly deformed. It is important to obtain a complete dejellification.

5. Rinse immediately at least five times with 100–200 ml of HSB per milliliter of eggs. Transfer the eggs to a large glass Petri dish for observation under a microscope. If, at this point, necrosis is visible in more than 20% of the eggs, do not proceed further. Transfer to 50-ml glass beakers.

All subsequent steps should be performed at 4°C, with precooled buffer solutions and tubes.

6. Transfer the eggs to a beaker or 50-ml Corning tube. Rinse very gently twice with cold XB-HS, then with cold XB-HS containing 10 µg/ml protease inhibitors, 100 µg/ml cytochalasin B.
7. Transfer the eggs to a cold Ultra-clear 10-ml Beckman tube and leave in ice for 5–10 minutes, depending on the egg volume.
8. Remove the excess buffer and immediately centrifuge at 17,000g for 30 minutes at 1°C in a swinging bucket rotor to break the eggs. In this protocol, eggs are not packed before crushing and the first centrifugation lasts longer than that for low-speed extracts. In these conditions, the membranes are more easily removed from the extract.
9. Remove the supernatant with a 5-ml syringe mounted with a 20G needle. Insert the needle just above the black pigment layer and collect the cytoplasmic extract leaving the yellow lipid top layer. Transfer to a cold Ultra-clear tube.
10. Add 10 µg/ml protease inhibitors, 10 µg/ml cytochalasin B, 1 mM DTT, 10 mM creatin phosphate, 10 µg/ml creatin kinase.
11. Transfer the supernatant to a cold 5-ml ultra clear tube. Do not exceed $\frac{1}{3}$ of the tube's volume in order to obtain a clear supernatant (slightly brown is fine). Overlay with paraffin oil. Centrifuge in a SW50 TI Beckman rotor, 40,000 rpm (192,000g) for 60 minutes at 2°C.
12. The pellet is composed of two layers: the bottom layer is golden and contains glycogen and ribosomes; it is covered by a grey layer of mitochondria and membranes (Figure 11-1). The high-speed extract should be collected with a 5-ml syringe mounted with a 20G needle. Insert the needle above the fluffy membrane layer and collect the high-speed extract, avoiding the yellow lipid layer.
13. If the extract is too turbid, centrifuge again, in the same conditions.
14. Freeze in liquid nitrogen in 50–100-µl aliquots in 0.7-ml Eppendorf tubes, and store at −80°C. High-speed extracts can be frozen and thawed several times without any detectable loss in their efficiency for chromatin assembly or single-strand DNA replication. They will remain stable for more than 1 year at −80°C.

Preparation of Nuclei from *Xenopus* Embryos for DNA Replication and Chromatin Studies

The isolation of nuclei from the embryos of *Xenopus laevis* presents a range of problems not encountered in other cell and tissue types. The large store of vitellus in the early embryo, the low number of classic interphasic nuclei in the rapidly dividing early embryo, and the changes in the physical constitution of the embryo during development can all cause difficulties in obtaining good-quality nuclei in sufficient numbers. We present here two methods that have been developed from previous methods in order to supply a large number of intact nuclei, from embryonic stages up to and including the tail-bud stage, suitable for use in genomic footprinting with LMPCR.

Nuclei from Early Embryos

When early embryos are crushed in the absence of buffer, the vitellus sediments to the bottom of the tube whereas nuclei remain in the supernatant. The viscosity and density of the extract is such that most nuclei do not sediment even after centrifugation at 17,000g for 10 min. This property can be exploited to isolate nuclei up to at least the gastrula–neurula stage of development.

1. Dejelly embryos in HSB 0.1×, cysteine 2%, pH 7.9.
2. Wash embryos several times with 0.1× Barth's solution.
3. Place embryos in an Ultra-clear centrifuge tube and remove as much excess buffer as possible. Centrifuge at 700g at 4°C briefly in order to pack, but not break, the embryos.
4. Rapidly remove excess buffer and centrifuge at 17,000g at 4°C for 10 minutes.
5. Remove the liquid layer by side-puncturing of the tube just above the black pigment layer with a 20G needle. This liquid layer is either homogeneous or is composed of two distinct layers, depending on the stage of development of the embryos used.
6. Add 10 volumes of buffer Nuc A and pass three times through a syringe with a 21G needle to remove further debris from the nuclei.
7. Centrifuge at 500g at 4°C for 10 minutes.
8. Resuspend the gray pellet in 10 ml of Nuc A and layer over 4 ml of 12.5% glycerol in the same buffer.
9. Centrifuge at 500g at 4°C for 10 minutes and resuspend the pellet in a buffer of choice. For DNase I-hypersensitive sites, use 10% glycerol, 0.25 M sucrose, 10 mM HEPES, pH 7.6, 60 mM KCl, 15 mM NaCl, 0.15 mM spermine, 0.5 mM spermidine, 1 mM DTT, 10 µg/ml protease inhibitors.

Nuclei from Neurula Stages and Onwards

After the neurula stage, the preceding method is no longer applicable. The viscosity of the extract changes, and the nuclei, which co-sediment with the vitellus after centrifugation, are surrounded by less cytoplasm.

Two previously published methods (Farzaneh and Pearson 1978; Gorski et al. 1986), based on sucrose cushion layers, have been adapted in order to yield a large number of high-quality nuclei. The method described here circumvents the difficulty associated with the removal of the nuclei from an interface between highly viscous layers, whilst ensuring a high yield through the use of detergent. The buffers have been, as far as possible, synchronized with those from the previous method so that the early and late embryonic stages can be compared. Sucrose buffers dissolve slowly and it is thus advisable to prepare them before starting the collection of the embryos.

1. Wash and dejelly embryos and place in a Potter homogenizer on ice.
2. Reconstitute Nuc B by mixing parts B1 with B2, and add 30 ml to the embryos.
3. Homogenize the embryos using a Potter homogenizer, with five strokes at 1500 rpm at 4°C.
4. Dilute the homogenate to 85 ml using cold Nuc B.
5. Prepare three Ultra-clear centrifuge tubes, each containing 10 ml of Nuc B.
6. Layer on 28 ml of homogenate and centrifuge at 24,000 rpm for 30 minutes at 4°C using a SW28 rotor.
7. Pour off the supernatant and resuspend the pellets in a total volume of 56 ml of a 9:1 (v/v) mixture of Nuc B and glycerol.
8. Layer the homogenate over two 10-ml cushions of Nuc B, as before, and centrifuge at 24,000 rpm for 30 minutes at 4°C using a SW28 rotor.
9. Resuspend nuclei in a buffer of choice. For DNase I-hypersensitive sites we use 10 mM HEPES, pH 7.6, 0.25 M sucrose, 60 mM KCl, 15 mM NaCl, 0.15 mM spermine, 0.5 M spermidine, 1 mM DTT, 10% glycerol.

Preparation of Demembranated Sperm Nuclei

The following protocol includes modifications made by Gurdon (1976) and Murray (1991).

1. Inject three male frogs with 100 units of human chorionic gonadotropin (Chorulon) 1–7 days before the preparation.
2. Frogs should be anesthetized in ice water until they no longer respond to manipulation (around 30 minutes). Lay the frog on its back on an ice bed and cut the skin and the peritoneal wall. The first cut should be made from top to bottom, at 1 cm to the right of the midline (to avoid any blood vessels), and other cuts should made on both sides to expose the body wall. Use a pair of forceps to push aside the intestine and the yellowish fat bodies. Locate the white testes on either side of the backbone. Collect the testes with scissors and wash them in HSB. The remaining fat and blood vessels should be removed from the testes.
3. Rinse the testes twice in SNB and place them in a Petri dish laid on ice.
4. Chop the testes with a sterile blade.
5. Transfer to a Dounce homogenizer (large pestle), together with 2 ml of SNB.
6. Homogenize with one or two slow strokes.
7. Filter the homogenate on sterile cheesecloth placed on a small funnel on the top of a 15-ml conical tube. Alternatively, centrifuge briefly (10 seconds, 100g) to remove large pieces of tissue.
8. Rinse with 5–10 ml of SNB.
9. Centrifuge at 1700g for 10 minutes at 4°C. If blood vessels have not been sufficiently removed from the testes, a thin red layer of reticulocytes will be visible. In this case, the pellet can be resuspended in 6 ml of Tris 10 mM, pH 7.6, EDTA 1 mM and left for 5 minutes at 4°C to break the reticulocyte cytoplasmic membrane.

10. Centrifuge at 1700g, 10 minutes at 4°C, and resuspend the pellet in SNB. If the pellet is difficult to resuspend, homogenize gently with a Dounce homogenizer (large pestle).
11. Centrifuge again and resuspend the pellet in 1 ml of SNB at room temperature.
12. When the tube is equilibrated at room temperature (20–22°C), add 50 µl of lysolecithin 10 mg/ml (prepared extemporaneously in water), and incubate 5 minutes. Rapidly check the progress of demembranation with aliquots stained with 1 µg/ml Hoechst.
13. Add 6 ml of cold SNB containing 30 mg/ml BSA.
14. Centrifuge at 1700g for 10 minutes and resuspend the pellet in 5 ml of SNB, 3 mg/ml BSA.
15. Centrifuge again and resuspend the pellet in 1 ml of SNB, 3 mg/ml BSA, 30% glycerol. Measure the concentration of nuclei using a hemocytometer and nuclei stained with Hoechst. Around 10^8 nuclei can be obtained from one frog.
16. Freeze in liquid nitrogen in 10-µl aliquots and store at −80°C. Do not refreeze any thawed aliquots.

Assays for DNA Replication in *Xenopus* Extracts

Xenopus egg extracts are very efficient in replicating DNA from sperm nuclei, or in complementary DNA strand synthesis from a single-strand DNA template (Table 11-1). In both cases, a 100% replication from the template DNA is routinely observed. With plasmid double-strand DNA molecules, the efficiency is lower, between 2% and 20% input DNA is replicated (values also obtained in vivo after microinjection). The extent of in vitro replication increases with the size of the DNA template (Blow and Laskey 1986), as observed in vivo (Méchali and Kearsey 1984). It is not yet clear why plasmid DNA molecules do not replicate as well as sperm nuclei, but because DNA replication is dependent on nuclear membrane assembly, it is generally admitted that incorrect assembly will result in less efficient DNA synthesis. The assembly of chromatin in higher order structures may be also less efficient in plasmids than in nuclei.

Like all biochemical reactions, DNA replication is dependent on template concentration. This is an important point to take into consideration when sperm nuclei are used as template, since a limiting quantity of DNA replication factors may change not only the rate of DNA synthesis but also the specificity of the reaction. Assuming that 1 µl of low-speed extract represents the soluble content of two eggs, and that DNA replication factors are not limiting up to the mid-blastula transition stage (8000 cells in 8 hours in vivo), the concentration of nuclei in the extract should not exceed 10,000–16,000 nuclei/µl. We routinely use 500–1000 nuclei/µl of extract, a concentration sufficient for most analyses.

DNA Replication Assay by TCA Precipitation

1. Prepare 2.1- or 2.5-cm Whatman GF/C glass filters which have been numbered with India ink.
2. Immediately before use, thaw an aliquot of the *Xenopus* egg extract (LSE) and put it on ice.
3. Add the following to a 0.5-ml Eppendorf tube:

Energy mix 20×	2.5 µl
LSE	50 µl
Sperm nuclei	25,000 nuclei (500 nuclei/µl of LSE)
α^{32}P-dATP or α^{32}P-dCTP	10 µCi

   Since the extract is extremely sensitive to dilution, avoid diluting it by more than 20%. Homogenize very gently and incubate at room temperature (20–23°C). In good extracts, DNA replication starts after 30–40 minutes and reaches 100% by 60–90 minutes (Figure 11-2).
4. Spot 2 µl of the mix on a Whatman GF/C glass filter to determine the total radioactivity of the extract (input). Leave it to dry, put it into scintillation vials, add scintillation fluid, and count.
5. Take 2–5 µl aliquots of the mix at 0 minutes and then every 10 minutes. Add the aliquot directly to 25 µl of Stop Mix solution supplemented with 0.6 mg/ml of proteinase K. Vortex vigorously and incubate all the kinetic points at 42°C for at least 1 hour.
6. Spot each sample on a Whatman GF/C glass filter and drop the filter into a large beaker containing cold 5% TCA, 2% pyrophosphate solution (20 ml/filter). Leave for at least 20 minutes at 4°C. Filters can be accumulated for several hours in the beaker.
7. Wash the filters four times in cold 5% TCA, 5 minutes per wash, and once in ethanol. Agitation is not necessary.
8. Leave all the filters to dry on tissue paper, and count the incorporated TCA-precipitable radioactivity (T).

Calculation of the Activity of the Extract

To calculate the activity of the extract, it is necessary to evaluate the quantity of deoxynucleotide triphosphates in the extract and the percentage of radioactivity incorporated.

1. *Quantity of deoxynucleotides in the reaction mix.* The extract contains stockpiles of deoxynucleotide triphosphate, including ATP which is estimated at 50–70 µM by isotope dilution under steady-state replication conditions (Méchali and Harland 1982; Blow and Laskey 1986). We did not observe any significant variations in this value for dATP using different extracts. The value of 60 pmol/µl of extract used in the assay is usual.
2. *Amount of DNA synthesized*:

   pg DNA synthesized = (% dATP incorporated) × (pmol dATP in the assay × 4 × 330

Example

In a reaction assay of 55 µl containing 50 µl of extract and 25,000 sperm nuclei, 20,000 cpm above the control are detected after TCA precipitation in a 5-µl aliquot. A 2-µl sample of the reaction assay (input) contains 0.72×10^6 cpm.

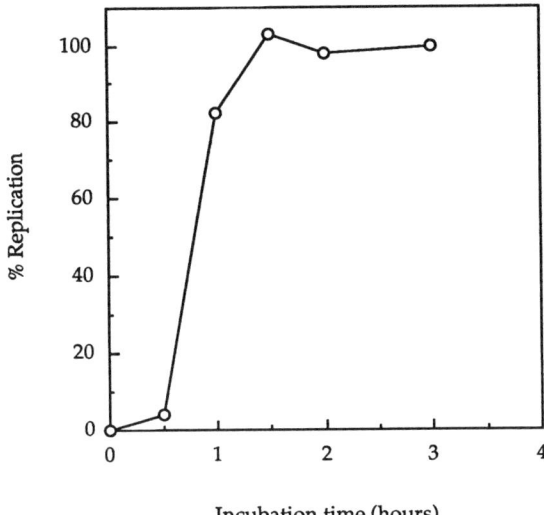

Figure 11-2 Kinetics of DNA replication afer incubation of sperm nuclei in a *Xenopus* egg extract. *Xenopus* sperm nuclei were introduced into low-speed extract (500 nuclei/μl). At the indicated times, aliquots were removed from the reaction mixture and the percentage of template DNA that replicated was calculated by the TCA-precipitation method.

$$\text{pmoles dATP in the assay: } 60 \times 50 = 3000$$

$$\text{Total radioactivity: } \frac{0.72 \times 10^6 \times 55}{5} = 20 \times 10^6 \text{ cpm}$$

$$\text{Total incorporated: } \frac{20{,}000 \times 55}{5} = 220{,}000 \text{ cpm}$$

$$\text{\% incorporated radioactivity: } \frac{220{,}000 \times 100}{20 \times 10^6} = 1.1\%$$

$$\text{pg DNA synthesized: } \frac{1.1 \times 3000 \times 4 \times 330}{100} = 43{,}560 \text{ pg}$$

Since one sperm nucleus contains 3 pg of DNA (*n* chromosomes), the DNA in the assay corresponds to 14,520 nuclei. There were 25,000 sperm nuclei in the assay, therefore, the amount of DNA synthesized is 58%.

DNA Replication Assay by Immunofluorescence

DNA replication can be analyzed by immunofluorescence with the nucleotide thymidine deoxynucleotide analogs BrdUTP (5-bromodeoxyuridine triphosphate) or biotin-dUTP (biotinylated deoxyuridine triphosphate). This method has several advantages. It is rapid, very sensitive, and detects sites of DNA replication (Figure 11-3). We nevertheless recommend the use of TCA-precipitation assays in parallel to determine the efficiency of DNA replication. A low level of synthesis may, for example, reflect repair synthesis, which will be interpreted as positive replication signals by immunofluorescence.

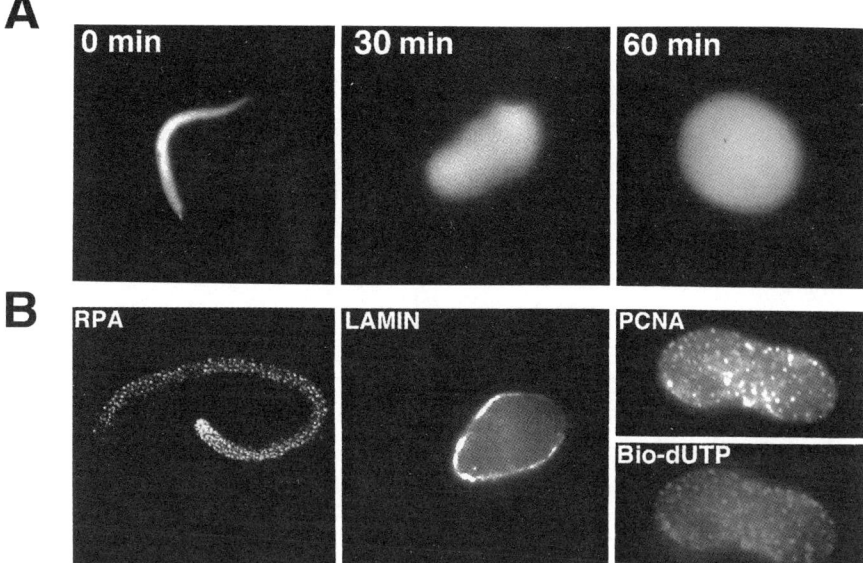

Figure 11-3 Changes in nuclear morphology in replicating extracts. (A) Demembranated sperm nuclei were incubated in a low-speed extract and examined by fluorescence with the Hoechst 33342 dye. (B) Prereplication centers are detected by immunolocalization of RPA protein after 15 minutes incubation. After 30 minutes, the nuclear envelope has formed and the lamina was detected around the chromatin with an antilamin anitobdy. Replication centers were detected by incorporation of biotin-dUTP and colocalized with PCNA staining.

1. Add to 50 µl of extract between 300 and 1000 sperm nuclei/µl and either 0.25 mM BrdUTP or 40 µM biotin-dUTP.
2. Every 30 minutes take a 10-µl aliquot of the reaction and fix for 1 hour at room temperature in 200 µl of XB containing 6% formaldehyde.
3. Spin down at 2500g through a 0.6 M sucrose cushion in XB onto coverslips. Coverslips can be coated with polylysine to increase the recovery but this treatment also increases immunofluorescence background, and we do not recommend it.
4. Incubate the coverslips in 4 N HCl in PBS for 30 minutes at room temperature (HCl denatures DNA and gives access to anti-BrdUTP antibody). Wash for 10 minutes at room temperature.
5. Postfix the nuclei for 4 minutes in −20°C methanol.
6. Incubate the coverslips in PBS for 10 minutes at room temperature and then for 1 hour in PBS, 1% BSA.
7. Wash in PBS, 0.1% Tween 20 for 10 minutes.
8. Load on the coverslips 50 µl of either streptavidin coupled with fluorescein or Texas red in PBS 1% BSA (diluted according to the supplier) and incubate for 1 hour at room temperature. If the nucleotide analog is BrdUTP, use 50 µl

of anti-BrdUTP antibody in PBS 1% BSA, and then a second antibody coupled with fluorescein or Texas red for 1 hour at room temperature.
9. Wash twice in PBS, 0.1% Tween 20 for 15 minutes at room temperature.
10. Stain DNA for 10 minutes at room temperature with either Hoechst dye 1 µg/ml in PBS, or propidium iodide 10^{-6} M.
11. To visualize the nuclear membrane, incubate the coverslips for 10 minutes in PBS containing 5 µg/ml DIOC (lipid dye) to obtain a green fluorescence.
12. Wash three times in PBS, 0.1% Tween 20 for 15 minutes at room temperature.
13. Mount the coverslips with Citifluor antifading agent and seal with nail polish before observation under a fluorescence microscope.

Protocol for Single-Strand DNA Replication and Chromatin Assembly

Single-stranded (SS) M13 DNA incubated in the egg extract is entirely converted into a double-stranded DNA form assembled in a minichromosome (Méchali and Harland 1982). Chromatin assembly occurs during the process of DNA replication in the extract (Almouzni and Méchali 1988a). Figure 11-4 shows the kinetics of replication analyzed by agarose gel electrophoresis.

This reaction can be also used to prepare double-stranded labeled DNA by a single addition of ^{32}P dATP to the reaction medium.

1. Prepare a reaction medium as follows, scaled to the number of assays to be performed.

Single-stranded DNA 500 µg/ml	0.4 µl
^{32}P-dATP 0.5 µCi/µl	1 µl (note A)
H$_2$O	0.6 µl
Total	2 µl reaction mix (note B)

2. Add to small 0.5-ml Eppendorf tubes:

Reaction mix	2 µl
Xenopus egg extract	18 µl
Total	20 µl (note C)

   Mix and incubate at room temperature (20–25°C) (note D).
3. Take a 2–5-µl aliquot to determine the extent of labeling by TCA precipitation (note E).
4. Take another 2–5 µl aliquot. Add 25 µl of Stop Mix and 1 volume of chloroform. Vortex at maximum speed for 20 seconds and centrifuge for 3 minutes at maximum speed in an Eppendorf centrifuge at room temperature. Take 18 µl from the aqueous phase and add the sample buffer for gel electrophoresis.
5. Analyze the reaction product on a 1% agarose gel run with Tris-acetate-EDTA buffer (note F).

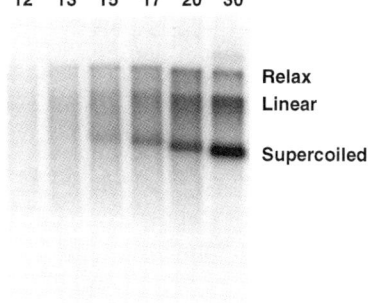

Figure 11-4 Replication of M13 ss DNA in a high-speed extract. M13 single-stranded DNA was added together with α^{32}P-dATP in high-speed egg extract. Aliquots were taken at the indicated times and analyzed by agarose gel electrophoresis. The first complete double-stranded forms were detected after 15 minutes at 20°C.

Notes

A. A small amount of ^{32}P-dATP (0.5 µCi per reaction) is sufficient to obtain >50,000 cpm incorporated in the assay. The values can be increased by increasing the amount of ^{32}P-dATP.

B. The endogenous ATP/Mg^{2+} normally present in the extract is sufficient for DNA replication and nucleosome assembly. The addition of exogenous ATP/Mg^{2+} improves the physiological spacing of the nucleosomes (Almouzni and Méchali 1988b).

C. The reaction medium can be diluted up to at least 10 µl if other components need to be included in the assay. In that case, the final concentration of single-stranded DNA and ATP/Mg^{2+} should be adjusted accordingly in the reaction mix. We recommend the use of the following dilution medium: 20 mM HEPES, pH 7.6, 70 mM potassium acetate, 6 mM MgCl$_2$, 1 mM ATP, 2 mM DTT, 5% sucrose, 1 mg/ml nuclease-free BSA.

D. >95% of DNA is replicated and assembled in a minichromosome after 90 minutes at 22°C. The physiological spacing of chromatin is slightly improved after 2 hours at 22°C. The reaction can be left for at least 6 hours at room temperature without affecting the final product. Do not incubate below 18°C.

E. The reaction can be quantified by TCA-precipitation of aliquots. The extent of incorporation of ^{32}P-dATP in double-stranded DNA is determined by spotting a 2–5-µl aliquot directly on a Whatman GF/C filter followed by 20 minutes incubation in cold 5% TCA, 2% sodium pyrophosphate, four washes in 5% cold TCA, and a final wash with ethanol. The filters can be left for several hours in cold 5% TCA if necessary. DNA synthesis is estimated as described in the section "Assays for DNA Replication in *Xenopus* Extracts" above, and the efficiency of DNA replication is calculated in relation to the input single-strand DNA.

F. The protocol of extraction given here is sufficient for an analysis of the DNA products on an agarose gel. It can be improved by treatment with proteinase K, followed by phenol and phenol chloroform extractions in larger volumes if the DNA has to be extensively purified.

Chromatin Assembly

A double-stranded plasmid DNA molecule incubated in the egg extract is assembled into a minichromosome containing evenly spaced nucleosomes, as occurs in vivo (Laskey et al. 1977; Almouzni and Méchali 1988b). Assembly of spaced chromatin also occurs concomitantly with DNA synthesis (Almouzni and Méchali 1988a). If the plasmid DNA molecule is circular-relaxed, it will be converted into a supercoiled form after incubation in the extract and extraction of the DNA.

Chromatin assembly is efficient because a large store of histones is present in the *Xenopus* egg, estimated at 120 ng, an amount sufficient to assemble 20,000 nuclei (Newport and Kirschner 1982). Using a high-speed extract and a plasmid molecule to titrate histones, the decrease in the supercoiling reaction is not detectable below 50 ng DNA/μl extract and at 80 ng DNA/μl the DNA molecule population is mainly relaxed (M. N. Prioleau and M. Méchali, unpublished data). Thus, 1 μl of extract contains enough histones to assemble 50–80 ng of DNA, a value in close agreement with the estimate made by Laskey et al. (1977).

Chromatin assembly can be monitored by a variety of procedures and we describe here two simple protocols. The first is based on the observation that the assembly of nucleosomes introduces negative superhelical turns into the DNA (supercoiling assay, Figure 11-5). The second is based on the differential sensitivity to micrococcal nuclease of nucleosomes and linker DNA (micrococcal nuclease assay).

Supercoiling Assay (First Method)

1. Prepare a reaction medium as follows, scaled to the number of assays to be performed.

Double-stranded DNA 500 μg/ml	0.4 μl
ATP/Mg^{2+}	1 μl
Sterile H$_2$O	0.6 μl
Total	2 μl reaction mix.

2. Add to small 0.5-ml Eppendorf tubes:

Reaction mix	2 μl
Xenopus egg extract	18 μl
Total	20 μl (note A)

   Incubate at room temperature (20–25°C) (note B).

3. Add 100 μl of Stop Mix containing 500 μg/ml proteinase K. Incubate for 1 hour at 37°C.

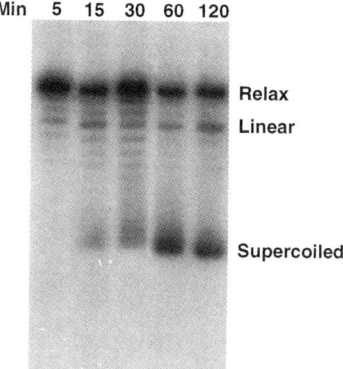

Figure 11-5 Chromatin assembly in a high-speed egg extract. Supercoiled bluescript DNA was added to a high-speed extract and aliquots were processed as described in the text for analysis by agarose gel electrophoresis. The supercoiled DNA is rapidly relaxed and then progressively supercoiled. The relaxation step is due to topoisomerase I and the supercoiling step is correlated with the assembly of nucleosomes from the endogenous histone pool.

4. Add 1 volume of phenol/chloroform saturated with TE. Vortex vigorously and centrifuge for 4 minutes at maximum speed in an Eppendorf centrifuge (room temperature).
5. Transfer the upper aqueous phase to a new tube and extract with 1 volume of chloroform/isoamyl alcohol. Centrifuge for 3 minutes in an Eppendorf centrifuge.
6. Transfer the aqueous phase to a new tube. Add 1/10 volume of 3 M sodium acetate and 2 volumes of cold ethanol and mix.
7. Centrifuge for 10 minutes. Dry the pellet and resuspend in 15 µl of TE containing 25 µg/ml DNase-free RNase. Incubate for 20 minutes at 37°C.
8. Analyze the reaction by Southern blot followed by hybridization (Figure 11-5) (note C).

Supercoiling Assay (Second Method)

It can be more convenient to analyze chromatin assembly on a molecule that is already radiolabeled, eliminating the need for Southern blot hybridization. Then 2–5-µl aliquots are sufficient to follow the reaction. In that case, the DNA can be labeled through the conversion of single-stranded DNA to the double-stranded form in the extract (see "Protocol for Single-Strand DNA Replication and Chromatin Assembly"). After extraction, a double-stranded labeled supercoiled molecule is obtained. Another procedure is to linearize the DNA at one restriction site, dephosphorylate with alkaline phosphatase, label with T4 polynucleotide kinase and recircularize with T4 DNA ligase (Razvi et al. 1983). If a radiolabeled DNA is used, aliquots of the reaction can be analyzed on an agarose gel after a single chloroform extraction.

Micrococcal Nuclease Digestion assay
Steps 1 and 2 are the same as steps 1 and 2 of the first supercoiling assay.
3. Add $CaCl_2$ to 3 mM final concentration and 30 units of micrococcal nuclease. Remove aliquots during digestion (2–15 minutes) at room temperature.
4. Each aliquot is adjusted to 30 mM EDTA, 0.5% SDS, 500 µg/ml proteinase K and extracted as described above (steps 3–7 of the first supercoiling assay).
5. Analyze the reaction products on a 1.5–2% agarose gel run in TBE, or in acrylamide gels for mononucleosome resolution (Almouzni and Méchali 1988b).

Notes
A. The reaction medium can be diluted up to 10 µl if other components have to be included in the assay. In that case, the final concentration of DNA and ATP/Mg^{2+} should be adjusted accordingly in the reaction mix. We suggest the following dilution medium: 20 mM HEPES, pH 7.6, 70 mM potassium acetate, 6 mM $MgCl_2$, 1 mM ATP, 2 mM DTT, 5% sucrose, 1 mg/ml nuclease-free BSA.
B. Chromatin assembly is a multistep process. The DNA is first relaxed by the topoisomerase I present in the extract in a reaction which takes place in less than 5 minutes at 20°C. It is then progressively assembled into chromatin in a reaction which takes place within 2–4 hours.
C. It is also possible to analyze directly the reaction products by ethidium bromide staining of the gel. In that case, the amount of DNA and egg extract in the assay must be increased. We suggest using 1 µg of DNA with 25 µl of *Xenopus* egg extract. A larger amount of DNA may deplete the histone pool.

Analysis of DNA Replication by Microinjection into *Xenopus* Eggs

The semiconservative DNA replication of small DNA molecules microinjected into unfertilized *Xenopus* eggs was first reported with Polyoma virus DNA (Laskey and Gurdon 1973). The replication of a wide range of DNA molecules occurs under strict cellular control (Harland and Laskey 1980; Méchali et al. 1983; Méchali and Kearsey 1984). The injected DNA molecules replicate once per cell cycle, and do not require specific sequences to initiate DNA synthesis. This is not a feature that is peculiar to the injected DNA since it has been demonstrated that, at least during the early developmental period, the endogenous genome initiates in an apparently random manner, with no preferred sites of initiation (Hyrien and Méchali 1993; Hyrien et al. 1995). The injected DNA molecules assemble into minichromosomes, from a large histone pool stored in the egg, and nuclei are assembled around the injected DNA (Forbes et al. 1983; Tredelenburg et al. 1986). One interesting feature of this system, as opposed to the in vitro system, is that several cycles of DNA replication can be observed after microinjection, in synchrony with the endogenous cell cycles. Initiation of DNA replication can still be detected in the injected DNA 8 hours after the injection (unpublished results). Supercoiled, linear, and circular molecules all replicate, but linear molecules are also ligated into concatemers, which complicates the analysis.

We will describe three main protocols for the analysis of DNA replication. Quantitative and qualitative analysis of DNA synthesis are, respectively, followed by TCA precipitation (Figure 11-6) and gel electrophoresis. The assay for semiconservative DNA replication combines the incorporation of BrdUTP and CsCl gradient analysis (Figure 11-7). This is an unambiguous and crucial assay for checking that DNA synthesis is indeed due to a replication process and not merely to a repair reaction. This protocol also makes it possible to ensure that DNA replicates only once per cell cycle. Localization of initiation sites of DNA replication can be analyzed by two-dimensional agarose gel electrophoresis (Figure 11-8).

Injection of Plasmid DNA Molecules

Protocols for microinjection have been described in several laboratories (Gurdon 1991; Harland and Laskey 1980; Newport and Kirschner 1982; Méchali and Kearsey 1984) and will not be described here in detail. DNA replication is proportional both to the size of the molecule in the 5–15 kbp range and to the

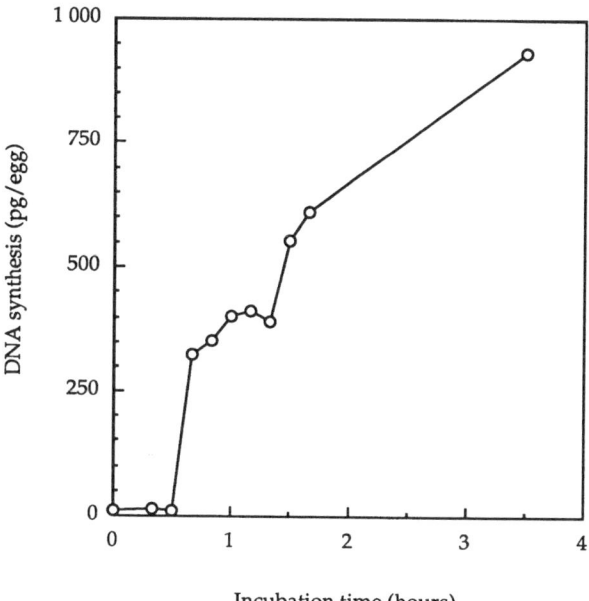

Figure 11-6 Replication of plasmid DNA microinjected into *Xenopus* eggs. Unfertilized eggs were microinjected with 10 ng of plasmid DNA and α^{32}P-dCTP. At different times, groups of 14 eggs were processed for acid-insoluble material precipitated on GF/C filters.

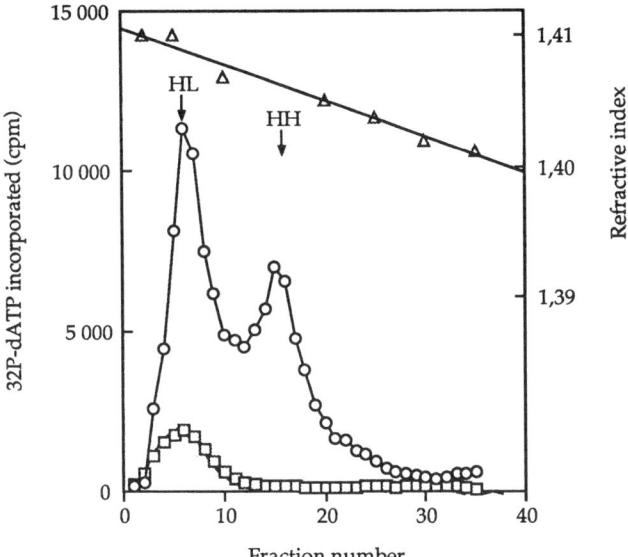

Figure 11-7 Semiconservative synthesis of λ DNA in *Xenopus* eggs. CsCl equilibrium density gradient of DNA extracted from eggs after 5 hours 40 minutes incubation. Eggs were injected with λ DNA, BrdUTP, and α^{32}P-dATP (circles). Replication of genomic endogenous DNA was followed by microinjection of BrdUTP and α^{32}P-dATP (squares). Refraction indexes of some fractions were noted on the graph (triangles). HL, heavy–light DNA, HH, heavy–heavy DNA.

amount of injected DNA, up to 30–50 ng per egg. Usually, 5 ng of DNA is injected per egg, in order to retain steady-state conditions.

Injected eggs are collected in Eppendorf tubes after incubation in batches of 15–30 eggs per point. Any remaining liquid is removed and tubes are frozen at −80°C if they are not processed immediately. DNA is injected together with 0.025 µCi ^{32}P-dATP for labeling experiments.

Analysis by TCA Precipitation and Agarose Gel Electrophoresis

1. Add to each tube containing frozen eggs, 20 µl per egg of Stop Mix solution supplemented with 600 µg/ml proteinase K just before use. Eggs should be homogenized in this solution while they are thawing, by means of a small pestle or an Eppendorf tip.
2. Incubate 1 hour at 37°C.
3. Add 1 volume of phenol/chloroform/isoamyl alcohol (25:24:1). Vortex vigorously and centrifuge for 4 minutes at maximum speed at room temperature in an Eppendorf-type centrifuge.
4. Take the upper aqueous phase and, if necessary, wash the phenolic phase with half its volume. Pool the aqueous phase with the previous one.
5. Perform a second phenol/chloroform extraction as in stage 3 and take the aqueous phase.

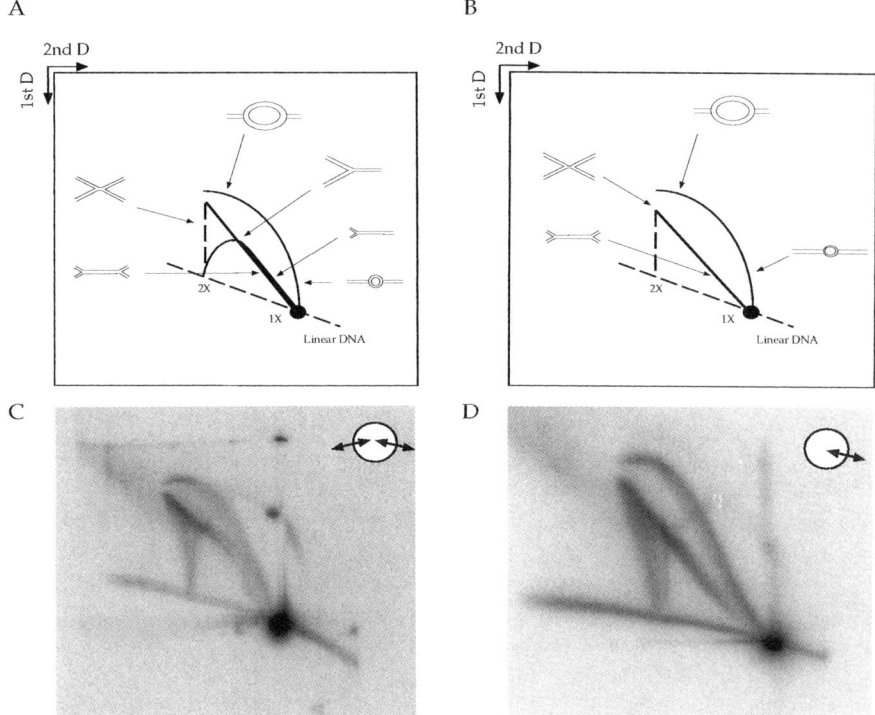

Figure 11-8 Two-dimensional gel electrophoresis analysis of plasmid DNA after microinjection into *Xenopus* eggs. Patterns of replication intermediates for an (A) double or a (B) single restriction enzyme digestion of a plasmid are shown. 5 ng of a plasmid DNA was co-injected with α^{32}P-dCTP into unfertilized eggs. After 70 minutes incubation, the DNA from 60 or 30 eggs was analyzed by 2D gel after (C) two or (D) one cut, respectively.

6. Take a 10-μl aliquot and spot on a Whatman GF/C glass filter for the total input radioactivity. Take a 100-μl aliquot and spot on a Whatman GF/C glass filter for TCA precipitation to measure the amount of DNA synthesis (see "DNA Replication Assay by TCA Precipitation" protocol above). Calculate the efficiency of DNA replication essentially as described in the "Calculation of the Activity of the Extract" protocol above, taking into account the fact that one egg contains 13 pmol of dATP (Woodland and Pestell 1972).
7. Take a 20-μl aliquot and process for agarose gel electrophoresis in Tris-acetate buffer to analyze the reaction products (Figure 11-6).

Analysis of DNA Replication by Cesium Chloride Density Gradients
Semiconservative DNA replication is monitored by incorporation of BrdUTP, a heavy analog of dTTP. One round of DNA replication results in a peak of heavy–light molecules, whereas two rounds or more of DNA replication gives both heavy–heavy and heavy–light DNA molecules (Figure 11-7). In contrast, if

DNA synthesis is due to a repair reaction, the DNA migrates as a smear between the positions of light–light (parental, unreplicated molecules) and heavy–light DNA.

1. Inject the DNA together with 0.05 µCi ^{32}P-dATP and 0.3 nmol BrdUTP. Around 50 eggs should be injected per experimental point.
2. Extract the DNA as described for TCA precipitation (cf. "Analysis by TCA Precipitation and Agarose Gel Electrophoresis" protocol, steps 1 to 6).
3. Precipitate the remaining aqueous phase with 2 volumes of ethanol and centrifuge at 10,000g for 10 minutes.
4. Dissolve the pellets in 30 µl of TE containing 100 µg/ml DNase-free RNase and incubate for 30–60 minutes at 37°C.
5. Remove the unincorporated deoxynucleotides by gel filtration on a 1-ml column, or by spin column, with TE as a buffer.
6. Adjust the DNA sample to 3 ml with TE containing 0.91 g/ml CsCl. This solution should be carefully layered on top of 3 ml of TE containing 1.13 g/ml CsCl and the gradients centrifuged to equilibrium in a Beckman 50 TI rotor for 38 hours at 36,000 rpm. The use of two layers of CsCl is not essential but produces a more rapid separation of the different reaction products. Do not run at temperatures below 20°C to avoid any risk of CsCl crystallization. If the concentration and volumes of CsCl gradient solutions have to be adjusted or the centrifugation conditions changed, check the allowable run speed conditions of the rotor when using CsCl solutions.
7. Collect the fractions of the gradient directly in a scintillation vial (35–40 fractions per gradient gives an optimum separation) and measure the radioactivity in each fraction. An example of separation is given in Figure 11-7.

Localization of Origins of DNA Replication by Two-Dimensional Gel Analysis

The conformation of a DNA molecule influences its electrophoretic mobility in an agarose gel. This can be frequently observed when comparing the mobilities of circular, linear, and supercoiled DNA in an agarose gel. Making use of this property, Brewer and Fangman (1987) developed a two-dimensional (2D) agarose gel electrophoresis to analyze intermediates of DNA replication.

Restriction fragments which are replicated can have three basic shapes depending on the site at which of DNA replication commences (Figure 11-8). Restriction fragments containing bidirectional origins produce a series of bubble-shaped molecules, whereas termination within fragments leads to X-shaped molecules (also called double-Y molecules). These two families of molecules generate, respectively, a bubble arc and a double-Y arc, which are maximally retarded when fragments are completely replicated. Restriction fragments which are replicated from an external origin assume a Y-like configuration during their replication. They all form a Y arc and are maximally retarded when they are half replicated.

Electrophoresis on the first dimension minimizes the effect of molecular shape and separates molecules according to their mass. These conditions can be obtained with a low concentration of agarose and when the gel is run at low voltage during approximatively 15 hours for a fragment in a range of 2–6 kbp. In contrast, in the second dimension the gel is run under conditions that emphasize the shape effect. To obtain this migration a higher concentration of agarose is necessary and the gel should be run at a higher voltage (5–6 V/cm) in the presence of 0.5 µg/ml of ethidium bromide in the gel and in the buffer. The migration must be carried out during 5 hours with ethidium bromide and at 4°C in order to stabilize the intermediates of DNA replication.

We will describe conditions for analysis of fragments within the 2–6 kpb range. For larger fragments (6–9 kb), the first and the second dimension have to be longer and a lower voltage should be used during the second dimension (Hyrien and Méchali 1992).

Injected eggs or embryos should be processed immediately since freezing and thawing may nick DNA molecules. The extraction procedure is also modified to maintain the integrity of DNA molecules.

1. Inject 5 ng of DNA per egg, together with 0.05 µCi ^{32}P-dATP if the analysis is done by labeling. Fifty eggs should be injected per experimental point.
2. Collect the eggs in an Eppendorf tube, remove any remaining solution, and homogenize immediately with 30 µl per egg of the Stop Mix 2D solution supplemented with 0.6 mg/ml proteinase K.
3. Transfer to 15-ml conical Falcon tubes and incubate at 37°C for at least 3 hours and at most overnight.
4. Add 1 volume of phenol saturated in TE. Vortex at maximum speed and centrifuge for 10 minutes at 2000g at room temperature.
5. Transfer the upper aqueous phase to a new 15-ml Falcon tube and wash the phenolic phase with half its volume of Stop Mix 2D solution.
6. Extract, twice, the total aqueous phase with 1 volume of phenol/chloroform/isoamyl alcohol (25:24:1).
7. Add 1 volume of chloroform/isoamyl alcohol (24:1). Vortex and centrifuge 5 minutes at 2000g at room temperature.
8. Transfer the aqueous phase to a 50-ml Corex tube. Add 2 volumes of cold ethanol.
9. Centrifuge for 20 minutes at 10,000g at 4°C.
10. Dry the pellet and resuspend it in 100 µl of TEN300 containing 20 µg/ml of RNase A. Incubate for 30 minutes at 37°C. The DNA can be further purified by gel filtration or precipitated by ethanol.
11. Digest the DNA with the appropriate restriction enzyme.
12. Adjust the concentration to 20 mM EDTA, sodium acetate 0.3 M, and precipitate with 2 volumes of cold ethanol.
13. Centrifuge at maximum speed for 20 minutes at 4°C, dry the pellet, and resuspend it in 20 µl of TENX.
14. Analyze the sample by neutral/neutral 2D gel electrophoresis, as follows.
15. Load the sample containing the intermediates of DNA replication on a 0.4% agarose gel in 1× TBE buffer and load a ladder of molecular weight (e.g., 1 kb ladder) in a separate lane. Run the gel at 0.8–1 V/cm for approximatively 15 hours at room temperature.
16. Stain the gel for 20 minutes in 1× TBE, 0.5 µg/ml ethidium bromide. Leave in the dark during the staining. Take a photograph.
17. Cut a lane of 10 cm, beginning 1 cm below the position of the linear fragment to be analyzed and extending up to twice the size of this fragment.
18. Pour a 1.1% agarose gel into 1× TBE, 0.5 µg/ml ethidium bromide equilibrated at 55°C. The new agarose gel should cover the old one.

19. Run the gel in a cold room (4°C) at 5–6 V/cm for 5 hours in 1× TBE buffer containing 0.5 µg/ml ethidium bromide. Recircularize the buffer from anode to cathode to prevent accumulation of the ethidium bromide at the anode pole, and keep the gel in the dark during the electrophoresis.
20. Take a photograph of the gel. If the DNA has been radiolabeled during DNA replication, dry the gel and expose by autoradiography. If the analysis is done by Southern blot, perform as follows.
21. Wash the gel twice for 10 minutes in 0.25 N HCl, rinse for 5 minutes in water, and transfer the gel to 0.4 N NaOH for 20 minutes.
22. Transfer the DNA to a nylon membrane (Hybond N$^+$, Amersham) by alkali capillary blotting in 0.4 N NaOH during 5 hours or overnight.
23. Prehybridize for 1 hour with Church solution at 65°C.
24. Add the denatured probe previously prepared with a random priming kit at 2×10^5–10^6 cpm/ml of Church solution. Hybridize overnight at 65°C.
25. Wash twice in 1× SSPE 0.1% SDS at 65°C.
26. Expose for 2 days on a phosphorImager cassette or an autoradiographic film. Figure 11-8 shows an example of such analysis using two kinds of restriction enzyme digestion.

References

Almouzni, G., and Méchali, M. (1988a). Assembly of spaced chromatin promoted by DNA synthesis in extracts from Xenopus eggs. EMBO J. 7(3):665–672.

Almouzni, G., and Méchali, M. (1988b). Assembly of spaced chromatin involvement of ATP and DNA topoisomerase activity. EMBO J. 7(13):4355–4365.

Almouzni, G. , Méchali, M., and Wolffe, A. P. (1990). Competition between transcription complex assembly and chromatin assembly on replicating DNA. EMBO J. 9:573–582.

Blow, J. J., and Laskey, R. A. (1986). Initiation of DNA replication in nuclei and purified DNA by a cell-free extract of Xenopus eggs. Cell 47:577–587.

Blow, J. J. , and Laskey, R. A. (1988). A role for the nuclear envelope in controlling DNA replication within the cell cycle. Nature 332:546–548.

Brewer, B. J., and Fangman, W. L. (1987). The localization of replication origins on ARS plasmids in S. cerevisiae. Cell 51:463–471.

Church, G. M., and Gilbert, W. (1984). Genomic sequencing. Proc. Natl. Acad. Sci USA 81:1991–1995.

Farzaneh, F., and Pearson, C. K. (1978). A method for isolating uncontaminated nuclei from all stages of developing Xenopus laevis embryos. J. Embryol. Exp. Morphol. 48:101–108.

Forbes, D. J., Kirschner, M. W., and Newport, J. W. (1983). Spontaneous formation of nucleus-like structures around bacteriophage DNA microinjected into Xenopus eggs. Cell 34:13–23.

Ford, C. C., and Woodland, H. R. (1975). DNA synthesis in oocytes and eggs of Xenopus laevis injected with DNA. Dev. Biol. 43:189–199.

Gorski, K., Carneiro, M., and Schibier, U. (1986). Tissue-specific in vitro transcription from the mouse albumin promoter. Cell 47:767–776.

Gurdon, J. B. (1976). Injected nuclei in frog oocytes: fate, enlargement, and chromatin dispersal. J. Embryol. Exp. Morphol. 36:523–540.

Gurdon, J. B., Birnstiel, M. L., and Speight, V. A. (1969). The replication of purified DNA introduced into living egg cytoplasm. Biochim. Biophys. Acta 174:614–628.
Gurdon, J. B. (1991). Nuclear transplantation in *Xenopus*. Methods Cell Biol. 36:299–309.
Harland, R. M., and Laskey, R. A. (1980). Regulated replication of DNA microinjected into eggs of *Xenopus laevis*. Cell 21:761–771.
Hutchison, C. J., Cox, R., Drepaul, R. S., Gomperts, M., and Ford, C. C. (1987). Periodic DNA synthesis in cell-free extracts of Xenopus eggs. EMBO J. 6:2003–2010.
Hyrien, O., and Méchali, M. (1992). Plasmid replication in Xenopus eggs and egg extract: a 2D gel electrophoresis analysis. Nucleic Acids Res. 20:1463–1469.
Hyrien, O., and Méchali, M. (1993). Chromosomal replication initiates and terminates at random sequences but at regular intervals in the ribosomal DNA of Xenopus early embryos. EMBO J. 12:4511–4520.
Hyrien, O., Maric, C., and Méchali, M. (1995). Transition in specification of embryonic metazoan DNA replication origins. Science 270:994–997.
Laskey, R. A., and Gurdon, J. B. (1973). Induction of polyoma DNA synthesis by injection into frogg-egg cytoplasm. Eur. J. Biochem. 37:467–471.
Laskey, R. A., Mills, A. D., and Morris, N. R. (1977). Assembly of SV40 chromatin in a cell-free system from Xenopus eggs. Cell 10:237–243.
Li, J. J., and Kelly, T. J. (1984). Simian virus 40 replication in vitro. Proc. Natl. Acad. Sci. USA 81:6973–6977.
Lohka, M. J., and Masui, Y. (1983). Formation in vitro of sperm pronuclei and mitotic chromosomes induced by amphibian ooplasmic components. Science 220:719–721.
Mahbubani, H. B., Paull, T., and Blow, J. J. (1992). DNA replication initiates at multiple sites on plasmid DNA in Xenopus egg extract. Nucleic Acids Res. 20:1457–1462.
Méchali, M., and Harland, R. M. (1982). DNA synthesis in a cell-free system from Xenopus eggs: priming and elongation on single-strand DNA in vitro. Cell 30:93–101.
Méchali, M., and Kearsey, S. (1984). Lack of specific sequence requirement for DNA replication in Xenopus eggs compared with high sequence specificity in yeast. Cell 38:55–64.
Méchali, M., Méchali, F., and Laskey, R. A. (1983). Tumor promoter TPA increases initiation of replication on DNA injected into Xenopus eggs. Cell 35:63–69.
Murray, A. (1991). Cell cycle extracts. Methods in Cell Biol. 36:581–605.
Murray, A. W., and Kirschner, M. W. (1989). Cyclin synthesis drives the early embryonic cell cycle. Nature 339:275–280.
Newport, J. (1987). Nuclear reconstitution in vitro: stages of assembly around protein-free DNA. Cell 48:205–217.
Newport, J., and Kirshner, M. (1982). A major developmental transition in early Xenopus embryos: I. Characterization and timing of cellular changes at the midblastula stage. Cell 30:675–686.
Prioleau, M. N., Huet, J., Sentennac, A., and Méchali, M. (1994). Competition between chromatin and transcription complex assembly regulates gene expression during early development. Cell 77:439–449.

Razvi, F., Gargiuolo, G., and Worcel, A. (1983). A simple procedure for parallel sequence analysis of both strands of 5'-labeled DNA. Gene 23:175–183.

Toyoda, T., and Wolffe, A. P. (1992). Characterization of RNA polymerase II-dependent transcription in Xenopus extracts. Dev. Biol. 153:150–157.

Trendelenburg, M. F., Oudet, P., and Montag, M. (1986). DNA injections into Xenopus embryos: fate of injected DNA in relation to formation of embryonic nuclei. J. Embryol. Exp. Morphol. 97:243–255.

Wolffe, A. (1989). Transcriptional activation of Xenopus class III genes in chromatin isolated from sperm and somatic nuclei. Nucleic Acids Res. 17:767–780.

Wolffe, A. (1994). The role of transcription factors, chromatin structure and DNA replication in 5S RNA gene regulation. J. Cell Sci. 107:2055–2063.

Woodland, H. R., and Pestell, R. Q. (1972). Determination of the nucleoside triphosphate contents of eggs and oocytes of *Xenopus laevis*. Biochem. J. 127:597.

Appendix: Buffers and Solutions

Extracts

High-Salt Barth (HSB) Solution
15 mM HEPES, pH 7.6
110 mM NaCl
2 mM KCl
1 mM KCl
1 mM $MgSO_4$
0.5 mM $Na2HPO_4$
2 mM $NaHCO_3$
Prepare an 8× stock and store at 4°C.

XB
10 mM HEPES, pH 7.7
100 mM KCl
0.1 mM CaCl2
1 mM MgCl2
5% Sucrose
1 mM DTT
Prepare a 2× stock and store at −20°C.

Cytochalasin B, 10 mg/ml
10 mg/ml Cytochalasin B in DMSO
Store at −20°C.

Protease Inhibitors, 10 mg/ml
10 mg/ml Leupeptin
10 mg/ml Pepstatin
10 mg/ml Aprotinin in DMSO
Store in aliquots at −20°C.

Energy Mix 20×
200 µg/ml Creatine kinase
200 mM Creatine phosphate

20 mM ATP
20 mM MgCl$_2$
Store in aliquots at −20°C.

Creatine Phosphate, 1 M
1 M Creatine phosphate in water
Store in aliquots at −20°C.

Creatine Kinase, 10 mg/ml
10 mg/ml Creatin kinase in HEPES 10 mM, pH 7.6
20% Glycerol
Store in aliquots at −20°C.

XB-HS
10 mM HEPES, pH 7.7
10 mM potassium acetate
5 mM MgCl$_2$
1 mM DTT
5% Sucrose

MMR
5 mM HEPES, pH 7.8
0.1 M NaCl
2 mM KCl
0.1 mM EDTA (added before Ca^{2+} and Mg^{2+})
1 mM MgCl$_2$
2 mM CaCl$_2$
Prepare a 10× stock and store at 4°C.

HSB-CSF
Same as HSB + 2 mM EGTA

Energy Mix-CSF 20×
Same as Energy Mix + 2 mM EGTA

XB-CSF
10 mM HEPES pH 7.7
100 mM KCl
1 mM MgCl$_2$
5% Sucrose
1 mM DTT
5 mM EGTA

Calcium Ionophore
5 mM in DMSO and store at −20°C.

Single-Strand DNA and Chromatin Assembly

Single-stranded M13 DNA or any other single-stranded molecule 200 to 500 μg/ml (stock solution).

ATP/Mg^{2+} 20×
60 mM ATP

100 mM MgCl$_2$
pH 7.5. Check the pH of the solution with a few microliters spotted on a pH paper.
Store at $-20°C$.

Stop Mix
40 mM EDTA
1% SDS
0.5% Triton ×100

Double-stranded M13 DNA 500 µg/ml in TE.

Preparation of Nuclei

Sperm Nuclei
Lysolecithin
10 mg/ml Lysolecithin in water
Prepare extemporaneously.

SNB
15 mM HEPES, pH 7.4
0.25 M Sucrose
75 mM NaCl
0.5 mM Spermidine
0.15 mM Spermine
1 mM DTT

BSA
BSA (nuclease-free) 100 mg/ml in Tris 10 mM, pH 7.6
NaCl 20 mM.

PBS
137 mM NaCl
2.7 mM KCl
43 mM Na$_2$HPO$_4$
1.4 mM KH$_2$PO$_4$

Embryonic Nuclei
Nuc A
10 mM HEPES, pH 7.6
15 mM KCl
1 mM EDTA
0.5 mM Spermidine
0.15 mM Spermine
0.5 mM DTT
10 µg/ml Protease inhibitors
0.4% Triton N-101

This latter is not essential, but it improves the yield. The point at which it is added should also be considered. Its addition or omission will depend on the requirements of the researcher.

Nuc B
Part B1
  10 mM HEPES, pH 7.6
  10% Glycerol
  2 M Sucrose
  15 mM KCl
  1 mM EDTA

Part B2
  0.5mM Spermidine
  0.15mM Spermine
  0.5M DTT
  10 µg/ml Protease inhibitors

Analysis of DNA Replication

TCA Precipitation
Stop Mix
30 mM EDTA
1% SDS

5% TCA
Keep at 4°C.

5% TCA
2% Sodium pyrophosphate
Keep at 4°C.

Glass fiber filters (e.g., Whatman GF/C)

Plasmid DNA Injected into Eggs or Embryos
Stop Mix 2D
20 mM Tris, pH 7.9
30 mM EDTA
1% Sarkosyl
300 mM NaCl
300 mM NaCl will stabilize DNA replication intermediates.

Proteinase K, 10 mg/ml
10 mg proteinase K per ml TE
Incubate for 30 minutes at 37°C
Store in aliquots at −20°C.

TENX
20 mM Tris, pH 7.9
1 mM EDTA
x mM NaCl (x refers to adjustable NaCl concentration)

RNase A, 10 mg/ml
10 mg DNase-free RNase A per ml TE

Sodium acetate
3 M Sodium acetate
Adjust to pH 4.0 with glacial acetic acid.

1× TBE
0.089 M Tris
0.089 M Boric acid
2 mM EDTA
pH 8.3
1× SSPE
0.15 M Sodium chloride
0.01 M Sodium phosphate
0.001 M EDTA

Ethidium Bromide, 10 mg/ml

Church Solution (Church and Gilbert 1984)
0.5 M Sodium phosphate, pH 7.2
7% SDS
1 mM EDTA
Filter and store at room temperature.

12

Isolation and Characterization of Masked mRNPs from *Xenopus* Oocytes

Mary T. Murray
Rolland Reinbold

Cytoplasmic mRNA exists in two functionally and physically distinct states. Actively translated mRNA is found in polyribosomes, whereas nontranslated mRNA is bound by protein and sequestered in messenger ribonucleoprotein particles (mRNPs). Translational activation of a nontranslated mRNP is referred to as mobilization. Four distinct types of mRNPs have been defined (Spirin 1996). The first type consists of experimentally derived polyribosomal mRNPs that can be released from polysomes by the addition of EDTA. The remaining classes of mRNPs are the classically defined nontranslating mRNPs. These are (1) free mRNPs that are found in all somatic cells and represent an excess of mRNA above the translational capacity, (2) nontranslatable mRNPs where translational initiation is blocked due to the binding of a specific repressor (e.g., ferritin nontranslated mRNPs), and (3) masked mRNPs that are stably stored in the cytoplasm of germ cells awaiting a developmental signal for mobilization to polysomes. Both the storage and the correct spatial temporal mobilization of masked mRNPs to polyribosomes are necessary for male and female germ cell maturation and the earliest stages of embryogenesis.

The vast majority of *Xenopus* oocyte poly(A)$^+$ mRNA is found in masked mRNPs complexed with oocyte-specific proteins (Darnbourgh and Ford 1981). The two most abundant mRNP proteins of fully grown oocytes were originally named mRNP3 and mRNP4 (Darnbourgh and Ford

1981), but are also called FRGY2a + b, p54/p56, and pp56/pp60. Using an mRNA-binding and UV-cross-linking assay, we identified soluble forms of mRNP3 + 4 that assemble exogenous mRNA into 40-200S mRNPs resembling endogenously masked mRNPs (Murray et al. 1991). Cloning and characterization of mRNP3 and mRNP4 demonstrated that the two polypeptides are pseudoallelic, having 87% amino acid identity (Murray et al. 1992). Their predicted molecular masses are 36 and 37 kDa, respectively; however, they run anomalously in SDS–PAGE in the range of 50–60 kDa (Deschamps et al. 1992). mRNP3 + 4 are germ-line members of the Y box multigene family of nucleic acid-binding proteins whose members all contain the evolutionarily conserved cold shock domain (CSD; Wolffe 1994). The CSD contains a motif related to RNP-1 (Landsman 1992) that mediates binding to either single-stranded DNA or RNA (Kolluri et al. 1992; Tafuri and Wolffe 1992; Ladomery and Sommerville 1994; Murray 1994). In addition to the CSD, mRNP3 + 4 contain four repeats of arginine-rich basic/aromatic islands (Murray et al. 1992) that bind RNA exclusively (Ladomery and Sommerville 1994; Murray 1994).

The abundance of mRNP3 + 4 in mRNPs suggests that they have a general role in mRNP structure and/or regulation (Darnbourgh and Ford 1981). In direct support of a functional role in translational regulation, these *Xenopus* mRNP polypeptides inhibit translation of β-globin mRNP in vivo following oocyte microinjection (Richter and Smith 1984) or in vitro in a wheat-germ translation system (Richter and Smith 1984; Kick et al. 1987). Furthermore, the expression of mRNP4 in somatic cells or oocytes causes a generalized inhibition of translation (Ranjan et al. 1993; Bouvet and Wolffe 1994). In vitro, recombinant mRNP4 represses translation of mRNA, provided a 30-fold excess of protein over mRNA is used (Matsumoto et al. 1996). However, a 320-kDa particle containing mRNP3 + 4 represses translation of particle-bound mRNA in vitro at nearly equimolar concentrations of particle to mRNA (Yurkova and Murray 1997). In addition to mRNP3 + 4, this 320-kDa particle contains nucleolin, suggesting it may be nuclear and participate in an early step of mRNP masking. (For more on the involvement of a nuclear step in masking, see chapter 13 in this volume). Further analysis of mRNPs assembled in vitro promises to provide insight into the mechanism of masking. With regard to the regulation of masking, we note that the 320-kDa particle has a protein kinase activity that phosphorylates mRNP3 + 4, and that mRNA-binding by mRNP3 + 4 is dependent on phosphorylation (Kick et al. 1987; Murray et al. 1991). With regard to the mobilization of mRNPs, the recent identification of an RNA helicase component of masked mRNPs (Ladomery et al. 1997), and the association of nucleoplasmin with mobilized mRNPs (Meric et al. 1997) hold promise for beginning to understand the role of mRNP proteins during unmasking.

Methods presented in this chapter can be used to purify mRNP proteins, analyze in vitro assembled mRNPs, and determine the translational status of a maternal mRNA. It is worth emphasizing that the translational status of any newly identified maternal mRNA must be investigated to determine if it is translated or masked throughout oogenesis. In the absence of a specific antibody, Northern blotting of RNA from fractionated mRNPs and polysomes can determine if an mRNA is translated. Since masked mRNAs are often subsequently mobilized for translation, it is useful to fractionate mRNPs and polysomal mRNA from different-stage oocytes, eggs, and embryos to determine when an mRNA is translated. The methods described here are generally applicable and can be adapted for analysis of masked mRNPs from other biological systems.

Xenopus oocyte mRNPs are primarily fractionated by size or density using either rate zonal or isopycnic centrifugation, respectively. The commonly employed methods are summarized in Table 12-1. Rate zonal centrifugation was directly adapted from polysomal analysis of somatic cells (Woodland 1974) and is widely used. Although the majority of mRNPs are 40–90S, mRNPs as large as 200S are observed that cosediment with polysomes. The identification of large mRNPs is verified by EDTA treatment, which has no effect on mRNPs but causes the release of ribosomes from polysomes. Alternatively, masked mRNPs and polysomes can be cleanly separated from one another on isopycnic gradients because the buoyant density of mRNPs is significantly less than that of polysomes. Isopycnic gradients of cesium salts are used to determine the buoyant density of *Xenopus* masked mRNPs, which is directly related to the protein content per mole of mRNA (Spirin 1969). Using this method, the mole ratio of protein to mRNA in 40–60S mRNPs is determined to be 4:1 (Richter and Smith 1984; Cummings and Sommerville 1988). A significant disadvantage in using cesium salt gradients is their high ionic strength; in the case of CsCl this necessitates prior fixation of the mRNP. In contrast, the nonionic nature of the iodinated aromatic compound Nycodenz® (Nycomed Pharma AS) allows isopycnic fractionation of mRNPs and polyribosomes for subsequent analysis of mRNA or protein (Houssais 1983). Nycodenz density gradient fractionation of mRNPs and polysomes is currently preferred over rate zonal fractionation because mRNPs are cleanly separated from polysomes (Tafuri and Wolffe 1993). Finally, UV-photo-cross-linking methods and fractionation of mRNPs by oligo(dT) affinity chromatography are also presented for analysis of mRNP proteins.

Preparation of Postmitochondrial Supernatant

The starting material for oocyte masked mRNP purification is either isolated oocytes or whole ovary depending on the inquiry. If relatively large amounts of purified mRNPs are required, it is most practical to start with whole ovary containing a mixture of different-stage oocytes. Selection for early-stage oocytes can be achieved by using immature ovaries from early postmetamorphic frogs (2–2.5″ in size). In instances where specific oogenic

Table 12-1 Fractionation of mRNPs and polysomes

Method	Gradient	Comments	References
Rate zonal centrifugation	Sucrose, 15–50% (w/v), continuous	RNA profile by A_{260} scanning during displacement; 40–200S mRNPs cofractionate with ribosomes and polysomes; use EDTA disruption of polysomes to identify large mRNPs	Woodland 1974; Darnbourgh and Ford 1981
	Glycerol, 15–40% (w/v), continuous	Same as above, except sometimes preferred or protein isolation	Cummings and Sommerville 1988
Isopycnic centrifugation	CsCl, 30–60% (w/v), preformed	Requires fixation of mRNP; RNA profile by A_{260} scanning	Richter and Smith 1984; Cummings and Sommerville 1988
	Nycodenz, 20–60% (w/v), step gradient	High UV absorbance of Nycodenz prevents RNA profile scanning; cleanly separates mRNPs from polyribosomes	Houssais 1983; Tafuri and Wolffe 1993

stages are required, oocytes should be isolated by manual defolliculation using watchmaker's forceps (for a detailed method and determination of oocyte stage, see Smith et al. 1991). Release of oocytes by collagenase treatment should be avoided as protein synthesis is severely affected and artifacts may result (Wallace and Misulovin 1978). Alternatively, eggs or embryos can be used as starting material.

If starting with whole ovary, thoroughly rinse ovary with ice-cold PB to remove all blood, transfer ovary lobes to a small sheet of aluminum foil using forceps, and leave behind any connective tissue and blood vessels. Ovary may be frozen in liquid nitrogen and stored at −80°C or used directly. Determine tissue mass and homogenize in an equal volume of ice-cold PB using a Dounce homogenizer with the L pestle. Transfer homogenate to a centrifuge tube chilled on ice and centrifuge 12,000g at 4°C for 10 minutes. Avoiding the top lipid layer, remove the supernatant; this is most easily done by puncturing the wall of the tube with a needle and syringe and removing the supernatant. Defolliculated oocytes are washed in cold PB, homogenized, and processed in a similar manner as the ovary. If eggs or embryos are used, they must first be dejellied and then transferred to cold PB. If one is interested in less tightly associated mRNP proteins, the lower ionic strength LSB buffer can be used (Cummings and Sommerville

1988); however, recovery of polysomes is significantly reduced in comparison with 0.3 M KCl PB (Woodland 1974).

Rate Zonal Fractionation of mRNPs

Prepare continuous gradients of 15–50% (w/v) sucrose or 15–40% (w/v) glycerol from equal volumes of the two stock solutions. Store gradients on ice while preparing the postmitochondrial supernatant (PMS). Up to 0.2 ml of PMS can be loaded per 4 ml gradient in an SW60 rotor (Beckman, Palo Alto) that is run at 55,000 rpm for 90 minutes at 4°C. For larger scale preparations, 0.5 ml of PMS can be fractionated per 12-ml gradient in SW41 tubes at 37,000 rpm for 3 hours, or 1.0 ml of PMS per 36-ml gradient in SW28 tubes at 17,000 rpm for 16 hours. Gradients are fractionated by displacement from the bottom of the tube using Fluorinert (Sigma, St. Louis) with continuous UV monitoring at 260 nm. The A_{260} profile is dominated by ribosomal subunits with the major peak at 80S (Woodland 1974), and in low-salt preparations, the 42S storage particle also is detected as a major peak (Cummings and Sommerville 1988).

EDTA Release of Ribosomes from Polyribosomal mRNA

Because large masked mRNPs cosediment with polysomes in rate zonal density gradients (Darnbourgh and Ford 1981), the presence of an mRNA in the polysomal fraction does not predict its translational status. Chelation of Mg^{2+} with EDTA releases ribosomes from polysomal mRNP without affecting masked mRNPs (Dolecki and Smith 1979; Cardinali et al. 1987). Experimentally, the PMS is prepared in the absence of cycloheximide, EDTA is added (35 mM final concentration), and the sample is incubated on ice for 10 minutes. The EDTA-treated sample is then applied to a sucrose or glycerol gradient that does not contain cycloheximide. It is not necessary to add EDTA to the gradients since EDTA-treated ribosomes do not reassociate with the polysome-derived mRNP. This control is essential when investigating an mRNA that appears to be associated with polyribosomes on glycerol or sucrose gradients, since approximately 70% of the mRNA that sediments in this region of the gradient is found in large masked mRNPs (Dolecki and Smith 1979).

Isolation of mRNPs by Cesium Salt Density Gradient Centrifugation

Isopycnic centrifugation with CsCl or $CsSO_4$ density gradients is used to determine the protein content of purified mRNPs or in vitro assembled mRNPs. Masked mRNPs that sediment at 40–60S in rate zonal fractionation have a density of 1.36 g/ml in CsCl (Richter and Smith 1984; Cummings and Sommerville 1988), which corresponds to 80% of the total mass being protein (Spirin 1969). When using CsCl gradients, mRNPs must first be fixed either by formaldehyde treatment (4% neutralized formaldehyde for 2 hours at 4°C; Spirin et al. 1965) or by UV photo-cross-linking (Marello et al. 1992), and then applied onto preformed gradients for centrifugation. Even though cesium density gradients have been used extensively to characterize masked mRNPs, they have been almost completely replaced by the adaptation of Nycodenz density gradients. If needed, details on the application of cesium density gradients can be found in original articles by Sommerville and coworkers (Kick et al. 1987; Cummings and Sommerville 1988; Marello et al., 1992).

Isolation of mRNPs and Polyribosomes by Nycodenz Isopycnic Fractionation

Nycodenz gradients are set up as 20–60% (w/v) step gradients in swing-out buckets to form a continuous density gradient during centrifugation. The high absorbance of Nycodenz prevents monitoring of RNA by UV; however, the location of mRNPs and polysomes is precisely determined from the fraction density. The density of each fraction is calculated from its refractive index (h) determined by an Abbe-type refractometer using the equation: ρ (g/ml) = $3.242\eta - 3.323$. The PB and LSB gradient buffers do not significantly affect the refractive index of Nycodenz; however, if other buffers are used, the refractive index may need to be corrected before the density can be calculated.

Nycodenz step gradients are composed of equal volumes of 60%, 50%, 40%, 30%, and 20% Nycodenz layered directly on top of each other, and the PMS is applied. The gradient size is determined by the sample: PMS from small numbers of hand-isolated oocytes or embryos (<100) can be fractionated on short gradients (e.g., 1.2.–1.8 ml in the TLS55 rotor, Beckman) whereas ovary preparations are separated on longer gradients (e.g., 5.0 ml in SW50.1 rotors). Because a continuous gradient is formed during centrifugation, the sample loading volume affects the final gradient. Therefore, consistent sample loading volumes are required to assure reproducible density gradients. Gradients are run at 150,000g (42,000 for TLS55 or 47,000 for SW50.1) for 20 hours at 4°C. Gradients may be fractionated by carefully aspirating from the top with a micropipette preset to the desired fraction volume, or fractionated by displacement with Fluorinert.

A Nycodenz gradient profile from an immature *Xenopus* ovary preparation is presented in Figure 12-1, with the protein and RNA content of each fraction shown with its corresponding density. The 5S ribosomal RNA storage particles concentrate at 1.15–1.18 g/ml. Masked mRNPs sediment at 1.18–1.23 g/ml with the peak of mRNP3+4. Free ribosomal subunits peak at 1.23–1.27 g/ml and polysomes are found in the densest fractions at 1.28–1.34 g/ml. The first mRNP fractions typically contain 5S ribosomal storage particles while the densest mRNPs overlap with the first ribosomal

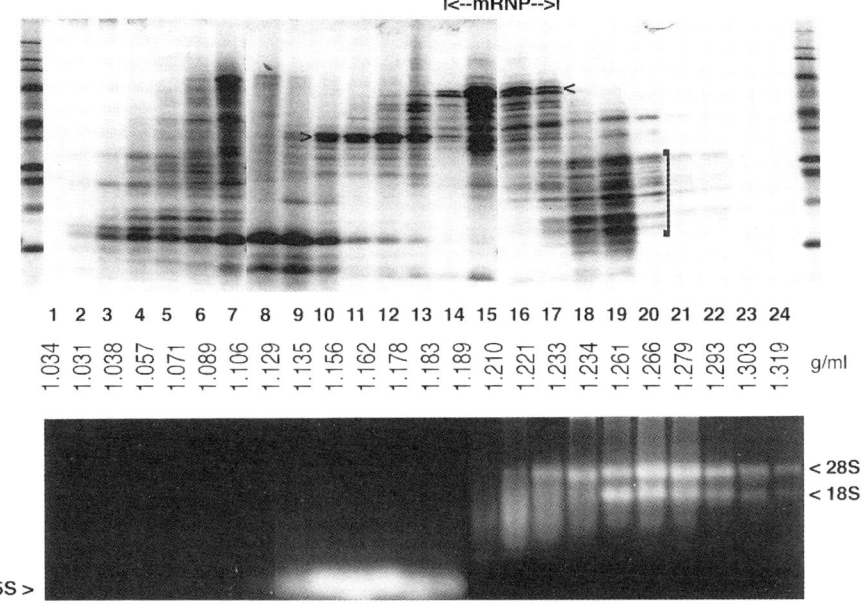

Figure 12-1 mRNP and polysome fractionation on Nycodenz density gradients. The protein profile of immature *Xenopus* ovary PMS fractionated on a 20–60% Nycodenz gradient is shown by Coomassie blue staining (upper panel), along with the corresponding RNA profile (lower panel) and density of each fraction. In the protein gel, TFIIIA (>), mRNP3 + 4 (<), and the ribosomal proteins (large bracket) are noted. Molecular mass markers are shown on the left and right (from top to bottom: 205, 116, 98, 66, 45, 36, 29, 24, and 14 kDa). In the RNA gel, 5S, 18S, and 28S ribosomal RNAs are noted. The final gradient fraction (number 25; 1.376 g/ml) is not shown, however, its protein and RNA content resembles fraction 24.

subunits. Northern blots of RNA from oocyte RNPs fractionated on similar Nycodenz gradients (Figure 12-2) show the distribution of mRNAs that are masked in the oocyte (FRGY1 and histone H1) in comparison with mRNAs that are translated (FRGY2 and TFIIIA). The gap in poly(A)$^+$ RNA observed in the Northern blot between mRNPs and polysomes (Figure 12-2) corresponds to the peak of free ribosomal subunits. Although the gradients shown in Figures 12-1 and 12-2 differ substantially, mRNPs and polysomes are cleanly separated in each case.

Recovery of RNA and Protein from Fractionated mRNPs

If RNA is to be recovered from sucrose or glycerol gradients, it is best to first precipitate the RNP particles from the fraction using 0.1 volumes of 3.0 M sodium acetate and 2.5 volumes of ethanol. The RNP particles are then dissolved in TNES, extracted once with phenol/chloroform, and once with

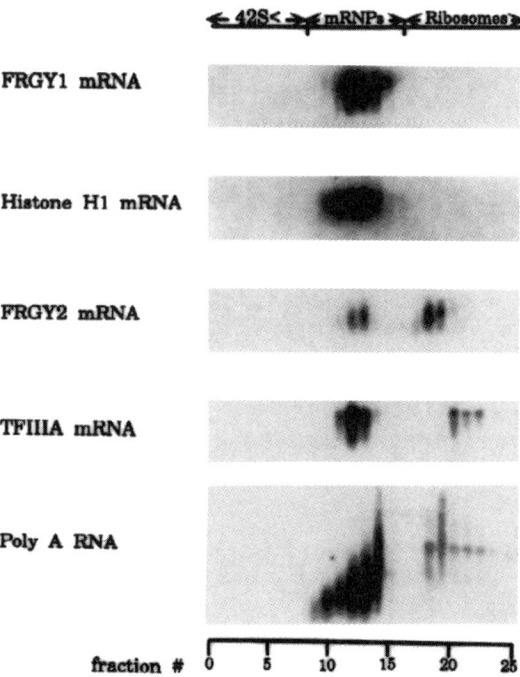

Figure 12-2 Northern blot analysis of the Nycodenz density gradient fractionated mRNPs and polysomes. PMS from stage III and IV oocytes was fractionated on 20–60% (w/v) Nycodenz gradient as described. The RNA in each fraction of the gradient was deproteinized and analyzed by Northern blotting with specific probes for the mRNAs shown on the left. In the lower panel, the blot was probed with poly(U). The location of 5S ribosomal RNA storage particles (42S), mRNPs, and ribosome subunits and polysomes (ribosomes) is indicated at the top. Reprinted with permission from Tafuri and Wolffe (1993).

chloroform, and the RNA ethanol is precipitated. If RNA is to be recovered from Nycodenz gradients, TNES can be added directly to an aliquot of the fraction, in equal volume, and the RNA purified by organic extraction and precipitation as above. RNA is processed for Northern blots using standard protocols.

If the mRNP protein is to be recovered for further use, the gradient material should first be removed by dialysis. However, analysis of the protein composition of sucrose, glycerol, or Nycodenz gradient fractions by SDS-gel electrophoresis can proceed directly by precipitation of aliquots from each fraction with the addition of tricholoroacetic acid to 10% final concentration. Samples are vortexed and kept on ice for 1 hour. Precipitates are recovered by centrifugation at 12,000g for 10 minutes, washed with 90% acetone, dried, and prepared for electrophoresis using standard protocols.

mRNP and ribosomal proteins are well separated on 18% acrylamide SDS-gels (Thomas and Kornberg 1975). For two-dimensional gel electrophoresis, NEPHGE (O'Farrell et al. 1977) gives the best separation of mRNP3+4 isoforms (Murray et al. 1991).

Photo-cross-linking of mRNPs

Photo-cross-linking of an RNA–protein complex results in covalent adduct formation at sites of amino acid proximity to a nucleic acid base. The aromatic residues phenylalanine, histidine, and tyrosine are the most photoreactive (Smith 1969). mRNP3+4 efficiently photo-cross-link to mRNA in vivo (Swiderski and Richter 1988) and in vitro (Murray et al. 1991; Marello et al. 1992). This is undoubtedly due to the presence of multiple phenylalanine and tyrosine residues in the four basic/aromatic islands of mRNP3+4 that bind RNA (Murray 1994). Masked mRNPs can be photo-cross-linked in vivo using stage 1 oocytes that do not contain yolk platelets or pigment granules (Swiderski and Richter 1988). More typically, in vitro transcribed radiolabeled mRNA is microinjected into oocytes for protein binding (Bouvet and Wolffe 1994) or incubated with cellular extracts for in vitro protein binding (Murray et al. 1991). High specific activity ^{32}P-radiolabeled mRNA is prepared by in vitro transcription using standard protocols (Wormington 1991). The use of 5-bromouridine-substituted RNA increases the efficiency of photo-cross-linking to mRNP3+4 more than 100-fold (Murray et al. 1991). In vitro transcription of 5-bromouridine-substituted RNA requires complete replacement of UTP because phage polymerases discriminate heavily against 5-bromoUTP (unpublished observation); in this case, the radiolabeled nucleotide triphosphate is CTP.

Photo-cross-linking of mRNA bound by mRNP3+4 from soluble extracts of oocytes, ovary, eggs, or embryos is highly efficient and well characterized (Murray et al. 1991). PMS is prepared from microinjected oocytes for UV irradiation, while in vitro binding samples are irradiated directly. Photo-cross-linking is best achieved with a 30-minute exposure with 312 nm light; however, a 90-second exposure with 254 nm light is also efficient. The photo-cross-linked mixture is then treated with RNase and the radiolabeled protein is separated by SDS-gel electrophoresis. Digestion of the photo-cross-linked RNA–protein complex using both RNase A and RNase T1 (0.34 mg/ml and 500 U/ml, respectively, at 37°C for 30 minutes) results in near-normal migration of the photo-cross-linked protein, with only a slight change in mobility detected by two-dimensional electrophoresis (unpublished observation). This suggests that only a few nucleotides remain bound to the polypeptide.

Purification of mRNPs by Affinity Chromatography

Masked poly(A)$^+$ mRNPs are bound by mRNP protein in the heterogeneous region of the mRNA, leaving the poly(A) tail accessible for hybridization to oligo(dT) cellulose (Darnbourgh and Ford 1981). In addition, *Xenopus* oocytes have a low concentration of poly(A) binding protein (PABP), so that most poly(A) is unbound in the oocyte (Stambuk and Moon 1992; Wormington et al. 1996). Oligo(dT)-cellulose affinity chromatography can provide an initial fractionation of mRNP proteins, but one must consider the possibility that polyribosomal mRNPs may also copurify due to the low endogenous concentration of PABP.

Poly(A)$^+$ mRNPs are bound to oligo(dT) cellulose in NTE equilibration buffer, then washed successively with 25% formamide buffer, 50% formamide buffer, and finally with a buffer that does not contain salt (Lindberg and Sunquist 1974; Darnbourgh and Ford 1981; Richter and Smith 1983). Under these conditions, the major mRNP proteins elute with the 50% formamide wash. The presence of 10 mM EDTA in the equilibration buffer is critical for assuring that retention is due to nucleic acid hybridization rather than direct protein binding to the matrix because mRNP3+4 readily bind single-stranded DNA in the presence of Mg^{2+} (Murray 1994). Poly(U) Sepharose can be used in place of oligo(dT) cellulose using the same formamide wash buffers (Cummings and Sommerville 1988); however, 90% formamide is used as the final elution buffer.

To investigate less tightly associated mRNP proteins, alternative elution protocols for oligo(dT) chromatography employ a salt gradient (0.2 M to 1.0 M NaCl; Ladomery et al. 1997). In this method, mRNA and mRNP3+4 are eluted either with water (or buffer without NaCl) or 60% buffered formamide (Cummings and Sommerville 1988; Marello et al. 1992).

ACKNOWLEDGMENTS We thank Drs. Markku Kurkinen, Allen Nicholson, and John Tomkiel (WSU) for helpful comments on the manuscript. This work was supported by the National Science Foundation (MCB-9513751).

References

Bouvet, P., and Wolffe, A. P. (1994). A role for transcription and FRGY2 in masking maternal mRNA within Xenopus oocytes. Cell 77:931–941.

Cardinali, B., Campioni, N., and Pierandrei-Amaldi, P. (1987). Ribosomal protein, histone and calmodulin mRNA are differentially regulated at the translational level during oogenesis of *Xenopus laevis*. Exp. Cell Res. 169:432–441.

Cummings, A., and Sommerville, J. (1988). Protein kinase activity associated with stored messenger ribonucleoprotein particles of Xenopus oocytes. J. Cell Biol. 107:45–56.

Darnbourgh, C. H., and Ford, P. J. (1981). Identification in *Xenopus laevis* of a class of oocyte-specific proteins bound to messenger RNA. Eur. J. Biochem. 113:415–424.

Deschamps, S., Viel, A., Garringos, M., Denis, H., and le Marie, M. (1992). mRNP4, a major mRNA-binding protein from Xenopus oocytes is identical to transcription factor FRGY2. J. Biol. Chem. 267:13799–13802.

Dolecki, G. J., and Smith, L. D. (1979). Poly(A)$^+$ RNA metabolism during oogenesis in *Xenopus laevis*. Dev. Biol. 69:217–236.

Houssais, J. F. (1983). Fractionation of ribonucleoproteins from eukaryotes and prokaryotes. In Iodinated Density Gradient Media, Rickwood, D., ed. Oxford, IRL Press, pp. 43–67.

Kick, D., Barrett, P., Cummings, A., and Sommerville, J. (1987). Phosphorylation of a 60 kDa polypeptide from Xenopus oocytes blocks messenger RNA translation. Nucleic Acids Res. 15:4099–4109.

Kolluri, R., Torrey, T. A., and Kinniburgh, A. J. (1992). A CT promoter element binding protein: Definition of a double-stranded and novel single-stranded DNA binding motif. Nucleic Acids Res. 20:111–116.

Ladomery, M., and Sommerville, J. (1994). Binding of Y-box proteins to RNA: involvement of different protein domains. Nucleic Acids Res. 22:5582–5589.

Ladomery, M., Wade, E., and Sommerville, J. (1997). Xp54, the Xenopus homologue of human RNA helicase p54, is an integral component of stored mRNP particles in oocytes. Nucleic Acids Res. 25:965–973.

Landsman, D. (1992). RNP-1, an RNA-binding motif is conserved in the DNA-binding cold shock domain. Nucleic Acids Res. 20:2861–2864.

Lindberg, U., and Sunquist, B. (1974). Isolation of messenger riobonucleoproteins from mammalian cells. J. Mol. Biol. 86:451–468.

Marello, K., LaRovere, J., and Sommerville, J. (1992). Binding of Xenopus oocyte masking proteins to mRNA sequences. Nucleic Acids Res. 21:5593–5600.

Matsumoto, K., Funda, M., and Wolffe, A. P. (1996). Translational repression dependent on the interaction of the Xenopus Y-box protein FRGY2 with mRNA. J. Biol. Chem. 271:22706–22712.

Meric, F., Matsumoto, K., and Wolffe, A. P. (1997). Regulated unmasking of in vivo synthesized maternal mRNA at oocyte maturation. A role for the chaperone nucleoplasmin. J. Biol. Chem. 272:12840–12846.

Murray, M. T. (1994). Nucleic acid-binding properties of the Xenopus oocyte Y box protein mRNP3 + 4. Biochem. 33:13910–13917.

Murray, M. T., Krohne, G., and Franke, W. W. (1991). Different forms of soluble cytoplasmic mRNA binding proteins and particles in *Xenopus laevis* oocytes and embryos. J. Cell Biol. 112:1–11.

Murray, M. T., Schiller, D. L., and Franke, W. W. (1992). Sequence analysis of cytoplasmic mRNA-binding proteins of Xenopus oocytes identifies a family of RNA-binding proteins. Proc. Natl. Acad. Sci. USA 89:11–15.

O'Farrell, P. Z., Goodman, H. M., and O'Farrell, P. H. (1977). High resolution two-dimensional electrophoresis of basic as well as acidic proteins. Cell 12:1133–1142.

Ranjan, M., Tafuri, S. R., and Wolffe, A. P. (1993). Masking mRNA from translation in somatic cells. Genes Dev. 7:1725–1736.

Richter, J. D., and Smith, D. L. (1983). Developmentally regulated RNA binding proteins during oogenesis of *Xenopus laevis*. J. Biol. Chem. 258:4864–4869.

Richter, J. D., and Smith, L. D. (1984). Reversible inhibition of translation by *Xenopus* oocyte-specific proteins. Nature 309:378–380.

Smith, K. C. (1969). Photochemical addition of amino acids to ^{14}C-uracil. Biochem. Biophys. Res. Comm. 34:345–357.

Smith, L. D., Weilong, X., and Varnold, R. L. (1991). Oogenesis and oocyte isolation. In *Methods in Cell Biology. Xenopus laevis: Practical Uses in Cell and Molecular Biology*, Kay, B. K., and Peng, H. B., eds. San Diego, Academic Press, pp. 45–60.

Spirin, A. S. (1969). Informosomes. Eur. J. Biochem. 10:20–35.

Spirin, A. S. (1996). Masked and translatable messenger ribonucleoproteins in higher eukaryotes. In *Translational Control*, Hershey, J. W. B., Mathews, M. B., and Sonenberg, N., eds. Cold Spring Harbor, NY, Cold Spring Harbor Laboratory Press, pp. 319–334.

Spirin, A. S., Belitsina, N. V., and Lerman, M. I. (1965). Use of formaldehyde fixation for studies of ribonucleoprotein particles by caesium chloride density-gradient centrifugation. J. Mol. Biol. 14:611–615.

Stambuk, R. A., and Moon, R. T. (1992). Purification and characterization of recombinant Xenopus poly(A)$^{+}$-binding protein expressed in a baculovirus system. Biochem. J. 287:761–766.

Swiderski, R. E., and Richter, J. D. (1988). Photocrosslinking of proteins to maternal mRNA in Xenopus oocytes. Dev. Biol. 128:349–358.

Tafuri, S. R., and Wolffe, A. P. (1992). DNA binding, multimerization, and transcription stimulation by the Xenopus Y box proteins in vitro. The New Biologist 4:349–359.

Tafuri, S. R., and Wolffe, A. P. (1993). Selective recruitment of masked maternal mRNA from mRNPs containing FRGY2(mRNP4). J. Biol. Chem. 268:24255–24261.

Thomas, J. O., and Kornberg, R. D. (1975). An octomer of histones in chromatin and free in solution. Proc. Natl. Acad. Sci. USA 72:2626–2630.

Wallace, R. A., and Misulovin, Z. (1978). Long-term growth and differentiation of Xenopus oocytes in a defined growth medium. Proc. Natl. Acad. Sci. USA 75:5534–5538.

Wolffe, A. P. (1994). Structural and functional properties of the evolutionarily ancient Y-box family of nucleic acid binding proteins. BioEssays 16:245–251.

Woodland, H. R. (1974). Changes in the polysome content of developing Xenopus laevis embryos. Dev. Biol. 40:90–101.

Wormington, M. (1991). Preparation of synthetic mRNAs and analyses of translational efficiency in microinjected Xenopus oocytes. In *Methods in Cell Biology. Xenopus laevis: Practical Uses in Cell and Molecular Biology*. Kay, B. K., and Peng, H. B., eds. San Diego, Academic Press, pp. 167–183.

Wormington, M., Searfoss, A. M., and Hurney, C. A. (1996). Overexpression of poly(A) binding protein prevents maturation-specific deadenylation and translational inactivation in Xenopus oocytes. EMBO J. 15:900–909.

Yurkova, M. S., and Murray, M. T. (1997). A translation regulatory particle containing the Xenopus oocyte Y box protein, mRNP3+4. J. Biol. Chem. 272:10870–10876.

Appendix: Solutions and Buffers

Stringent precautions against RNase contamination of solutions and glassware should be taken. All chemicals should be of the highest quality, preferably certified RNase free. Stock solutions of dithiothreitol (DTT) and cycloheximide cannot be autoclaved but should be sterilized by filtering through 0.2 mm sterile filters. Equipment for gradient preparation and fractionation that cannot be autoclaved should be treated with RNase Zap (Ambion, Inc., Austin) following the manufacturer's procedure. Nycodenz is distributed in the USA by Life Technologies Inc. (Gaithersburg).

Polysome Buffer (PB)
Core PB is 300 mM KCl, 2 mM $MgCl_2$, 10 mM Tris-HCl, pH 7.5, 0.5% (v/v) Nonidet P-40.
Autoclave for 15 minutes at 15 psi and store at 4°C.
For final PB, add DTT to 1 mM final concentration just before use.*

Low-Salt Buffer (LSB)
Core LSB is 50 mM NaCl, 2 mM $MgCl_2$, 10 mM Tris-HCl, pH 7.5, 0.5% (v/v) Nonidet P-40.
Autoclave for 15 minutes at 15 psi and store at 4°C.
For final LSB add DTT to 5 mM final concentration just before use.*

Glycerol Gradient Stock Solutions
15% (w/v) and 40% (w/v) in PB or LSB.
Prepare in core PB or LSB buffer, autoclave for 15 minutes at 15 psi and store at 4°C.
Before using, add DTT to the appropriate final concentration.*

Sucrose Gradient Stock Solutions
15% (w/v) and 50% (w/v) in PB or LSB.
Prepare in core PB or LSB buffer, autoclave for 15 minutes at 15 psi and store at 4°C.
Before using add DTT to the appropriate final concentration.*

Nycodenz Gradient Stock Solution
60% (w/v) in PB or LSB.
Prepare in core PB or LSB buffer, autoclave for 15 minutes at 15 psi and store at 4°C.
Before using add DTT to the appropriate final concentration.*

RNA Extraction Buffer (TNES)
100 mM Tris-HCl, pH 7.5, 300 mM NaCl, 10 mM EDTA, 2% SDS.
Autoclave for 15 minutes at 15 psi and store in aliquots at room temperature.

---

*Cycloheximide (10 µg/ml), RNase inhibitor (100 U/ml placental ribonuclease inhibitor or 30 U/ml PRIME, 5 Prime-3 Prime Inc., Boulder), and/or protease inhibitors (5 µg/ml leupeptin, 2 µg/ml aprotinin, 0.5 mM phenylmethylsulfonyl fluoride) may also be added depending on the experiment.

Oligo(dT) Equilibration Buffer (NTE)
0.2 M NaCl, 50 mM Tris-HCl, pH 7.8, 10 mM EDTA, 0.2% Nonidet-40.
Autoclave for 15 minutes at 15 psi and store in aliquots at room temperature.

Oligo (dT) Cellulose Wash Buffer 1
25% formamide in NTE.
Deionize formamide and mix with sterile NTE, store in aliquots at $-20°C$.

Oligo(dT) Cellulose Wash Buffer 2
50% formamide in NTE.
Deionize formamide and mix with sterile NTE, store in aliquots at $-20°C$.

Oligo(dT) Cellulose Elution Buffer
10 mM Tris-HCl, pH 7.5, 1 mM EDTA, 0.1% SDS.
Autoclave for 15 minutes at 15 psi and store in aliquots at room temperature.

Poly(U) Sepharose Elution Buffer
90% formamide, 10 mM Tris-HCl, pH 7.5, 1 mM EDTA, 0.1% SDS.
Deionize formamide and mix with sterile oligo(dT) elution buffer. Store in aliquots at $-20°C$.

13

Complications and Problems in the Assay of Transcription in *Xenopus* Oocytes, Eggs, and Embryos

Alan P. Wolffe

Xenopus oocytes and embryos provide extraordinarily useful experimental systems for the assay of transcription and translation. Practically any metazoan promoter can facilitate transcription following microinjection into an oocyte nucleus, and most mRNAs are translated following microinjection into oocyte cytoplasm. An increasing number of investigators make use of the capacity of the oocyte to assemble exogenous DNA into chromatin to explore how the packaging of regulatory elements by the histones contributes to the control of gene expression (Almouzni and Wolffe 1993; Landsberger and Wolffe 1995; Wong et al. 1995; Kass et al. 1997; Wong et al. 1997). Likewise, the packaging of mRNA into ribonucleoprotein complexes within the oocyte nucleus and cytoplasm influences translational fate (Bouvet et al. 1995; Matsumoto et al. 1996; Meric et al. 1996, 1997). Importantly, the transcription of an mRNA in the oocyte nucleus can influence the translation of that mRNA in the cytoplasm (Bouvet and Wolffe 1994). The discovery of these novel regulatory principles creates several complications and problems for the analysis of gene expression in the *Xenopus* oocyte system. Nevertheless, recognition of the issues involved can provide a great deal of useful information concerning how a particular gene might be controlled in vivo.

Many of the regulatory principles controlling gene expression in the oocyte also operate during *Xenopus* embryogenesis; however, the analysis of transcriptional mechanisms becomes complicated by the imposition of developmentally regulated transitions in chromatin structure (Dimitrov et al. 1993, 1994). An example of this is the capacity of certain histone modifications, such as acetylation, to influence transcription changes during development (Almouzni et al. 1994). There are also developmentally regulated constraints on the activities of the transcriptional machinery (Wolffe and Brown 1987; Almouzni and Wolffe 1995; Landsberger and Wolffe 1997). Once again, recognition of these complications can provide many opportunities in elucidating molecular mechanisms of gene control.

Assays for Transcription in Oocytes

Microinjection of many diverse genes into oocyte nuclei will result in their transcription by the appropriate RNA polymerase (Brown and Gurdon 1977; McKnight and Kingsbury 1982; Reeder et al. 1983). For transcription directed by RNA polymerase II, the method of choice for determining both transcription start site and the abundance of a particular mRNA that is being synthesized is primer extension. This technique has the advantage of providing direct quantitation of the abundance of accurately initiated transcripts. The use of reporter genes, such as chloramphenicol acetyl transferase (CAT), β-galactosidase, and luciferase, is helpful because these cDNA sequences allow the use of primers to quantitate mRNAs that do not crosshybridize with endogenous mRNAs in the oocyte. However, as we will discuss later, the transcription of mRNA in vivo will influence translational fate in the cytoplasm. Therefore assays for CAT, β-galactosidase, and luciferase activity may not reflect the level of mRNA synthesized by a particular promoter. Moreover, the deletion of the normal gene sequence might remove important transcriptional regulatory elements downstream of the start site of transcription (Wong et al. 1995).

The transcriptional control of templates injected into oocyte nuclei is sensitive to the assembly of DNA into chromatin. Therefore, the mass of DNA can influence transcriptional efficiency (Landsberger and Wolffe 1995; Landsberger et al. 1995), as can the pathway of chromatin assembly (Almouzni and Wolffe 1993; Wong et al. 1995). The more DNA that is injected into an oocyte, the less that DNA will be assembled into chromatin and the more promiscuously will the promoter be utilized (Landsberger et al. 1995). Double-stranded DNA is assembled into chromatin with relatively slow kinetics compared with single-stranded DNA undergoing second-strand synthesis in the oocyte nucleus (Almouzni and Wolffe 1993). Therefore, transcription of double-stranded DNA is relatively promiscuous compared with replicating single-stranded templates. These results can be illustrated using the following protocols and experiments.

Oocyte Injections, RNA and DNA Analysis methods

Stage 6 oocytes are collected and defolliculated in 0.2% collagenase (Sigma) in calcium-free OR2 (82.5 mM NaCl, 2.5 mM KCl, 1 mM $MgCl_2$, 1 mM Na_2HPO_4, 500 mM HEPES, pH7.8). After defolliculation, follicle-free oocytes are selected under a microscope and washed free of collagenase in OR2. They are then incubated overnight at 18°C. The following day, healthy oocytes retaining clear animal and vegetal pole pigmentation are collected and injected with 30 nl of a DNA solution containing either double-stranded or single-stranded templates at different concentrations (as indicated in the figure legends). The oocytes are then allowed to recover at 18°C for varying periods of time, but typically an overnight incubation of 8–10 hours is sufficient to allow abundant transcripts to accumulate. The oocytes are then collected and homogenized in 20 mM Tris-HCl (pH 8.0, 10 µl per oocyte). This homogenate is processed for both DNA and RNA analysis. From half of the sample, RNA is extracted using the RNAzol TM method (TM, Cinna Scientific). The RNA equivalent to three to four oocytes is analyzed by primer extension (Toyoda and Wolffe 1992). Transcripts are annealed with 0.05 pmol of $5'$ ^{32}P-labeled primer (2×10^5 cpm) in 10 µl of 20 mM Tris-HCl, pH 8.0, 0.2 mM EDTA, 0.25 M KCl at 65°C for 5 minutes and at 55°C for 20 minutes, before cooling to room temperature. The annealed primer is elongated using 2 U of avian myeloblastosis virus reverse transcriptase (Promega) in 30 µl of 20 mM Tris-HCl, pH 8.0, 80 mM KCl, 8 mM $MgCl_2$, 80 µg/ml of actinomycin D (BMB), 10 mM DTT, 0.4 mM of each of four deoxynucleoside $5'$ triphosphates (BMB), and 10 U of human placental RNase inhibitor (Gibco BRL) at 37°C for 1 hour. The reaction is stopped by the addition of 150 µl of ethanol. The nucleic acid is precipitated and washed in 70% ethanol, dried, and dissolved in 10 µl of formamide dye buffer. The solution is boiled for 2 minutes, chilled on ice, and then electrophoresed on a 6% polyacrylamide-urea gel. The particular primer used depends on the transcribed sequence. It is important to have an extension product that is clearly resolved from the radiolabeled primer. Extension products of 80–180 bp, dependent on the particular primer chosen, are typical. For DNA analysis, the samples are incubated for 1 hour at 37°C in 15 mM EDTA, 20 mM Tris (pH 8.0), 0.5% SDS, and 500 µg/ml proteinase K. The DNA is extracted twice with phenol/chloroform (1:1) and ethanol-precipitated. After RNase A treatment, the samples are subjected to 1% agarose gel electrophoresis, transferred on a Hybond N membrane (Amersham), and hybridized against a nick-translated hsp70 CAT plasmid utilizing the Rapid-hyb buffer (Amersham). The probes are prepared utilizing a random primed labeling kit (Biolabs). Assay of the mass and topology of DNA provides an essential control for efficient and successful microinjection of DNA templates into the oocyte nucleus and for the degree of chromatin assembly. Each nucleosome assembled on a closed circular plasmid DNA molecule introduces a single negative superhelical turn. Therefore, the degree of supercoiling indicates the number of nucleosomes assembled onto the template.

Illustrative Results

Xenopus have a heat shock response (Wolffe et al. 1984) that can be assayed in oocyte nuclei by heat-inducible transcription of the Xenopus hsp70 promoter (Landsberger and Wolffe 1995; Landsberger et al. 1995). However, observation of this response depends on the efficiency of chromatin assembly. Early attempts to observe heat-inducible transcription in oocytes failed because of inadequate chromatin assembly (Bienz 1984). Microinjection of 3 ng of the hsp70 template into oocytes, together with a control human cytomegalovirus immediate/early enhancer/promoter (CMV), reveals strong heat-inducible transcription (Figure 13-1, compare lanes 1 and 2). Transcription of the hsp70 promoter is so strong under heat shock conditions (34°C) that it competes with the CMV promoter for components of the basal transcriptional machinery (compare lanes 1 and 2). As increasing amounts of nonspecific DNA are injected to titrate out chromatin components, heat-inducible transcription is lost (Figure 13-1, lanes 3–6). The *Xenopus* hsp70 promoter becomes constitutively active at 18°C and 34°C (Bienz 1984; Landsberger et al. 1995). The reduction in chromatin assembly is documented by the Southern blot (Figure 13-1B), showing the progressive decline in fully supercoiled form I DNA with the microinjection of increasing amounts of nonspecific DNA (compare lanes 1 and 4). Therefore, chromatin assembly can exert a strong influence on the transcriptional regulation of templates microinjected into Xenopus oocytes. Experimental conclusions can differ dramatically depending on the mass of template injected (Bienz 1984; Landsberger and Wolffe 1995).

The influence of the pathway of chromatin assembly on transcription can be illustrated using the *Xenopus* thyroid hormone receptor βA promoter (TRβA, Figure 13-2A). This particular experiment shows that the microinjection of single-stranded DNA, followed by replication-coupled chromatin assembly, leads to a dramatic chromatin-mediated repression of transcription (Figure 13-2B, lanes 3 and 4), whereas the microinjection of double-stranded DNA results in high levels of basal transcription (Figure 13-2B, lanes 1 and 2). These differences occur even though the DNA recovery demonstrates approximately equal masses of DNA, with comparable DNA topologies, in the oocyte nucleus (Figure 13-2C, lanes 3–6). Obviously, a much wider range of gene regulation will potentially be achieved from the chromatin template that initially gives a low level of basal transcription before gene activation (Wong et al. 1995). It is important to note that for both the hsp70 and the TRβA promoter, double-stranded DNA microinjected into the oocyte retains substantial transcriptional activity even though it is packaged into chromatin (Figures 13-1A and 13-2B). This is due to the slow kinetics of nucleosome assembly on nonreplicative

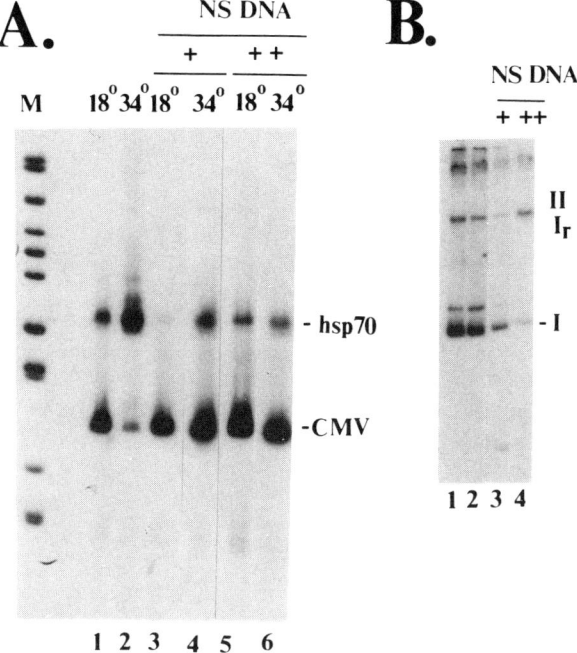

Figure 13-1 The *Xenopus* hsp70 promoter responds to heat shock of *Xenopus* oocytes under conditions that promote chromatin assembly. *Xenopus* oocytes were injected with 3 ng of a mixture of double-stranded wild-type hsp70 CAT and CMV constructs. They were then incubated for 4 hours at 18°C before exposure to heat shock at 34°C for 2 hours or maintenance at 18°C throughout this time, as indicated. RNA and DNA were extracted. Increasing amounts of bacteriophage λ DNA (nonspecific, NS DNA) were mixed with the 3 ng of template DNAs. (A) Transcription at 18°C (lanes 1, 3, and 5) or 34°C (lanes 2, 4, and 6), as indicated, from hsp70 or CAT promoters. Lanes 1 and 2 have no NS DNA, lanes 3 and 4 have 6 ng (+) and lanes 5 and 6 have 12 ng (++) (Almouzni and Wolffe 1993). (B) DNA analysis. DNA samples were resolved on a 1% agarose gel and blotted. Hybridization to the hsp70 construct is shown. Lane 1 shows DNA isolated from the nuclei of oocytes maintained at 18°C, lane 2 shows DNA isolated from nuclei heat shocked at 34°C, lane 3 shows DNA in the presence of 6 ng of NS DNA (+) at 18°C, lane 4 shows DNA in the presence of 12 ng of NS DNA (++) at 18°C. The various DNA forms are indicated: double-stranded form I (closed circular supercoiled), form II (nicked circle), form I_r (closed circular relaxed).

substrates relative to replicating DNA. These slow kinetics allow the basal transcriptional machinery to compete effectively with histones for association with regulatory DNA (Almouzni and Wolffe 1993). For these regulated promoters (hsp70 and TRβA) the high levels of basal transcription obtained using double-stranded templates is not representative of their normal

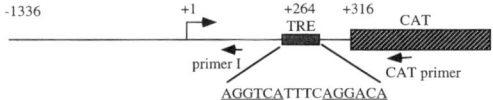
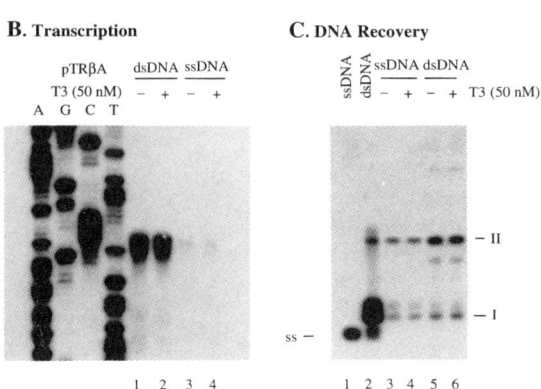

Figure 13-2 Efficient T3-independent transcription of double-stranded but not single-stranded TRβA promoter when injected into *Xenopus* oocytes. (A) Structure of the TRβA promoter construct. The transcriptional start site in oocytes is designated as +1 (see B). A thyroid hormone response element (TRE) identified is located at +264 (underlined). Also indicated are the relative positions of primer I (+107) and CAT primer (+397), which were used for primer extension analysis. (B) Replication-coupled chromatin assembly effectively represses the transcription from the TRβA promoter. Forty oocytes were injected into the nuclei with double-stranded (ds; 2.3 ng/oocyte) or single-stranded (ss; 1.15 ng/oocyte) pTRβA templates. The injected oocytes were then divided into two groups with equal numbers of oocytes and incubated at 18°C overnight with (+) or without (−) 50 nM of T3. The transcription from the TRbA promoter was analyzed by primer extension using primer I. DNA sequencing markers using the TRβA promoter and end-labeled primer I indicate the +1 as the start site. (C) Recovery of DNA after either ss or ds DNA; are injected into oocytes. DNAs recovered from the groups of oocytes in (B) were resolved by a 1% agarose gel, blotted, and hybridized with random primer-labeled pTRβA. (Lane 1) Uninjected ss DNA; (lane 2) uninjected ds DNA. Double-stranded form I (closed circle supercoiled), form II (nicked circular), and ss DNA (ss) are indicated.

physiological activity state in the chromosomes of oocytes or somatic cells. Other promoters that are constitutively active in the oocyte chromosome may have high levels of transcription in oocytes, reflecting the imposition of tissue-specific control (see Cho and Wolffe 1994); however, this is the exception rather than the general rule.

Problems with the Use of Reporter Genes in Oocytes

In somatic cells, the use of reporter genes such as CAT, β-galactosidase, and luciferase has greatly simplified the assay of promoter activity. These assays rely upon the assumption that a linear relationship exists between the synthesis of mRNA encoding the various enzymes and the translation of that mRNA into protein. In *Xenopus* oocytes, this is not necessarily true (Bouvet and Wolffe 1994; Matsumoto et al. 1996; Meric et al. 1996, 1997).

The transcription of a particular mRNA from plasmid DNA molecules microinjected into *Xenopus* oocyte nuclei leads to very inefficient translation compared with the same mRNA microinjected into *Xenopus* oocyte cytoplasm (Bouvet and Wolffe 1994). In Figure 13-3A, CAT mRNA is either synthesized in vivo following microinjection of a template in which the CMV regulatory elements are fused to a CAT reporter gene into the nucleus, or it is produced from capped and polyadenylated CAT mRNA that is synthesized in vitro and microinjected into oocyte cytoplasm (Krieg and Melton 1984). The efficiency with which CAT protein is synthesized is shown for both mRNA synthesized in vivo and that synthesized in vitro (Figure 13-3B). CAT activity was also assayed using the methodology of Nordeen et al. (1987). These results clearly indicate that the use of CAT assays to quantitate the activity of a promoter microinjected into Xenopus oocyte nuclei is insensitive, at best, and that a linear relationship between mRNA abundance and translation does not necessarily occur (Meric et al. 1996). Comparable results have been obtained with all the commonly used reporter genes (F. Meric and A. P. Wolffe, unpublished observations).

Work from the Kingsman laboratory has also suggested that particular promoter sequences and pre-mRNA splicing can also influence translational fate (Braddock et al. 1994; Gunkel et al. 1995). Therefore considerable caution must be used in determining the significance of individual promoter mutations for transcriptional activity if the assays rely only on the translation of mRNA synthesized in vivo as an indicator of mRNA abundance.

Assays for Transcription in Eggs and Embryos

In *Xenopus*, oogenesis is characterized by vigorous transcriptional activity; however, maturation of the oocyte into an unfertilized egg leads to transcriptional quiescence. This physiological transition is recapitulated on DNA templates injected into the oocyte nucleus (Landsberger and Wolffe 1997). The molecular basis for this phenomenon involves a significant increase in the efficiency of chromatin assembly in the egg compared with the oocyte. Therefore, the dependence of transcription on the mass of DNA present becomes more extreme in eggs and embryos in comparison with oocytes (Figure 13-4). Just as in the oocyte, titration of chromatin components through the injection of a high DNA mass correlates with higher levels of constitutive transcription.

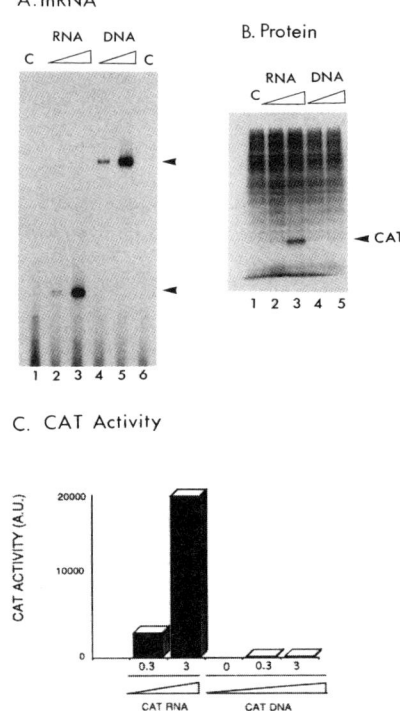

Figure 13-3 In vivo transcribed mRNA is masked. Oocytes are injected with in vitro transcribed CAT mRNA into the oocyte cytoplasm or with plasmid DNA encoding for CAT mRNA into the nucleus. After a 16-hour incubation, oocytes were injected with [^{3}H]arginine and [^{3}H]lysine and incubated for another 6 hours. (A) Primer extension analysis of the steady-state level of CAT mRNA in oocytes injected with 0.3 ng and 3 ng of CAT mRNA (lanes 2 and 3, respectively), or 0.3 ng and 3 ng of CAT DNA (lanes 4 and 5, respectively). Oocytes injected with amino acids alone were used as control (lane C). Arrows indicate transcripts. (B) Radiolabeled proteins synthesized in these oocytes were resolved by SDS–PAGE. The CAT protein is indicated with an arrow. (C) CAT activity corresponding to the oocytes injected in (A) is shown.

Following fertilization, the cells of the embryo undergo 11 rapid cleavage cycles in the absence of significant transcription. When the embryo reaches the mid-blastula transition (MBT), zygotic transcription of many genes is activated (Newport and Kirschner 1982; Rupp and Weintraub 1991). Specialized programs of gene activity are subsequently established within particular cell lineages, depending on the segregation and accumulation of particular transcription factors. Many investigators have sought to explore the expression of exogenous DNA templates in embryos or to manipulate

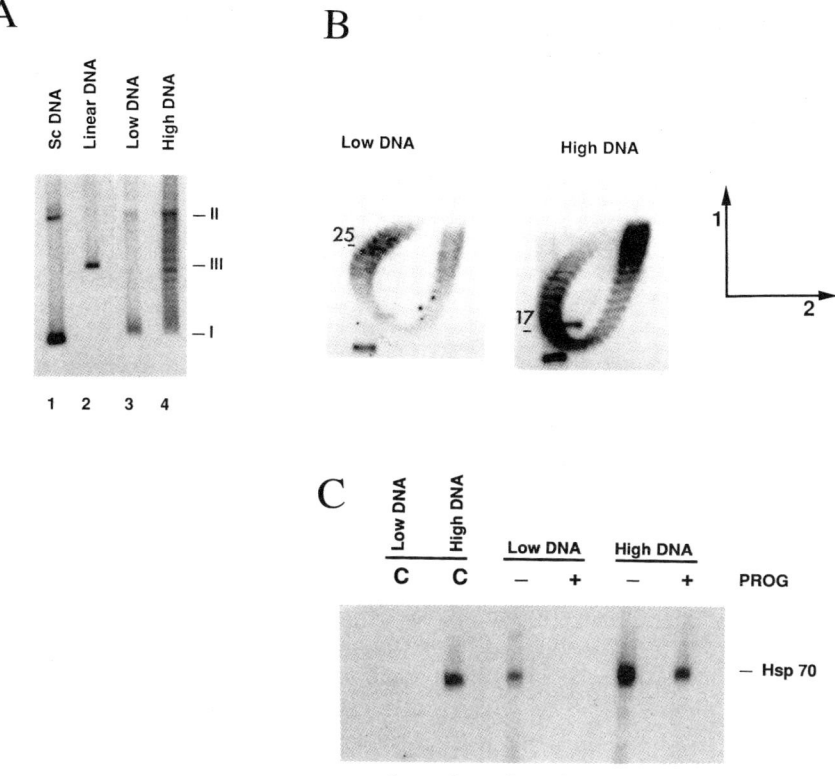

Figure 13-4 The titration of repressive chromatin components alleviates the transcriptional repression obtained after maturation. (A, B, and C) Oocyte nuclei were injected with either a low mass (1 ng) or a high mass (15 ng) of HSPCAT DNA (Landsberger and Wolffe 1997). (A) DNA was extracted from the injected oocytes and resolved on a 1% agarose gel. After transfer, DNA was probed with a random primed labeled HSPCAT vector. Markers of supercoiled and linear DNA are resolved, respectively, in lanes 1 and 2. Lane 3 shows the DNA extracted from oocytes injected with a low mass. Lane 4 contains DNA purified from oocytes injected with a high mass. The positions of supercoiled (I), nicked closed circular (II), and linear DNA (III) are indicated. (B) The purified DNAs (low DNA and high DNA) were resolved on a two-dimensional agarose gel containing chloroquine (Landsberger and Wolffe 1997). DNA was transferred and probed as described in (A). The arrows indicate the two dimensions of electrophoresis. In each panel, the number indicates the average number of negative supercoils. (C) RNA was extracted from the injected oocytes and analyzed by primer extension. Lanes 1 and 2 (C, control) show the amount of transcripts accumulated in the first 2 hours after the DNA injection in the absence of progesterone. In lanes 3 and 5, oocytes were incubated in the absence of progesterone. Lanes 4 and 6 contain oocytes incubated in the presence of progesterone. The position of accurately initiated transcripts from the hsp70 (Hsp70) promoter is indicated.

their expression through expression of transcription factors (Sargent and Mathers 1991; Vize et al. 1991). The results of these studies are, again, influenced by the process of chromatin assembly and the transcriptional environment of the developing embryonic cells (Almouzni and Wolffe 1995).

Embryo Injections, RNA and DNA Analysis

Embryo injection is carried out during the first 30 minutes of the first cell cycle. Zygotes are prepared by artificial fertilization and dejellied in 2% cysteine. Injection is of 300 pg of template DNA, supplemented with 25 ng of λ DNA as indicated, in a total volume of 40 nl of injection buffer. During the injection process, embryos are positioned on a special slide chamber (Medical Specialties, Baltimore, MD) in 1 × MMR (100 mM NaCl, 2 mM KCl, 2 mM $CaCl_2$, 1 mM $MgCl_2$, 5 mM HEPES-KOH, pH 7.4, in 5% Ficoll). The embryos are transferred to 0.1 × MMR at 23°C until they reached the desired stage.

To assess the level of transcript accumulated per DNA template, the injected embryos are collected and suspended in 20 mM Tris-HCl (pH 8; 15 ml per embryo). Half of the sample is made up to 30 mM EDTA, 1% SDS, and processed for DNA analysis; the remaining sample is processed using the RNAzol TM method (TM Cinna Scientific) to isolate RNA, followed by a LiCl precipitation to remove glycoproteins and yolk contaminants (Maniatis et al. 1982). Additional treatment with RNase-free DNase I (Boehringer, Indianapolis, IN) is performed. Generally, the RNA equivalent of five embryos is analyzed for the presence of specific transcripts by primer extension (Toyoda and Wolffe 1992) on 6% polyacrylamide gels containing 8 M urea in TBE buffer. For DNA analysis, the template is treated with proteinase K (500 µg/ml) for ≥2 hours. After deproteinization by phenol-extraction and ethanol-precipitation, the samples are treated with RNase A, phenol-extracted, and DNA-precipitated with ethanol. The samples are then subjected to 1% agarose gel electrophoresis in TAE buffer. After electrophoresis, the gels are stained with ethidium bromide and genomic DNA replication can be estimated by visualization under UV light. Gels are then processed for blotting to probe for the injected DNA templates and to control for the exact recovery and topological state of the template (Maniatis et al. 1982). Probes were made using a random primed labeling kit (Boehringer). Quantitation of transcripts and DNA recovery were carried out with a PhosphorImager (Molecular Dynamics).

Illustrative Results

Embryos are much more fragile than oocytes and every care should be taken not to perturb normal development. The quantity of DNA typically injected (300 pg) is much less than the amount present at the MBT per embryo (24 ng). Under these conditions, normal embryonic development ensues and the

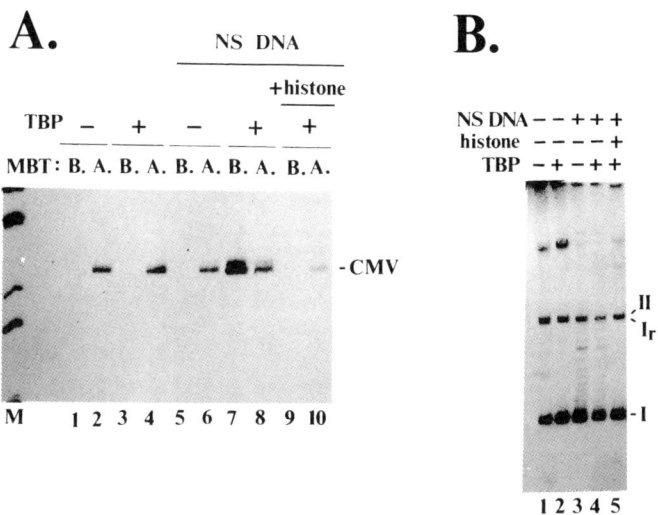

Figure 13-5 CMV transcription in the embryo, the effect of exogenous TBP on transcription, and the consequences of the titration of chromatin with nonspecific DNA and core histone supplementation. Closed circular, double-stranded supercoiled CMV template DNA (300 pg) was injected into each embryo. After injection, before the first cleavage, the embryos were incubated in 1 × MMR, 5% Ficoll and transferred after ~1 hour in 0.1 × MMR. Embryos were collected either before MBT (B), after 2 hours incubation at 22°C, or after MBT (A) after 8 hours incubation at 22°C. Titration of chromatin assembly with nonspecific DNA (NS) was achieved using 25 ng of bacteriophage λ DNA per embryo. Exogenous TBP (~10 ng) was injected with the CMV template and core histones (10 ng of each pair of H2A/H2B and H3/H4) were injected as indicated. (A) For transcription analysis, RNA isolated from the equivalent of five embryos was analyzed by primer extension. The markers were end-labeled DNA fragments, generated by the digestion of pBR322 with HpaII. The bands above and below the CMV primer extension product are 110 and 90 nucleotides long, respectively. (B) DNA was extracted before the MBT (2 hours of incubation), resolved on a 1% agarose gel, transferred to nitrocellulose, and the filter hybridized with radiolabeled CMV template DNA. The different forms of DNA are indicated, circular supercoiled (I), relaxed (I_r) and nicked (II).

injected genes are brought under the same temporal regulation of transcription as the endogenous genes (Krieg and Melton 1985; Lund and Dahlberg 1992; Almouzni and Wolffe 1995; Figure 13-5). The transcription of DNA microinjected into the embryo is sensitive to both the DNA mass injected and the abundance of components of the basal transcriptional machinery (Figure 13-5). Elevation of DNA mass injected (to 25 ng) through the injection of nonspecific bacteriophage λ DNA does not itself activate transcription of the reporter CMV template (300 ng). However the addition of exogenous TATA-binding protein (TBP) does activate transcription prior to

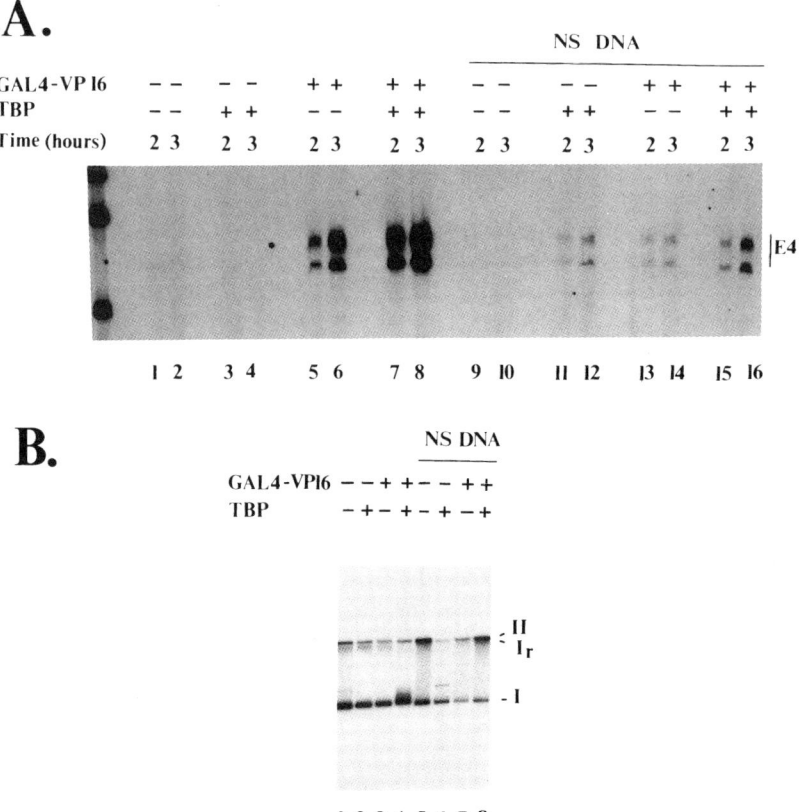

Figure 13-6 The influence of the ratio of DNA to cytoplasm, exogenous TBP, and GAL4-VP16 on the transcription of double-stranded templates containing the E4 promoter before the MBT. Closed circular, double-stranded templates (300 pg) containing the minimal E4 promoter with five GAL4-VP16 binding sites were injected into the embryo. Exogenous TBP (~10 ng) or GAL4-VP16 (~10 ng), or both together, were injected with the E4 template as indicated. Titration of chromatin assembly with non-specific DNA (NS) was achieved using 25 ng of λ DNA per embryo. After 2 or 3 hours of incubation at 22°C, embryos were collected and processed for RNA isolation. (A) Transcription from the E4 promoter was assayed by primer extension. (B) DNA was extracted [from the embryos whose transcription properties are shown in (A)] after 3 hours of incubation, resolved on a 1% agarose gel, transferred to nitrocellulose, and the filter hybridized with radiolabeled E4 template DNA. The different topological forms of DNA are indicated: circular supercoiled (I), relaxed (I_r), and nicked (II).

the MBT (Figure 13-5A, lane 7). This activation depends on the titration of the histones by nonspecific DNA because the injection of additional core histones restores transcriptional repression prior to the MBT, even in the presence of TBP (Figure 13-5, lane 9). This result indicates that the sensitivity of particular promoters to the basal transcriptional machinery depends on the mass of DNA microinjected. An alternate strategy for the activation of transcription on exogenous templates prior to the MBT is to introduce binding sites for a strong transcriptional activator, such as GAL4-VP16, near the TATA box and to express the protein in embryos following fertilization (from injected mRNA; Figure 13-6). However, the capacity of the transcriptional activator to function in embryos is also dependent on the mass of DNA introduced into the embryo (Figure 13-6A, compare lanes 5–8 with lanes 13–16). In this case, a large mass of DNA (25 ng), sufficient to titrate histones, compromises GAL4-VP16 function, indicating that additional constraints on activator function (aside from the titration of chromatin components) contribute to transcriptional quiescence during early Xenopus embryogenesis (Almouzni and Wolffe 1995). These constraints might reflect the lack of essential coactivators necessary to facilitate transcription in the early embryo, or the presence of inhibitors of their activities. The embryo potentially provides a useful in vivo system for the definition of such regulatory proteins and the characterization of the molecular mechanisms by which they work.

Summary

The *Xenopus* oocyte, egg, and embryo provide powerful in vivo systems for the analysis of gene control. It is important to recognize that each system contains cell(s) in a distinct physiological state. These states differ from each other in terms of the properties of the transcriptional machinery and the chromatin proteins that package DNA. Variations in the mass of DNA injected into oocytes, eggs, or embryos will lead to major changes both in basal transcription and in the capacity to respond to sequence-specific transcription activators. A key aspect of these studies is the necessity of assaying accurately initiated mRNA levels directly by primer extension. Use of CAT, β-galactosidase, and/or luciferase to quantitate transcriptional activity cannot be encouraged due to the influence of transcriptional control elements, splicing, and the transcription process itself on translational fate.

ACKNOWLEDGMENTS I thank Drs. Geneviève Almouzni, Nicoletta Landsberger, Funda Meric, and Jiemin Wong for carrying out the experiments described in this chapter. I am grateful to Ms. Thuy Vo for manuscript preparation.

References

Almouzni, G., and Wolffe, A. P. (1993). Replication coupled chromatin assembly is required for the repression of basal transcription in vivo. Genes Dev. 7, 2033–2047

Almouzni, G., and Wolffe, A. P. (1995). Constraints on transcriptional activator function contribute to transcriptional quiescence during early Xenopus embryogenesis. EMBO J. 14:1752–1765.

Almouzni, G., Khochbin, S., Dimitrov, S. and Wolffe, A. P. (1994). Histone acetylation influences both gene expression and development of Xenopus laevis. Dev. Biol. 165:654–669.

Bienz, M. (1984). Xenopus hsp70 genes are constitutively expressed in injected oocytes. EMBO J. 3:2477–2483.

Bouvet, P., and Wolffe, A. P. (1994). A role for transcription and FRGY2 in masking maternal mRNA in Xenopus oocytes. Cell 77:931–941.

Bouvet, P., Matsumoto, K., and Wolffe, A. P. (1995). Sequence specific RNA recognition by the *Xenopus* Y-box proteins: an essential role for the cold shock domain. J. Biol. Chem. 270:28297–28303.

Braddock, M., Muckenthaler, M., White, M. R. H., Thorburn, A. M., Sommerville, J., Kingsman, A. J., and Kingsman, S. M. (1994). Intron-less RNA injected into the nucleus of *Xenopus* oocytes accesses a regulated translation control pathway. Nucleic Acids Res. 22:5255–5264.

Brown, D. D., and Gurdon, J. B. (1977). High fidelity transcription of 5S DNA injected into *Xenopus* oocytes. Proc. Natl. Acad. Sci. USA 74:2064–2068.

Cho, H. and Wolffe, A. P. (1994). Characterization of the *Xenopus laevis* B4 gene: an oocyte specific vertebrate linker histone gene containing introns. Gene 143:233–238.

Dimitrov, S., Almouzni, G., Dasso, M., and Wolffe, A. P. (1993). Chromatin transitions during early *Xenopus* embryogenesis: changes in histone H4 acetylation and in linker histone type. Dev. Biol. 160:214–227.

Dimitrov, S., Dasso, M. C., and Wolffe, A. P. (1994). Remodeling sperm chromatin in Xenopus laevis egg extracts: the role of core histone phosphorylation and linker histone B4 in chromatin assembly. J. Cell Biol. 126:591–601.

Gunkel, N., Braddock, M., Thornburn, A. M., Muckenthaler, M., Kingsman, A. J., and Kingsman, S.M. (1995). Promoter control of translation in Xenopus oocytes. Nucleic Acids Res. 23:405–412.

Kass, S. U., Landsberger, N., and Wolffe, A. P. (1997). DNA methylation directs a time-dependent repression of transcription initiation. Curr. Biol. 7:157–165.

Krieg, P. A., and Melton, D. A. (1984) Functional messenger RNAs are produced by SP6 in vitro transcription of cloned cDNAs. Nucleic Acids Res. 12:7057–7070.

Krieg, P.A., and Melton, D. A. (1985). Developmental regulation of a gastrula-specific gene injected into fertilized *Xenopus* eggs. EMBO J. 4:3463–3471.

Landsberger, N., and Wolffe, A. P. (1995). The role of chromatin and *Xenopus* heat shock transcription factor (XHSF1) in the regulation of the *Xenopus* hsp70 promoter *in vivo*. Mol. Cell. Biol. 15:6013–6024.

Landsberger, N., and Wolffe, A. P. (1997). Remodeling of regulatory nucleoprotein complexes on the *Xenopus* hsp 70 promoter during meiotic maturation of the *Xenopus* oocyte. EMBO J 16:4361–4373.

Landsberger, N., Ranjan, M., Almouzni, G., Stump, D., and Wolffe, A.P. (1995). The heat shock response in *Xenopus* oocytes, embryos and somatic cells: a regulatory role for chromatin. Dev. Biol. 170:62–74.

Lund, E., and Dahlberg, J. E. (1992). Control of 4-8S RNA transcription at the midblastula transition in *Xenopus laevis* embryos. Genes Dev. 6:1097–1106.

Maniatis, T., Fritsch, E. F., and Sambrook, J. (1982). *Molecular Cloning*. New York, CSH Press.

Matsumoto, K., Meric, F., and Wolffe, A.P. (1996). Translational repression dependent on the interaction of the *Xenopus* Y-box protein FRGY2 with mRNA: the role of the cold shock domain, the tail domain and selective RNA sequence recognition. J. Biol. Chem. 271:22706–22712.

McKnight, S. L., and Kingsbury, R. (1982). Transcriptional control signals of a eukaryotic protein coding gene. Science 217:316–324.

Meric, F., Matsumoto, K., and Wolffe, A. P. (1997). Regulated unmasking of *in vivo* synthesized maternal mRNA at oocyte maturation: a role for the chaperone nucleoplasmin. J. Biol. Chem. 272:12840–12846.

Meric, F., Searfoss, A. M., Wormington, M., and Wolffe, A. P. (1996). Masking and unmasking maternal mRNA: the role of polyadenylation, transcription, splicing and nuclear history. J. Biol. Chem. 271: 30804–30810.

Nordeen, S. K., Green III, P. P., and Fowlkes, J. L. (1987). A rapid, sensitive, and inexpensive assay for chloramphenicol acetyltransferase. DNA 6:173–178.

Newport, J. W., and Kirschner, M. W. (1982). A major developmental transition in early *Xenopus* embryos. 1. Characterization and timing of cellular changes at the mid blastula stage. Cell 30:675–686.

Reeder, R. H., Roan, J. G. and Dunaway, M. (1983). Spacer regulation of *Xenopus* ribosomal gene transcription: competition in oocytes. Cell 35:449–456.

Rupp, R. A. W., and Weintraub, H. (1991). Ubiquitous Myo D transcription at the mid-blastula transition precedes induction-dependent Myo D expression in presumptive mesoderm of *X. laevis*. Cell 65:927–937.

Sargent, T. D., and Mathers, P. H. (1991). Analysis of class II gene regulation. Methods Cell Biol. 36:347–365.

Toyoda, T., and Wolffe, A. P. (1992). Characterization of RNA polymerase II dependent transcription in *Xenopus* extracts. Dev. Biol. 153:150–157.

Vize, P. D., Melton, D. A., Hemmati-Brivanlou, A., and Harland, R. M. (1991). Assays for gene function in developing *Xenopus* embryos. Methods Cell Biol. 36:367–387.

Wolffe, A. P., and Brown, D. D. (1987). Differential 5S RNA gene expression in vitro. Cell 51:733–740.

Wolffe, A. P., Perlman, A. J., and Tata, J. R. (1984). Transient paralysis by heat shock of hormonal regulation of gene expression. EMBO J. 3:2763–2770.

Wong, J., Shi, Y.-B., and Wolffe, A. P. (1995). A role for nucleosome assembly in both silencing and activation of the Xenopus TRβA gene by the thyroid hormone receptor. Genes Dev. 9:2696–2711.

Wong, J., Shi, Y.-B., and Wolffe, A. P. (1997). Determinants of chromatin disruption and transcriptional regulation instigated by the thyroid hormone receptor: hormone regulated chromatin disruption is not sufficient for transcriptional activation. EMBO J. 16:3158–3171.

14

Analysis of Localized RNAs in *Xenopus* Oocytes

Malgorzata Kloc
Laurence D. Etkin

Localization of RNA is an important strategy used to control the spatial pattern of protein synthesis within a cell, to specify individual cell lineages, and to establish cellular polarity (Jeffery 1988; Lipshitz 1995; St Johnston 1995). Localized RNAs are found in a variety of somatic cell types, such as fibroblasts and neurons. In addition, localized RNAs are commonly detected in both invertebrate and vertebrate oocytes and embryos. A large number of RNAs are localized at the vegetal cortex in *Xenopus laevis* oocytes (Rebagliati et al. 1985; Kloc et al. 1993; Mosquera et al. 1993). These RNAs are likely to play important roles in axial patterning and in the specification of the germ cell lineage (Kloc et al. 1993, 1996; Kloc and Etkin 1994, 1995a,b; Forristall et al. 1995).

In recent years, numerous laboratories have focused on the identification, localization patterns, and mechanisms of localization of RNAs in *Xenopus* oocytes. The methodology used to study these processes involve the detection of endogenous RNAs through in situ hybridization with radioactively or nonradioactively labeled probes. Additionally, to study the mechanisms involved in RNA localization, in vitro synthesized transcripts are injected into the oocyte. These RNAs may be either labeled with radioactive ribonucleotides and detected by autoradiography or labeled with ribonucleotides derivatized with digoxigenin or biotin and detected using a simple color reaction. In this chapter, we present the techniques commonly used for the analysis of localized RNAs in *Xenopus* oocytes. Many of these are modifications of existing protocols, while others are techniques that have been developed in our laboratory.

Analysis of Endogenous RNA in Oocytes

Frogs and Manipulation of the Ovary

Wild type and albino *Xenopus laevis* were purchased from either Xenopus Express, Nasco, or Xenopus I. Frogs are anesthetized in 0.25% solution of 3-aminobenzoic acid ethyl ester (Sigma A-5040) for 30 minutes. We use pre- and postmetamorphic froglets (stages 63–66; Nieuwkoop amd Faber 1967) to obtain oogonia in the earliest stages of differentiation, such as those in mitotic nests and pre-stage 1 (Dumont, 1972). At these stages, ovaries are located on top of the kidneys; they are lobulated, completely transparent, and barely visible without a dissecting microscope (Figure 14-1a,b). To use oocytes and oogonia from these ovaries, it is best to remove both kidneys with the ovaries attached. These can be manipulated much more easily through the fixation and/or whole mount in situ hybridization procedures. The ovaries are treated with collagenase (1 mg/ml of collagenase type II, Sigma, cat# C6885) in 1× OR2 solution (Ca^{2+}- and Mg^{2+}-free) for 30 minutes at 28°C to loosen the layers of connective tissue and follicular cells without affecting the integrity of the whole ovary. The ovaries attached to the kidneys are then used for subsequent procedures, such as whole-mount in situ hybridization.

To obtain stage 1 to early stage 2 oocytes we use 2.5–5-cm young frogs, and for stage 2–3 oocytes we use adult female frogs (Figure 14-2a,b). After surgery, ovaries are placed in 1× OR2 solution (Ca^{2+}- and Mg^{2+}-free), and ovarian sacs are opened using the forceps and then torn into small pieces. These are transferred to a large Petri dish containing 50 ml of collagenase solution (see above) per ovary. After 30 minutes of digestion (28°C in an incubator shaker), pieces of ovary are pipetted a few times through a wide-mouth plastic transfer pipette to free them from the surrounding follicular cells. After an additional 30 minutes of collagenase treatment, oocytes are washed in five changes (100 ml each) of 1× Barth's solution. The washing step is very important if the oocytes are to be used for injection. It not only prevents toxicity from the traces of collagenase, but also removes dead oocytes and proteins released from broken oocytes which prevent the oocytes from sticking to the injection plate. Oocytes can be sized either manually or by sieving through different-sized sieves (Fisher brand stainless-steel sieves with 80–500 μm mesh) We find that stage 4 through stage 6 oocytes are more sensitive to collagenase treatment. For this reason, in all experiments which require the culturing of oocytes of stage 4 and older, we use manually defolliculated oocytes.

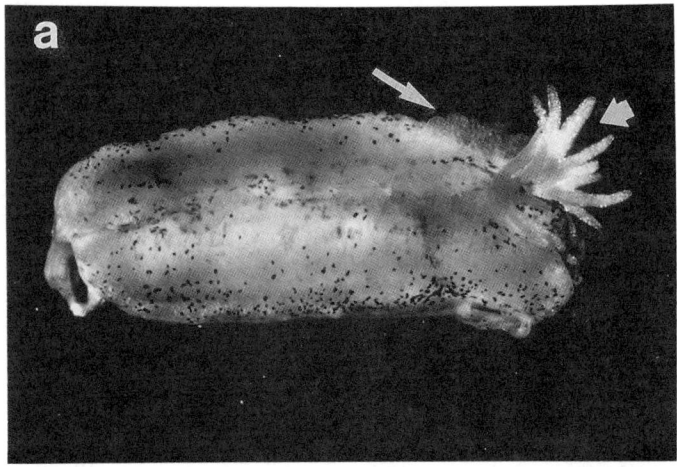

Figure 14-1 Ovaries from postmetamorphic froglets (stage 66). (a) Two kidneys with a portion of the fat body (arrowhead) and transparent, barely visible, ovaries (arrow). (b) High magnification of transparent lobulated ovary on the top of the kidney.

Whole-Mount In Situ Hybridization to Analyze Endogenous RNAs in Oocytes

In situ hybridization is commonly used to detect the spatial pattern of expression of genes during development. We have used this procedure to detect the distribution of a variety of localized RNAs during oogenesis. The following procedure has been used extensively in our laboratory to detect such localized RNAs. It is a modification of the procedure used by Harland (1991).

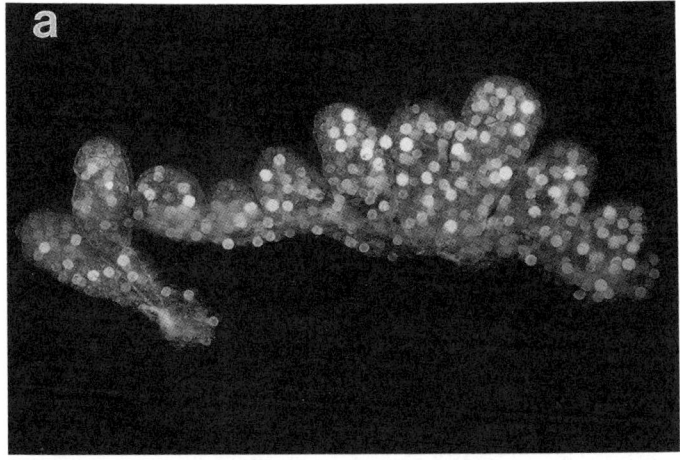

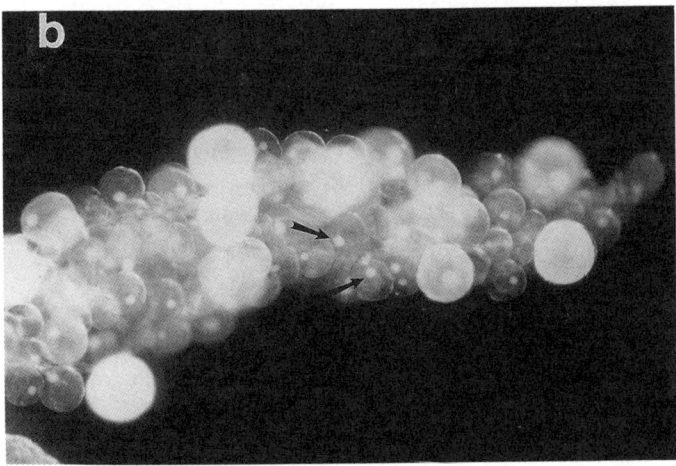

Figure 14-2 Ovary from 4-cm frog. (a) Whole ovary with stage 1 (transparent), and stage 2 (opaque) oocytes. (b) Higher magnification of a fragment of the same ovary. The nuclei and mitochondrial clouds (arrows) are clearly visible in stage 1 oocytes.

Preparation of DNA Template for In Vitro Synthesis of Antisense RNA Probe This step is crucial for the successful in vitro synthesis of high quality RNA. For all reactions, we use RNase- and DNase-free biotechnology-grade water and biotechnology-grade stock solutions from BioWhittaker (Walkersville, MD). Plasmid DNA is purified using Qiagen plasmid purification kit (Qiagen, Santa Clarita, CA) according to the manufacturer's protocol. Twenty micrograms of plasmid DNA in 50 µl of 1× restriction buffer is then completely linearized at the 5' of the insert using an appropriate enzyme. To the linearized DNA add proteinase K (2 µl of 10 mg/ml proteinase K to each 50 μl of DNA in restriction buffer) and incubate for 30 minutes at 50°C. After incubation, extract the DNA twice with phenol/chloroform/isoamyl alcohol (25:24:1 ratio) and once with chloroform/isoamyl alcohol (24:1 ratio), and precipitate overnight at −20°C with 1/10 volume of 3 M sodium acetate and 2.5 volumes of 100% ethanol. Centrifuge the DNA and wash the DNA pellet twice in 70% ethanol, dry, and dissolve it in biotechnology-grade water. After measuring the OD at 260 nm adjust the DNA concentration to 0.5 mg/ml.

In Vitro Synthesis of Labeled RNA This procedure is based on the in vitro transcription procedure developed by Melton et al. (1985). We routinely start with 3 μg of linearized template DNA to make RNA and incorporate digoxigenin-11-UTP, biotin-16-UTP or fluorescein-12-UTP (all from BMB, Indianapolis, IN) modified ribonucleotides to label the RNA.

In a sterile 1.5-ml Eppendorf tube, combine: 6 μl of DNA template (0.5 mg/ml), 6 μl of appropriate labeling mix (see Appendix), 6 μl of 10× transcription buffer (the buffer is supplied with the RNA polymerase, from Pharmacia or BMB), 6 μl of 0.1 M DTT, 27 μl of biotechnology-grade water, 3 μl of RNase inhibitor (Pharmacia or BMB), and 6 μl of appropriate RNA polymerase (T7, T3, or Sp6). The final reaction volume is 60 μl. Incubate the reaction mixture for 2 hours at 37°C. After incubation, add to the reaction 6 μl of RNase-free DNAase (BMB), incubate for an additional 15 minutes at 37°C, add 3 μl of 0.5 M EDTA, and precipitate RNA overnight at −20°C with 7.5 μl of 4 M LiCl and 225 μl of 100% ethanol. Centrifuge the precipitated RNA for 30 minutes at 4°C, wash twice with 1 ml of 70% ethanol, dry, and dissolve in 50 μl of water. Hydrolyze the probe by adding an equal volume of 80 mM sodium bicarbonate and 120 mM sodium carbonate mixture. Incubate at 60°C for 30 minutes — the objective is to hydrolyze the probe into 100–300-nt fragments. Precipitate the RNA overnight at −20°C with 12.5 μl of 4 M LiCl, 1 μl of 20 mg/ml glycogen, and 2.5 volumes of 100% ethanol. We find that further purification of the probe is unnecessary. Centrifuge and wash the pellet twice in 70% ethanol, then dry, resuspend the pellet in 60 μl of water, and measure the OD at 260 nm. Expected yield from 3 μg of template is approximately 30 ug of RNA (0.5 μg/μl). RNA probe can be frozen and stored for many months at −20°C.

Whole-Mount In Situ Hybridization We present a reliable protocol for whole-mount in situ hybridization. This protocol has been modified from Harland (1991) and works equally well for oocytes of different stages and for whole pre- and postmetamorphic ovaries. All procedures are performed at room temperature (if not indicated otherwise) with shaking in 6-, 12- or 24-well sterile tissue culture plates. The choice of size of the plates depends on the number of oocytes used for hybridization. Up to 200 oocytes can be hybridized simultaneously in one well of a six-well plate. The volumes of solutions used for each sized plate are: 5 ml per well for 6-well plates, 2 ml per well for 12-well plates and 1.5 ml per well for 24-well plates.

1. Fix defolliculated oocytes or ovaries attached to kidneys in MEMFA (see Appendix) for 1–2 hours. If you are going to perform in situ hybridization on the same day, proceed to step 2, otherwise transfer oocytes into a scintillation vial containing 100% high-purity methanol and store them at $-20°C$. Oocytes can be stored this way for many months. However, before the stored oocytes can be used for in situ hybridization, they have to be rehydrated by transferring stepwise from 100% methanol through successive washes of methanol of decreasing concentrations (90%, 70%, and 50%, 10 minutes each). Finally, wash once with water, and once with PBS/0.1% Tween 20, for 10 minutes each, and proceed to step 3 of the in situ protocol.
2. Remove fixative (never leave oocytes completely dry, always leave a few drops of fluid) and wash oocytes in PBS/0.1% Tween 20, three times, for 5 minutes each.
3. Treat oocytes for 7 minutes with 10 µg/ml of proteinase K in PBS/Tween.
4. Wash oocytes twice, for 5 minutes each, in 0.1 M triethanolamine (see Appendix).
5. Change to fresh triethanolamine solution and add acetic anhydride (Sigma). For 5 ml of solution add 12.5 µl of anhydride. After 5 minutes, repeat this step but double the amount of added acetic anhydride (25 µl to 5 ml), and incubate for 5 minutes.
6. Wash oocytes in PBS-Tween, three times, for 5 minutes each.
7. Postfix oocytes in MEMFA for 20 minutes.
8. Wash in PBS-Tween, five times, for 5 minutes each.
9. To the last PBS-Tween wash, add $\frac{1}{4}$ volume of hybridization buffer (see Appendix).
10. After 1 minute, replace with fresh hybridization buffer and incubate for 10 minutes at 50°C.
11. Replace with fresh hybridization buffer and prehybridize at 50°C, for 6 hours. We find that 6 hours prehybridization completely eliminates background. To prevent excessive evaporation put tissue culture plates inside a small Zip-lock plastic bag.
12. Replace with fresh hybridization buffer and add denatured RNA probe at final concentration of 1–5 µg/ml. Denature RNA by heating for 5 minutes at 80°C and placing immediately on ice.
13. Hybridize overnight at 50°C. Put the plate inside a Zip-lock bag.
14. Next day, remove the the plate from the plastic bag, collect the hybridization buffer containing the RNA probes in sterile Eppendorf tubes, and replace with fresh hybridization buffer for 10 minutes at 50°C. The collected probe can be frozen at $-20°C$ and recycled up to five or six times; however, remember to denature it again before every hybridization.
15. Replace the hybridization buffer with the 1:1 mixture of fresh hybridization buffer and 2× SSC/0.3% CHAPS (see Appendix); incubate at 50°C for 10 minutes.
16. Replace with 1:4 mixture of hybridization buffer and 2× SSC/ 0.3% CHAPS for 10 minutes at 50°C.
17. Wash twice, for 20 minute each, in 2× SSC/0.3% CHAPS at 37°C.

18. Incubate with 20 μg/ml of RNase in 2× SSC/0.3% CHAPS at 37°C for 20 minutes.
19. Wash in 2× SSC/0.3% CHAPS at room temperature for 10 minutes.
20. Wash in 0.2× SSC/0.3% CHAPS, twice, for 30 minutes each at 50°C.
21. Wash in PBS/0.1%Tween/0.3% CHAPS, twice, for 10 minutes each at 50°C.
22. Wash in PBS/0.1% Tween for 10 minutes.

Now samples are ready for antibody incubation. All steps of antibody incubation protocol also require shaking (if not indicated otherwise).

23. Wash oocytes with malate buffer (see Appendix) for 15 minutes.
24. Replace malate buffer with 2% blocking solution (see Appendix) and incubate for 1 hour.
25. Replace with fresh 2% blocking solution containing appropriate antibody, and incubate overnight at 4°C. We usually use alkaline phosphatase-conjugated antidigoxigenin antibody at 1:5000 dilution. Antifluorescein antibody or alkaline phosphatase-conjugated strepavidin are used at 1:2000 dilution. Equally good results are achieved when horseradish peroxidase conjugated antibodies are used, but in that case decrease the dilution of antibody to 1:1000. It should be noted that in our experience antibiotin antibodies conjugated with alkaline phosphatase produce a high background. Therefore, we prefer to detect biotin-labeled RNAs with alkaline phosphatase conjugated strepavidin.
26. On the next day replace antibody with malate buffer and wash the oocytes in five changes of this buffer, for 1 hour each. This step is very important; washing for too short a period results in increased background.
27. Replace the malate buffer wash with alkaline phosphatase (or horseradish peroxidase) reaction buffer (see Appendix) for 5 minutes.
28. Change to the appropriate color substrate solution (see Appendix). *Do Not Shake*. It is not necessary to develop color in the dark or to use inhibitors to block endogenous alkaline phosphatase or horseradish peroxidase. Monitor the intensity of color under the dissecting microscope. Since all color products are partially soluble in either water or alcohol, we prefer to overstain oocytes a little. This is especially true with red color, which is much more soluble in alcohol. The time of the color development depends on the concentration of the probe used for in situ hybridization and on the abundance of the RNA. Average time for the development of a good intensity of color is usually 30 minutes to 1 hour. After this time, the background staining slowly increases.
29. To stop the color reaction, replace the staining solution with MEMFA fixative. Fix oocytes overnight for 1–3 days in cold (4°C) conditions, preferably with shaking. Again, use a longer time for red color reactions. Fixation is necessary to stabilize the color precipitate.
30. Transfer oocytes to 100% methanol. Impurities can be detrimental for color precipitate (especially red color) so use high-purity methanol. After a few hours in methanol, oocytes become partially transparent and can be easily photographed at this point (Figure 14-3). They can be stored in methanol, in cold (4°C) conditions, in tightly capped scintillation vials for years without losing the color.

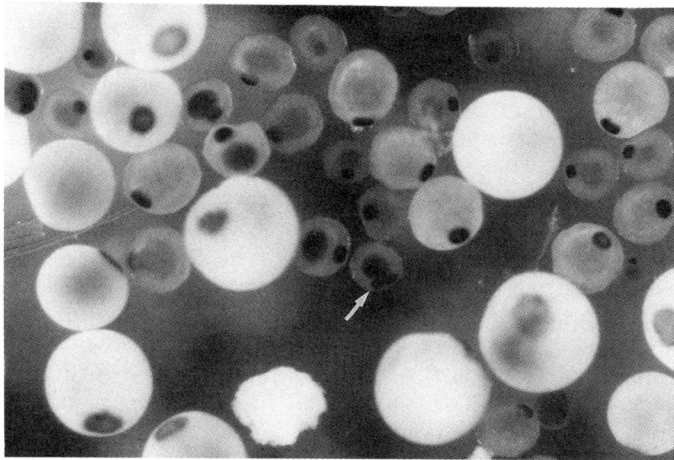

Figure 14-3 Whole-mount in situ hybridization with Xlsirts antisense digoxigenin-labeled RNA probe. Oocytes were stained with NBT/X phosphate, postfixed in MEMFA and photographed in methanol. In the transparent stage 1, oocytes, nuclei and mitochondrial clouds are stained; in one of the oocytes, two clouds are present and both are stained (see arrow).

Use of a Single Probe with One Color Detection for Whole-Mount In Situ Hybridization Our preferred choice for in situ hybridization is to label a single RNA probe with digoxigenin-11-UTP and use antidigoxigenin alkaline phosphatase antibody followed by a color reaction using NBT/X phosphate staining solution (see Appendix). This substrate gives a very stable purple-color precipitate which, after incubation in methanol, changes to a dark blue color.

Use of Two or More Probes for Whole-Mount In Situ Hybridization This method allows the investigator to visualize two or more different RNA species in the same oocyte and requires the preparation of differently labeled antisense RNA probes. For example, one probe can be labeled with digoxigenin and a second with fluorescein or biotin (consult the beginning of this section for probe preparation). The procedure is as follows.

> Proceed through the whole-mount *in situ* protocol to step 11. At step 12, add two or more denatured, differently labeled RNA probes to the hybridization buffer, each at a final concentration of 1–5 μg/ml. Proceed as usual until step 24. At step 25, add antidigoxigenin alkaline phosphatase-conjugated antibody at 1:5000 dilution and proceed to step 26. At step 27, use the alkaline phosphatase reaction buffer for NBT/X phosphate. At step 28, stain oocytes with NBT/X phosphate substrate. After the development of the color to the desired intensity, stop the color reaction by fixing oocytes overnight in MEMFA, at 4°C. Wash oocytes in several five 5-minute changes of PBS-Tween. Since even prolonged

fixation does not fully destroy alkaline phosphatase activity, inactivate the residual alkaline phosphatase with a 10-minute incubation in 0.1 M glycine-HCl, pH 2.2 with 0.1% Tween-20 at room temperature (Hauptmann and Gerster 1994). Wash oocytes four times, for 10 minute each, in PBS-Tween. Replace PBS-Tween with malate buffer, and wash for 5 minutes. Now, proceed again to step 24 of the whole-mount procedure. At step 25, add antifluorescein or strepavidin (depending how you labeled your second RNA probe) alkaline phosphatase- conjugated antibody at 1:2000 dilution. Proceed again to step 26 of the whole-mount in situ protocol. At steps 27 and 28, use alkaline phosphatase reaction buffer from the Vector red kit (see Appendix) supplemented with 0.3 M NaCl to enhance the red color (Chiu et al. 1996). After the development of the red color to the desired intensity, stop the color reaction by MEMFA fixation. Finally, after 3 days in MEMFA, transfer oocytes to 100% methanol. This should result in the signals from the first RNAs being stained blue and those from the second RNAs red.

Preparation of Sections from Whole-Mounts The method of whole-mount in situ hybridization allows for excellent three-dimensional analysis of the distribution of RNA in oocytes. However, to obtain more detailed information regarding the subcellular localization and association with specific subcellular structures, it is optimal to examine sectioned material (Figure 14-4). We have developed a fast method in which we section oocytes after whole-mount in situ hybridization, instead of performing in situ hybridization on sectioned material. Below, we describe the protocol for this posthybridization method.

Single or double whole-mount in situ oocytes or small ovaries attached to kidneys, which were either fixed in MEMFA for 1–3 days or were fixed in MEMFA and stored in methanol, can be used in this procedure. Transfer oocytes or ovaries to fresh 100% methanol and incubate overnight at −20°C in a tightly capped scintillation vial. Transfer the specimen to fresh 100% methanol for a few minutes. Replace the methanol with two 1-hour changes of Histoclear (National Diagnostic, Atlanta, GA). Remember to use glass vials at this point since Histoclear dissolves plastic. Incubate oocytes, with shaking, at room temperature. Replace the last Histoclear change with melted paraplast, and incubate oocytes in an oven at 60–65°C for 30 minutes. Replace with two changes of fresh paraplast for 30 minutes each change, then embed oocytes in paraplast.

Section at 10 µm and place sections on silane-coated slides from Polysciences (Warrington, PA). Dry slides for 1–2 days in a warm room at 37°C. Deparaffinize slides in two 10-minute changes of Histoclear (do not use Histoclear II because it does not dissolve paraplast well). Mount the slides in Permount.

Whole-Mount In Situ Hybridization Combined with Detection of Protein by Whole-Mount Immunostaining It is possible to detect both RNA and protein in the same oocyte using a combination of in situ hybridization and immunostaining. This method allows for simultaneous visualization of endogenous RNA and protein components of the oocyte, and permits us

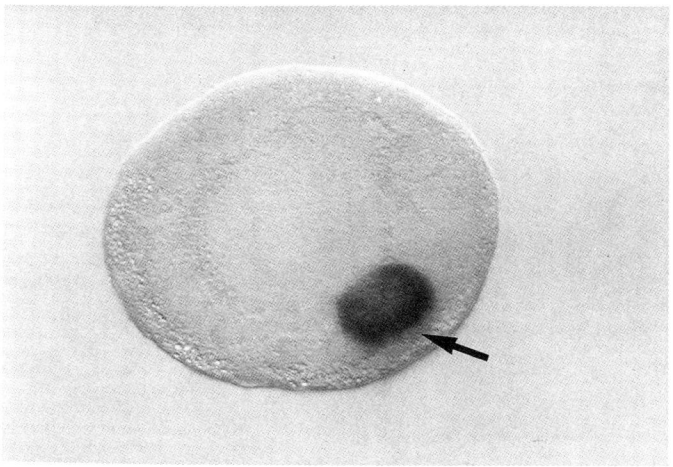

Figure 14-4 Section of stage 1 oocyte after in situ hybridization. Ten-micrometer paraffin section of whole-mount in situ hybridization of a stage 1 oocyte hybridized with Xlsirt antisense digoxigenin-labeled probe. There is staining in the mitochondrial cloud (arrow).

to analyze the relationships between RNA and a particular protein, or even a specific cell organelle. Below, we give a detailed protocol for this method.

Fix defolliculated oocytes in MEMFA; go through the regular whole-mount in situ protocol (see "Whole-Mount In Situ Hybridization" above), but since we want to detect protein, skip the proteinase K treatment (step 3). Stain whole mounts as usual with NBT/X phosphate staining solution. Stop the color reaction by overnight incubation in MEMFA at 4°C. Wash the oocytes in five changes of PBS-Tween, for 5 minutes each. Inactivate alkaline phosphatase by incubation in glycine-Tween (see "Use of Two or More Probes for Whole-Mount In Situ Hybridization" above). Wash oocytes in four changes, 10 minutes each, of PBS-Tween. Proceed to the whole-mount in situ protocol starting from step 23. At step 25, use appropriate antibody for immunostaining. The working dilution of the antibody has to be determined experimentally for each particular antibody, but we usually use a 1:50 dilution for the majority of commercially available monoclonal antibodies. Go through step 26. Now come back again to step 25. This time, use antimouse (if you used a monoclonal antibody for immunostaining) or antirabbit (if you used a polyclonal antibody for immunostaining) alkaline phosphatase-conjugated antibody, both at 1:2000 dilution. The following day, go through steps 26–28. At steps 27 and 28, use Vector red alkaline phosphatase reaction buffer supplemented with 0.3 NaCl to enhance the red color (see "Use of Two or More Probes for Whole-Mount In Situ Hybridization" above). At step 28, use the Vector red kit for staining. After development of the color (it is better to overstain a little than to understain oocytes), postfix oocytes in MEMFA for 3 days and transfer to 100% methanol. At this point, oocytes can be photographed as whole mounts or, for more detailed analysis they can be sectioned according to the protocol given in "Preparation of Sections from Whole Mounts" above.

Analysis of Exogenous RNA in Injected Oocyte

An important area of investigation is to define the *cis*-acting elements found in RNAs that localize to various regions of the oocyte [e.g., Vg1 (Rebagliati et al. 1985), Xcat2 (Mosquera et al. 1993), Xlsirts (Kloc et al., 1993)]. Early work in this area utilized radioactively labeled RNAs that were synthesized in vitro, injected, and analyzed on sections by autoradiography. The shortcomings of this approach were that it was time consuming, limited to analysis on sectioned material, and that it had limited resolution. We have developed an approach whereby one can inject RNAs labeled with fluorochromes and detect these at various time points after injection by a simple color reaction on whole mounts or sections. In addition, using this approach in conjunction with confocal microscopy, we can detect the fate of the injected RNAs in living oocytes. These procedures are discussed below.

Culturing of Oocytes

Different-staged oocytes that were manually defolliculated are stored in 1× Barth's solution. Culture in this solution does not support oocyte growth and oocytes are unable to localize RNA. Fortunately, Wallace and Misulovin (1978) formulated a more complex medium that can support oocyte growth in culture for many weeks (Wallace et al. 1973, Wallace et al. 1980a,b). We routinely use this medium for culturing injected oocytes for RNA localization experiments. The composition of the medium is 50% Leibovitz's L-15 medium (with L-glutamine) (Gibco), 15 mM HEPES NaOH, 100 µg/ml of gentamicin, 50 µg/ml of tetracycline, 50 U/ml of nystatin, and 1 µg/ml of porcine insulin (50× stocks solutions of gentamicin, tetracycline, and nystatin in 100% ethanol are stored frozen at $-20°C$). Just before use, we add purified vitellogenin to a final concentration of 5 mg/ml. Previous studies of RNA localization in *Xenopus* oocytes (Melton 1987; Yisraeli and Melton 1988; Yisraeli et al. 1990) showed that vitellogenin is necessary for the proper localization of Vg1 RNA to the vegetal cortex. Instead of using purified vitellogenin, some researchers supplement the culture medium with vitellogenin-rich frog serum. We find that the purified vitellogenin gives much better results than frog serum. Below, we give a relatively simple protocol for purification of vitellogenin from hormone-induced frogs.

Preparation of Vitellogenin

The protocol described below is a modification of a method published by Wiley et al. (1979). Our protocol, although shorter, gives high-quality vitellogenin which supports growth and differentiation of oocytes as indicated by oocyte morphology, rate of the survival, and proper RNA localization in injected oocytes. We usually prepare vitellogenin from 20 frogs. Xenopus Express (64 S. Jackson Street, Beverly Hills, FL 34465) sells so-called 'poor-quality frogs' which are much cheaper then regular frogs and can be used for this purpose. To induce frogs to produce vitellogenin, inject 1.5 mg of water-soluble 17-β-estradiol (Sigma) into the dorsal lymph sacs of female frogs. Reinject frogs after 3 days, and wait 10 days from the first injection before bleeding. To bleed frogs, first anesthetize, then cut a triangular opening in the sternum and gently pull out the heart. Make a small incision in the heart and collect blood into a 50-ml Falcon sterile tube containing 50% Dulbecco calcium-free saline (Gibco) with 0.07 M sodium citrate (pH 7.6). Use 1 ml of saline for the blood collected from two frogs. Keep collecting tubes on ice at all times. After collecting the blood from all frogs, centrifuge the tubes in a benchtop clinical centrifuge at $2500\,g$ for 15 minutes at 4°C. Carefully collect the supernatant, and recentrifuge it to remove cell debris. After the second centrifugation, transfer the supernatant into a Beckman centrifuge tube. To each 5 ml of this serum add 20 ml of 20 mM EDTA and 1.6 ml of 0.5 M $MgCl_2$, mix by inversion, and centrifuge at 5000 rpm in a J20 Beckman rotor for 15 minutes. After centrifugation, decant and dispose of the supernatant and keep the green vitellogenin pellet. Dissolve the pellet in buffer A (1 M NaCl and 50 mM Tris, pH 7.5) using 3 ml of buffer A for the vitellogenin collected from every two frogs. We dissolve the vitellogenin pellet in a centrifuge tube by putting a small stirring bar inside the tube and placing it in an ice-filled beaker or in the cold room on a magnetic stirrer. It takes up to one hour to dissolve the pellet. Stirring should be done very slowly to prevent excessive denaturation of vitellogenin. After dissolving the vitellogenin pellet, dialyze the light green solution overnight in the cold room against four 1-liter changes of buffer A. Dialyzing is necessary to remove Mg^{+2} ions. We do our dialysis in disposable dialyzing cassettes (Slide-A-Lyzer) from Pierce (Rockford, IL). The vitellogenin solution should be dialyzed again overnight, in the cold, this time against four 1-liter changes of 50 mM Tris, pH 7.5. Recover the vitellogenin from the dialyzing cassettes, measure OD at 280 nm, and freeze 500 μl aliquots at −20°C. Frozen vitellogenin can be stored for long periods without any obvious loss of activity. Usually, our vitellogenin concentrations range from 25 to 90 mg/ml depending on the frogs. We generally collect around 25 ml of purified vitellogenin from 20 frogs.

To determine the final vitellogenin concentration, we measure vitellogenin OD at 280 nm, multiply the reading by the dilution factor, and divide by 0.75 (coefficiency factor). This will give the concentration of vitellogenin expressed in mg/ml. To this value add a correction factor of 2.3 mg/ml. The sum of these two values represents the final concentration of vitellogenin preparation (Wallace et al. 1980b).

Injection of Exogenous RNA into Oocytes

In Vitro Synthesis of Sense RNA for Injection   The preparation of template for the synthesis of sense RNA for injection is identical to that described in the section "Preparation of DNA Template for In Vitro Synthesis of Antisense RNA Probe" above. The only difference is that the template has to be linearized downstream of the insert. Also, the in vitro synthesis of labeled sense RNA is virtually identical to synthesis of antisense RNA (consult section "In Vitro Synthesis of Labeled RNA" above for detailed protocol). One can choose freely between digoxigenin, biotin, fluorescein, rhodamine, or Texas red to label the RNA.

After the synthesis of sense RNA, remove the DNA template by treatment with DNase, and precipitate the RNA with LiCl and ethanol. Recover and wash the RNA pellet (see section "In Vitro Synthesis of Labeled RNA") but *do not* hydrolyze the RNA. After drying, dissolve the RNA pellet in 60 µl of biotechnology-grade water and measure the OD at 260 nm. Check the quality of RNA by electrophoresis on a 1% agarose formaldehyde gel and by performing a slot blot followed by staining with an antibody against derivatized labeled RNA. Adjust the concentration of RNA to 100 ng/µl and freeze aliquots at −20°C. For injection, mix 4 µl of this RNA with 1 µl of RNase inhibitor just before use. *Do not* refreeze or re-use RNA mixed with RNase inhibitor.

Injection of RNA   For injections of stage 1 and 2 oocytes, we use specially prepared injection dishes that affix the oocytes to the bottom of the plate. Corning 60-mm tissue culture dishes (cat# 25010) are coated with a solution of 1% protamine sulfate (Sigma cat# P-4830 dissolved in water) for 1 minute at room temperature, rinsed twice in distilled water, and then air-dried. Plates are stored at room temperature and can be used for a few weeks. Stage 1 and 2 oocytes stick firmly to the bottom of the dish; however, this technique does not work well for oocytes older then stage 2. Oocytes are placed into the coated dish in 1× Barth's solution. After 1–2 minutes, they stick to the bottom of the dish. Oocytes older than stage 2 are injected in 72-well injection plates (one oocyte per well). To prevent oocytes from moving inside the well, remove as much buffer as possible from the well before the injection, and replace it immediately after the injection.

We use glass micropipettes with a tip diameter of 1–5 µm and inject 100–500 pg of RNA per oocyte depending on the oocyte size. For RNA localization studies, we inject RNA into the oocyte germinal vesicle (GV); however, there is always some leakage from the GV into the cytoplasm. After the injection, oocytes are transferred to the Leibovitz culturing medium containing vitellogenin (see "Culturing of Oocytes" above).

Detection of Injected RNA by Whole-Mount Analysis   Below, we describe a simple protocol for detection of injected RNA in whole oocytes. This protocol is very similar to the in situ hybridization procedure ("Whole-Mount In Situ Hybridization" section above). All steps are performed in sterile tissue culture plates, with shaking, at room temperature unless indicated otherwise.

1. Fix oocytes injected with labeled RNA in MEMFA for 1–2 hours.
2. Wash in PBS/01.% Tween-20 twice, for 5 minutes each.
3. Treat oocytes for 7 minutes with proteinase K in PBS/0.1% Tween (add 5 µl of 10 mg/ml proteinase K stock solution to 5 ml of PBS-Tween).
4. Wash once, for 5 minutes in 0.1 M triethanolamine.
5. Incubate once for 10 minutes in fresh 0.1 M triethanolamine with acetic anhydride (add 25 µl of acetic anhydride to 5 ml of triethanolamine).
6. Wash once, for 5 minutes, with PBS-Tween.
7. Postfix oocytes for 20 minutes in MEMFA.
8. Wash twice, for 5 minutes each, in PBS-Tween.
9. Wash once, for 5 minutes, in malate buffer.
10. Block for 30 minutes to 1 hour in 2% blocking solution.
11. Change to fresh blocking solution containing appropriate antibody [antidigoxigenin, antifluorescein alkaline phosphatase-conjugated antibody, or streptavidin conjugated with alkaline phosphatase (see section "Whole-Mount In Situ Hybridization" above for recommended dilutions of antibodies)]. Incubate overnight at 4°C.
12. Next day, wash oocytes six times, for 15 minutes each in malate buffer.
13. Replace the last wash with alkaline phosphatase reaction buffer and wash for 5 minutes.
14. Stain oocytes in NBT/X phosphate or other color substrate of your choice in the appropriate reaction buffer. Monitor color development under the dissecting microscope. To stop the color reaction and to stabilize the color precipitate, fix oocytes in MEMFA, in cold (4°C) conditions, for 1–3 days. Change into 100% methanol, and photograph.

Analysis of Injected RNA in Sectioned Oocytes Whole-mount analysis, in many cases, is not sufficient for detailed analysis of the subcellular distribution of injected RNA. Therefore, we often analyze sectioned material from injected oocytes. There are two options: (1) section oocytes stained using the whole-mount method and (2) use postsection staining that involves fixation of injected oocytes, embedding in paraplast, sectioning, and then staining the labeled injected RNA on sections.

Sectioning of Whole-Mounts The first method uses the same principles as described in the section "Preparation of Sections from Whole Mounts" above. However, we found that, contrary to the in situ hybridization whole mounts, the whole mounts of injected oocytes (especially stage 1 and early stage (2) do not section well. This problem can be overcome by several modifications to the regular embedding protocol.

1. After staining labeled RNA, fix the oocytes in MEMFA (see "Whole-Mount In Situ Hybridization" above).

2. Transfer oocytes to a glass scintillation vial with 100% methanol overnight.
3. Next day, change to fresh 100% methanol for 10–15 minutes.
4. Replace methanol with xylene. Keep the oocytes in xylene for 1–3 days. This step is especially important for yolkless, and thus very soft, oocytes of stage 1 and early stage 2. Older oocytes with an ample amount of yolk are much harder and they do not need such long clearing in xylene. We find that treatment with xylene does not dissolve strong color precipitate.
5. From xylene, transfer the oocytes to melted paraplast, and infiltrate at 60–65°C for 1 hour.
6. Change to fresh paraplast overnight.
7. Next day, change to fresh paraplast, infiltrate oocytes for 1 more hour, and embed.

Samples prepared according to this protocol should section well.

Postsection staining of oocytes injected with RNA

1. Fix injected oocytes in MEMFA for 1–2 hours.
2. Transfer to 100% methanol overnight in a glass scintillation vial.
3. Transfer to fresh methanol for few minutes.
4. Replace methanol with xylene for young yolkless oocytes or with Histoclear for oocytes of stage 3 and up. Histoclear, which is a nontoxic equivalent of xylene, does not harden the tissue, and works very well for all hard samples like older oocytes and frog embryos that are rich in yolk; however, it is not good for soft tissues like young oocytes. Keep young oocytes in xylene for 1–3 days; keep older oocytes in two changes, 1 hour each, of Histoclear.
5. Replace xylene or Histoclear with melted paraplast. Incubate first for 1 hour at 60–65°C, in paraplast, then overnight in a second change of paraplast, then for 1 hour in a third change of paraplast, and then embed. Section as usual on precoated slides (see "Preparation of Sections from Whole Mounts"). From now on, one has to use an RNase-free environment [i.e., either baked glass or plastic Coplin jars washed with RNase Zap (Ambion, Austin, TX) and biotechnology-grade water and stock solutions].
6. Deparaffinize sections in two 10-minute changes of Histoclear.
7. Hydrate through decreasing concentrations of ethanol (2× 100%, 90%, 70%, and 50%, then water for 5 minutes each).
8. Proceed to the whole-mount staining protocol (see "Whole-Mount In Situ Hybridization" above), starting from step 2. The only difference between staining of labeled RNA in whole mounts and that on sections is that staining on sections demands much higher volumes of solutions (thus, it is much less economical); otherwise, the duration and number of washes remain the same for both protocols.
9. After the staining, fix sections in MEMFA for 1–3 days, and 3 days for red color.
10. Dehydrate sections through increasing concentrations of ethanol (50%, 70%, 90%, and 2× 100% for 5 minutes each).
11. Clear in two changes, of Histoclear, 10 minutes each, and mount in Permount.

Simultaneous Detection of Exogenous and Endogenous RNAs This is a very powerful method which allows the investigator to compare directly the distribution and/or the behavior of the exogenous injected RNA and the endogenous RNA in the same oocyte. This procedure is described below.

Inject oocytes with labeled RNA and go through the whole-mount protocol (see "Detection of Injected RNA by Whole-Mount Analysis" above). Stain oocytes in NBT/X phosphate staining solution, and fix overnight in MEMFA. Wash the oocytes five times, for 5 minutes each, in PBS/0.1% Tween. Proceed through the "Whole-Mount In Situ Hybridization" protocol above starting from step 4. At step 12, remember that your RNA probe has to be labeled differently than the injected RNA. In step 28, use Vector Red staining substrate supplemented with 0.3 M NaCl in order to intensify the red color.

Antisense RNA probe has the potential to hybridize not only with the endogenous RNA, but also with the exogenous injected RNA. We believe that staining of injected RNA blocks, either partially or completely, its hybridization to the probe. However, it is still important to use darker color substrate (e.g., NBT/X phosphate) to visualize injected RNA, and lighter color substrate (e.g., Vector red) for detection of the hybridized RNA probe. This way, even if there is some cross-hybridization of the probe with injected RNA, in the place where exogenous (blue) and endogenous (red) RNAs overlap you will see darker blue color, and in the place where only endogenous RNA is present you will see only the red color. If one uses opposite combination of colors (lighter for injected RNA and darker for in situ hybridization), then the endogenous RNA will stain blue and the place of overlap will stain darker blue, and it will be impossible to distinguish between endogenous and exogenous RNA. After the last staining, fix oocytes in MEMFA for 3 days, and transfer to 100% methanol for storage. At this point, you can section oocytes according to the "Preparation of Sections from Whole Mounts" protocol above.

Confocal Analysis of Injected RNA in Living Oocytes The beauty of confocal analysis lies in the possibility of observing the injected RNA in living oocytes from the moment of injection through the time at which it is localized. In addition, by the coinjection of two RNAs labeled with different fluorochromes, one can compare the behavior of different RNA species in the same living oocyte.

All methods of obtaining oocytes, in vitro synthesis of labeled RNA, and injection and culturing of oocytes are identical to those described in "Analysis of Exogenous RNA in Injected Oocytes" above. The difference is that injected RNA has to be labeled with fluorochrome such as fluorescein-12-UTP (BMB), Cascade blue-7-UTP, tetramethylrhodamine-5-UTP, or Texas red-5-UTP (all from Molecular Probes, Eugene, OR; see Appendix for the composition of labeling mixes). In coinjection experiments, two different RNAs are labeled with two different compatible fluorochromes (i.e., fluorochromes which can be viewed under different filter conditions, such as fluorescein and Texas red, fluorescein and rhodamine, Cascade blue and fluorescein, Cascade blue and rhodamine, or Cascade blue and Texas red).

After the injection, oocytes are placed in a plastic Petri dish filled with Leibovitz culturing medium supplemented with vitellogenin. The Petri dish has had the bottom replaced with a #1 coverslip. The following protocol was used by Kloc et al. (1996) for detection of injected Vg1, Xcat2, and Xlsirt RNAs. The oocytes were viewed using an MRC-1024 laser scanning confocal microscope (Bio-Rad) equipped with an inverted Nikon Diaphot 200 microscope using either a Nikon 20× Fluor objective lens (0.75 NA) or a 60× PlanApo oil-immersion objective lens (1.4 NA). Images were collected with a krypton/argon laser using an intensity setting of 10% laser power to minimize potential damage to the oocytes. The krypton/argon laser has three excitation lines so that three fluorophores and, therefore, up to three different RNAs, can be viewed (either simultaneously or sequentially) using appropriate filters.

Detection of Proteins by Immunostaining

Immunostaining of Sections

We worked out a very reliable and short protocol for immunostaining of sectioned oocytes.

1. Fix oocytes with 4% formalin in PBS for 1–2 hours at room temperature or overnight at 4°C, preferably with shaking. Transfer to 100% methanol overnight at −20°C. Replace methanol with fresh methanol for a few minutes, then change to Histoclear or xylene, infiltrate with paraplast, and embed (for details consult "Postsection staining of oocytes injected with RNA" above).
2. Deparaffinize and rehydrate 10 µm-sections as described in steps 6 and 7 of "Postsection staining of oocytes injected with RNA" above.
3. Wash sections twice, for 5 minutes each, in PBS/0.05% Tween. Block for 30 minutes in 0.1% BSA in PBS/0.05% Tween in a Coplin jar.
4. Prepare the humidity chamber. On the bottom of a large dish, place a few paper towels wetted with PBS-Tween. Make a stand for slides from two plastic pipettes placed parallel on the bottom of the dish.
5. Take one slide from the blocking solution, place it horizontally on the stand, and pipette 200–1000 µl of the first antibody (monoclonal or polyclonal) in 1% BSA in PBS/0.05% Tween. The dilution of antibody has to be established experimentally, but we usually start from a 1:50 dilution. Repeat with all slides. It is important not to let the slides dry between consecutive steps. Cover the dish with Saran wrap and incubate the slides for 30 minutes. Collect the antibody and freeze at −20°C for future reuse.
6. Immediately transfer the slides into a Coplin jar filled with PBS/0.05% Tween and wash twice, for 10 minute each, with shaking.

7. Return the slides to the humidity chamber, and pipette 200–1000 μl of alkaline phosphatase-conjugated secondary antibody (antimouse or antirabbit), cover with Saran Wrap and incubate for 30 minutes. Collect and freeze the antibody.
8. Transfer the slides to PBS/0.05% Tween (wash twice, for 10 minutes each).
9. Transfer the slides to the humidity chamber and pipette staining solution (any alkaline phosphatase substrate can be used — see Appendix).
10. Develop the color to the desired intensity. Transfer the slides to 50% ethanol for 3 minutes. Change to 70%, 90%, and 2× 100% ethanol for 5 minutes each. Clear in two changes of Histoclear, for 10 minutes each, and mount in Permount.
11. To use secondary antibody conjugated with horseradish peroxidase, instead of alkaline phosphatase, slides have to be bleached in order to remove endogenous peroxidase activity. After deparaffinization and rehydration, bleach slides for 15 minutes in 6% H_2O_2 in PBS/0.05% Tween. Wash in two changes of PBS-Tween, for 10 minutes each and proceed to the blocking step. In the step when you add secondary antibody, add horseradish-conjugated antimouse or antirabbit antibody at 1:50 dilution. Stain the slides in horseradish peroxidase substrate (see Appendix), dehydrate, clear, and mount in Permount.

ACKNOWLEDGMENTS This work was supported by grants from NIH and NSF. We would like to thank Dr. Agnes Chan for her critical reading of the manuscript and Dr. Carolyn Larabell (Lawrence Berkeley Laboratories) for help in developing the confocal protocol.

References

Chiu, K., Sullivan, T., and Bursztajn, S. (1996). Improved in situ hybridization: color intensity enhancement procedure for the alkaline phosphatase/fast red system. Biotechniques 20:964–968.

Dumont, J. (1972). Oogenesis in *Xenopus laevis* (Daudin) stages of oocytes in laboratory maintained animals. J. Morphol. 136:153–180.

Forristall, C., Pondel, M., Chen, L., and King, M. L. (1995). Patterns of localization and cytoskeletal association of two vegetally localized RNAs Vg1 and Xcat2. Development 121:201–208.

Gurdon, J. B. (1968). Changes in somatic cell nuclei inserted into young and maturing amphibian oocytes. J .Embryol. Exp. Morphol. 20:401–414.

Harland, R. (1991). *In situ* hybridization: an improved whole mount method. In *Methods in Cell Biology*, Vol. 36, Peng, H. B., and Kay, B., eds. San Diego, Academic Press, pp. 685–695.

Hauptmann, G., and Gerster, T. (1994). Two-color whole-mount *in situ* hybridization to vertebrate and *Drosophila* embryos. Trends Genet 10:266.

Hemmati-Brivanlou, A., and Harland, R. (1989). Expression of an engrailed-related protein is induced in the anterior neural ectoderm of early *Xenopus* embryos. Development 106:611–617.

Jeffery, W. (1988). The role of cytoplasmic determinants in embryonic development. In *Developmental Biology: A Comprehensive Synthesis*, Vol. 5, Browder, L., ed. New York, Academic Press, pp. 3–56.

Kloc, M., and Etkin, L. (1994). Delocalization of Vg1 mRNA from the vegetal cortex in *Xenopus* oocytes after destruction of Xlsirt RNA. Science 265:1101–1103.

Kloc, M., and Etkin, L. (1995a). Two distinct pathways for the localization of RNAs at the vegetal cortex in *Xenopus* oocytes. Development 121:287–297.

Kloc, M., and Etkin, L. D. (1995b). Genetic pathways involved in the localization of RNA in *Xenopus* oocytes. In *Localized RNAs*, Lipshitz, H. D., ed. Austin, TX, R. G. Landes Co., pp. 149–154.

Kloc, M., Spohr, G., and Etkin, L (1993). Translocation of repetitive RNA sequences with the germ plasm in *Xenopus* oocytes. Science 262:1712–1714.

Kloc, M., Larabell, C., and Etkin, L. (1996). Elaboration of the messenger transport organizer pathway for localization of RNA to the vegetal cortex of *Xenopus* oocytes. Dev. Biol. 180:119–130.

Lipshitz, H. D. (1995). Introduction. In *Localized RNAs*, Lipshitz, H. D., ed. Austin, TX, R. G. Landes Co., pp. 1–5.

Melton, D. A. (1987). Translocation of a localized maternal mRNA to the vegetal pole of *Xenopus* oocytes. Nature 328:80–83.

Melton, D. A. Kreig, P. A. Rebegliatti, M. R., Maniatis, T., Zinn, K., and Green, M. (1985). Efficient *in vitro* synthesis of biologically active RNA hybridization probes from plasmids containing bacteriophage SP6 promoter. Nucleic Acids Res. 12:7035–7056.

Mosquera, L., Forristall, C., Zhou, Y., and King, M. L (1993). A mRNA localized to the vegetal cortex of *Xenopus* oocytes encodes a protein with a nanos-like zinc finger domain. Development 117:377.

Nieuwkoop, P. D. and Faber, J. (1967) *Normal table of Xenopus laevis*, Amsterdam, (Dauden) North Holland Pub.

Rebagliati, M. R., Weeks, D. L., Harvey, R. P., and Melton, D. A. (1985). Identification and cloning of localized maternal mRNA from Xenopus eggs. Cell 42:769–777.

Speel, E. J. M., Jansen, M. P. H. M., Ramaekers, F. C. S., and Hopman, A. H. N. (1994). A novel triple-color detection procedure for brightfield microscopy, combining in situ hybridization with immunocytochemistry. J. Histochem. Cytochem. 42:1299–1307.

St Johnston, D. (1995). The intracellular localization of messenger RNAs. Cell 81:161–170.

Wallace, R. A., and Misulovin, Z. (1978). Long term growth and differentiation of Xenopus oocytes in a defined growth medium. Proc. Natl. Acad. Sci. USA 75:5534–5538.

Wallace, R. A., Jared, D. W., Dumont, J. N., and Sega, M. W. (1973). Protein incorporation by isolated amphibian oocytes. J. Exp. Zool. 184:321–334.

Wallace, R., Misulovin, Z., and Etkin, L. D. (1980a). Full grown oocytes from *Xenopus laevis* resume growth when placed in culture. Proc. Natl. Acad. Sci. USA 78:3078–3082.

Wallace, R. A., Misulovin, Z., and Wiley, H. S. (1980b). Growth of anuran oocytes in serum-supplemented medium. Reprod. Nutr. Dev. 20:699–708.

Wiley, H. S., Opresko, L., and Wallace, R. A. (1979). New method for the purification of vertebrate vitellogenin. Anal. Biochem. 97:145–152.
Yisraeli, J. K., and Melton, D. A. (1988). The maternal RNA Vg1 is correctly localized following injection into *Xenopus* oocytes. Nature 336:592–595.
Yisraeli, J. K., Sokol, S., and Melton, D. A. (1990). A two step model for the localization of maternal mRNA in *Xenopus* oocytes: involvement of microtubules and microfilaments in translocation and anchoring of Vg1 mRNA. Development 18:289–298.

Appendix

10 × Barth's Solution (From Gurdon 1968)
0.88 M NaCl, 25.65 g
0.01 M KCl, 0.375 g
0.1 M HEPES, 11.9 g
0.008 M $MgSO_4 \times 7H_2O$, 1.0 g
0.003 M $Ca(NO_3)_2 \times 4H_2O$, 0.39 g
0.004 M $CaCl_2 \times 2H_2O$, 0.30 g
Add water to 500 ml, pH 7.6.
Sterilize and store at 4°C.

10 × OR2 (Modified from Wallace et al. 1973)
0.8 M NaCl, 24.11 g
0.025 M KCl, 0.932 g
0.01 M Na_2HPO_4, 0.710 g
0.038 M NaOH, 0.76 g
0.05 M HEPES, 5.96 g
Add water to 500 ml, pH 7.8.
Sterilize and store at 4°C.

10 × MEM (Hemmati-Brivanlou and Harland 1989)
1 M MOPS, pH 7.4
20 mM EGTA
10 mM $MgSO_4$
Sterilize and store at 4°C.

MEMFA
1 ml 10× MEM
1 ml 37% Formalin
8 ml Water
Mix just before use.

Triethanolamine Hydrochloride (Sigma, cat# T-1502)
For 0.1 M solution, add 9.28 g to 500 ml of water, pH 8. Sterilize.

20 × SSC
3 M NaCl
0.3 M Na_3 citrate $\times 2H_2O$
pH 7.0
Sterilize.

Malate Buffer
0.1 M Maleic acid
0.15 M NaCl
pH 7.5
Sterilize.

10% Blocking Solution
Dissolve 50 g of Blocking reagent (BMB, cat# 1096 176), with stirring, on a hot plate in 500 ml of malate buffer.
Autoclave, aliquot, and freeze at $-20°C$ in 50-ml Falcon sterile tubes.

2% Blocking Solution
10 ml 10% Blocking solution
40 ml Malate buffer
Store frozen at $-20°C$.

In Situ Hybridization Buffer
50% Deionized formamide
1 mg/ml Yeast RNA
1× Denhardt's Solution
0.1% CHAPS
5× SSC
100 µg/ml Heparin
5 mM EDTA
0.1% Tween 20
Mix just before use.

Chemicals Used For In Situ Hybridization Buffer
Deionized Formamide
To 500 ml of formamide add 50 g of ion exchange resin (ICN, cat# 150-330 or Biorad, cat# 143-7424).
Stir at room temperature for 30 minutes.
Filter twice through Whatman #1 filter paper.
Aliquot into 50-ml sterile Falcon tubes.
Store frozen at $-20°C$.

50 × Denhardt's Solution
0.5 g Polyvinylpyrrolidone (PVP)
0.5 g Bovine serum albumin (BSA)
0.5 g Ficoll 400
Add water to 50 ml.
Store frozen at $-20°C$.

Heparin, Sodium Salt (Sigma, cat# H-3393)
Make 10 mg/ml solution in water.
Store frozen at $-20°C$.

Yeast RNA (BMB, cat# 109 223)
Dissolve 5 g of RNA in water (adjust to pH 7.0 to dissolve).

After extracting once with phenol/chloroform/isoamyl alcohol, and once with chloroform/isoamyl alcohol, precipitate with 1/10th volume of 3 M Na acetate and 3 volumes of 100% ethanol for 20 minutes at −20°C.
Centrifuge the wash pellet twice with 70% ethanol.
Dissolve in water.
Measure OD at 260 nm.
Adjust concentration to 10 mg/ml.
Store frozen at −20°C.

Digoxigenin, Fluorescein, or Biotin 10 × Labeling Mix
CTP, GTP, ATP, 3.3 mM each
UTP, 2.2 mM
Digoxigenin or fluorescein or Biotin, 1.1 mM

For 100 µl of 10× labeling mix combine:
   100 mM CTP, 3.3 µl
   100 mM UTP, 2.2 µl
   100 mM GTP, 3.3 µl
   100 mM ATP, 3.3 µl
   10 mM Digoxigenin-11-UTP or 10 mM fluorescein-12-UTP or 10 mM biotin-16-UTP (from BMB), 11 µl each
Add water to 100 µl.
Store frozen at −20°C.

Texas Red, Rhodamine, or Cascade Blue 10 × Labeling Mix
For 100 µl of 10× NTP mix combine:
   100 mM CTP, 3.3 µl
   100 mM UTP, 2.2 µl
   100 mM GTP, 3.3 µl
   100 mM ATP, 3.3 µl
Add water to 100 µl.
Store frozen at −20°C.

To label RNA, combine:
   6 µl DNA template (0.5 µg/ml)
   6 µl 10× NTP mix
   6 µl Appropriate 10× transcription buffer
   6 µl 0.1 M DTT
   6.6 µl 1 mM stock of Texas red-5-UTP or 1 mM stock of tetramethylrhodamine-5-UTP or 1 mM stock of Cascade blue-7-UTP (all from Molecular Probes)
   20.4 µl Water
   3 µl RNase inhibitor
   6 µl Appropriate RNA polymerase

Alkaline Phosphatase Substrates
NBT(4-Nitroblue tetrazolium chloride) from BMB, cat# 1 383 213
X-phosphate = BCIP (5-bromo-4-chloro-3indolyl-phosphate 4-toluidine salt) from BMB, cat# 1 383 221
Vector red kit, cat# SK 5100, Vector black kit, cat# SK 5200, Vector blue kit, cat# SK 5300, all from Vector Laboratories, Burlingame, CA

Alkaline Phosphatase Color Reaction Solutions
1. *For NBT/X-phosphate substrate*
   100 mM Tris-HCl, 100 mM NaCl, 50 mM $MgCl_2$, pH 9.5
   To 10 ml of buffer add 45 µl of NBT and 35 µl of X-phosphate.
2. *For Vector red kit*
   100 mM Tris-HCl supplemented with 0.3 M NaCl, pH 8.2
   To 10 ml of buffer add 4 drops of reagent 1, 4 drops of reagent 2, and 4 drops of reagent 3 from the kit.
3. *For Vector black kit*
   100 mM Tris-HCl, pH 9.5
   To 10 ml of buffer add 4 drops of reagent 1, 4 drops of reagent 2, and 4 drops of reagent 3 from the kit.
4. *For Vector blue kit*
   100 mM Tris-HCl, pH 8.2
   To 10 ml of buffer add 4 drops of reagent 1, 4 drops of reagent 2, and 4 drops of reagent 3 from the kit.

Horseradish Peroxidase Substrates and Color Reaction Buffers
1. *DAB* (3,3'-Diaminobenzidine) *kit* from BMB, cat# 1 718 096. Kit contains metal-enhanced DAB and peroxidase buffer. Follow manufacturer's protocol.
2. *Vector SG kit* from Vector Laboratories, cat# SK-4700. Follow manufacturer's protocol.
3. *PO-TMB* (Horseradish peroxidase tetramethylbenzidine) *green substrate* (According to Speel et al. 1994).
   Dissolve 100 mg of sodium tungstate (Sigma cat# S-0765) in 100 mM of citrate buffer pH 5.0.
   Adjust pH to 5–5.5 with HCl.
   Just before use, dissolve 20 mg of dioctyl sodiumsulfosuccinate (Sigma cat# D-0885) and 6 mg of 3,3',5,5'-tetramethylbenzidine (Sigma cat# T-5525) in 2.5 ml of 100% ethanol at 80°C.
   Add this mixture, together with 10 µl of 30% H_2O_2, to sodium tungstate solution.

Citrate buffer pH 5.0: mix 20.5 ml of 0.1 M solution of citric acid (21.1 g in 1000 ml) with 29.5 ml of 0.1 M solution of sodium citrate (29.41 g $C_6H_5O_7Na_3 \times 2H_2O$ in 1000 ml).

15

In situ Hybridization Techniques with *Xenopus* Embryos

Thomas Hollemann
Frank Panitz
Tomas Pieler

In situ hybridization techniques have developed into a tool of primary importance in the analysis of gene expression characteristics in normal and experimentally manipulated *Xenopus* embryos. The advent of a standard in situ hybridization protocol making use of chemically modified antisense RNA probes (Hemmati-Brivanlou et al. 1990) has led to the description of spatial and temporal expression characteristics for a plethora of genes which are differentially transcribed during *Xenopus* embryogenesis.

In addition to the basic protocol that details this technique, we describe use of double-staining whole-mount in situ hybridization, which allows for the simultaneous analysis of two different transcripts, thereby establishing the exact correlation of their expression characteristics. In order to be able to reconstruct the three-dimensional pattern of gene expression, useful information must come not only from sectioned embryos that had previously been subjected to whole-mount in situ hybridization, but also from direct in situ hybridization on embryonic sections, helping to circumvent problems that could arise from a difficulty of the RNA probes to penetrate to the inner cell layers of an intact embryo (Strahle et al. 1994; Ryan et al. 1996). Finally, the basic in situ hybridization procedure combined with fast, PCR-based cloning/sequencing protocols has enabled the simultaneous screening of larger numbers of cDNA clones for their expression characteristics (Gawantka et al. 1995). We describe our recipe for the processing of large numbers of samples, including the design and use of a simple, yet helpful, apparatus for this purpose.

Whole-Mount In Situ Hybridization: The Basic Protocol

Preparation of Embryos

Best results for whole-mount in situ hybridization are obtained using albino embryos, but pigmented embryos are equally suitable for in situ hybridization on sections. For staging, albinos have to be stained with Nile blue after dejellying (saturated solution in 50 mM phosphate buffer, pH 7.8, filtered through a 3MM filter). Prior to stage 12.5, it is not necessary to remove the vitelline membrane. However, if later stages have to be analyzed, the membrane should be removed manually with a pair of good forceps (e.g., Dumont No. 5). If it turns out to be too difficult to remove the membrane manually, the membrane can be loosened by proteinase K treatment: 5 µg/ml in 0.1× Modified Barth's Solution (MBS: 88 mM NaCl, 1 mM KCl, 0.41 mM $CaCl_2$, 0.33 mM $Ca(NO_3)_2$, 0.82 mM $MgSO_4$, 2.4 mM $NaHCO_3$, 10 mM HEPES, pH 7.4) for 5–10 minutes. The extent of proteolysis has to be monitored under the microscope; if membranes start to lift, the digestion must be stopped immediately by washing with excess 0.1× MBS. If any broken embryos are found, discard the whole batch; neurula-stage embryos, in particular, are very sensitive to proteinase K treatment. Staged embryos are collected in pools and washed once with distilled water prior to fixation. Fixation is performed in 5-ml glass vials with plastic screw caps (< 250 embryos) in MEMFA buffer (0.1 M MOPS, pH 7.4, 2 mM EGTA, 1 mM $MgSO_4$, and 3.7% formaldehyde) on a rotator for 1–1.5 hours. Fresh formaldehyde solution must be used, but freshly prepared paraformaldehyde may replace it. The embryos are transferred to 100% ethanol and can be stored up to 1 year at $-20°C$.

In Situ Hybridization

Synthesis of RNA Probes

1. Digoxigenin- or fluorescein-labeled antisense probes are prepared using the Stratagene in vitro RNA Transcription Kit supplemented with RNasin (Promega) and digoxigenin- or fluorescein-UTP (Boehringer Mannheim). The following reaction mixture is allowed to react for 2–4 hours at 37°C: 5µl 5× RNA polymerase buffer, 4 µl of 2.5 mM nucleotide triphosphate mix (NTP: 10 µl of 10 mM ATP, 10 µl of 10 mM CTP, 10 µl of 10 mM GTP, 6.5 µl of 10 mM UTP and 3.5 µl of 10 mM digoxigenin- or fluorescein-UTP), 1 µl of 750 mM DTT, 9.5 µl of RNase-free H_2O, 0.5 µl of 40 U/µl of RNasin (Promega), 4 µl of linearized DNA template (1 µg), and 1µl 10 U/µl RNA polymerase. After RNA synthesis, digest the DNA template by adding 1 µl of RNase-free DNase (Boehringer Mannheim) and incubate for 20 minutes at 37°C. The length of the probe should

be in the range 150–4000 bp. Best results for the chromogenic reaction are obtained with RNA probes of 1000–2500 bp. It is not necessary to partially digest longer probes.

2. Purify the RNA with the help of the RNeasy Kit (Qiagen). Elute the RNA in a small volume (25 µl H$_2$O) and check 1/10 of the RNA on a freshly prepared 1% agarose-gel in 1× TBE. Suspend the remaining RNA in 1 ml of hybridization buffer [50% formamide (Merck, Germany), 5× SSC (20× SSC: 3 M NaCl, 0.3 M sodium citrate, pH 7.0, with NaOH), 1 mg/ml Torula RNA (Sigma, USA), 100 µg/ml Heparin (Sigma, USA), 1× Denhardt's (100× Denhardt's: 2% BSA (Fluka, Germany), 2% PVP (Sigma, USA), 2% Ficoll 400 (Pharmacia, Sweden), 0.1% Tween 20 (Sigma, USA), 0.1% CHAPS (Sigma, USA), 10 mM EDTA (Merck, Germany, in DEPC-H$_2$O)] and store at 20°C.

Hybridization

1. Carefully collect embryos of the stages to be analyzed in a 5-ml glass vial. Up to 50 embryos can be processed in a single vial.
2. Rehydrate the embryos by successive 5-minute incubations in (a) 5 ml of 100% ethanol, (b) 75% ethanol/25% water, (c) 50% ethanol/50% water, (d) 25% ethanol/75% Ptw (1× PBS, 0.1% Tween 20), and (e) wash four times, each in 100% Ptw, on a rocking platform with gentle agitation.
3. A proteinase K step is included to increase the permeability for the RNA probes and for the antibodies. Incubate the embryos in 10 µg/ml (Ptw) proteinase K for 10–30 minutes depending on the room temperature (a good guideline at 20°C is : –20 minutes for embryos that are still enclosed in a vitelline membrane, –15 minutes for neurula-stage embryos, and –30 minutes for tail-bud stage embryos; if a complete collection of stages is used, incubate for 20 minutes). Wash the embryos three times with 5 ml of Ptw in order to remove the proteinase K.
4. Following the proteinase K treatment, the embryos are incubated with acetic anhydride to neutralize positive charges which would otherwise bind nucleic acids by ion exchange (Hayashi et al. 1978). Rinse the embryos twice with 5 ml 0.1 M triethanolamine (Sigma, USA), pH 7.5, horizontally on a nutator. Add 12.5 µl of acetic anhydride to the embryos in 5 ml of triethanolamine and rock the vials for 5 minutes on a nutator. Add another 12.5 µl of acetic anhydride and rock again for 5 minutes.
5. Wash the embryos five times with 5 ml of Ptw prior to refixation. Refix the embryos for 20 minutes with 5 ml of 4% formaldehyde in 1× Ptw. Rinse the embryos five times with 5 ml of 1× Ptw for 5 minutes each.
6. For prehybridization, remove all but 1 ml of Ptw and add 250 µl of hybridization buffer. Wait until all embryos have settled. Remove the buffer and rinse twice with 500 µl of hybridization buffer. Incubate the embryos in hybridization buffer for 10 minutes at 65°C, with gentle shaking, in a water bath. Renew the hybridization buffer and prehybridize for at least 6 hours at 60°C with gentle shaking in a water bath.
7. Bring 500 µl of the RNA probe to 70°C, exchange with the prehybridization buffer, and hybridize overnight at 60°C with gentle shaking in a water bath.

Washing

8. Replace the RNA probe solution with 500 µl of prewarmed hybridization buffer and incubate at 60°C for 10 minutes (the probe can be reused up to three times; in some cases, reused probes show lower background). Add 1 volume of prewarmed 4× SSC and incubate at 60°C for 20 minutes. Wash the embryos three times with prewarmed 2× SSC at 60°C for 20 minutes each.
9. In order to reduce background staining, RNase treatment is performed by replacing the 2× SSC with 2× SSC including 20 µg/ml of RNase A (Sigma R-5000) and 10 U/ml of RNase T1 (Sigma R-8251), followed by incubation at 37°C for 30 minutes. RNase A stock solutions are prepared by dissolving RNase A at 10 mg/ml in TE (10 mM Tris-HCl, pH 7.5, 1 mM EDTA) and heating to 100°C for 10 minutes; store aliquots at −20°C. Following RNase treatment, wash the embryos with 2× SSC at room temperature for 10 minutes, and repeat twice with 0.2× SSC at 60°C for 30 minutes.

Antibody Incubation

10. Rinse the embryos twice at room temperature for 15 minutes with 5 ml of maleic acid buffer (MAB: 100 mM maleic acid, pH 7.5, 150 mM NaCl) and rock gently. Replace MAB with 2.5 ml of MAB/2% BMB (BMB: Boehringer Mannheim Blocking Reagent; dissolve by heating to obtain 10% stocks and autoclave, store in aliquots at −20°C).
11. Replace the 2.5 ml of MAB/2% BMB with 2.5 ml of MAB/2% BMB/20% lamb serum at room temperature and incubate for at least 1 hour (prepare lamb serum by heating to 55°C for 30 minutes; store in aliquots at −20°C).
12. Replace the solution by 1 ml of MAB/2% BMB/20% lamb serum and include a 1:2000 to 1:5000 dilution of the affinity-purified sheep antidigoxigenin (antifluorescein) antibody coupled to alkaline phosphatase (Boehringer Mannheim) and incubate at room temperature for 4 hours, with gentle rocking.
13. Wash the embryos three times with 5 ml of MAB at room temperature for 5 minutes each. Transfer the embryos to a 50-ml Falcon tube and wash overnight with 50 ml of MAB at 4°C, rocking gently.
14. Transfer the embryos to a glass vial. Wash twice with 5 ml of MAB at room temperature for 5 minutes.

Chromogenic Reaction

15. Rinse the embryos twice with 5 ml of alkaline phosphate buffer (APB: 100 mM Tris-HCl, pH 9.5, 100 mM NaCl, 50 mM $MgSO_4$, 0.1% Tween 20) in a total volume of 5 ml at room temperature for 5 minutes. The staining reaction is then performed in 1 ml of ABP plus 4.5 µl of nitro blue tetrazolium [NBT; 75 mg/ml in dimethylformamide (Boehringer Mannheim, Germany)] and 3.5 µl of 5-bromo-4-chloro-3-indolyl phosphate [BCIP; 50 mg/ml in dimethylformamide (Boehringer Mannheim, Germany)]. The chromogenic reaction is allowed to continue in the dark. For abundant transcripts, the reaction is very fast and will be visible within 5 minutes. For less abundant transcripts, the reaction can take up to 4–6 hours by which time background staining usually begins to become a significant problem. It can be reduced, if the colour reaction is performed in the coldroom at 6°C. If high

background is still a problem, use Purple AP Substrate (Boehringer Mannheim, Germany). The advantage of the Purple AP Substrate is that background staining is not observed. Staining can even proceed overnight without background problems. The disadvantage of the Purple AP Substrate is that the stain tends to be more diffusible than NBT/BCIP and this staining procedure is therefore not suitable for subsequent sectioning of stained embryos. For the chromogenic reaction with Purple AP Substrate, the embryos are washed first with MAB, then twice with water, which is then replaced with 1 ml of the Purple AP Substrate solution at room temperature. The chromogenic reaction with the Purple AP Substrate is approximately five times slower than the NBT/BCIP reaction.

16. The staining reaction is stopped by rinsing the embryos several times with Ptw. The contrast of the NBT/BCIP stain can be enhanced by carefully rinsing the embryos with methanol for 1 minute. Afterwards, the embryos are rehydrated by rinsing with 75% methanol/25% water, 50% methanol/50% water, 25% methanol/75% Ptw, and four washes with 100% Ptw for 30 seconds each. The embryos can then be stored in MEMFA at 4°C.

Double-Staining Whole-Mount In Situ Hybridization

Reliable methods for double-staining are available, and, in principle, it is possible to stain embryos for up to three different target gene transcripts. The probes are synthesized as described above, using either digoxigenin-, fluorescein-, or biotin-modified nucleotides. The hybridization procedure is essentially the same as described above. Both probes are used simultaneously during the hybridization. Because all the antibodies are coupled to alkaline phosphatase, the antibody and chromogenic reactions for each probe must be carried out consecutively.

Steps 1–2. As described above.
3. The proteinase K treatment is performed for 30 minutes at 20°C.
Steps 4–6. As described above.
7. Preheat 500 µl of a mixture of both RNA probes to 70°C before incubation of the embryos. Hybridize overnight at 60°C, with gentle shaking, in a water bath.
Steps 8–11 as described above.
12. It is advisable to start with the analysis of the less abundant transcript because the second staining is usually less pronounced. There is no difference in using Digoxigenin or Fluorescein-labeled probes; both antibodies (Boehringer Mannheim) work equally well.
Steps 13–15 as described above.
16. As a first stain, we recommend Fast red (Boehringer Mannheim) because the stain is less sensitive to heating than NBT/BCIP. The reaction is stopped by rinsing the embryos several times with Ptw. The primary alkaline phosphate is deactivated by heating at 65°C for 10 minutes in 5 mM EDTA. Restart the protocol at step 10.
As an alternative to Fast red or NBT/BCIP, the Vectastain II detection kit

(Vector Laboratories, USA) can be used, although the contrast is not as good as with NBT/BCIP. The Vectastain II stains in a brownish color.

Nonradioactive In Situ Hybridization on Sections

For the analysis of genes expressed in the deeper cell layers of the embryo, lack of penetrance for the RNA probes utilized can disturb the results obtained. In such a situation, in situ hybridization analysis can be performed directly on sections prepared from *Xenopus* embryos.

Preparation of Embryos

Embryos are fixed in MEMFA for 2 hours and then stored in ethanol at $-20°C$. After dehydration, embryos are embedded in paraffin.

Pretreatment of the Slides

Superfrost plus glass slides (Menzel, Germany) can be directly used, without pretreatment. Alternatively, regular glass slides can be prepared for mounting as follows:

1. Wash slides, in a slide holder, with 1 N HCl at room temperature for 10 minutes. Wash the slides once with 70% ethanol. Rinse twice with water and dry the slides at 70°C overnight.
2. Submerse slides in acetone for 10 minutes. Then submerse the slides in freshly prepared 2% (v/v) TESPA (Sigma, USA) in acetone at room temperature for 1 minute. Rinse once with acetone and twice with water. Bake the slides at 160°C overnight.

Embedding and Mounting
3. Carefully collect staged embryos in a 5-ml glass vial. Wash twice with ethanol at room temperature for 20 minutes. Replace ethanol with xylene. Wash twice with xylene at room temperature for 20 minutes.
4. Heat paraffin (Polysciences, USA) to 60°C. Replace xylene with the prewarmed paraffin. Wash twice with paraffin by incubations at 60°C for 1 hour.
5. Fill disposable plastic tissue-embedding molds (Polysciences, USA), approximately 5 mm deep, with freshly molten paraffin. Place the molds on a warming plate at 60°C to keep the paraffin in a liquid state, or let only a thin bed solidify. Transfer no more than one embryo per mold by use of a prewarmed pipette. To maintain orientation, use prewarmed forceps or glass needles. Let the paraffin solidify at room temperature for 30 minutes.
6. Remove the plastic mold. Carefully melt the upper side of the paraffin block and fix it quickly onto a small ($3 \times 3 \times 1$ cm) wooden block holder.
7. Cut ribbons of 5–12.5 µm sections (Supercut Jung RM 2065, Leica, Germany).
8. To remove any wrinkles, float the ribbons on 50% ethanol/50% degassed RNase-free water and transfer them directly onto superfrost plus or TESPA-coated slides. Dry the slides at 40°C overnight. Up to three lanes of sections can be fixed onto one slide.

Dewaxing and Rehydration

9. Transfer the slides, in a slide holder, to xylene and incubate at room temperature for 10 minutes; repeat twice. Rehydrate the slides by 2-minute incubations at room temperature in 100% ethanol, 95% ethanol in water, 80% ethanol in water, 70% ethanol in water, 40% ethanol in Ptw, and finally in 100% Ptw.
10. Refix in 1× PBS/4% formaldehyde at room temperature for 20 minutes.

Prehybridization

11. Rinse the slides with 2× SSPE (20× SSPE: 3.6 M NaCl, 0.2 M NaH_2PO_4 × $1H_2O$, pH 7.4 (NaOH), 20 mM EDTA).
12. Incubate the slides in 3 µg/ml proteinase K in 0.1 M Tris-HCl, pH 7.5, 10 mM EDTA at 37°C for 30 minutes. Rinse the slides twice with 2× SSPE to remove the proteinase K.
13. Treat the slides with 0.2 M HCl for 15 minutes at room temperature to remove all basic proteins that are bound to RNA and therefore prevent hybridization. Rinse the slides twice with 2× SSPE.
14. Transfer the slides to 0.1 M triethanolamine, pH 8.0, and add 0.25% (v/v) acetic anhydride while agitating the slides carefully on a horizontal shaker at room temperature. After 10 minutes, add another 0.25% (v/v) acetic anhydride and agitate the slides again for 10 minutes. Rinse the slides three times with 2× SSPE.
15. For prehybridization and hybridization, place the slides horizontally in an air-tight humidified chamber. For prehybridization, place 300 µl of hybridization buffer onto each slide. Incubate at 65°C for 2–3 hours.
16. For hybridization, heat 500 µl of the probe to 70°C before using it to replace the hybridization buffer. Do not use coverslips. Hybridize overnight at 65°C in a humidified chamber.

Washing

17. Replace the probe with 500 µl of prewarmed hybridization buffer and incubate at 65°C for 10 minutes. Rinse the slides with 500 µl of 2× SSPE at 65°C. Drain the slides, add 300 µl of one part hybridization buffer and one part 2× SSPE including 0.3% CHAPS, and incubate at 65°C for 10 minutes. Drain the slides, add 500 µl of 2× SSPE containing 0.3% CHAPS, and incubate at room temperature for 30 minutes. Wash the slides with 500 µl 2× SSPE at room temperature for 30 minutes.
18. Replace the SSPE with 500 µl of 20 µg/ml RNase A (Sigma R-5000) in 4× SSPE at 37°C for 60 minutes. Rinse the slides with 2× SSPE at 65°C. The final high-stringency washing is done in 500 µl of 50% formamide, 2× SSPE at 65°C for 45 minutes. Drain the slides, add 500 µl of 2× SSPE containing 0.3% CHAPS, and incubate at room temperature for 10 minutes.

Antibody Incubation

19. Rinse the slides three times at room temperature for 10 minutes with buffer TN (TN: 100 mM Tris-HCl, pH 7.5, 150 mM NaCl). Replace the TN with 500 µl of TNBL (TNBL: TN, 1% BMB, 20% lamb serum) and incubate at room temperature for 2 hours.

286 Methods for Studying *Xenopus* Oocytes and Embryos

20. Replace the solution with 200 µl of TNBL, include a 1:1000 to 1:4000 dilution of the affinity-purified sheep antidigoxigenin antibody coupled to alkaline phosphatase (Boehringer Mannheim), and incubate at room temperature for 1 hour. Rinse the slides three times, for 10 minutes each, with buffer TN.

Chromogenic Reaction

21a. For abundant target transcripts, use nitro blue tetrazolium/5-bromo-4-chloro-3-indolyl phosphate (NBT/BCIP) in alkaline phosphate buffer (ABP). The slides are rinsed twice with APB at room temperature for 10 minutes. Replace the buffer with 1 ml of APB including 4.5 µl of NBT (75 mg/ml in dimethylformamide) and 3.5 µl BCIP (50mg/ml in dimethylformamide). The chromogenic reaction is allowed to continue in the dark and will be visible within 2–3 hours.

b. For less abundant target transcripts, use NBT/BCIP in polyvinyl alcohol buffer [PVA: 10% (w/v) PVA MG30000-70000 (Sigma P-8136), 100 mM Tris-HCl, pH 9.5, 100 mM NaCl; dissolve at 90°C]. The addition of polyvinyl alcohol enhances the alkaline phosphate chromogenic reaction and stabilizes the stain; therefore, this increases the sensitivity without background problems (De Block and Debrouwer 1993). The slides are rinsed twice with APB at room temperature for 10 minutes. Place the slides in a vertical staining dish or a staining trough filled with PVA/NBT/BCIP buffer (PVA buffer including 5 mM $MgCl_2$, 0.2 mM NBT, 0.2 mM BCIP). The chromogenic reaction is allowed to proceed at 30°C in the dark overnight.

22. The reaction is stopped by rinsing the slides several times with Ptw. The contrast of the NBT/BCIP stain can be enhanced by carefully rinsing the slides with methanol. The slides are rehydrated by rinsing with 75% methanol/25% water, 50% methanol/50% water, and 25% methanol/75% water, and then washed four times, briefly, in 100% Ptw. The slides are fixed in MEMFA and mounted in PBS glycerol.

Processing of Large Numbers of Samples: A Helpful Novel Apparatus

Random screening of cDNA libraries for genes with suggestive expression characteristics has become a popular method for the identification of novel genes with regulatory activities in early *Xenopus* embryogenesis. This approach implies processing of high numbers of whole-mount samples by hybridization analysis. To this end, we have developed a simple PCR-based probe synthesis protocol and designed an easy-to-use and inexpensive apparatus for the whole-mount in situ hybridization procedure.

Antisense Probe Synthesis

Template Synthesis

1. An aliquot of the PCR-amplified insert from a single cDNA clone is directly used as DNA template for antisense RNA synthesis. Any cDNA library that

is unidirectional and contains an antisense promoter can be used in PCR-based probe synthesis. To obtain single clones, the cDNA library must be plated to single plaques or single colonies, respectively.
2. Single clones are collected into 96-well microtiter plates (Greiner, Germany) in 100 µl of SM buffer for phages, or 100 µl of LB containing the appropriate antibiotic for bacteria, and eluted or incubated overnight.

PCR Reaction
3. The PCR is performed using a 96-well plate PCR cycler (UNOII, Biometra, Germany). For each position, 1 µl of liquid culture or phage eluate is transferred to a 96-well PCR plate using an eight-channel pipette (Brand, Germany). The volume of a single PCR reaction is 25 µl. A premix is prepared for 100 reactions (2.5 ml in total) including 1× PCR buffer, 1.5 mM $MgCl_2$, 0.2 mM dNTPs, 0.15 mM 5' primer, 0.15 mM 3' primer, and 25 U/ml Taq-Polymerase (Perkin-Elmer). The PCR premix is pipetted in 25 µl aliquots using a multipipette (Eppendorf, Germany). The primers were designed such that the amplified product contains no sense-promoter but a nested antisense-promoter. For λ-ZAP-Express libraries, the following primers can be utilized: downstream-primer 5'-CGCGCCTGCAGGTC-GACACTA, upstream-primer 5'- GCAAGGCGATTAAGTTGGGTA. The cycling parameters are initial denaturation for 3 minutes at 95°C, followed by 40 cycles (1 minute, 95°C; 45 seconds, 58°C; 3 minutes, 72°C), and a final polymerization step (10 minutes, 72°C). The amplification products are checked by running a 2.5 µl aliquot from each reaction on a 1.5% agarose gel in 1× TBE.
4. For antisense RNA probe synthesis, 5 µl of each amplification product is transferred directly to a 96-well PCR plate, serving as DNA template for the in vitro transcription reactions. The volume of one in vitro transcription reaction is 25 µl. A premix was prepared for 100 reactions (2 ml in total): 1× transcription buffer, 30 mM DTT, NTPs (0.4 mM ATP, 0.4 mM CTP, 0.4 mM GTP, 0.26 mM UTP, 0.14 mM fluorescein UTP), 0.5 U/µl RNasin, RNase-free water, and 0.5 U/µl RNA-Polymerase. The premix is pipetted in 20-µl aliquots using a multipipette (Eppendorf, Germany). These in vitro transcription reactions are incubated in a 96-well plate PCR cycler at 37°C for 4 hours.
5. In contrast to the basic protocol (see above), the DNA template is not digested with DNase. To remove nonincorporated nucleotides, the RNAs are purified using a 96 gel filtration kit (MoBiTec, Germany). This kit was designed to remove label from primers, or dye terminators from sequencing reactions, but it works just as well for RNA purifications. The 96-well gel filtration block is first centrifuged for 2 minutes at 1500g in a table-top centrifuge with a microplate carrier block to pack the beads. The gel filtration material is then equilibrated with 100 µl of 1× STE buffer (1× STE: 150 mM NaCl, 10 mM Tris-HCl, 10 mM EDTA in RNase-free H_2O). Before loading, each in vitro transcription reaction (25 µl) is diluted with 25 µl 2× STE, loaded onto the 96-well gel filtration block, and centrifuged for 2 minutes at 1500g. One tenth of the eluate should be analyzed on a freshly prepared 1.5% agarose gel in 1× TBE. Because of the incorporated fluorescein nucleo-

tides, the RNA is detectable under UV light without using ethidium bromide. The RNAs are then transferred from the 8 × 12 microtiter plate to four 4 × 6-well tissue culture plates (Nunc, USA) and dissolved in 750 µl of hybridization buffer.

Preparation of Embryos

6. Collection of the embryos is carried out as described above. Three embryonic stages and two embryos of each stage are used for every clone in the screen (two early- to mid-gastrula-stage embryos, two mid- to late-neurula stage embryos, and two late tail-bud to early tail-tip embryos).

Hybridization Conditions
The buffers and solutions are the same as described in the basic protocol. The volumes are approximately three times as large as in the basic protocol.

7. Carefully collect approximately 200 embryos for each of the stages to be analyzed into a 15-ml polystyrene vial (Falcon 2095). Rehydrate the embryos as described above in 15-ml liquid volumes on a rocking platform with gentle agitation.
8. Incubate the embryos in proteinase K at room temperature (gastrula-stage embryos for 15 minutes, neurula-stage embryos for 10 minutes, and tail-bud-stage embryos for 30 minutes). Wash the embryos extensively (at least three times) with 15 ml of Ptw to remove the proteinase K.
9. Rinse the embryos twice with triethanolamine "head over head" on a nutator (leave no air in the vials). Add 2 × 40 µl of acetic anhydride to the rinsed embryos and rock the vials for 5 minutes on a nutator.
10. Wash the embryos five times with Ptw before refixation. Refix the embryos for 20 minutes with 15 ml of 4% formaldehyde in 1× Ptw. Rinse the embryos five times with 1× Ptw for 5 minutes each.
11. For prehybridization, remove most of the Ptw and add 5 ml of hybridization buffer. Wait until all embryos have settled. Remove all buffer and wash twice with 5 ml of hybridization buffer. Incubate the embryos for 20 minutes at 65°C, with shaking, in a water bath. Exchange the hybridization buffer and prehybridize for at least 6 hours at 60°C while shaking.
12. Preheat the 4 × 6-well tissue culture plates containing the RNA probes (step 5) covered with a lid at 60°C in the water bath. If the 3′-poly(A)-tail derived p(U) of the antisense RNA probe is longer than 30 base pairs, add excess unmodified p(U) to a final concentration of 1 µg/ml to compete for unspecific hybridization.
13. Transfer the prehybridized embryos to small, warm glass dishes filled with hybridization buffer. Place the "in situ helpers" (Figure 15-1) into 4 × 6-well tissue culture plates containing 1ml hybridization buffer in each well. Very carefully aliquot the embryos into the individual chambers of the apparatus. Transfer the "in situ helpers" loaded with embryos to the 4 × 6-well tissue culture plates containing the RNA probes and hybridize in the water bath at 60°C, while shaking, overnight.
14. Put the "in situ helpers" back into 4 × 6-well tissue culture plates containing 1 ml of prewarmed (60°C) hybridization buffer in each well.

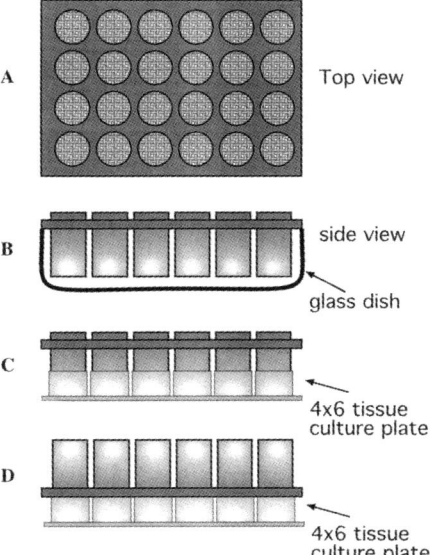

Figure 15-1 An apparatus for the simultaneous processing of 24 different RNA probes in whole-mount in situ hybridization analysis with *Xenopus* embryos. (A) Top view. The apparatus is made out of aluminum; the tubes, which are closed by a nylon grid at the bottom, fit exactly into standard 4 × 6 tissue culture plates. (B) For washing steps, the apparatus is placed into a glass dish. (C) For hybridization and antibody incubation, the apparatus is placed into a 4 × 6 tissue culture plate. (D) For the chromogenic reaction, the embryos are transferred to a 4 × 6 tissue culture plate. See text for further details.

Washing

15. All the washing is done in glass dishes that fit to the "in situ helpers." A dish volume is 250–300 ml. Wash the embryos three times with prewarmed 2× SSC at 60°C for 20 minutes each.
16. RNase treatment is carried out by replacing the 2× SSC with 2× SSC including 20 µg/ml RNase A and incubating at 37°C for 30 minutes. Wash the embryos twice with 0.2× SSC at 60°C for 30 minutes, rocking gently.

Antibody Incubation

The incubations in expensive buffers and antibody solutions were done in 4 × 6-well tissue culture plates to reduce buffer volumes.

17. Rinse the embryos twice at room temperature for 15 minutes with maleic acid buffer, rocking gently.
18. Place the "in situ helpers" into 4 × 6-well tissue culture plates containing 1 ml of MAB/2% BMB in each well and incubate for 30 minutes. Change to MAB/2% BMB/20% lamb serum for at least 1 hour, rocking gently.

19. Change to 0.5 ml of MAB/2% BMB/20% lamb serum and include a 1:2000 to 1:5000 dilution of the antidigoxigenin (antifluorescein) antibody coupled to alkaline phosphatase and incubate for 4 hours, rocking gently.
20. Place the "in situ helpers" back into glass dishes filled with MAB. Wash the embryos four times with MAB for 5 minutes each wash. Wash overnight with MAB at 4°C, rocking gently.
21. Wash the embryos twice with MAB for 5 minutes each wash.
22. Carefully transfer the embryos to 4 × 6-well tissue culture plates.

Chromogenic Reaction
23. The embryos are rinsed twice with alkaline phosphate buffer. The embryos are developed in 1 ml of ABP plus NBT/BCIP. Because the speed of the chromogenic reactions differs from sample to sample, one has to stop the reactions individually.
24. The reaction is stopped by rinsing the embryos several times with water. The contrast of the NBT/BCIP stain can be enhanced by carefully rinsing the embryos with methanol for 1 minute. Afterwards, the embryos are rehydrated as described above. The embryos can be stored in MEMFA at 4°C.

References

De Block, M., and Debrouwer, D. (1993). RNA-RNA in situ hybridization using digoxigenin-labeled probes: the use of high-molecular-weight polyvinyl alcohol in the alkaline phosphatase indoxyl-nitroblue tetrazolium reaction. Anal. Biochem. 215:86–89.

Gawantka, V., Delius, H., Hirschfeld, K., Blumenstock, C., and Niehrs, C. (1995). Antagonizing the Spemann organizer: role of the homeobox gene Xvent-1. EMBO J., 14:6268–6279.

Hayashi, S., Gillam, I. C., Delaney, A. D. and Tener, G. M. (1978). Acetylation of chromosome squashes of *Drosophila melanogaster* decreases the background in autoradiographs from hybridization with [^{125}I]-labeled RNA. J. Histochem. Cytochem. 26:677–679.

Hemmati-Brivanlou, A., Frank, D., Bolce, M. E., Brown, B. D., Sive, H. L., and Harland, R. M. (1990). Localization of specific mRNAs in *Xenopus* embryos by whole-mount in situ hybridization. Development 110:325–330.

Ryan, K., Garrett, N., Mitchell, A., and Gurdon, J.B. (1996). Eomesodermin, a key early gene in *Xenopus* mesoderm differentiation. Cell 87:989–1000.

Strahle, U., Blader, P., Adam, J., and Ingham, P. W. (1994). A simple and efficient procedure for non-isotopic in situ hybridization to sectioned material. Trends Genet. 10:75–76.

16

Visualizing Endogenous and Exogenous Proteins in *Xenopus laevis* (and Other Organisms)

Timothy Carl
Michael W. Klymkowsky

The history of cell and developmental biology has been greatly influenced by our ability to see within the cell and within the organism as a whole. A close, high-resolution look at a cell poses a number of obstacles. Perhaps the most significant is that most cellular components do not efficiently absorb visible light, and so are difficult to see and distinguish from one another. Two generalized approaches have been used to deal with this problem. The first is the use of "phase-based" optical systems, in which differences in refractive index are used to visualize subcellular organelles. This method is limited to relatively thin, optically "simple" specimens. In all but the smallest organisms, such as *Caenorhabditis elegans*, or at higher magnifications, phase-based images can be difficult to interpret. The second approach is to "stain" specific biological molecules so that they absorb or emit light of specific wavelengths, rendering them visible. Until ~25 years ago, such staining was based primarily on the differential binding affinities of various relatively low-molecular-weight dyes to biological macromolecules (see Pearse 1972). While the major classes of biological macromolecules (e.g., carbohydrates, nuclei acids, proteins, lipids) can be distinguished in this manner, the limitations of "generic" staining methods are severe. For example, proteins with dramatically different functions are often, from a chemical perspective, quite similar to one another and not easily distinguished by "generic stains." One alternative to "generic staining" has been to exploit the enzymatic properties of specific proteins (see Pearse 1972) and a number of enzymes can be localized using such "histochemical" techniques. The limitations of histochemistry, however, are painfully obvious: most polypeptides do not have discrete enzymatic activities.

Two breakthroughs have been particularly important in allowing us to visualize the subcellular localization of biological macromolecules: antibody-based detection of specific polypeptides and in situ hybridization-based detection of specific nucleic acids. Initially, these methods were limited to cultured cells or tissue sections. To visualize the spatial distribution of proteins within an intact embryo required the rather tedious and time-consuming process of serial reconstruction of sections. A major breakthrough was stimulated by the work of Mitchison and Sedat (1983); their demonstration that "whole-mount" methods could be used in *Drosophila* embryos to generate a global view of nucleic acid and protein expression spurred our own work to develop similar methods for *Xenopus* (Figure 16-1; Dent et al. 1989). Although whole-mount methods have their limitations, which we will discuss, they provide a mass of data quickly and allow higher resolution studies to focus on specific regions of the embryo, with some assurance that the major sites of RNA/protein expression have been identified.

One weakness of both in situ hybridization and immunocytochemistry is that they "work" only on fixed tissues. To circumvent this limitation, workers have turned to fluorescently modified versions of specific proteins, which can be introduced into the living cell. Particularly in the case of cytoskeletal proteins, this approach was been quite useful (see Kreis and Birchmeier 1982). Originally, this approach involved chemically modifying purified proteins, which were then injected back into cells. The technical problems associated with such methods can be quite daunting. The polypeptide of interest may be difficult to purify in adequate amounts, inherently unstable, or inactivated by the conjugation process. The introduction of the green fluorescent protein (GFP) of *Aequorea victoria*, and its variants, as vital markers by Chalfie et al. (1994) has rejuvenated this approach. Also, GFP has become a powerful tool in both lineage and promoter analysis with organisms. Recent studies indicate that single GFP molecules can be imaged, opening up even more interesting possibilities (see Dickson et al. 1997; Pierce et al. 1997).

The goal of this chapter is to review the methods involved in generating and using GFP-based reagents, their expression in cultured cells and embryos, and the use of whole-mount immunocytochemistry within the developing vertebrate embryo. Our main purpose is to provide a simple, "how-to" guide for the relative novice. Those interested in whole-mount staining of bone and cartilage should refer to Klymkowsky and Hanken (1991). A very useful web site has been prepared by David Gard (http://froglab.biology.utah.edu/html/techniques.html; see also Gard et al. 1995) which describes, in detail, techniques associated with using confocal microscopy to visualize subcellular detail in the *Xenopus* oocyte and early embryo.

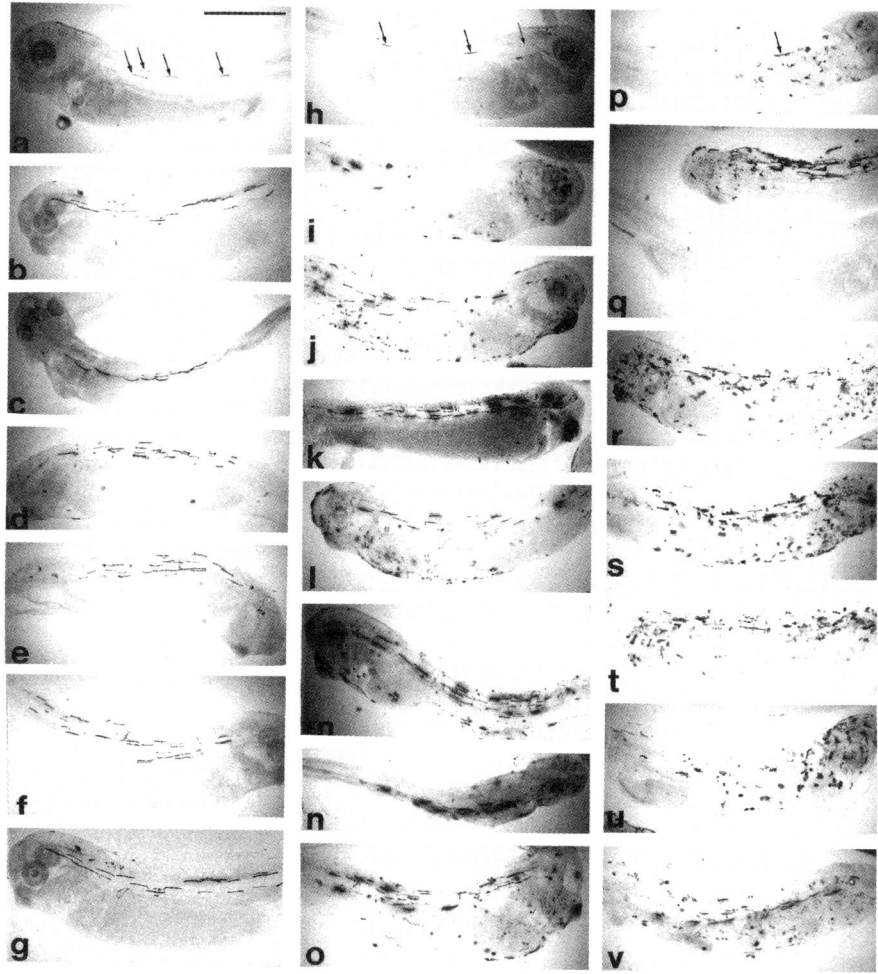

Figure 16-1 Visualization of exogenous proteins by whole-mount staining in *Xenopus*. The development of whole-mount immunocytochemistry makes visualizing the global pattern of protein expression in embryos relatively simple. In this example, fertilized eggs were injected with plasmids in which the cytoskeletal actin promoter of *X. borealis* was used to drive the expression of either (panels a–g) a myc-tagged form of *Xenopus* desmin (Cary and Klymkowsky 1994), (panels h–o) a myc-tagged *Xenopus* vimentin (Dent et al. 1992), or (panels p–v) the *E. coli lacZ* gene (β-galactosidase; supplied by Richard Harland, UC Berkeley). Embryos were fixed in Dent's fixative, bleached, and reacted with either the anti-myc antibody 9E10 or an anti-*lacZ* antisera, and then stained with appropriate peroxidase-conjugated secondary antibodies. After reaction with DAB, the embryos were dehydrated and viewed "cleared" in BABB. Isolated muscle cells, expressing the exogenous proteins, can be seen in the myotome (arrows in parts a, h, and p). The overall impression is that desmin accumulates to high levels preferentially in muscle cells, whereas the vimentin and *lacZ* accumulate in a wider range of cell types. These images are taken from the previously unpublished work of Robert Cary and Mike Klymkowsky.

GFP Protein and its Uses in *Xenopus*

The original GFP sequence encodes an ~28 kDa polypeptide (Prasher et al. 1992). The nascent polypeptide is not itself fluorescent, but rather it undergoes an isomerization reaction within the cell (Cody et al. 1993); typically, this reaction is relatively slow. The native protein has a complex emission spectra, with two absorption maxima: a stronger one at 395 nm (near UV) and a weaker one at 477 nm (blue). In the light-emitting glands of the jellyfish, GFP is excited by light emitted by luciferase or a Ca^{2+}-activated photoprotein, such as aequorin (see Cody et al. 1993). Since its original isolation, a number of mutational changes have been introduced into the GFP sequence to increase its transcriptional and translational efficiency and the efficiency of its activation within the cell. New versions of GFP are constantly being introduced. The simplest of these mutations introduce a single amino acid change of Ser_{65} to either a Thr or an Ala (Heim et al. 1995; Heim and Tsien 1996). The $Ser_{65} \rightarrow Thr$ mutation eliminates the 395 nm absorption maxima of the wild-type protein and leads to a single excitation peak at 490 nm; its emission peak is six-fold higher than the wild-type protein (Heim et al. 1995; Heim and Tsien 1996). In addition, these mutations increase the rate at which the nascent GFP polypeptide becomes fluorescent. A second set of useful variants are the "blue fluorescent proteins" (BFP) which carry mutations at other sites (see Heim et al. 1995; Heim and Tsien 1996; Zernicka et al. 1996, and references therein).

GFP has a compact structure (Ormo et al. 1996) characterized as an 11-stranded β-barrel. It is now widely appreciated that the GFP moiety can be appended to a large number of different polypeptides, such that it (the GFP moiety) folds correctly and generates a fluorescent chimeric polypeptide within the cell. In many cases, these "green" chimeric polypeptides behave quite normally. That said, GFP is much larger than typical "epitope tags" which range between 10 and 15 amino acids in length. Therefore, care must be taken to establish whether the chimeric "green" protein retains its original activities. In particular, it is possible that the presence of the GFP moiety can generate dominant negative changes in protein structure. An example from our own experience comes from studies of the intermediate filament proteins vimentin and desmin. We have epitope-tagged vimentin and desmin with the "myc-epitope" (MEQKLISEEDL — recognized by the monoclonal antibody 9E10; Evan et al. 1985) and found that it assembles normally into 10-nm filaments, both within cells and in vitro (Figure 16-2A; Dent et al. 1992; Cary and Klymkowsky 1994; H. Herrmann, personal communication). In collaboration with Harald Herrmann (German Cancer Research Center), we have examined the behavior of vimentin and desmin tagged with GFP, rather than the myc-epitope. The presence of GFP leads to the formation of nonfilamentous aggregates (Figure 16-2B) and poisoned filament formation by the wild-type protein when present at a ratio of 1:20

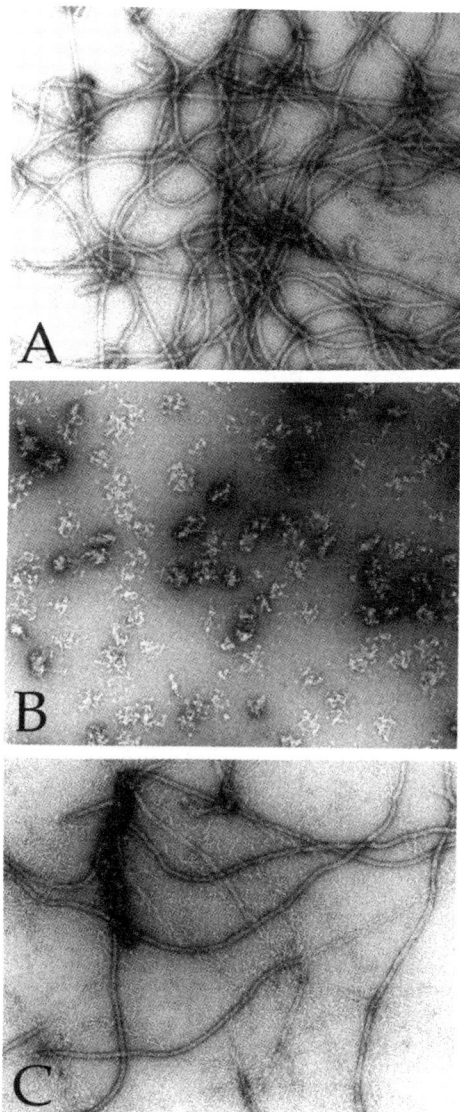

Figure 16-2 Effects of GFP on vimentin assembly. (A) *Xenopus* vimentin, expressed in bacteria and purified to homogeneity, assembles into 8–10 nm diameter filaments in vitro. (B) In contrast, chimeric polypeptides with GFP appended to the C-terminus of vimentin fail to assemble into filaments under the same conditions. (C) Moreover, addition of 1 part in 20 of GFP–vimentin to a solution of vimentin dramatically disrupts the efficiency of vimentin polymerization into filaments; the filaments that do form often fray at their ends into protofibrils. These images are taken from previously unpublished work of Harald Herrmann (German Cancer Research Center) and Mike Klymkowsky.

(GFP to wild type; Figure 16-2C). In a similar vein, the position of the GFP tag can produce profound changes in the behavior of the chimeric polypeptide. For example, the same region of the cytoplasmic tail of human β4 integrin behaves quite differently depending upon whether the GFP moiety is appended to its N- or C-terminus (T. A. Holwell et al., manuscript submitted).

Finally, it is well worth remembering that, within the cell, many polypeptides are unstable and subject to proteolytic processing. Proteolysis can disconnect the GFP group from the rest of the polypeptide, generating an inaccurate view of the chimeric polypeptide's distribution within the cell. This is particularly true given the compact structure of GFP. It is therefore prudent to plan to construct both N- and C-terminal versions of GFP-tagged polypeptides and to monitor their behavior within the cell and embryo by immunoblot analysis.

Intracellular Distribution of GFP

The GFP is small enough to enter nuclei passively. In cultured cells expressing GFP alone (or with a short epitope tag), green fluorescence is seen in both the cytoplasm and the nucleus. In the cytoplasm, green fluorescence can often be seen to be excluded from subregions, which presumably correspond to the endoplasmic reticulum, mitochondria, and other intracellular organelles. There is more free water volume in nuclei than the cytoplasm (Century et al. 1970; Paine, 1984), so that the local GFP level is expected to be higher in the nucleus than it is in the cytoplasm. Coupling the GFP moiety to various polypeptides leads to the redistribution of fluorescence from this "base" state. For example, GFP coupled to the intermediate filament proteins (IFP) vimentin and desmin is exclusively cytoplasmic, whereas GFP coupled to transcription factors (e.g., mouse LEF-1 or XTcf-3) is almost completely nuclear. It should be remembered, however, that if the chimeric polypeptide is proteolytically processed within the cell, fluorescence can appear in either the cytoplasm or the nucleus due to the "released" GFP moiety, rather than to the distribution of the chimeric polypeptide.

GFP Plasmids

Expression plasmids containing the GFP sequence with a number of mutations to optimize expression in various biological systems are available from various commercial vendors. The plasmids we use are based on the pCS2mt plasmid developed by Dave Rupp (Rupp et al. 1994), a description of which can be found linked to the *Xenopus* home page (vize222.zo.utexas.edu/ Marker_pages/PlasMaps/CS2.html). We have inserted the $S_{65} \rightarrow T$ and $S_{65} \rightarrow A$ mutated forms of GFP into these plasmids, such that sequences can be inserted via EcoRI and Xba I restriction sites (see spot.colorado.

edu/~klym/plasmids.html). Using these plasmids, we have prepared constructs tagged at the N-terminus with "myc-GFP" or, at the C-terminus, with GFP alone (typically together with an N-terminal 6 myc sequence of the pCS2mt plasmid. Another very useful plasmid appears to be RN3P and its green and blue derivatives (Zernicka et al. 1996). Capped RNA can be synthesized from these plasmids and injected into fertilized *Xenopus* eggs (10–20 nl or a 0.5 ng/nl solution), using standard pressure injection methods. Green fluorescence is visible beginning around the mid-blastula transition and grows stronger over the next day, persisting into relatively late embryonic stages (Zernicka et al. 1996; Rubenstein et al. 1997) (Figure 16-3).

The pCS2 plasmids also contain a cytomegalovirus (CMV) promoter which works efficiently in a number of *Xenopus* and mammalian cell lines. Having made transgenic tadpoles with the plasmid, however, it is clear that the CMV promoter does not drive expression uniformly in all tissues; higher level expression is found in the ciliated cells of the epidermis and the myotome (C. Dufton and M. W. Klymkowsky, unpublished observations; see Kroll and Amaya 1996).

Plasmid DNA can be injected directly into the nuclei of cultured cells using a simple apparatus consisting of a micromanipulator and a large glass-barrel syringe, with pressure applied using a rubber band. We use standard miniprep DNA prepared with Wizard® minipreps (Promega). As isolated, however, the plasmid DNA will rapidly clog microneedles. To circumvent this problem, we treat 30 µl of miniprep DNA with a Qiagen "PCR clean-up column." The column is eluted with 32 µl of water (2 µl of water are lost in the column) into 30 µl of 2× injection buffer (150 mM KCl, 10 mM Tris, pH 7.8, 1 mM EDTA, 0.05% sodium azide). This solution flows through the microneedle like water. Normally, we inject this solution neat, but it can be diluted if the expression levels obtained are too high.

Following intranuclear injection, green fluorescence is typically visible to the naked eye by ~1–1.5 hours (Figure 16-3). Using the Quantum Biotechnology, Inc. "Blue Fluorescent Protein" plasmid, we found that blue fluorescence appeared much more slowly, typically taking ~12 hours to become visible and longer to become prominent.

To visualize GFP fluorescence, any standard epifluorescence microscope can be used. We carry out epifluorescence visualization using a Zeiss IM35 inverted microscope using a 75 W xenon lamp, a Leitz stereomicroscope equipped with epifluorescence optics and a 100 W mercury lamp, or a Molecular Dynamics confocal microscope using the setting specified for viewing fluorescein. By a day after injection, green fluorescent cells can be easily visualized using a 6× or 16× lens (Figure 16-3). A filter set designed for the blue protein can be purchased from Omega Optical Co. (Brattleboro, VT 05302); however, filter sets designed for DAPI have also been used. The GFP is fairly resistant to photobleaching, but BFP is reportedly more sensitive (Heim and Tsien 1996). To suppress bleaching, 10 µM Troxol (Aldrich

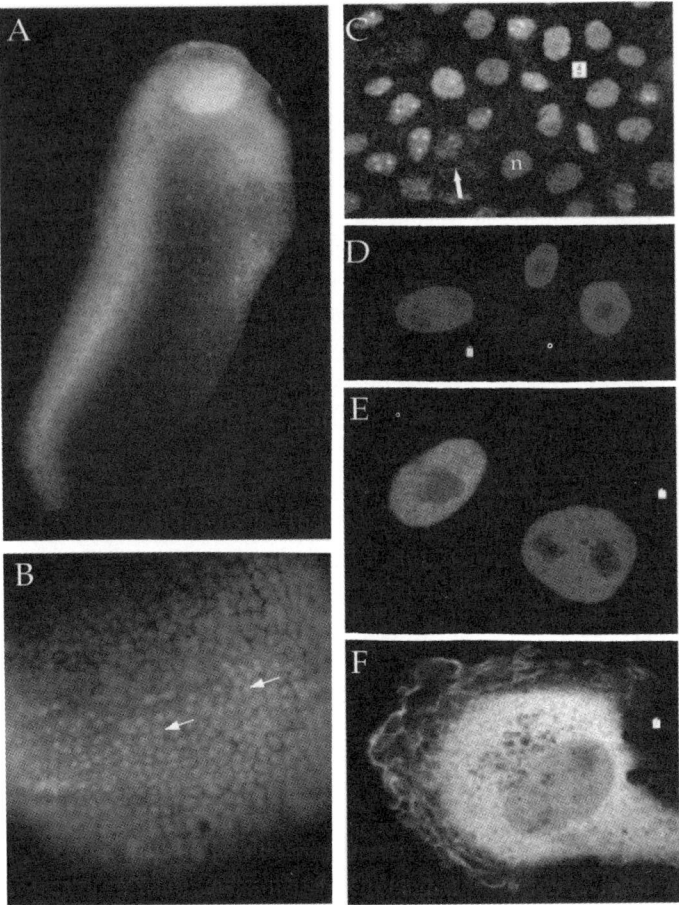

Figure 16-3 Green embryos. Using GFP-tagged polypeptides, it is simple to follow protein expression and intracellular distribution with cells and embryos. (A and B) Mutated forms of the adhesion junction/Wnt-signaling polypeptide plakoglobin are visualized in living (A) neural and (B) gastrula stage *Xenopus* embryos. Part (A) is a late-neural-stage embryo, while part (B) is a gastrula-stage embryo. In the gastrula-stage-embryo (B), the chimeric plakoglobin–GFP polypeptide can be seen to accumulate in the nuclei (arrows). (C) GFP-tagged forms of human plakophilin-1 (original cDNA supplied by Werner Franke, German Cancer Research Center), a desmosomal protein (Hatzfeld 1997), accumulates in the nuclei in later stage ($\sim$30) *Xenopus* embryos, and can also be seen to be weakly associated with cell–cell junctions, where desmosomes are known to be present. (D) In cultured *Xenopus* A6 cells, a full-length, GFP-tagged form of plakophilin is also found primarily in the nuclei. To define the region involved in plakophilin's nuclear localization, we have constructed GFP-tagged versions consisting of (E) the polypeptides N-terminal "head" domain or (F) its C-terminal "armadillo repeat" domain. The N-terminal head domain appears to be primarily responsible for the nuclear targeting of plakophilin-1. These images are taken from the previously unpublished work of Adam Rubenstein, John Merriam, Joshua McElwee, and Mike Klymkowsky.

Chemical Co.) is added to the media (from a 10 mM stock made up in DMSO). Typically, we fix cells in either 100% methanol, 70% acetone/ 30% methanol, or 3.7% formaldehyde in PBS, followed by extraction in acetone/methanol. Fixed cells (or sections) can be mounted in an appropriate media to suppress photobleaching—we use a solution of polyvinyl alcohol (Airvol®)/glycerol/Tris-HCl, pH 8.0, supplemented with propyl gallate (see Klymkowsky and Hanken, 1991). More recently, we have switched to secondary antibodies conjugatd to Alexa 488 or Alexa 594 fluorochromes (Molecular Probes, Inc.). Since GFP and Alexa-conjugates do not bleach rapidly, we routinely omit the n-propyl gallate, from the mounting media.

Viewing GFP-Expressing Embryos

GFP-expressing embryos can be visualized with little, if any, apparent damage to the embryo. Low-power lenses (2.5×, 4× or 6×) are particularly useful for visualizing the entire embryo (Figure 16-3). Late-stage embryos can be quite active and should be anesthetized. Place the embryos in a 0.03% solution of ethyl *m*-amino-benzoate tricaine methanesulfonate (TMS) made up in 10% Holtfreter's solution for 5–10 minutes prior to viewing. Care should be take to minimize exposure to TMS and the embryos should be washed and returned to standard solution (10% MRS) as soon as possible. To hold eggs, embryos, and tadpoles, we use specially constructed brass slides (Figure 16-4). In these slides, the living eggs, embryos, or tadpoles are placed between two glass coverslips, such that the entire slide can be flipped over to view either side of the specimen. High-resolution, oil-immersion lenses can also be used. When approaching the "non-wax-attached" side of the slide, however, care must be taken to move oil-immersion lenses slowly, so that the coverslip is not pulled away by the surface tension between the lens face and the coverslip. Living embryos can be removed from this slide after viewing without harmful side-effects. This slide system is also useful for viewing with BABB-cleared embryos. Fixed embryos should not be left for prolonged periods (i.e., a few days) in the slide.

According to Zernicka et al. (1996), embryos can be fixed for 2 hours in 4% formaldehyde, dehydrated, and embedded in wax as described by Kintner and Brockes (1984), and still preserve green fluorescence. We have had better success in visualizing green fluorescence in embryos fixed for 20–30 minutes in MEMFA (see below). In cultured cells, fixation with methanol, acetone/methanol or formaldehyde followed by acetone/methanol all preserved green fluorescence without significant diminution.

Whole-Mount Immunocytochemistry

The basics of this method have not changed significantly since its introduction in 1988 at the 2nd International *Xenopus* Meeting in New Orleans (see

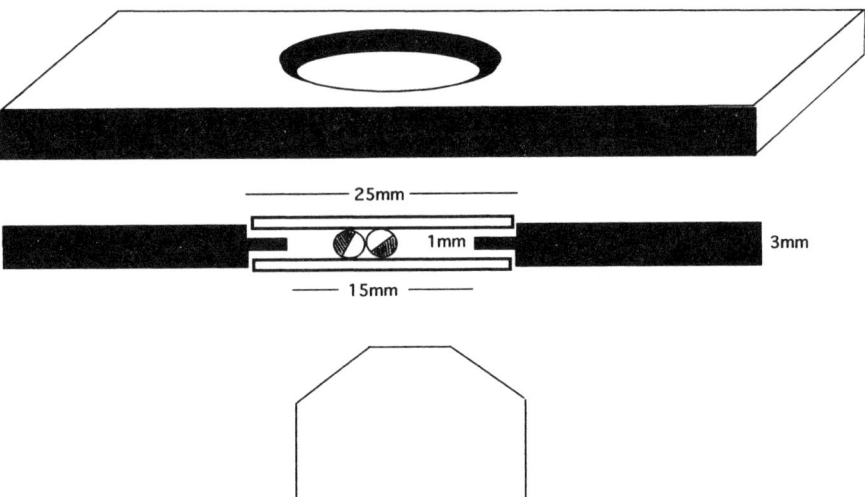

Figure 16-4 Slide schematic. For visualizing living or whole-mount stained embryos, we use custom-designed brass slides. These are the same dimensions as standard glass microscope slides (1 × 3 in), in which a central hole of 1.5 cm diameter has been machined. On either side, and centered on this hole, 2.5-cm indentations have been machined, leaving a 1-mm-thick inner ring. Using melted paraffin wax, #1 thickness coverslips (22 mm diameter) can be attached to one side of the slide. Eggs, embryos, and tadpoles are placed in the well created by the coverslip and slide, the well is filled with media or mounting material, and a second coverslip is placed above. If excess media/mounting material is removed, the slide can be safely flipped over, and the embryos examined on either side. The 1-mm space holds the embryos firmly.

Dent et al. 1989; Hemmati-Brivanlou and Harland 1989; Klymkowsky and Hanken 1991). Fixation can be carried out using either Dent's fixative (80% methanol/20% DMSO) or 3.7% formaldehyde in either phosphate-buffered saline or MEMFA (the fixative used for whole-mount immunocytochemistry). Since aldehyde fixatives do not permeabilize the plasma membrane, formaldehyde fixation must be followed by a permeabilization step—we use Dent's fixative. Embryos are fixed for 2–12 hours at room temperature in Dent's fixative or for 2 hours in aldehyde-based fixatives, followed by an overnight fixation in Dent's fixative to permeabilize the embryo (see below). If your ultimate goal is ultrastructural detail, i.e., electron microscopy, you will need to use a fixative that includes glutaraldehyde and the permeabilization step will be omitted (see below; Cary and Klymkowsky 1994, 1995). Ultrastructural analysis is not compatible with whole-mount staining, and you will need to prepare and stain sections (see below).

If the antigen you are attempting to localize is highly concentrated within a tissue, you will find that whole-mount staining is restricted to the tissue periphery. For example, in staining embryos with antibodies against skeletal muscle myosin, we found that staining of the interior of the myotome was much less intense than that seen at its periphery (Cary and Klymkowsky 1995). Essentially, the high concentration of myosin in myotomal muscle cells acts to absorb the antibody. Similarly, proteins that are expressed at high levels in the epidermis (keratins, β-catenin, etc.) will also act to absorb specific antibodies, inhibiting staining of more interior regions. While staining under these conditions can be informative, it *must* be complemented with section-based methods (see below). In addition to the absorption problem, there is also an imaging problem. Dense surface staining of an embryo obscures interior details. While confocal analysis can be useful for superficial layers, sectioning is required for imaging deeper regions of the embryo.

Choosing Fixatives

The choice of fixatives must be determined empirically. Some antibodies work much better in one type of fixative versus another. For example, the anti-"pan keratin" antibody C11 (sold by Sigma Chemical Co.) reacts very nicely with *Xenopus* keratin filaments in alcohol-fixed tissues, but its reactivity is abolished by aldehyde-based fixatives (see Gard et al. 1997). Alternatively, we have found that staining certain antiplakoglobin antibodies is better preserved by formaldehyde fixation, whereas others are best used in alcohol-based fixatives.

Assuming the target polypeptide is present in cultured cells, fixation sensitivities and appropriate dilutions of primary and secondary antibodies can be rapidly determined. Fixation times are reduced to 5–10 minutes (followed by a 5–10 minute alcohol or detergent extraction in the case of aldehyde-based fixatives). Coverslips are then rinsed briefly with PBS, incubated in primary antibody for 20–30 minutes at room temperature, then rinsed and incubated for 30 minutes in fluorochorome-conjugated or peroxidase-conjugated secondary antibodies. In the case of fluorochrome-conjugated secondary antibodies, the cells (typically grown on a glass coverslip) are mounted on a slide and examined. Normally, we find that an antibody dilution that generates a good signal on cultured cells also works well for whole-mount immunocytochemistry.

If no convenient cell culture system is available to determine fixation sensitivity and antibody titer, these have to be defined using embryos. It is particularly common to use antibodies generated against mammalian proteins in preliminary studies of proteins in amphibian embryos. Care must be taken to recognize "fortuitous" cross-reactivities. For example, we have found that some guinea pig antiplakoglobin antibodies fail to recognize *Xenopus* plakoglobin, but react in a highly specific manner with certain neuronal structures in the tadpole-stage embryo (Figure 16-5). Determining antibody titer and specificity is critical. We have found that different

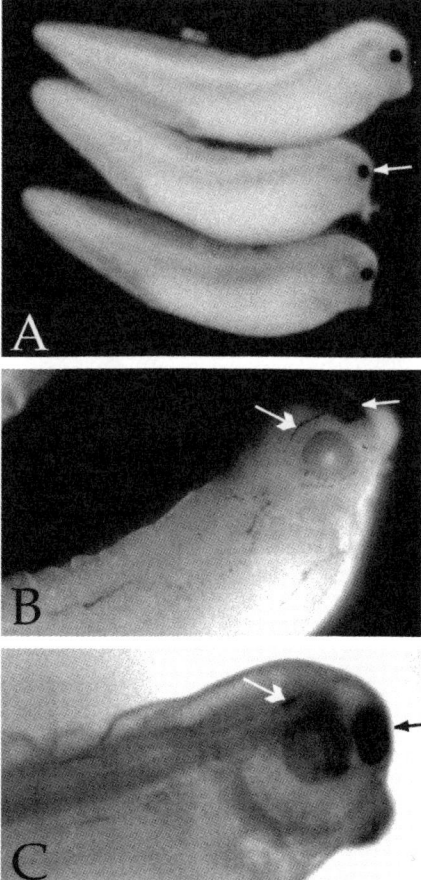

Figure 16-5 The specificity of whole-mount staining. In the course of studies on the distribution of plakoglobin in *Xenopus* embryos, we have examined a number of antiplakoglobin antibodies. In this case, a guinea pig antiplakoglobin antisera (generously supplied by Pam Cowin) was found to stain cells associated with the lateral line system. (A and B) Embryos stained with this antisera reveal an intense reaction with the neuromast, a discrete structure located anterior to the eye (arrow in part A). Higher magnification images reveal that the antisera also stains axonal tracts and varicosities (bold headed arrow in part B). This pattern of staining is more visible in embryos *before* they have been cleared. (C) A cleared embryo. These images are taken from the previously unpublished work of Mike Klymkowsky.

batches of secondary antibody can differ rather dramatically in their effective whole-mount working dilution, even though they appear to be much more uniform with respect to their working dilutions for other use, e.g., immunoblotting.

Bleaching

A major problem faced by those who want to image cells within an embryo is that embryos are not transparent. Given the size of vertebrate embryos and the refractive index of cytoplasm, even the nonpigmented zebra fish embryo is really translucent rather than transparent, particularly when it is examined with high-resolution optics. In the case of *Xenopus* and a number of other amphibians, there is also the matter of animal hemisphere pigmentation and the appearance of pigmented cells later in development. Two approaches have been used to circumvent the problems posed by pigment. The first is the use of albino *Xenopus*, which are available from most commercial suppliers; albino eggs can be fertilized by normal sperm for studies of early embryogenesis (pre stage 30); if later stages are to be examined, the albino eggs must be fertilized with sperm from albino males.

The second approach is to bleach the embryos. Typically, embryos are treated with hydrogen peroxide (we use 10% H_2O_2 in Dent's fixative). Bleaching proceeds faster in embryos fixed with formaldehyde and can be further speeded up by exposing the bleaching embryos to a bright (e.g. halogen) lamp. Bleaching also inactivates endogenous peroxidase activity, which is critical when peroxidase-conjugated secondary antibodies are to be used for staining. It is important to check whether immunoreactivity is abolished or altered by bleaching. In our experience, the antigenicity and intracellular distribution are rarely destroyed by bleaching.

Fixation Artifacts and Pitfalls

An important point regarding fixation that is generally not well appreciated (and rarely discussed) is that while fixation is commonly assumed to "lock down" cellular components more or less instantaneously and permanently, this is often not the case. Proteins have been found to redistribute rather dramatically during and after fixation. A particularly illustrative set of examples has been provided by Melan and Sluder (1992), who found dramatic changes in the intracellular distribution of proteins induced by detergent-extraction and fixation (Figure 16-6). Of particular interest was the cell-type specific redistribution of proteins which were exclusively cytoplasmic in the living cell to a primarily nuclear distribution in the fixed cell. In a complementary set of observations, Mertens et al. (1996) found that the desmosomal protein plakophilin-2 is normally localized to nuclei in cells that do not form desmosomes (see Figure 16-3C–F), but can be almost completely extracted from nuclei in fixed tissues by prolonged incubation in the presence of detergent. Similarly, we have found that under certain conditions exogenous plakoglobin, which accumulates in the nuclei of cultured *Xenopus* A6 cells (and embryonic cells), can be extracted from nuclei following fixation (data not shown). It is now clear that a number of proteins, previously assumed to be exclusively cytoplasmic, turn up to localize

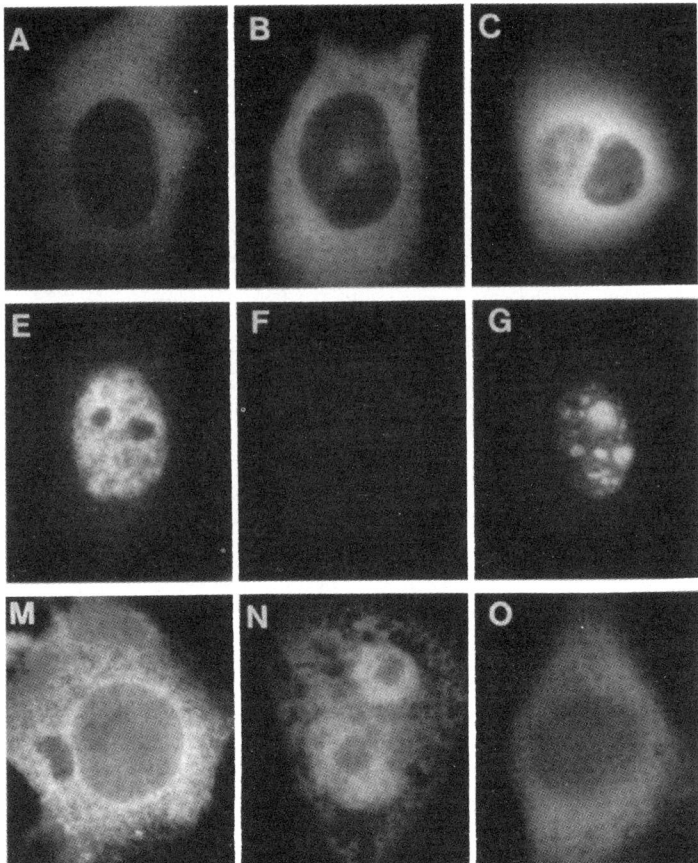

Figure 16-6 Artifacts associated with fixation. In this figure, taken from the work of Melan and Sluder (1992), we see the distribution of fluorescein-conjugated BSA in living (A) PtK$_1$, (B) CHO, and (C) 3T3 cells. In each case, the protein is cytoplasmic. If cells are extracted with 0.1% Triton X-100 prior to formaldehyde fixation, the fluorescent BSA is seen to accumulate in the nuclei in (E) PtK$_1$ and (G) 3T3 cells but not in (F) CHO cells. In cells fixed with 3.7% paraformaldehyde for 30 minutes *before* permeablization with 0.1% Triton, nuclear BSA can be seen in (M) PtK$_1$ and (N) CHO cells, but not in (O) 3T3 cells. These results should instil some caution in a facile interpretation of immunofluorescence data. The image was scanned (600 lines per inch optical resolution) and manipulated using Adobe Photoshop/Illustrator. Original image © The Company of Biologists, used with permission.

to nuclei under certain conditions and in certain cell types (see Figure 16-7D; see Funayama et al. 1995; Karnovsky and Klymkowsky 1995; Gottardi et al. 1996; Keon et al. 1996; Mertens et al. 1996; Rubenstein et al. 1997).

On the other hand, it is a well-known artifact that certain secondary antibody reagents will "nonspecifically" stain nuclear components. Moreover, the level of such nonspecific staining can be dramatically different under

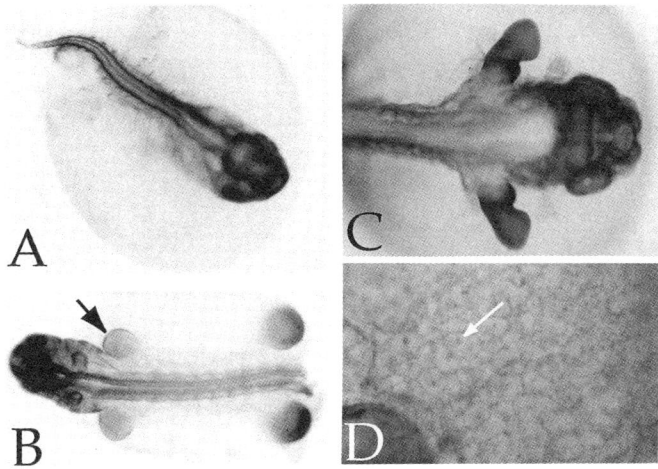

Figure 16-7 Staining of larger embryos. The embryos of *Eleutherodacytlus coqui* are much larger than those of *Xenopus laevis* (~5 mm in diameter versus ~1 mm). Yet they can be easily stained using the whole-mount methods described here. Here we present images of (A) a stage 5 *E. coqui* embryo stained with a monoclonal antibodies against HNK-1 (obtained from Becton Dickinson, San Jose, CA), (B) a stage 5^+ embryo stained with a polyclonal antibody against BMP-4 (a generous gift of the Genetics Institute) and (C and D) a stage 6 embryo stained with antibodies against β-catenin (supplied by Pierre McCrea; McCrea et al. 1993). The embryos have been cleared. In the case of the anti-β-catenin-stained embryo, it is evident that β-catenin is localized to nuclei in the cells that cover the ventral yolk mass (arrow in part D). β-catenin is also highly concentrated in the limbs, where it may participate in Wnt signal processing. These images are taken from the previously unpublished work of Tim Carl, James Hanken, and Mike Klymkowsky.

different fixation conditions. This makes it particularly critical that each batch of secondary reagent be tested against cells or embryos to determine that it does not display unwanted cross-reactivities under the fixation conditions used (see above). All in all, great care (and some caution) must accompany immunocytochemical localization of proteins in order to avoid erroneous (and occasionally preconceived) conclusions.

Clearing

The main characteristic of a clearing agent is that it must match the refractive index (n_D) of the specimen. In addition, it must not bleach out the stains used in visualization reactions or noticeably alter the structure of the embryo. Clearing agents have a long history (see Pearse 1972; Klymkowsky and Hanken 1991). For *Xenopus*, the most effective appears to be BABB, a 1:2 mixture of benzyl alcohol and benzyl benzoate (both available from Aldrich Chemical Co.). Andrew Murry developed BABB while he was working in Marc Kirschner's lab, and its introduction certainly stimulated our own work on the development whole-mount methods. *Xenopus* embryos, and the embryos of a number of other species, are rendered completely transparent by BABB (Figures 16-1 and 16-7; Dent and Klymkowsky 1989; Klymkowsky and Hanken 1991). The drawbacks of using BABB are that (1) it is extremely irritating if you get it on your skin or in your eyes, and it is very difficult to remove — gloves should be worn at all times when using it; (2) it renders embryos brittle; and (3) it will solubilize many chromogenic substrates (Dent et al. 1989). It is compatible with DAB-based peroxidase visualization and fluorescent reagents. Both components of BABB are harmful, and the proper care should be taken when handling and disposing of this solution.

Clearing embryos makes visualizing the intraembryonic distribution of immunostaining more clearly visible (Chu and Klymkowsky 1989; Dent et al. 1989) provided that the staining is not dense (see Figure 16-1). Clearing can also generate image confusion by effectively superimposing different regions of the embryo, and by the internal detail made visible (Figures 16-5 and 16-7). In particular, internal cavities can be stained weakly and apparently nonspecifically. Such staining is not visible in the uncleared embryo, but adds to the overall noise in the cleared embryo (Figure 16-5), particularly when the specific signal is relatively weak. Confocal microscopy can be helpful in this regard, but the large size of the embryo makes high-resolution imaging of internal regions difficult, if not impossible, due to the limited working distance of most high-resolution lenses. In part, this problem can be circumvented by preparing sections of the embryo (see below). In particular, the studies by Gard and colleagues on the cytoskeletal systems of the *Xenopus* oocyte utilize this approach and have led to a number of stunning images (Gard 1994; Gard et al. 1997).

Incubation and Washing Conditions

We dilute antibodies into 20% calf serum, 10% DMSO, 70% TBS, and 0.05% thimoserol (as a perservative, do not use sodium azide as it inactivates peroxidases). Embryos are incubated overnight, with gentle rocking, at room temperature. In the morning, embryos are washed three times in rapid succession in TBS and then washed three more times for 2 hours each with TBS. At the end of the day, they are placed into secondary antibody solution and again incubated overnight, washed, and then staining is visualized.

Visualization Reagents

We normally use either horseradish peroxidase-, fluorescein-, or Alexa 488 (green) or Alexa 594 conjugated secondary reagents to visualize antibody staining in the embryo. We find that fluorochrome conjugates work well at a

1:200 dilution; peroxidase conjugates work in the range of 1:100 to 1:2000 dilution. In the case of peroxidase-conjugated antibodies, embryos are placed in a solution containing 0.5 mg/ml DAB and 0.02% H_2O_2 — the reaction is monitored visually and stopped when the staining intensity seems appropriate. Typically, reactions are performed for 15–60 minutes. Very rapid and intense staining (less than 5 minutes) indicates that primary and secondary antibody concentrations are too high and should be reduced. The peroxidase reaction is stopped by removing the DAB solution and washing the embryos in TBS, and then dehydrating the embryos through a methanol series. Embryos can be stored indefinitely in 100% methanol. Once in methanol, the embryos can be cleared in BABB and examined (Figures 16-1 and 16-7). Alternatively, embryos can be washed in TBS and visualized prior to clearing (Figure 16-5).

Fluorescently stained embryos can be viewed directly using epifluorescence microscopy (see Merriam et al. 1997). If fluorochrome-conjugated secondary reagents are to be used, it is probably better to avoid bleaching the embryo, since the pigment acts to shield out some of the intense autofluorescence associated with the underlying yolk. Alternatively, confocal microscopy can be used to minimize autofluorescence. Embryos stained with fluorescein-conjugated secondary reagents can be mounted in one of the commercially available antifade reagents or our own home-made mounting media (see "Appendix: Solutions") and visualized using the type of slide described in Figure 16-4. Alternatively 0.8-mm depression slides can be used, with a coverslip attached over the depression.

A recently introduced variation to the DAB-based visualization strategy is the use of "conjugated tyramide" substrates (NEN). These reagents can be used to generate a fluorescent signal from a peroxidase-conjugated secondary reagent and are said to offer higher sensitivity and make it possible to reduce primary and secondary antibody concentrations. In preliminary studies, we have found these reagents to work well and they are likely to prove quite useful. The newly introduced family of Alexa-conjugated antibodies, from Molecular Probes, Inc.) are also a real improvement over fluorescein, Texas Red and rhodamine-conjugated antibodies. They are brighter, do not bleach, and are significantly less expensive than the tyramide reagents.

Working Method for Whole-Mount Immunocytochemistry
1. If a jelly coat is present, dejelly the embryo chemically (2% cysteine, pH 7.8–8.0). We normally dejelly *Xenopus* embryos following fertilization. Wash with 20% MRS and maintain in 10% MRS until the desired embryonic/developmental stage is reached.
2. Fix overnight at room temperature in Dent's fixative or for 2 hours in MEMFA, followed by overnight fixation in Dent's fixative. (At this stage, embryos can be placed into 100% methanol and stored indefinitely).

3. If appropriate, bleach embryos using Dent's bleach (10% H_2O_2 in Dent's fixative). To speed bleaching, place embryos in clear glass vials, under a bright light, such as the sun or a halogen lamp.
4. Rehydrate specimens through a methanol/TBS series (1:1, 1:3; for 5 minutes each) into TBS.
5. Incubate specimens overnight in primary antibody diluted into 20% calf serum/10% DMSO/TBS/0.05% thimerosal (as a preservative). Appropriate dilution of the primary antibody must be determined empirically. Typically, ascites fluids can be diluted up to 1:2500 or higher. Polyclonal antibodies typically range from 1:10 to 1:10,000.
6. Wash over the course of the next day in TBS (typically five or six changes).
7. Incubate overnight in secondary antibody. Again, dilution of secondary antibody must be determined empirically.
8. Wash over the course of the next day in TBS (typically five changes, 1–2 hours between changes).

Alternative Routes
1. Fluorescently stained specimens can be mounted and examined. Typically, we use specially manufactured brass slides (Figure 16-4) . A glass coverslip is attached, using wax, to one side of the slide and embryos are placed on that coverslip along with the mounting media. A second coverslip is applied to the top, taking care to avoid air bubbles and to remove excess mounting media. The embryo is sealed between the two coverslips, and normally the slide can be viewed in either orientation. If oil-immersion lenses are used, care must be taken with the non-wax-sealed coverslip, to avoid pulling it away from the slide. Surface tension is normally sufficient to hold the coverslip in place. Because of the design, these slides will securely hold the embryo in place and both sides of the embryo can be examined. One disadvantage is that in later stage embryos, the geometry of the slide constrains views to the lateral surfaces of the embryo.
2. Peroxidase-stained specimens are reacted with DAB solution or tyramide reagents (see manufacturer's directions) until the desired staining intensity is achieved. When using DAB, wear gloves as DAB is carcinogenic. Bleach can be used to destroy DAB.
3. Stop the DAB reaction by washing with TBS supplemented with 0.05% Na azide. At this point, the embryo can be examined directly (without clearing) or dehydrated and cleared. Dehydrate through a methanol series (3:1, 1:1, 1:3; 5 for minutes each) into 100% methanol.
4. Transfer into a BABB series (or through a Histoclear series for paraffin embedding). Sometimes, direct transfer from methanol to BABB will cause wrinkling of the embryo, if this is the case, transfer through a methanol/BABB series into 100% BABB.

Sectioning and Immunostaining

There are times when whole-mount methods are simply inadequate. Typically, proteins that are expressed throughout the embryo generate very poor images, and their intracellular distribution can be clearly seen only in superficial layers. Under these conditions, the embryo must be sectioned and stained. An extremely nice example of a section-based analysis in *Xenopus* is the study by Fagotto and Gumbiner (1994; Figure 16-8). They fixed embryos in Dent's fixative and then embedded them in 15% cold-water-fish gelatin (Sigma Chemical Co., G-7765) and 15% sucrose or in 7.5% porcine gelatin (Bloom 300, Sigma Chemical Co.) and 15% sucrose, and stored then at 4°C. For sectioning, embryos were frozen using Tissue-Tek and a dry ice/ethanol bath and sectioned using a cryostat. Sections were postfixed in acetone, rehydrated, blocked in 5% nonfat milk and incubated in primary and secondary antibodies. For their studies, sections were rinsed once in PBS containing 0.1% Eriochrome black to suppress yolk autofluorescence, as described by Torpey et al. (1992), and mounted for microscopy. We have found, however, that the Eriochrome blocking step can often be omitted.

Cryostat Sectioning Method (after Fagotto and Gumbiner 1994)
1. Fix embryos with a fixative compatible with the antibodies to be used; Dent's fixative, 100% methanol, and MEMFA are all good choices.
2. Wash briefly in PBS and soak in either 15% cold-water-fish gelatin/15% sucrose in PBS overnight or in 7.5% porcine gelatin (Bloom 300)/15% sucrose in PBS for 6 hours at 37°C.
3. Embed in 15% sucrose/7.5% gelatin, chill to 4°C (you can hold the samples here for a few days).
4. Remove sucrose/gelatin solution and add Tissue-Tek OCT (Miles Scientific, Naperville, IL 60566) compound, allow to equilibrate for 30 minutes.
5. Prepare a block using Tissue-Tek. By controlling temperatures, it is possible to position your sample on the block in the desired orientation. Typically, this is done using a stereomicroscope and forceps. Once in position, refreeze the block, either in a −80°C freezer or using a dry ice/ethanol bath.
6. Place the block into the cryostat and allow it to equilibrate to the cutting temperature (i.e., −17°C). This normally takes 30–40 minutes. Trim OCT with a razor blade, so that the specimen resides within a trapezoidal block. If the block is not trimmed properly it will be difficult to obtain good sections — see Presnell and Schreibman (1997) for a more exhaustive description of sectioning techniques and methods.
7. Prepare 12–14 μm thick sections using a cryostat at −17°C and collect onto precoated or frosted glass slides (Colorfrost[R]/Plus, Fisher Scientific Co. — these work very well!) and store until you are ready to stain them.
8. Warm the slides and allow them to dry at room temperature.
9. Extract for 2 minutes in 100% acetone.
10. Rehydrate in PBS and block by incubation in 5% low-fat milk in PBS (or the blocking reagent that is often supplied by the manufacturers of various staining kits). If you are planning to use peroxidase-conjugated antibodies, it is advisable to inactivate endogenous peroxidase by treating the rehydrated sections with 1% H_2O_2 in PBS for 15 minutes at room temperature *prior* to blocking.

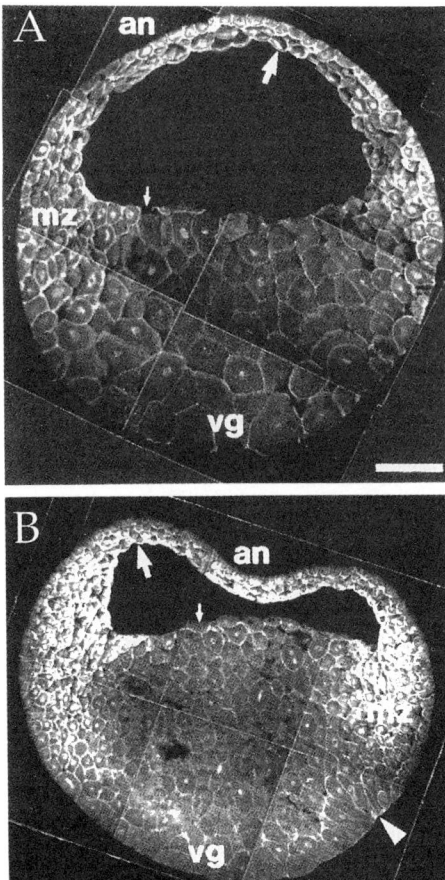

Figure 16-8 Using sections to visualize endogenous proteins. This pair of sections, taken from the work of Fagotto and Gumbiner (1994), show a beautiful example of the use of sectioning to visualize the global distribution of a protein. an, animal pole; vg, vegetal pole. Sections were stained with an anti-β-catenin antibody (McCrea et al. 1993) and fluorescein-conjugated secondary antibody. (A) β-catenin is seen associated with the cell periphery and nuclei in blastula-stage embryos. (B) A similar pattern of staining is seen in gastrula-stage embryos. As is evident from these images, and as stated by Fagotto and Gumbiner (1994), "No obvious difference in the intensity of the β-catenin staining could be detected between dorsal and ventral regions." The image was scanned (600 lines per inch optical resolution) and manipulated using Adobe Photoshop/Illustrator. Original image © The Company of Biologists, used with permission)

11. Incubate in primary antibody (2 hours at 30–37°C) (overnight incubation can also be tested).
12. Rinse in 1% low-fat milk in PBS and incubate with secondary antibody (2 hours at 30–37°C).

13. (*Optional*) Rinse in 1% low-fat milk in PBS and then in 0.1% Eriochrome black. The Eriochrome black step can often be omitted particularly if the samples are to be examined using a confocal microscope, which suppresses much of the yolk autofluorescence (see, e.g., Merriam et al. 1997).
14. Mount the slide using an "antifade" mounting media and examine.

Ultrastructural Methods

If ultrastructural information is required, plastic sections must be prepared. A number of methods have been used over the years. For example, in our studies of the behavior of chimeric forms of the IFPs vimentin and desmin in the *Xenopus* myotome (Cary and Klymkowsky 1994), embryos were fixed in 4% paraformaldehyde, 0.1% glutaraldehyde in 10 mM $MgCl_2$, 10 mM $CaCl_2$, and 150 mM sodium cacodylate, pH 7.4, overnight at 4°C. Fixed embryos were then rinsed for 30 minutes in 150 mM glycine, 150 sodium cacodylate, pH 7.4, and dehydrated through an ethanol series. Dehydrated embryos were infiltrated with LR White resin (20% resin for 1 hour, 50% resin for 2 hours, 66% resin for 4 hours, and then 100% resin overnight at 4°C). After replacing the resin twice, the embryos were placed in gelatin capsules and the resin polymerized in a vacuum oven for 36 hours at 58°C. Sections (2.5 μm thick and taken with a glass knife) were stained with the 9E10 monoclonal anti-myc epitope antibody (used at a higher than normal concentration of 50 μg/ml) and fluorescein-conjugated secondary antibody to reveal cells expressing the myc-tagged polypeptides. If a cell expressing the mutant protein was found, that section was recovered, resectioned for electron microscopy, and stained with immunogold reagents (see Cary and Klymkowsky 1994, 1995).

This rather tedious process was made necessary by the highly mosaic nature of expression of exogenous DNA in the early embryo (Figure 16-1). The problem can be circumvented, to some extent, using confocal microscopy. It is the development of transgenic technology for *Xenopus*, however, that promises to greatly simplify ultrastructural studies of this type. It should be possible to go directly to the EM level section, assuming that uniform expression of the exogenous protein can be confirmed at the light level (e.g., using GFP).

ACKNOWLEDGMENTS We thank François Fagotto, Barry Gumbiner, Melissa Melan, and Greenfield Sluder for permission to reprint figures from their work. We thank Joe Heilig for the use of his cryostat and his advice on cryosectioning. We thank Pam Cowin for guinea pig antiplakoglobin antibodies and the Genetics Institute for anti-BMP antibodies. Our own work has been supported over the last decade by the American Cancer Society, the Muscular Dystrophy Association, the American Heart Association, the National Institutes of Health, and the National Science Foundation. We thank Harald Herrmann, Bob Evans, Bob Boswell, James Hanken, and our many other colleagues for their support and encouragement over the years.

References

Cary, R. B., and Klymkowsky, M. W. (1994). Differential organization of desmin and vimentin in muscle is due to differences in their head domains. J. Cell. Biol. 126:445–456.

Cary, R. B., and Klymkowsky, M. W. (1995). Disruption of intermediate filament organization leads to structural defects at the intersomite junction in *Xenopus* myotomal muscle. Development 122:1041–1052.

Century, T. J., Fenicheck, I. R., and Horowitz, S. B. (1970). The concentration of water, sodium and potassium in the nucleus and cytoplasm of amphibian oocytes. J. Cell Sci. 7:5–13.

Chalfie, M., Tu, Y., Euskirchen, G., Ward, W. W., and Prasher, D. C. (1994). Green fluorescent protein as a marker for gene expression. Science 263:802–805.

Chu, D. T. W., and Klymkowsky, M. W. (1989). The appearance of acetylated α-tubulin during early development and cellular differentiation in *Xenopus*. Dev. Biol. 136:104–117.

Cody, C. W., Prasher, D. C., Westler, W. M., Prendergast, F. G., and Ward, W. W. (1993). Chemical structure of the hexapeptide chromophore of the *Aequorea* green-fluorescent protein. Biochemistry 32:1212–1218.

Dent, J. A., and Klymkowsky, M. W. (1989). *The Cell Biology of Fertilization*, Shatten, H. and Shatten, G., eds, New York, Academic Press.

Dent, J. A., Polson, A. G., and Klymkowsky, M. W. (1989). A whole-mount immunocytochemical analysis of the expression of the intermediate filament protein vimentin in *Xenopus*. Development 105:61–74.

Dent, J. A., Cary, R. B., Bachant, J. B., Domingo, A., and Klymkowsky, M. W. (1992). Host cell factors controlling vimentin organization in the *Xenopus* oocyte. J. Cell. Biol. 119:855–866.

Dickson, R. M., Cubitt, A. B., Tsien, R. Y., and Moerner, W.E. (1997). On/off blinking and switching behavior of single molecules of green fluorescent protein. Nature 388:355–358.

Evan, G. I., Lewis, G. K., Ramsay, G., and Bishop, J. M. (1985). Isolation of monoclonal antibodies specific for human c-myc proto-oncogene product. Mol. Cell. Biol. 5:3610–3616.

Fagotto, F., and Gumbiner, B. M. (1994). β-catenin localization during *Xenopus* embryogenesis: accumulation at tissue and somite boundaries. Development 120:3667–3679.

Funayama, N., Fagotto, F., McCrea, P., and Gumbiner, B. M. (1995). Embryonic axis induction by the armadillo repeat domain of b-catenin: evidence for intracellular signaling. J. Cell. Biol. 128:959–968.

Gard, D. L. (1994). γ-tubulin is asymmetrically distributed in the cortex of *Xenopus* oocytes. Dev. Biol. 161:131–140.

Gard, D. L., Cha, B. J., and Schroeder, M. M. (1995). Confocal immunofluorescence microscopy of microtubules, microtubule-associated proteins, and microtubule-organizing centers during amphibian oogenesis and early development. Curr. Topics Dev. Biol. 31:383–431.

Gard, D. L., Cha, B. J., and King, E. (1997). The organization and animal–vegetal asymmetry of cytokeratin filaments in stage VI *Xenopus* oocytes is dependent upon F-actin and microtubules. Dev. Biol. 184:95–114.

Gottardi, C. J., Arpin, M., Fanning, A. S., and Louvard, D. (1996). The junction-associated protein, zonula occludens-1, localizes to the nucleus before the maturation and during the remodeling of cell–cell contacts. Proc. Natl. Acad. Sci. USA 93:10779–10784.

Hatzfeld, M. (1997). Band 6 protein and cytoskeletal organization. In *Cytoskeletal–membrane interactions and signal transduction*, Cowin, P., and Klymkowsky, M. W., eds. Austin, TX, R. G. Landes.

Heim, R., and Tsien, R. Y. (1996). Engineering green fluorescent protein for improved brightness, longer wavelengths and fluorescence resonance energy transfer. Curr. Biol. 6:178–182.

Heim, R., Cubitt, A. B., and Tsien, R. Y. (1995). Improved green fluorescence. Nature 373:663–664.

Hemmati-Brivanlou, A., and Harland, R. M. (1989). Expression of an engrailed-related protein is induced in the anterior neural ectoderm of early *Xenopus* embryos. Development 106:611–617.

Karnovsky, A., and Klymkowsky, M. W. (1995). Anterior axis duplication in *Xenopus* induced by the over-expression of the cadherin-binding protein plakoglobin. Proc. Natl. Acad. Sci. USA 92:4522–4526.

Keon, B. H., Schafer, S., Kuhn, C., Grund, C., and Franke, W. W. (1996). Symplekin, a novel type of tight junction plaque protein. J. Cell Biol. 134:1003–1018.

Kintner, C. R., and Brockes, J. P. (1984). Monoclonal antibodies identify blastemal cells derived from dedifferentiating limb regeneration. Nature 308:67–69.

Klymkowsky, M. W., and Hanken, J. (1991). Whole-mount staining of *Xenopus* and other vertebrates. In *Xenopus laevis: Practical Uses in Cell and Molecular Biology, Methods in Cell Biology*, Vol. 36, Kay, B. K. and Peng, H. B. eds, Academic Press, Inc., pp. 419–441.

Kreis, T. E., and Birchmeier, W. (1982). Microinjection of fluorescently labeled proteins into living cells, with emphasis on cytoskeletal proteins. Int. Rev. Cytol. 75:209–227.

Kroll, K. L., and Amaya, E. (1996). Transgenic *Xenopus* embryos from sperm nuclear transplantations reveal FGF signaling requirements during gastrulation. Development 122:3171–3183.

McCrea, P. D., Brieher, W. M., and Gumbiner, B. M. (1993). Induction of a secondary body axis in *Xenopus* by antibodies to β-catenin. J. Cell Biol. 123:477–484.

Melan, M. A., and Sluder, G. (1992). Redistribution and differential extraction of soluble proteins in permeabilized cultured cells: implications for immunofluorescence microscopy. J. Cell. Sci. 101:731–743.

Merriam, J., Rubenstein, A., and Klymkowsky, M. W. (1997). Cytoplasmically-anchored plakoglobin induces a Wnt-like phenotype in *Xenopus*. Dev. Biol. 185:67–81.

Mertens, C., Kuhn, C., and Franke, W. W. (1996.). Plakophilins 2a and 2b: Constitutive proteins of dual localization in the karyoplasm and the desmosomal plaque. J. Cell Biol. 135:1009–1025.

Mitchison, T. J., and Sedat, J. (1983). Localization of antigenic determinants in whole *Drosophila* embryos. Dev. Biol. 99:261–264.

Ormo, M., Cubitt, A. B., Kallio, K., Gross, L. A., Tsien, R. Y., and Remington, S. J. (1996). Crystal structure of the *Aequorea victoria* green fluorescent protein. Science 273:1392–1395.

Paine, P. L. (1984). Diffusing and non-diffusing proteins in vivo. J. Cell Biol. 99:188s–195s.

Pearse, A. G. E. (1972). *Histochemistry, Theoretical and Applied.* New York, Churchill Livingstone.

Pierce, D. W., Hom-Booher, N., and Vale, R. D. (1997). Imaging individual green fluorescent proteins. Nature 388:338.

Prasher, D. C., Eckenrode, V. K., Ward, W. W., Prendergast, F. G., and Cormier, M. J. (1992). Primary structure of the *Aequorea victoria* green-fluorescent protein. Gene 111:229–233.

Presnell, J. K., and Schreibman, M. P. (1997). *Humanson's Animal Tissue Techniques*, 5th ed., Baltimore and London, The Johns Hopkins University Press.

Rubenstein, A., Merriam, J., and Klymkowsky, M. W. (1997). Localizing the adhesive and signaling functions of plakoglobin. Dev. Genet. 20:91–102.

Rupp, R. A. W., Snider, L., and Weintraub, H. (1994). *Xenopus* embryos regulate the nuclear localization of XMyoD. Genes Dev. 8:1311–1323.

Torpey, N. P., Heasman, J., and Wylie, C. C. (1992). Distinct distribution of vimentin and cytokeratin in *Xenopus* oocytes and early embryo. J. Cell Sci. 101:151–160.

Zernicka, G. M., Pines, J., Ryan, K., Siemering, K. R., Haseloff, J., Evans, M. J., and Gurdon, J. B. (1996). An indelible lineage marker for *Xenopus* using a mutated green fluorescent protein. Development 122:3719–3724.

Appendix: Solutions

Dent's Fixative
20% Dimethyl sulfoxide (DMSO) in methanol

MEMFA
0.1 M MOPS, pH 7.4 (with NaOH)
2 mM EGTA
1 mM MgSO$_4$
3.7% Formaldehyde (1/10 volume of concentrated stock, 37%)

Make up a solution of 1 M MOPS, 20 mM EGTA, and 10 mM MgSO$_4$, pH 7.4 (10× salts).

This solution can be autoclaved and stored at room temperature. It may go yellow with exposure to light but this does not affect its performance. To make working solution, add 1 volume of 10× salts, 1 volume of 37% formaldehyde, and 8 volumes of water. Working stock is stable at room temperature for ∼2 weeks.

Dent's Bleach
1 part 30% hydrogen peroxide, 2 parts Dent's fixative

Diaminobenzidine (DAB)
Make stock of 10 mg/ml in distilled water and store in 1-ml aliquots at $-20°C$ until needed.

DAB Reaction Mixture
0.5 mg/ml DAB (diluted 1:20 from stock) in Tris-buffered saline
0.02% H_2O_2

Tris-Buffered Saline (TBS)
20 mM Tris, pH 7.5
150 mM NaCl

BABB
1 part Benzyl alcohol, 2 parts benzyl benzoate

Modified Ringer's Saline (MRS)
6.3 g NaCl
0.15 g KCl
0.2 g $MgCl_2$
0.17 g $NaHCO_3$
1.2 g HEPES
0.3 g $CaCl_2$
Add distilled water to 1 liter, pH to 7.8.

0.03% TMS
Ethyl *m*-amino-benzoate tricaine methanesulfonate (Sigma # E1626)
0.03% Ethyl m-amino-benzoate tricaine methanesulfonate, buffered to pH 7.0 with sodium bicarbonate

Holtfreter's Solution (100%)
60 mM NaCl
0.6 mM KCl
0.9 mM $CaCl_2$
0.2 mM $NaHCO_3$

Mounting Media
Dissolve 10 g of Airvol 205 (polyvinyl alcohol; Air Products, Inc., Allentown, PA) in 40 mL of 50 mM Tris-HCl, pH 8.0, for 24–48 hours at 37°C.
Add 20 ml of glycerol and 1.2 g of *n*-propyl gallate.
Aliquot and store. (Omit propyl gallate when using Alexa conjugated secondary antibodies.)

17

Qualitative Analysis of Differential Gene Expression during Early *Xenopus* Embryogenesis by Whole-Mount In Situ Hybridization and RT–PCR

L. Lynn McGrew
Stefan Hoppler
Randall T. Moon

During the embryonic development of an animal such as *Xenopus laevis*, cells are produced through many cycles of cell division. In order to form a developing embryo, these cells have to be organized. Early events set up the embryonic axes and the germ layers, but the germ layers are further patterned to position different organs which themselves consist of different tissues. When scientists started to study the processes during development which pattern the embryo, they had to assay the effects of their experimental manipulation on differentiated tissues which were often not recognizable until considerably later in development. This posed the question of whether the observed effects were a direct result of the experimental manipulation or whether they were cumulative and an indirect consequence of the experiment. With the advent of molecular biology, molecular markers became available which allowed the study of patterning as a dynamic process. Molecular markers are gene products, either RNA transcripts or proteins, the expression of which is restricted to a subpopulation of cells and is indicative of a certain tissue or a certain cell type. The products of these marker genes can range from tissue-specific transcription factors, e.g., *goosecoid* (Cho et al. 1991) to particular structural genes characteristic for certain organs, e.g., *Xag-1* for the cement gland, (Sive et al. 1989).

We would like to illustrate how molecular markers can be used to study the dorsoventral patterning of the developing mesoderm and the anteroposterior patterning of the developing central nervous system (CNS) of *Xenopus*. Molecular markers are useful in that they can predict early patterning events prior to the time when organized morphology is established. We have also found it often sufficient to analyze the expression of marker genes by qualitative methods since their transcription normally becomes upregulated many-fold above background in the tissues and cells in which they are expressed. Two methods of analysis of gene transcription have been used very successfully in recent years to investigate early patterning in *Xenopus* development: the whole-mount RNA in situ hybridization technique which was originally adapted for *Xenopus* embryos by Richard Harland (Harland 1991), and the reverse transcription–polymerase chain reaction (RT–PCR) method which was developed for *Xenopus* by Ralph Rupp and Hal Weintraub (Rupp and Weintraub 1991). We provide well-established, straightforward protocols which are intended for inexperienced users who would like to study molecular markers during *Xenopus* development.

The two methods are best used in conjunction in order to assess the effects of experimental manipulation on marker gene expression. The RNA in situ hybridization analysis provides accurate spatial information about the expression of marker genes in the context of a whole embryos whereas RT–PCR is very powerful for the analysis of gene expression in the more controlled experimental environment of embryonic explants. By simultaneously using multiple molecular markers, as with the panel of RT–PCR markers described here, one can gain a similar overview. It should also be stated that not all regulation of gene expression is at the level of transcription, and if antibodies recognizing the protein product of a marker gene are available, the antibody staining technique can give valuable additional information. For an antibody staining protocol, see Moon and Christian, 1989, and references therein. Furthermore, marker gene expression can sometimes be transient and misleading; we therefore encourage the comparison of any molecular data with morphological and histological analysis of the differentiated phenotype (for histology preparation, see Kelly et al. 1991).

Molecular Markers for the Dorsoventral Patterning of the Developing Mesoderm

It has been suggested that the gastrula marginal zone can be divided into four sectors based on mesodermal marker gene expression (Dosch et al. 1997): a dorsal sector, two dorsolateral sectors on either side of the embryo, and a lateroventral sector. Gene expression in the marginal zone is best viewed from the vegetal pole of mid-gastrula embryos (late stage 10 to stage 11) in order to see all four of these sectors. Although final gene expression may not precisely map to these sectors, they can serve as a useful framework for describing the early expression of genes in the marginal zone.

However, marginal zone markers further differ depending on whether their expression domain is closer to the blastopore lip or further towards the animal pole, and whether they are expressed only in the deeper cells, only in the surface cell layer, or in both tissues. When genes are expressed in the surface cell layer, the staining tends to have a grainy appearance because the expression in single cells can actually be perceived. The expression staining of genes that are only expressed in the deep cell layer appears much more diffuse because the cellular resolution is obscured by the surface tissue through which the staining is viewed. However, in order to determine more precisely the expression of genes in different cell layers, it is often better to view in situ staining in tissue sections (Lemaire and Gurdon 1994; Vodicka and Gerhart 1995). Dorsoventral patterning of the marginal zone can also be studied with RT–PCR in either marginal zone explants (Steinbeisser et al. 1995) or in the experimental system of animal cap explants (Hoppler et al. 1996). In Table 17-1, we provide RT–PCR primer sequences for all of the following mesodermal markers.

One of the earliest markers for the dorsal side of the embryo is the expression of *Xnr3*, *Xenopus* nodal related 3 (Smith et al. 1995). This gene is exclusively expressed in the surface cell layer, which explains the grainy appearance of the staining after whole-mount in situ hybridization (Figure 17-1). *Xnr3* is also a good early marker for the dorsal marginal zone when used with RT–PCR (Yang-Snyder et al. 1996). Another early dorsal marker is *siamois* (Lemaire et al. 1995), which is expressed in the deep layer of cells and is successfully used in RT–PCR (Brannon and Kimelman 1996; Yang-Snyder et al. 1996). *Goosecoid* (*gsc*) is a classical marker for the dorsal mesoderm (Cho et al. 1991). Its expression starts in the dorsal sector of the marginal zone before gastrulation (stage 9) but expands during the early stages of gastrulation into the dorsolateral sectors (stages 10 and 11; Figure 17-1). *gsc* is also a very reliable dorsal marker when used in RT–PCR (Hemmati-Brivanlou et al. 1994). Two further markers expressed in the dorsal marginal zone are *noggin* (Smith and Harland 1992) and *chordin* (Sasai et al. 1994). *Chordin* expression stays close to the dorsal lip and expands laterally into the dorsolateral sector, whereas *noggin* appears to stay within the dorsal sector. *Noggin* has been successfully used with RT–PCR (Smith and Harland 1992). Another marker which is exclusively expressed in the dorsal sector is *Xnot* (von Dassow et al. 1993); Figure 17-1). *Xnot* expression fills the dorsal sector and has very sharp borders towards the dorsolateral sector where it marks the precursors of the notochord.

Although *XmyoD* expression comprises both the dorsolateral sectors and the lateroventral sector, at an early stage, stage 10 (Frank and Harland 1991), it becomes more and more confined to the dorso-lateral sectors during mid-gastrulation, stage 11 (Figure 17-1; Hopwood et al. 1989). At this stage, *XmyoD* expression marks the cells that will later form the somites. *XmyoD* has also been used successfully as a marker in RT–PCR (Rupp and

Table 17-1 RT–PCR primers for dorsoventral mesoderm

Markers	Forward and reverse primers	Expression	Reference
Xnr3	F 5′-TCCACTTGTGCAGTTCCACAG-3′ R 5′-ATCTCTTCATGGTGCCTCAGG-3′	Dorsal, early	Yang-Snyder et al. 1996
Siamois	F 5′-CTCCAGCCACCAGTACCAGATC-3′ R 5′-GGGGAGAGTGGAAAGTGGTTG-3′	Dorsal, early	Yang-Snyder et al. 1996
gsc	F 5′-ACAACTGGAAGCACTGGA-3′ R 5′-TCTTATTCCAGAGGAACC-3′	Dorsal	Hemmati-Brivanlou et al. 1994
Noggin	F 5′-AGTTGCAGATGTGGCTCT-3′ R 5′-AGTCCAAGAGTCTGAGCA-3′	Dorsal	Smith and Harland 1992
Xpo	F 5′-CACTTAGGGATTGGTCTCAGGAGTC-3′ R 5′TGAGGGAGGGCTATGGTCTAGG-3′	Ventral	Hoppler et al. 1996
Evx-1 (Xhox-3)	F 5′-ACCTATATGATGAGCCACGAG-3′ R 5′-GCCGTTGGTTGAGGTACTAAGAC-3′	Ventral	Hoppler et al. 1996
Xwnt-8	F 5′-AGATGACGGCATTCCAGA-3′ R 5′-TCTCCCGATATCTCAGCA-3′	Ventral	Smith and Harland 1991

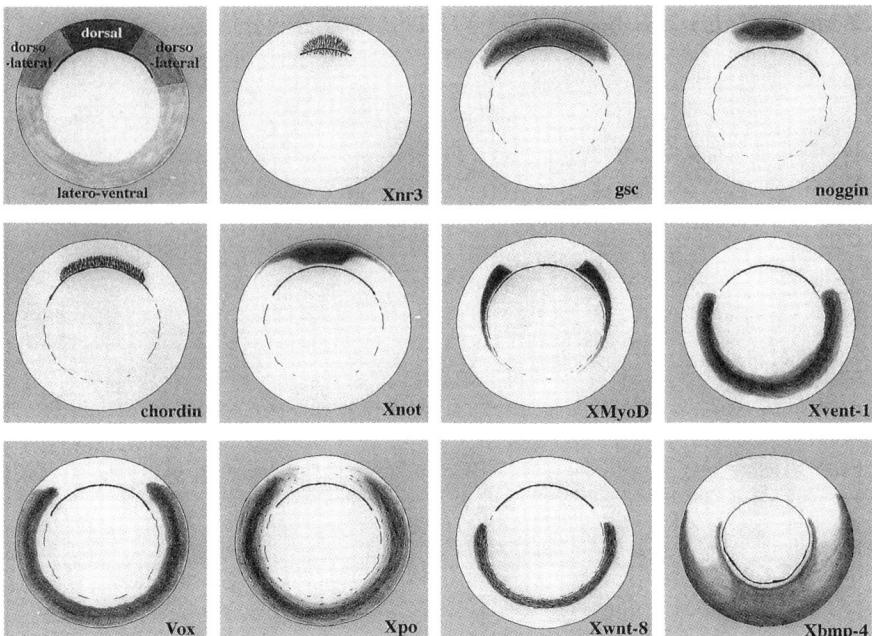

Figure 17-1 Sectors of gene expression and expression of different marker genes in the marginal zone. Embryos are shown from the vegetal pole at late stage 10 of development, apart from *Xnr3* expression, which is shown in a early stage 10 embryo, and *Xbmp-4* expression, which is shown in a late stage 11 embryo. See text for details.

Weintraub 1991). Another marker for the dorso-lateral mesoderm is the expression of *Myf-5*, which becomes restricted to the dorsolateral sector at an early stage 10 (Dosch et al. 1997).

A comparison of the expression patterns of the two closely related genes *Xvent-1* (Gawantka et al. 1995) and *Vox* (Schmidt et al. 1996), also called *Xvent-2*, (Onichtchouk et al. 1996), clearly illustrates the division of the gastrula marginal zone into sectors as described above (Figure 17-1). *Xvent-1* is exclusively expressed in the lateroventral sector. *Vox (Xvent-2)* is expressed in the lateroventral sector as well as in the dorsolateral sector. A further difference between the expression of these genes is that *Xvent-1* is expressed in a relatively narrow band very close to the blastopore lip, but *Vox (Xvent-2)* expression keeps its distance from the blastopore lip and extends towards the animal pole. The expression of *Xpo*, *Xenopus*-posterior (Sato and Sargent 1991; Amaya et al. 1993), is similar to that of *Vox (Xvent-2*; see Figure 17-1 for comparison) in that both are expressed in the lateroventral and the dorsolateral sectors, keep their distance from the blastopore lip, and extend to the animal pole. In addition, however, *Xpo* also appears to be expressed in some cells of the surface layer in the whole marginal zone, including the dorsal sector, as suggested by the speckled staining. Because it is expressed much higher in the lateroventral sector than in the dorsal sector, *Xpo* can also be used as a marker for ventral marginal tissue with RT-PCR when relatively few amplification cycles are performed (Hoppler et al. 1996). A ventral marker which has only been successfully used with RT-PCR and not with whole-mount in situ hybridization is *Evx-1* (previously called *Xhox-3*; Ruiz i Altaba and Melton 1989). *Xwnt-8* expression (Christian et al. 1991) is very similar to that of *Xvent-1* (see Figure 17-1 for comparison) in that both are expressed close to the blastopore and only in the lateroventral sector, but the expression staining of *Xwnt-8* has a somewhat more spotty appearance than that of *Xvent-1*. This could suggest that, as well as in the deep cells, *Xwnt-8* is expressed in some cells of the surface layer; however, it has also been suggested that some *Xwnt-8* transcripts localize to the nuclei of cells (Smith and Harland 1991).

Xbmp-4 is rather a late ventral marker. Early *Xbmp-4* expression is ubiquitous (Dale et al. 1992), then it becomes highly upregulated around the animal pole, but it is not until later stages of gastrulation (stage 11.5, Figure 17-1) that the marginal zone expression becomes established in the lateroventral sector (Fainsod et al. 1994; Schmidt et al. 1995). Similar to *Vox* (*Xvent-2*) and *Xpo*, *Xbmp-4* is expressed at a distance from the blastopore lip. However, unlike *Vox* (*Xvent-2*) and *Xpo*, in the lateral marginal zone *Xbmp-4* expression levels appear lower in cells further away from the blastopore lip before apparently increasing again in cells closer to the animal pole.

Molecular Markers for the Anteroposterior Patterning of the Developing Central Nervous System

Neural induction starts during the period of development known as gastrulation, whereby global cell movement serves to reposition the three primary germ layers. During this time, the dorsal mesoderm involutes toward the interior of the embryo, migrating along the inner surface of the blastocoel roof until it reaches a position underlying the presumptive neural ectoderm. During neural induction, signals emanating from the dorsal mesoderm induce the overlying ectoderm to become neural tissue and direct the surrounding mesoderm to form the body axis. Current students of development accept this view based on the classic experiments performed by Hilda Mangold and Hans Spemann in 1924. They took the region of the embryo which initiates gastrulation (the dorsal lip) and transplanted it to the ventral side of a second embryo, 180° away from the endogenous dorsal lip. Remarkably, a complete second axis formed at the site of the transplant. Because the donor tissue was pigmented and the host tissue was not, they could tell that the ectopic axis was not derived exclusively from the transplanted tissue. In fact, the secondary nervous system was *organized* from host tissues that would normally form epidermis. Hence, the name Spemann's Organizer was coined for the dorsal lip region.

These experiments make an important point: all of the instructions for a nervous system could be found in this small region of the embryo. Historically, two routes have been proposed to explain how this information is transmitted from the dorsal mesoderm to the dorsal ectoderm to form a patterned nervous system. The first was uncovered when the inductive capacity of the dorsal mesoderm was tested from different-stage neurula embryos (Hamburger 1988). Using the Einsteck method (for review, see Phillips 1991), which mimics the endogenous apposition between the two germ layers, mesoderm from anterior or posterior positions along the axis was inserted into the blastocoel of an early gastrula embryo. The induced neural tissue, when assayed by histology, was roughly equivalent to the anteroposterior level of the mesoderm. This experiment demonstrates that information could be passed vertically, where the type of mesoderm dictates the type of neural tissue formed. But the regional pattern in the induced neural ectoderm was not as precise as in a normal embryo. This result can be explained when examined from the context of the developing embryo. Gastrulation is a dynamic time when the mesoderm has the power to induce surrounding tissues while it is undergoing migration. It is understandable that the tissues have a better chance of creating the correct alignment along the anteroposterior axis if the final territories cover a broad area. But this scenario necessitates a second signal that can fine-tune the induced neural tissue, presumably by patterning events that take place within the plane of the induced neural tissue. Pieter Nieuwkoop addressed this issue in a set of experiments that examined the role of purely planar signals. In a gastrula embryo, flaps of competent ectoderm were inserted along different positions of the presumptive anteroposterior axis. Histologically, he found that the uppermost portion of the flap contained the most anterior neural tissue, although each flap contained a partial range of anteroposterior neural structures. Furthermore, only the caudally positioned flaps contained posterior tissue, whereas all of the flaps contained anterior tissue. These results led Nieuwkoop to propose a two-signal model, where neural induction and patterning are under the control of at least two distinct signals. The first signal induces competent ectoderm to become anterior-type neural tissue, and would originate within the dorsal mesoderm. The second signal serves to caudalize the induced neural tissue, and is present either in the posterior dorsal mesoderm or within the plane of the neural ectoderm.

We are fortunate that the molecular era has produced several candidate proteins that help explain the mechanisms underlying Nieuwkoop's two-signal model. Noggin (Smith et al. 1993), chordin (Sasai et al. 1994), and follistatin (Hemmati-Brivanlou et al. 1994) are all secreted proteins, expressed in the dorsal mesoderm during gastrulation. Each induces neural tissue of basically anterior phenotype fitting Nieuwkoop's first signal. Members of the Wnt (McGrew et al. 1995) and FGF (Doniach 1995) family can function to posteriorize the induced neural tissue without the ability to directly induce neural tissues by themselves. Thus, the Nieuwkoop prediction of at least two physically separate signals involved in the process of neural induction and patterning can be experimentally studied with modern molecular techniques.

Using the two-signal hypothesis as a guideline, we can divide our summary of molecular markers into two parts: those that define the onset of neural induction and those markers that demonstrate the subsequent patterning of the neural tissue into regionally distinct territories along the anteroposterior axis. We will not discuss the dorsal–ventral patterning of the neural tube in this chapter as three recent reviews cover this well (Lumsden and Krumlauf 1996; Placzek and Furley 1996; Tanabe and Jessell 1996).

When competent animal cap ectoderm is dissected from a stage 8–9 embryo and treated in isolation with either noggin (Lamb et al. 1993), follistatin (Hemmati-Brivanlou and Melton, 1994; Hemmati-Brivanlou et al. 1994), or chordin (Sasai et al. 1995), the tissue will become neural, as indicated by the presence of several general neural markers and the absence of dorsal mesoderm as assayed by muscle actin (Stutz and Spohr 1986; see Figure 17-3B). *N-CAM* is a cell adhesion molecule expressed throughout the neurons and glial cells of the CNS. Neural plate expression can first be detected by in situ hybridization at stage 16 and persists exclusively in the neural tube throughout later stages (Kintner and Melton 1987). *NF-M* encodes neurofilament-M and is a pan-neuronal marker that comes on only after neural tube closure when the first motor neurons differentiate by stage 20 (Sasai et al. 1995). *NRP-1* encodes a neural specific RNA-binding protein that is first expressed maternally, followed by expression throughout the open neural plate at stage 13. During later stages, it continues to be strongly expressed in the anterior neural folds, with less expression posteriorly, in two broad stripes that extend around the blastopore (Knecht et al. 1995).

Two families of proteins have been characterized for their involvement in the generation of neural pattern. Both Xwnt-3a (McGrew et al. 1995) and eFGF (Isaacs et al. 1992; Pownall et al. 1996) are secreted proteins that are expressed at the right time (during gastrula and early neurula) and in the right place (in the posterior dorsal mesoderm and throughout the dorsal ectoderm) to play a role as patterning agents. Functional tests are available that coexpress both the anterior neuralizing agents with a prospective patterning agent to assay the type of neural tissue produced (see Figure 17-3B). When regional neural markers are examined (Figure 17-2), neuralized animal caps express only anterior markers (*Xanf-1/2, Otx-A/2*) in the absence of either Xwnt-3a (McGrew et al. 1995) or bFGF (a closely related FGF family member; see Doniach 1995 for review). In contrast, when these are added to neuralized caps, both anterior and posterior (*En-2, Krox-20, and HoxB9*) regional markers are expressed.

The following is a general list of anteroposterior regional neural markers. They are useful due to their discrete and persistent expression patterns

324   Methods for Studying *Xenopus* Oocytes and Embryos

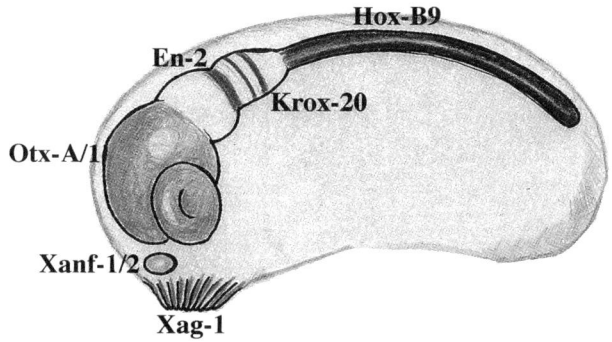

Figure 17-2 Regional gene expression of neural specific markers along the anteroposterior axis of an early tail-bud-stage embryo.

within different territories along the anteroposterior axis (Figure 17-2; Tables 17-2 and 17-3 for the corresponding RT–PCR primers). But a word of caution is advised. It is not unusual for regionally restricted markers to begin expression in widespread domains that may consist of more than one tissue layer. Therefore, it is critical to assay gene expression at the appropriate period of development. For example, studies of the regional identities along the open neural plate require a unique set of markers distinct from the ones mentioned here.

We start at the anterior end of the embryo with a marker of an ectodermally derived organ, the cement gland. Although *Xag-1* is not a neural marker, it is useful as a positional marker because the cement gland develops from the most anterior ectoderm and has been shown to be induced by the adjacent neural ectoderm (Sive et al. 1989), so it serves as a positive indicator of anterior neural induction. *Xanf-1* is a homeobox-containing gene expressed in the presumptive neural ectoderm during the late gastrula stage where it is progressively restricted to the extreme anterior edge of the neural plate. In mid- to late-neurula-stage embryos (stages 15–17), it is confined to the mediolateral anterior ridge, and by the end of neurulation, it is localized exclusively to the prospective anterior pituitary (Zaraisky et al. 1995). *Xanf-2* is 82% identical at the amino acid level to *Xanf-1* and shows a similar pattern of expression by in situ hybridization (Mathers et al. 1995). *Otx-A* is a member of the *orthodenticle* family of homeobox genes and was originally reported to be directly induced by noggin in naive animal cap ectoderm (Lamb et al. 1993). Before gastrulation, *Otx-A* is expressed throughout the marginal zone, then becoming restricted to a large anterior domain of the open neural plate. Later, during neural tube closure, it is restricted to parts of the midbrain, forebrain, and retina (Lamb et al. 1993). *Otx-2* is closely related to *Otx-A* but its expression starts later, during early gastrulation, where it is expressed in the prechordal mesoderm. By stage 10.5, expression begins in the overlying ectoderm, where it is main-

Table 17-2 RT–PCR primers for anteroposterior neural ectoderm

Markers	Forward and reverse primers	Expression	Reference
N-CAM	F 5′-CACAGTTCCACCAAATGC-3′ R 5′-GGAATCAAGCGGTACAGA-3′	Pan-neural	Hemmati-Brivanlou and Melton 1994
NF-M	F 5′-GCGGGTACCTTCTAATAGTCAC-3′ R 5′-GGCTTGGCTGTGGTTCTGAAGG-3′	Pan-neural	Sasai et al. 1995
NRP-1	F 5′-ACAACTGGAAGCACTGGA-3′ R 5′-TCTTATTCCAGAGGAACC-3′	Pan-neural	Hemmati-Brivanlou et al. 1994
Xag-1	F 5′-CTGACTGTCCGATCAGAC-3′ R 5′-GAGTTGCTTCTCTGGCAT-3′	Cement gland	Lai et al. 1995
Xanf-1	F 5′-TCGCTTGTGATCTCCCCATAC-3′ R 5′-AGAGGTCCAAGGCTCTATCAGG-3′	Anterior pituitary	Unpublished
Xanf-2	F 5′-ACTGACCTACAAGAGAGAAC-3′ R 5′-AGTGCATCATTGTTCCACAG-3′	Same as Xanf-1	Lai et al. 1995
Otx-A	F 5′-CATCGGACATAAAGCAGCTCATC-3′ R 5′-CTTTCCCTCCTCTGTTTCCTGG-3′	Forebrain and retina expression (tail bud)	Lai et al. 1995
Otx-2	F 5′-GGAGGCCAAAACAAAGTG-3′ R 5′-TCATGGGGTAGGTCCTCT-3′	Same as Otx-A	Blitz and Cho 1995
En-2	F 5′-CGGAATTCATCAGGTCCGAGATC-3′ R 5′-GCGGATCCTTTGAAGTGGTCGCG-3′	Midbrain/hindbrain border	Hemmati-Brivanlou and Melton 1994
Krox-20	F 5′-AACCGCCCAGTAAGACC-3′ R 5′-GTGTCAGCCTGTCCTGTTAG-3′	Rhombomeres 3 and 5 of the hindbrain	Hemmati-Brivanlou and Melton 1994
Hox-B9 (XlHbox-6)	F 5′-TACTTACGGGCTTGGCTGGA-3′ R 5′-AGCGTGTAACCAGTTGGCTG-3′	Spinal cord	Hemmati-Brivanlou and Melton 1994

Table 17-3 Control RT–PCR primers

Markers	Forward and reverse primers	Expression	Reference
Ef-1α	F 5'-CAGATTGGTGCTGGATATGC-3' R 5'-ACTGCCTTGATGACTCCTAG-3'	Loading control	Hemmati-Brivanlou and Melton 1994
Muscle actin	F 5'-GCTGACAGAATGCAGAAG-3' R 5'-TTGCTTGGAGGAGTGTGT-3'	Dorsal mesoderm	Hemmati-Brivanlou and Melton 1994
Histone H4	F 5'-CGGGATAACATTCAGGGTATCACT-3' R 5'-ATCCATGGCGGTAACTGTCTTCCT-3'	Loading control	Blitz and Cho 1995

tained in register between the anterior tip of the chordal mesoderm and the anterior neural plate and prospective cement gland (Blitz and Cho 1995). Upon neural tube closure, *Otx-2* expression is detected in the forebrain, midbrain, and developing eye anlage (Blitz and Cho 1995). *En-2* is another homeobox-containing gene whose expression was first examined using En-2-specific antibodies (monoclonal 4D9; Hemmati-Brivanlou and Harland 1989). At the open neural plate stages (stages 14–15), it is expressed as a stripe across the anterior neural plate, localized to the neural ectoderm. By early tail-bud (stage 20), the En-2 protein is limited to the posterior border of the midbrain and anterior border of the hindbrain (Hemmati-Brivanlou and Harland 1989). Although two transcripts are known to encode the En-2 protein, the antibody used in this study only recognizes the product of one of these clones (Hemmati-Brivanlou and Harland 1989). *Krox-20* is a zinc-finger gene exclusively expressed in the presumptive neural ectoderm and thought to demarcate the future segmentation of the hindbrain. Transcripts are first detected at stages 12.5–13 as two narrow stripes of expression, on the anterior neural plate (Bradley et al. 1992). This localized expression persists when morphologically it is distinguished as rhombomeres 3 and 5 of the hindbrain. *Krox-20* is also expressed in the premigratory and post-migratory neural crest aligned with rhombomere 5. *Hox-B9* (previously known as *Xlhbox-6*) was originally identified as a marker of neural induction in ectoderm/mesoderm conjugates (Sharpe et al. 1987). Hox-B9-specific antibodies have shown that the protein is expressed, upon neural tube closure, along the entire length of the spinal cord exclusive of the hindbrain (Wright et al. 1990).

Xenopus In Situ Hybridization Protocol

This is a simple and reliable protocol for inexperienced users which has been directly developed from the original *Xenopus* whole-mount in situ protocol (Harland 1991). This protocol will be useful for the study of gene expression during many stages of *Xenopus* development, although we have mainly used it during gastrula and neurula stages. This generic protocol has recently been adapted for a range of special applications, of which we would like to mention a few in order to guide the interested reader to the relevant publications. To allow whole-mount in situ hybridizations to be performed on embryonic explants, baskets have been designed which allow a more gentle transfer of explants (and embryos) between different incubation solutions (Knecht et al. 1995; Bradley et al. 1996). It has been noticed that the standard in situ hybridization protocol, as described here, is not adequate to detect reliably gene expression close to the vegetal pole. However, conditions have been worked out recently which allow the detection of endodermal gene expression close to the vegetal pole (Sasai et al. 1996). Different protocols have been developed for double in situ hybridization, which allows the comparison of expression of two different marker genes in the same embryo (Knecht et al. 1995; Bradley et al. 1996). In order to more reliably determine the expression of marker genes in different layers of tissue of the embryo, in situ hybridization staining can be viewed in tissue sections (Lemaire and Gurdon 1994; Vodicka and Gerhart 1995).

Reagents

Template DNA
The template DNA has to be cloned into a vector with viral promoter sequences (SP6, T3, or T7) in the antisense orientation, suitable vectors include pT7TS, pSP6T, pBluescript, and pCS2+.

NTP mixture
2.5 mM GTP, 2.5 mM ATP, 2.5 mM CTP, 1.6 mM UTP, 0.9 mM digoxigenin-UTP (Boeringer Mannheim).

10 × MEM Salts Stock Solution
1 M MOPS, 20 mM EGTA, 10 mM $MgSO_4 \cdot 7H_2O$. Adjust with NaOH to pH 7.4. Filter and store wrapped in aluminum foil at 4°C.

MEMFA
1× MEM salts with 4% formaldehyde. Always make up fresh in a small volume, e.g., 50 ml.

10 × PBS Stock Solution
Final 1× is as follows: 137 mM NaCl, 2.7 mM KCl, 4.3 mM $Na_2HPO_4 \cdot 7H_2O$, 1.4 mM KH_2PO_4; should give pH 7.3. Autoclave, and store at room temperature.

Ptw
1× PBS with 0.1% Tween 20. Keeps for weeks at room temperature.

20× SSC
3 M NaCl and 0.3 M sodium citrate. Adjust pH to 7, autoclave, and store at room temperature.

50× Denhardt's
1% Ficoll (Type 400, Pharmacia), 1% polyvinylpyrrolidone, 1% BSA (Fraction V, Sigma) in distilled water. Store at −20°C.

Hybridization Buffer
50% Formamide, 5× SSC, 1 mg/ml Torula RNA, 100 µg/ml heparin, 1× Denhardt's, 0.1% Tween, 0.1% CHAPS, 5 mM EDTA in distilled water.

5× MAB
100 mM maleic acic, 150 mM NaCl. Adjust pH to 7.5 with NaOH. Store at room temperature.

10% BMB in MAB
1× MAB (pH 7.5), 10% Boehringer Mannheim blocking agent. Autoclave, aliquot in 50 ml and store at −20°C. (The already diluted 2% BMB in MAB can also be stored at −20°C for at least several weeks.)

Alkaline Phosphate Buffer
Always prepare fresh by combining 8.8 ml of distilled water with 1 ml of 1M Tris-HCl (pH 9.5), 500 μl of 1M $MgCl_2$, 200 μl of 5M NaCl, and 100 μl of 10% Tween.

Staining Solution
Always prepare fresh by combining 8.7ml of distilled water with 1 ml of 1M Tris-HCl (pH 9.5), 500 μl of 1M $MgCl_2$, 200 μl of 5M NaCl, 100 μl of 10% Tween, 35 μl of 4-Nitro blue tetrazolium chloride (NBT, Boehringer Mannheim, 300 mg in 3 ml), and 35 μl of 5-bromo-4-chloro-3-indolyl-phosphate (BCIP, Boehringer Mannheim, 150 mg in 3 ml).

Murray's Clear
1 part benzyl alcohol, 2 parts benzyl benzoate.

Digoxigenin Probe Synthesis

1. For in vitro transcription, the plasmid with the template sequence (at least 5 μg!) first has to be linearized by digesting to completion with a restriction endonuclease which cuts 3′ of the template in the antisense orientation and preferentially produces a 5′ overhang. Following complete digestion, which can be tested by running an aliquot on an analytical agarose gel, the restriction enzyme should be inactivated by phenol extraction or heat inactivation, depending on the enzyme.

2. Assemble the transcription reaction at room temperature to a final volume of 50 μl. In the following order, add to a reaction tube, first the distilled water in the amount to bring the final volume up to 50 μl, then 10 μl of 5× transcription buffer for the polymerase used, 2.5 μl of 0.2 M DTT, 10 μl of NTP mix, approximately 5 μg of linearized template DNA, and 1 μl of the RNase inhibitor RNasin; then mix the components in the tube, add 90 units of polymerase (SP6, T7, or T3), mix again, and incubate for 2 hours at 37°C.

3. In order to digest the template DNA, add RQ DNase (1 unit per microgram of template DNA) and incubate at 37°C for another 10 minutes.

4. The enzymes, after the reaction, must never be inactivated by phenol extraction, since phenol sequesters the digoxigenin moiety. Enzyme inactivation is achieved by adding SDS to a final concentration of 1%, EDTA to a final concentration of 20 mM, Tris-Cl (pH 7.5) to a final concentration of 20 mM, and NaCl to a final concentration of 0.1 M.

5. The probe is separated from incorporated NTPs by runing over Sephadex G50 columns and then recovered by precipitation in 0.1 volume of sodium acetate and 2 volumes of ethanol. The pellet is resuspended in 20 μl of distilled water. An aliquot of the RNA probe should be analyzed by agarose gel electrophoresis in MOPS buffer. We keep our probes at −70°C and have successfully used probes almost 2 years after their synthesis.

6. Probes can be hydrolyzed by incubating in 40 mM sodium bicarbonate, 60 mM sodium carbonate for 30–60 minutes at 60°C; in our experience, however, the signal improves with hydrolyzation for only a few probes, and we therefore normally work with unhydrolyzed probes.

Preparation of Embryos

The in situ hybridization technique can be performed on either albino embryos or pigmented embryos. The staining is clearly more visible in albino embryos; however, the expression can often be analyzed better in pigmented embryos because it appears to be easier to orient the gene expression in the morphological context of the embryo. Hence, whereas expression staining in albinos is often more photogenic, we find in situ hybridization on pigmented embryos is more informative.

Because in situ hybridization probes can be trapped in the internal cavities of embryos and cause strong unspecific staining, which might obscure specific staining, the embryos have to be injured immediately before fixation to create a perforation between large body cavities and the outside in an area of the embryo which is not in the focus of interest. Mesodermal gene expression is studied at early gastrulation stages and is often viewed from the vegetal pole (see Figure 17-1). The blastopore roof on the animal pole of the embryo can easily be cut with either a fine glass or tungsten needle to allow the washing solutions used at later stages in the protocol to reach the blastocoel. Neural gene expression is studied at the neural plate stage and viewed from the dorsal side. An opening to the archenteron is created by using a pair of forceps like a pair of scissors to cut a cleft just ventral to the closed blastopore. (Contrary to older protocols, we did not find it necessary to remove the vitelline membrane carefully at this stage. We find that the injured membrane always becomes completely removed by the subsequent procedures of the protocol, presumably the proteinase K digest.)

The embryos are then immediately fixed in freshly made MEMFA for 1 hour (at most, 2 hours) and incubated in ethanol overnight at 4°C or at least 2 hours at room temperature. At this stage, the embryos can be stored in ethanol at 4°C for at least several weeks before further processing. We even found that long ethanol incubations reduce background staining.

In Situ Hybridization Staining Protocol

Hybridization
1. Rehydrate and warm up embryos to room temperature by incubating the embryos for 5 minutes each in 75% MeOH/25% H_2O, 50% MeOH/50% H_2O, 25% MeOH/75% Ptw, and three times in 100% Ptw.
2. Incubate embryos with proteinase K (20 µg/ml proteinase K in Ptw) for about 10 minutes at room temperature under constant agitation.
3. Rinse once in Ptw, wash twice in triethanolamide (0.1 M, pH 7.8), incubate for 5 minutes each, then replace triethanolamide and add acetic anhydride (2.5 µl per 1 ml) incubate for 5 minutes, add additional acetic anhydride (2.5 µl per 1 ml) and incubate for another 5 minutes. Wash twice for 5 minutes in Ptw.

4. Refix in 4% formaldehyde in Ptw for 20 minutes. Then wash at least three times in Ptw for 5 minutes each.
5. Replace last wash of Ptw with 500 µl of hybridization buffer, wait until embryos settle to the bottom. Prehybridize embryos by replacing the hybridization buffer and incubating the embryos at 60°C for at least 1 hour. (We usually prehybridize for at least 3 hours to overnight.)
6. Replace with 500 µl of hybridization buffer with digoxigenin antisense RNA probe (1–5 µg/ml), incubate at 60°C for at least 6 hours. (We hybridize overnight.)
7. Carefully remove hybridization solution, replace with fresh hybridization buffer, and incubate for 20 minutes at 60°C. (Save hybridization solution, because it can be stored at −20°C for about 1 year and reused up to 10 times; this may even reduce background staining with some probes.)
8. Wash embryos three times in 2× SSC for 20 minutes each at 60°C. Then, wash them twice in 0.2× SSC for 30 minutes each at 60°C.

Antibody Labeling
9. Wash twice in MAB for 10 minutes at room temperature. Then wash once in MAB/2% BMB for 15 minutes.
10. Both the embryos and the antibody (anti DIG Fab fragments 1:2000) are preblocked in MAB/2% BMB/20% Goat (or sheep) serum (heat inactivated) for 1 hour at room temperature.
11. Remove the blocking solution from the embryos, replace it with the preblocked antibody solution, and incubate for 1 hour at room temperature (or at 4°C overnight).
12. Wash six times during a total of at least 3 hours with MAB at room temperature. (The last wash can go overnight at 4°C.)

Color Reaction
13. Wash twice in freshly prepared alkaline phosphate buffer for 5 minutes each.
14. Replace buffer with freshly prepared staining solution, and incubate until specific staining is strong and background staining is coming up. (The incubation times vary considerably between different probes, but normally exceed 1 hour. On long reactions, it is often advisable to replace the staining solution every 2 hours to reduce background staining.)
15. Stop the staining reaction by removing the staining solution and adding MEMFA. Fix in MEMFA for 3–5 hours at room temperature under constant agitation.
16. Wash off background staining and dehydrate embryos by replacing the staining solution with methanol. Replace methanol several times, then leave the embryos in MeOH for at least 24 hours to wash off the last traces of background staining.
17. Stained embryos can be stored in methanol at 4°C for several months. Pigmented embryos can be bleached for several hours in 70% methanol/3% H_2O_2/27% H_2O under fluorescence-emitting light at room temperature. In order to see internal staining, embryos can be cleared in minutes by incubation in Murray's clear, but staining can desolve in Murray's clear.

Reverse Transcription–Polymerase Chain Reaction Protocol

Reagents

The following stock solutions are described in Maniatis et al., 1989. 3.0 M NaOAC, (pH 5.2); 29:1 acrylamide/bis-acrylamide (USB US32799); TBE electrophoresis buffer and gel loading buffer; 1 M dithiothreitol (DTT).

Guanidine Homogenization Buffer
Dissolve 47.44 g of guanidine thiocyanate (Boehringer Mannheim, 1685 929) in 55.6 ml of glass-distilled water. Heat to 60°C to dissolve, add 3.34 ml of 750 mM sodium citrate (Baker 3646-05) and 5 ml of 10% N-laurylsarcosine (Sigma L-5125). Adjust to a final volume of 100 ml and store at room temperature. The stock solution is stable for 6 months. Just before use, add 36 µl of 2-mercaptoethanol (Sigma M-6250) per 5 ml of the stock solution.

Phenol (USB, US75829)
Both water-saturated phenol and phenol equilibrated with Tris-HCl (pH 8) are used in this protocol.

Sevag
24:1 Chloroform: isoamyl alcohol

10× DNase Buffer
400 mM Tris-HCl (pH 8.0), 100 mM NaCl, 60 mM $MgCl_2$, 1 mM $CaCl_2$. Add glass-distilled H_2O to a final volume of 1 ml.

2.5 mM dNTP Mix
Add 10 µl of each of the following: 100 mM dATP, 100 mM dTTP, 100 mM dGTP, 100 mM dCTP (Pharmacia Biotech, pH 7.5). Adjust to a final volume of 400 µl with glass-distilled H_2O, and store at −20°C.

PCR Primers
Available from Gibco BRL Custom Primers. Dilute primers to 100 ng/µl, which is approximately 10 pmol for a primer 20 nucleotides in length (follow manufacturer's directions for reconstitution and calculation of concentration). Store primer pairs as a mixture of the forward (5′) and reverse (3′) primers, each at a concentration of 100 ng/µl.

Preparation of Embryonic Explants

Marker gene expression can be studied in whole embryos with RT–PCR. However, RT–PCR is particularly useful for studying gene expression in the controlled experimental system of embryonic explants. The animal cap assays are very powerful for investigating the activity of molecules in mesoderm induction (Slack *et al.,* 1987; Smith 1987; Kimelman and Kirschner 1987) and subsequent dorsoventral patterning (Christian et al. 1992; Hoppler et al. 1996). The animal caps (ectodermal explants) should be cut relatively small to exclude the marginal zone which contains endogenous mesoderm-inducing and -patterning factors. Marginal zone explants are used for studying the activity of factors that pattern the marginal zone along the dorsoventral axis (Steinbeisser et al. 1995). Relatively wide ventral marginal zone explants can be cut to assay for ectopic dorsal or reduced ventral gene expression, but dorsal marginal explants should be narrow and exclude the dorsolateral mesoderm. Dorsal marginal explants can be used to assay for reduced dorsal and ectopic ventral marker gene expression.

In order to study the individual components of neural induction and patterning one needs to decide which type of explant is best for the question at hand. Keller explants are useful because they rely on endogenous signals to drive induction and patterning (Figure 17-3A). They provide plenty of raw material that can be examined either by in situ hybridization or RT–PCR. For a detailed description of their generation and applications, see Doniach et al. 1992 and Figure 17-3A. The second method takes advantage of the neural-inducing properties of proteins such as noggin. In order to obtain pure anterior neural tissue, noggin RNA can be injected at the 2–4 cell stage and ectodermal caps cut at stage 8–9, then cultured in isolation until the appropriate stage (Figure 17-3B). A more rigorous test of direct induction involves the application of purified protein (available for noggin or bFGF) to ectodermal caps cut at the same stage from uninjected embryos. Dissection of animal cap ectoderm is straightforward as long as the cap is cut from the upper third of the embryo. This avoids contamination of marginal zone material which will lead to indirect patterning effects due to the presence of dorsal mesoderm tissue. All of these manipulations require specialized culture conditions which are described in detail in Peng (1991).

Preparation of RNA

1. Homogenize embryos or explants in 10–20 volumes of guanidine homogenization buffer (Chomczynski and Sacchi 1987) using a P200 pipette tip or Eppi pestle in a 1.5-ml Eppendorf tube. Embryos are roughly equivalent to 5 µl of volume and a single cap explant is probably one-fifth that amount. Do the initial homogenization in a small volume (approximately 200 µl/20 explants) and then increase to the final desired volume (400 µl/ 20 explants or four embryos). The samples are stable at $-70°C$ for several days.

2. Thaw samples on ice, add 40 µl (0.1 volume) of 3 M NaOAC, and if low RNA yield is expected (e.g., when using explants), add 10 µg of carrier yeast tRNA (Gibco BRL, 16051-039). Add 200 µl (or 0.5 volumes) of water-saturated phenol, vortex well and place on ice for approximately 20 minutes. Spin briefly, add 100 µl (0.25 volumes) of Sevag, vortex, and spin in a microfuge at room temperature for 5 minutes.

3. Recover the aqueous phase (avoiding the interface) and repeat the phenol/Sevag extraction 1×, followed by a Sevag extraction alone. Precipitate nucleic acid by the addition of 2.5 volumes of 100% EtOH, and incubate at $-70°C$ for 20 minutes.

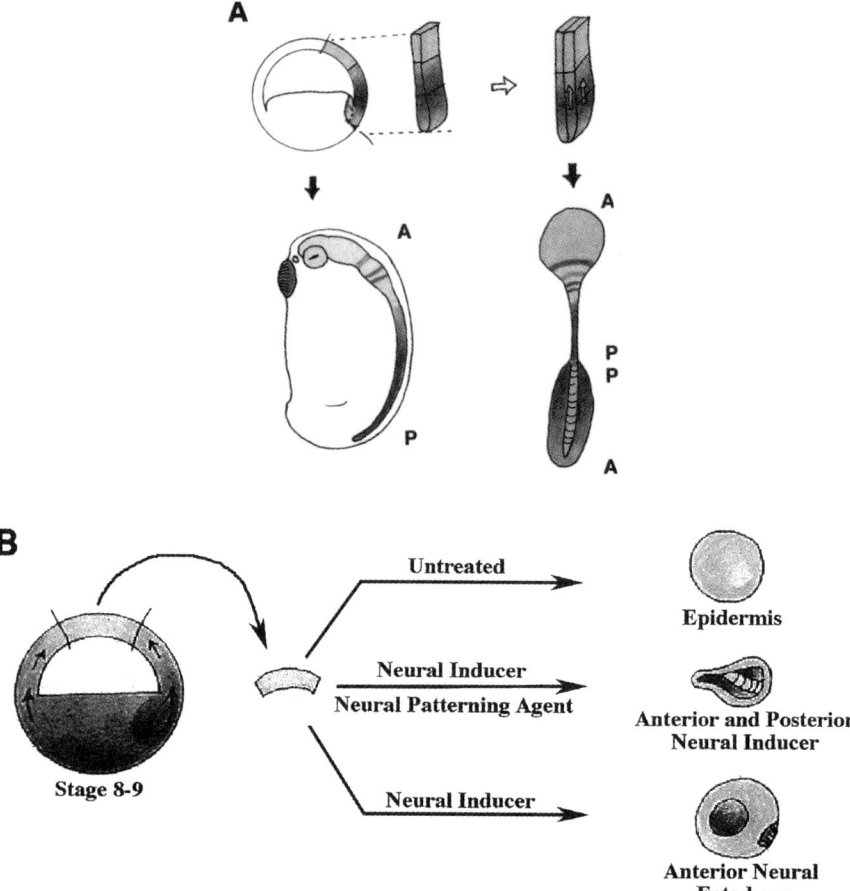

Figure 17-3 (A) Preparation of Keller explants to study regional neural induction in the absence of vertical signals as found in the whole embryo. Two explants are dissected from stage 10 gastrula embryos and sandwiched together with their inner surfaces apposed to one another. A, anterior; P, posterior. (B) Animal cap explants. Competent animal cap ectoderm is dissected from stage 8–9 embryos and cultured in the presence or absence of factors that induce and/or pattern neural tissue.

4. Pellet the RNA at 4°C in a microfuge (13,000–15,000 rpm) for 10 minutes. Wash the pellet with 70% EtOH, and dry at room temperature. Overdrying the pellet will make it impossible to resuspend.

5. Resuspend the pellet in 77 μl of glass-distilled H_2O.

6. DNase digest the samples by the addition of 10 μl of 10× DNase buffer, 10 μl of 0.1 M DTT, 1 μl of RNasin Ribonuclease Inhibitor (40 μ/μl; Promega N2511), 2 μl of RQ1 RNase-free DNase (1 U/μl; Promega M6101). Incubate for 30–60 minutes at 37°C.

7. Increase total volume to 200 µl with glass-distilled H_2O, add 20 µl (0.1 volume) of 3 M NaOAC, phenol/Sevag extract 2× using Tris-buffered phenol. Sevag extract 1×, and ethanol-(EtOH) precipitate.
8. Pellet the RNA and wash 2× with 70% EtOH. Air-dry briefly and resuspend in 34 µl of glass-distilled H_2O.

cDNA Synthesis

We use the Life Sciences, Inc. (St. Petersberg, Florida) cDNA Synthesis Kit to synthesize our first-strand cDNA. The manufacturer's instructions are followed using approximately 5–10 µg of total RNA for each 25-µl reaction volume. One embryo contains approximately 5 µg of total RNA, but recovery rate reduces that by 25–50%.

1. Add 1 µl of 0.5 µg/µl $p(dT)_{12-18}$ to 33 µl of sample from above. Denature at 70°C for 10 minutes. Anneal on ice for 10 minutes. Quick spin to pellet mix and split sample for −RT control (17 µl each).
2. Make a master mix for each sample consisting of the following reagents supplied by the manufacturer:
   1 µl of 0.25 M DTT
   1 µl of RNasin
   5 µl of 5× Reaction buffer (triturate to mix before aliquotting)
   1 µl AMV reverse transcriptase
   8 µl total volume
   Store on ice until use.
3. Add 8 µl of master mix to each +RT sample (17 µl) and mix gently. Add 8 µl of glass-distilled H_2O to each −RT control sample. Incubate reactions at 41°C for 60 minutes. The cDNA is stable indefinitely at −20°C.

PCR Amplification

Usually, explants are analyzed for changes in gene expression using a panel of RT–PCR markers. In order to simplify the amount of pipetting required to accommodate this many samples, master mixes are made for each cDNA sample, which contain all of the components except the primers. This method also decreases any chance of pipette error that could affect the even distribution of cDNA in each reaction. Primers are added to the PCR reaction tubes followed by an aliquot of the appropriate cDNA master mix. PCR amplifications, 50 µl, are carried out in a Perkin-Elmer GeneAmp 9600 cycler. We use either Perkin-Elmer or Promega enzymes and buffers with equal success.

1. cDNA master mix per reaction:
Water (up to 50 µl total volume)
5 µl of 10× PCR buffer (either Perkin-Elmer or Promega containing $MgCl_2$)
2µl of cDNA (directly from the first-strand synthesis)
4 µl of 2.5 mM dNTP mix
1 µl of 1.0 uCi/µl $[\alpha-^{32}P]$ dCTP (3000 Ci/mmol, diluted from 10.0 uCi/µl with H_2O)
0.2 U of Taq DNA polymerase (either Perkin-Elmer or Promega)

2. Aliquot 4 µl of 100 ng/µl primer pair mix directly into reaction tubes. Then aliquot 48 µl of each cDNA master mix into the corresponding sets of primers.

3. The number of cycles used will vary for individual primers and new primer sets should be tested on control samples (cDNA made from whole-embryo RNA at the appropriate stage) prior to use. In order to determine whether amplification is occurring in the exponential phase, aliquots are removed from a control PCR reaction every couple of cycles over a 10-cycle range, usually starting at approximately 20 cycles. Ideally, the signal should continue to double while in the exponential phase. All of the neural primers described here are detectable after 24 cycles, and the control or loading markers are detectable after 20 cycles. If the optimal number of cycles is surpassed, high-molecular-weight bands are common due to contamination of genomic DNA. Annealing times and temperatures are also primer dependent and should be tested using control cDNA.

We have found the following PCR program to give reproducible results using a minimum amount of starting material:

   a. Initial denature for 4 minutes at 94°C.

   b. For 28 cycles: denature for 1 minute at 94°C, anneal for 1.5 minutes at 55°C, and extend for 1 minute at 72°C.

   c. Final extension for 5 minutes at 72°C.

For a complete description of PCR conditions, we recommend the text, *PCR Protocols: A guide to Methods and Applications* (Innis et al. 1990).

Gel Elecrophoresis

1. After PCR amplification, add 0.1 volume of TBE loading dye.
2. Run approximately 5–10 µl of each reaction on a 5% TBE-polyacrylamide gel until the BPB runs off the gel. Dry the gel and place on X-ray film for several hours to overnight, depending on the intensity of the expected signal.

ACKNOWLEDGMENTS The authors wish to thank Jeff Brown for discussion of technical protocols and comments on the manuscript, Minh Sum for help with the preparation of figures, and Cheng-Jung Lai for his artistic talents.

References

Amaya, E., Stein, P. A., Musci, T. J., and Kirschner, M. W. (1993). FGF signaling in the early specification of mesoderm in *Xenopus*. Development 118:477–487.

Blitz, I. L., and Cho, K. W. Y. (1995). Anterior neuroectoderm is progressively induced during gastrulation: the role of the *Xenopus* homeobox gene *orthodenticle*. Development 121:993–1004.

Bradley, L. C., Snape, A., Bhatt, S., and Wilkinson, D. G. (1992). The structure and expression of the *Xenopus* Krox-20 gene: conserved and divergent patterns of expression in rhombomeres and neural crest. Mech. Dev. 40:73–84.

Bradley, L., Wainstock, D., and Sive, H. (1996). Positive and negative signals modulate formation of the *Xenopus* cement gland. Development 122:2739–2750.

Brannon, M., and Kimelman, D. (1996). Activation of *siamois* in the *wnt* pathway. Dev. Biol. 180:344–347.

Cho, K. W. Y., Blumberg, B., Steinbeisser, H., and De Robertis, E. M. (1991). Molecular nature of Spemann's organizer: the role of the *Xenopus* homeobox gene *goosecoid*. Cel 67:1111–1120.
Chomczynski, P., and Sacchi, N. (1987). Single-step method of RNA isolation by acid guanidinium thiocyanate-phenol-chloroform extraction. Anal. Biochem. 162:156–159.
Christian, J. L., McMahon, J. A., McMahon, A. P., and Moon, R. T. (1991). *Xwnt-8*, a *Xenopus Wnt-1/int-1*-related gene responsive to mesoderm inducing factors, may play a role in ventral mesodermal patterning during embryogenesis. Development 111:1045–1056.
Christian, J., Olsen, D., and Moon, R. T. (1992). *Xwnt-8* modifies the character of mesoderm induced by bFGF in isolated *Xenopus* ectoderm. EMBO J. 11:33–41.
Dale, L., Howes, G., Price, B. M. J., and Smith, J. C. (1992). Bone morphogenetic protein 4: a ventralizing factor in early *Xenopus* development. Development 115:573–585.
Doniach, T. (1995). Basic FGF as an inducer of anteroposterior neural pattern. Cell 83:1067–1070.
Doniach, T., Phillips, C. R., and Gerhart, J. C. (1992). Planar induction of anteroposterior pattern in the developing central nervous system of *Xenopus laevis*. Science 257:542-545.
Dosch, R., Gawantka, V., Delius, H., Blumenstock, C., and Niehrs, C. (1997). Bmp-4 acts as a morphogen in dorsoventral mesoderm patterning in *Xenopus*. Development 124:2325–2334.
Fainsod, A., Steinbeisser, H., and de Robertis, E. M. (1994). On the function of *Bmp-4* in patterning the marginal zone of the *Xenopus* embryo. EMBO J. 13:5015–5025.
Frank, D., and Harland, R. M. (1991). Transient expression of *XMyoD* in nonsomite mesoderm of *Xenopus* gastrulae. Development 113:1387–1393.
Gawantka, V., Delius, H., Hirschfeld, K., Blumenstock, C., and Niehrs, C. (1995). Antagonizing the Spemann organizer: role of the homeobox gene *Xvent-1*. EMBO J. 14:6268–6279.
Hamburger, V. (1988). *The Heritage of Experimental Biology, Hans Spemann and the Organizer*. Oxford, Oxford University Press.
Harland, R. M. (1991). *In situ* hybridization: an improved whole mount method for *Xenopus* embryos. In *Methods in Cell Biology*, Vol. 36, Kay, B. K., and Peng, H. J., eds. San Diego, Academic Press, pp. 685–695.
Hemmati-Brivanlou, A., and Harland, R. M. (1989). Expression of an engrailed-related protein is induced in the anterior neural ectoderm of early *Xenopus* embryos. Development 106:611–617.
Hemmati-Brivanlou, A., and Melton, D. A. (1994). Inhibition of activin receptor signaling promotes neuralization in Xenopus. Cell 77:273–281.
Hemmati-Brivanlou, A., Kelly, O. G., and Melton, D. A. (1994). Follistatin, an antagonist of activin, is expressed in the Spemann organizer and displays direct neuralizing activity. Cell 77:283–295.
Hoppler, S., Brown, J. D., and Moon, R. T. (1996). Expression of a dominant-negative Wnt blocks induction of MyoD in *Xenopus* embryos. Genes Dev. 10:2805–2817.

Hopwood, N. D., Pluck, A., and Gurdon, J. B. (1989). A *Xenopus* mRNA related to *Drosophila twist* is expressed in response to induction in the mesoderm and the neural crest. Cell 59:893–903.

Innis, M. A., Gelfand, D. H., Sninsky, J. J. and White T. J., eds. (1990). *PCR Protocols: A Guide to Methods and Applications.* New York, Academic Press.

Isaacs, H. V., Tannahill, D., and Slack, J. M. W. (1992). Expression of a novel FGF in the *Xenopus* embryo. A new candidate inducing factor for mesoderm formation and anteroposterior specification. Development 114:711–720.

Kelly, G. M., Eib, D. W., and Moon, R. T. (1991). Histological preparation of *Xenopus laevis* oocytes and embryos. Meth. Cell Biol. 36:389–414.

Kimelman, D., and Kirschner, M. (1987). Synergistic induction of mesoderm by FGF and TGF-β and the identification of an mRNA coding for FGF in the early *Xenopus* embryo. Cell 51:869–877.

Kintner, C. R., and Melton, D. A. (1987). Expression of *Xenopus* N-CAM RNA in ectoderm is an early response to neural induction. Development 99:311.

Knecht, A. K., Good, P. J., Dawid, I. B., and Harland, R. M. (1995). Dorsal-ventral patterning and differentiation of noggin-induced neural tissue in the absence of mesoderm. Development 121:1927–1935.

Lai, C. J., Ekker, S. C., Beachy, P. A., and Moon, R. T. (1995). Patterning of the neural ectoderm of *Xenopus laevis* by the amino terminal product of *hedgehog* autoproteolytic cleavage. Development 121:2349–2360.

Lamb, M. T., Knecht, A., Smith, W. C., Stachel, S. E., Econmides, A. N., Stahl, N., Yancopolous, G. D., and Harland, R. (1993). Neural induction by the secreted polypeptide noggin. Science 262:713–718.

Lemaire, P., and Gurdon, J. B. (1994). A role for cytoplasmic determinants in mesoderm patterning: cell-autonomous activation of the *goosecoid* and *Xwnt-8* genes along the dorsoventral axis of early *Xenopus* embryos. Development 120:1191–1199.

Lemaire, P., Garrett, N., and Gurdon, J. B. (1995). Expression cloning of *siamois*, a *Xenopus* homeobox gene expressed in dorsal-vegetal cells of blastulae and able to induce a complete secondary axis. Cell 81:85–94.

Lumsden, A., and Krumlauf, R. (1996). Patterning the vertebrate neuraxis. Science 274:1109–1115.

Maniatis, T., Fritsch, E. F., and Sambrook, J. (1989). *Molecular Cloning. A Laboratory Manual.* Cold Spring Harbor Laboratory Press, New York.

Mathers, P. H., Miller, A., Doniach, T., Dirksen, M.-L., and Jamrich, M. (1995). Initiation of anterior head-specific gene expression in uncommitted ectoderm of *Xenopus laevis* by ammonium chloride. Dev. Biology 171:641–654.

McGrew, L. L., Lai, C. J., and Moon, R. T. (1995). Specification of the anteroposterior neural axis through synergistic interaction of the Wnt signaling cascade with *noggin* and *follistatin*. Dev. Biol. 172:337–342.

Moon, R. T., and Christian, J. L. (1989). Microinjection and expression of synthetic mRNAs in *Xenopus* embryos. Technique 1:76–89.

Onichtchouk, D., Gawantka, V., Dosch, R., Delius, H., Hirschfeld, K., Blumenstock, C., and Niehrs, C. (1996). The *Xvent-2* homeobox gene is part of the *BMP-4* signalling pathway controlling dorsoventral patterning of *Xenopus* mesoderm. Development 122:3045–3053.

Peng, H. B. (1991). Appendix A: solutions and protocols. In *Xenopus laevis*: Practical Uses in Cell and Molecular Biology, Vol. 36, Kay, B. K., ed. New York, Academic Press, pp. 657–662.

Phillips, C. R. (1991). Neural Induction. *Xenopus laevis*. Practical Uses in Cell and Molecular Biology. *Methods in Cell Biology, Volume 36*, Academic Press.

Placzek, M., and Furley, A. (1996). Neural development: patterning cascades in the neural tube. Curr. Biol. 6:526–529.

Pownall, M. E., Tucker, A. S., Slack, J. M. W., and Isaacs, H. V. (1996). *eFGF, Xcad3* and *Hox* genes form a molecular pathway that establishes the antero-posterior axis in *Xenopus*. Development 122:3881–3892.

Ruiz i Altaba, A., and Melton, D. A. (1989). Bimodal and graded expression of the *Xenopus* homeobox gene *Xhox3* during embryonic development. Development 106:173-183.

Rupp, R. A., and Weintraub, H. (1991). Ubiquitous MyoD transcription at the midblastula transition precedes induction-dependent MyoD expression in presumptive mesoderm of *X. laevis*. Cell 65:927–937.

Sasai, Y., Lu, B., Steinbeisser, H., Geissert, D., Gont, L. K., and De Robertis, E. M. (1994). Xenopus chordin: a novel dorsalizing factor activated by organizer-specific homeobox genes. Cell 79:779–790.

Sasai, Y., Lu, B., Steinbeisser, H., and De Robertis, E. M. (1995). Regulation of neural induction by the Chd and Bmp-4 antagonistic patterning signals in *Xenopus*. Nature 376:333–336.

Sasai, Y., Lu, B., Piccolo, S., and DeRobertis, E. M. (1996). Endoderm induction by the organizer-secreted factors chordin and noggin in Xenopus *animal* caps. EMBO J. 15:4547–4555.

Sato, S. M., and Sargent, T. D. (1991). Localized and inducible expression of *Xenopus posterior* (*Xpo*), a novel gene active in early frog embryos, encoding protein with a 'CCHC' finger domain. Development 112:747–753.

Schmidt, J., Francois, V., Bier, E., and Kimelman, D. (1995). *Drosophila* short gastrulation induces an ectopic axis in *Xenopus*: evidence for conserved mechanism of dorsal-ventral patterning. Development 121:4319–4328.

Schmidt, J. E., von Dassow, G., and Kimelman, D. (1996). Regulation of dorsal-ventral patterning: the ventralizing effects of the novel *Xenopus* homeobox gene *Vox*. Development 122:1711–1721.

Sharpe, C. R., Fritz, A., De Robertis, E. M., and Gurdon, J. B. (1987). A homeobox-containing marker of posterior neural shows the importance of predetermination in neural induction. Cell 50:749–758.

Sive, H. L., Hattori, K., and Weintraub, H. (1989). Progressive determination during formation of the anteroposterior axis in *Xenopus laevis*. Cell 58:171–180.

Slack, J. M. Darlington, B. G., Heath, J. K. and Godsave, S. F. (1987). Mesoderm induction in early *Xenopus* embryos by heparin-binding growth factors. Nature 326:197–200.

Smith, J. (1987). A mesoderm-inducing factor is produced by a *Xenopus* cell line. Development 99:3–14.

Smith, W. C., and Harland, R. M. (1991). Injected Xwnt-8 RNA acts early in Xenopus embryos to promote formation of a vegetal dorsalizing center. Cell 67:753–765.

Smith, W. C., and Harland, R. M. (1992). Expression cloning of *noggin*, a new dorsalizing factor localized in the Spemann organizer in *Xenopus* embryos. Cell 70:829–840.

Smith, W. C., Knecht, A. K., Wu, M., and Harland, R. M. (1993). Secreted *noggin* protein mimics the Spemann organizer in dorsalizing *Xenopus* mesoderm. Nature 361:547–549.

Smith, W. C., McKendry, R., Ribisi, S., and Harland, R. M. (1995). A nodal-related gene defines a physical and functional domain within the Spemann organizer. *Cell* **82**, 37–46.

Steinbeisser, H., Fainsod, A., Niehrs, C., Sasai, Y., and De Robertis, E. M. (1995). The role of gsc and *BMP-4* in dorsal-ventral patterning of the marginal zone in *Xenopus*: a loss-of-function study using antisense RNA. EMBO J. 14:5230–5243.

Stutz, F., and Spohr, G. (1986). Isolation and characterization of sarcomic actin genes expressed in *Xenopus laevis* embryogenesis. J. Mol. Biol. 187:349–361.

Tanabe, Y., and Jessell, T. M. (1996). Diversity and pattern in the developing spinal cord. Science 274:1115–1123.

Vodicka, M. A., and Gerhart, J. C. (1995). Blastomere derivation and domains of gene expression in the Spemann Organizer of *Xenopus laevis*. Development 121:3505–3518.

von Dassow, G., Schmidt, J. E., and Kimelman, D. (1993). Induction of the *Xenopus* organizer: expression and regulation of *Xnot*, a novel FGF and activin-regulated homeobox gene. Genes Dev. 7:355–366.

Wright, C. V. E., Morita, E. A., Wilkin, D. J., and De Robertis, E. M. (1990). The Xenopus XlHbox 6 homeo protein, a marker of posterior neural induction, is expressed in proliferating neurons. Development 109:225–234.

Yang-Snyder, J., Miller, J. R., Brown, J. D., Lai, C. J., and Moon, R. T. (1996). A *frizzled* homolog functions in a vertebrate *Wnt* singaling pathway. Curr. Biol. 6:1302–1306.

Zaraisky, A. G., Ecochard, V., Kazanskaya, O. V., Lukyanov, S. A., Fesenko, I. V., and Duprat, A. M. (1995). The homeobox-containing gene *XANF-1* may control development of the Spemann organizer. Development 121:3839–3847.

18

Studying the Function of Maternal mRNAs in *Xenopus* Embryos: An Antisense Approach

M. V. Zuck
C. C. Wylie
J. Heasman

This chapter describes a method that has been used successfully to study the roles of a number of maternal mRNAs in *Xenopus* embryos (Weeks et al. 1991; Heasman et al. 1992, 1994; Kofron et al. 1997). The aim of the technique is to study the roles of maternal genes by creating a milieu in which the genes are not fully active. This is achieved by reducing the maternal pool of RNA for the gene of interest and studying the effect that depletion has on development. Depletion of the maternal mRNA of interest is accomplished by injection of antisense oligonucelotides (ODNs) into oocytes and subsequent fertilization of these oocytes in order to study the effects of depletion. While these underexpression experiments are technically more demanding than overexpression experiments, they are more rewarding in that they can reveal directly the functions of the gene of interest.

Underexpression experiments require injection of antisense ODNs complementary to the target mRNA into fully grown ovarian oocytes. Oocytes contain an endogenous RNase H activity that cleaves RNA/DNA duplexes; therefore, mRNA bound to the ODNs is cleaved by RNase H and then broken down further by other nucleases (Dash et al. 1987). A single injection of ODN at a single time point effectively reduces maternal RNA levels as, from the time the oocyte is fully grown until the embryo reaches the 4000-cell stage, no new RNA is synthesized. Therefore, new mRNA is not produced until synthesis of RNA from the embryonic genome begins (zygotic

Developmental northern blot of α-catenin mRNA

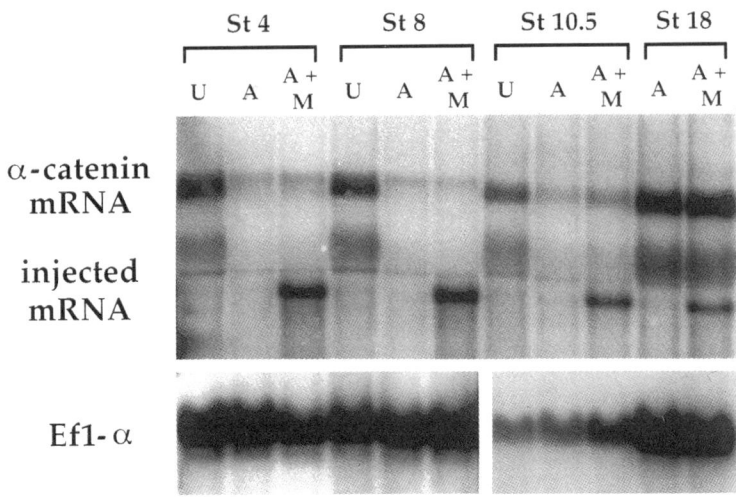

Figure 18-1 A northernl blot of α-catenin expression in embryos at cleavage stage (stage 4, Nieuwkoop and Faber), blastula stage (stage 8 — approximately 4000-cell stage), gastrula stage (stage 10.5), and neurula stage (stage 18). Lanes marked U represent uninjected control embryos. Lanes marked A represent embryos injected with antisense ODNs as oocytes, and lanes marked A + M received antisense ODN injections as oocytes followed by synthetic mRNA injections. Embryos at stages before zygotic synthesis (stage 4 and stage 8) exhibit significant depletion, as do embryos for some time after the 4000-cell stage, i.e., depletion remains significant after the 4000-cell stage (approximately stage 8) until at least stage 10.5. EF1-α expression patterns are shown as a loading control. Synthetic α-catenin mRNA is shown as well. The mRNA injected for the rescue experiment is much smaller because it does not contain its full-length 5'- and 3'UTRs (see text for details) (Kofron et al. 1997).

synthesis), and zygotic synthesis of the mRNA of interest does not begin until after the 4000-cell stage (see Figure 18-1). Thus, injection of ODN into the oocyte effectively removes mRNA until the 4000-cell stage. However, the continual synthesis of zygotic mRNA makes antisense depletion a poor choice for targeting zygotic mRNAs, as ODNs would need to be added continually to prevent depleted mRNA being replaced (see Figure 18-1). This portion of the host transfer technique — the depletion of RNA from the oocyte — is adapted from the original studies in *Xenopus* oocytes by the laboratories of Walder, Coleman, and Weeks (see Shuttleworth and Colman 1988; Shuttleworth et al. 1988; Dagle et al. 1990, 1991; Weeks et al. 1991).

The second part of the underexpression experiment involves the fertilization of oocytes taken from one frog and then transferred into a different frog, using a method devised by Subtelny et al. (1961) and Brun (1975), and refined in our lab (Holwill et al. 1987). Although it would seem simpler to bypass the process of removing ovary from one frog only to place it into another, we do not inject ODNs directly into fertilized eggs. Briefly, ODNs are more toxic in eggs, are not as effective in degrading mRNAs, and may be inherited in a mosaic fashion, making results difficult to interpret (Woolf et al. 1990; Heasman et al. 1992).

Underexpression has been utilized in our laboratory to study the functions of cytoskeletal proteins, adhesion proteins, signaling molecules, and transcription factors in *Xenopus* embryos, and a schematic overview of the entire underexpression technique is seen in Figure 18-2. In some cases, specific functions of the proteins were revealed, while for other molecules, no phenotype was seen when the mRNA was depleted (Heasman et al. 1992, 1994; Kofron et al. 1997; data not shown).

The disadvantage of the underexpression technique is that, in cases where no phenotype results, it is extremely difficult to determine whether the lack of effect is due to an insufficient depletion of the mRNA or protein (e.g., if the protein has a very long half-life), or whether the molecule is redundant in function. For this reason, it is important to analyze the level of gene function as accurately as possible. Therefore, when an assay exists to test gene activity (e.g., a kinase assay for a maternal kinase), it should be used to determine the level of gene function after underexpression. Also, when antibodies are available, Western analysis and/or immunostaining on ODN-injected oocytes should be utilized to show that depletion of the mRNA does lead to a reduction in the level of protein. This reduction would be expected to be greatest at the late blastula stage, when the maternal pool of protein will be most exhausted and before zygotic synthesis of RNA is underway (e.g., Figure 18-3). Underexpression is a powerful tool for analysis of gene function in *Xenopus laevis*, but it does have limitations.

Methods

Identifying Antisense ODNs that Deplete the Target mRNA

1. Selection of Six to Eight ODNs We choose ODNs of 18 bases in length based on experiments of Dagle et al. 1990) complimentary to either the 5′ untranslated region (5′UTR), the 3′UTR, or the coding region of the sequence of the mRNA. Parts of the sequence are chosen that do not have the same base occurring three or more times consecutively and which do have a balance of purines and pyrimidines. Utilizing these restrictions, six to eight ODNs are chosen at random, synthesized, and desalted (Biosynthesis, Inc.). Subsequently, the ODNs are resuspended in sterile (0.2 μm) filtered distilled water at a concentration of 1 mg/ml and stored in 10-μl aliquots in a −80° freezer until just before use.

mRNA Depletion Experiment
ODN injection into manually defolliculated *Xenopus* oocytes

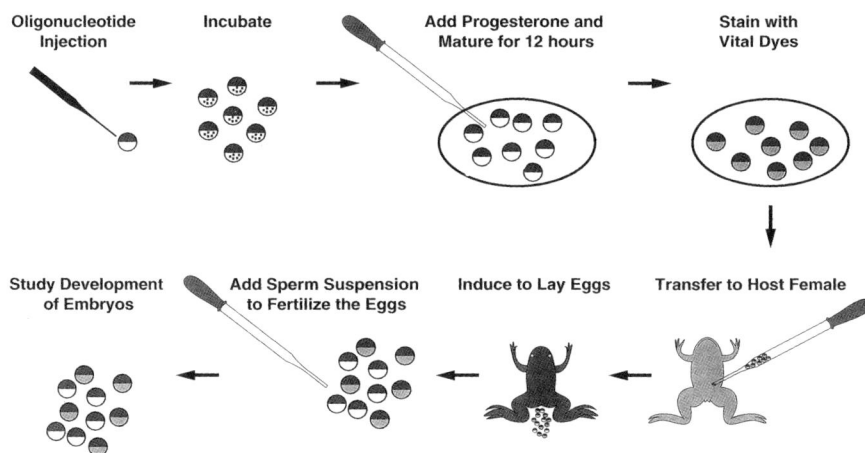

Figure 18-2 A schematic representation of the underexpression experiment.

2. Injection of ODNs and Incubation of Oocytes Small numbers of full-grown stage 6 oocytes are then manually defolliculated from pieces of ovary in oocyte culture medium (OCM, Appendix A, for further details see "Defolliculation and Injection" under "The Host Transfer Technique") and stored at 18°C for use. The six test ODNs are spun in an Eppendorf centrifuge at 20,000 rpm for 10 minutes at 4°C and placed on ice. Each ODN is injected into approximately five oocytes per dose in doses of 5 and 10 ng. Injections are accomplished using glass needles pulled in a moving coil microelectrode puller (model 753, Campden Instruments Ltd., Genetic Research Instrumentation Ltd.). The glass needle is broken off at

Western blots from ODN depletions

Figure 18-3 Western blots of (A) α-catenin protein levels in oocytes and (B) plakoglobin protein levels in late blastula embryos reveal a reduction in protein as well as in mRNA in embryos and oocytes injected with ODNs (Kofron et al. 1997).

the very tip and the needle is calibrated on a high-pressure injection system (Medical Systems, Inc., PLI-100) by collecting the volume of ten 1-second injections into a 1-μl capillary ("microcaps," disposable 1-μl micropipettes, Drummond). Only needles that conform to a volume of 2–10 nl per 1-second injection are used. Needles are attached to a micromanipulator (Leitz) and oocytes are then injected in the equatorial zones while in OCM under a dissecting microscope (Leica Wild M8) and incubated overnight to deplete the mRNA thoroughly. Recently, for vegetally localized mRNAs, we have found vegetal rather than equatorial injections deplete more successfully.

3. Freezing Oocytes and Performing Northern Analysis The next day, injected oocytes are frozen for Northern blot analysis along with uninjected controls. Northern blot analysis is carried out utilizing procedures described in Kofron et al. (1997), isolating RNA from 2–5 oocytes and loading 1–2.5 oocyte equivalents per lane (depending on the abundance of the mRNA of interest). Figure 18-4A is an example of such a Northern blot in which seven ODNs are tested in this fashion to determine which deplete plakoglobin mRNA.

It is interesting that, although all seven ODNs are completely complementary to the target mRNA, not all of them are effective in depleting the RNA (Figure 18-4A). This represents a common result for depletion of a message by a number of ODNs, and presumably reflects the fact that, as a result of folding or protein binding, only parts of the mRNA are available for the annealing of ODNs.

Oligodeoxynucleotide Modification

After testing the ODNs, an effective ODN or several ODNs: is/are chosen; a rough guideline for an effective ODN is one which depletes the target mRNA to less than 20% of the level of the uninjected control at a 10-ng dose, as seen by northern analysis. The ODNs so selected are generally resynthesized in a modified form and HPLC purified. The modification suggested by the experiments of Baker et al. (1990) and Dagle et al. (1990) is used, where three or four of the 5'-most and 3'-most phosphodiester bonds are replaced by phosphorothioate bonds, leaving at least eight unmodified bonds in the center of the ODN (e.g., Kofron et al., 1997).

Modified oligodeoxynucleotides provide two major advantages over unmodified ODNs. The modified ODNs can deplete the message more effectively than the unmodified ODNs at a much lower concentration (e.g., compare the 2-ng depletion by modified ODN #7 in Figure 18-4B with the 10-ng unmodified ODN #7 depletion in Figure 18-4A). Second, the modified ODNs provide a more reliable and reproducible depletion. However, it has been our experience that the phosphorothioate ODNs tend to reach a toxic dose at around 5–10 ng. Both the toxicity of the modified ODNs and their ability to further deplete mRNA probably result

346 Methods for Studying *Xenopus* Oocytes and Embryos

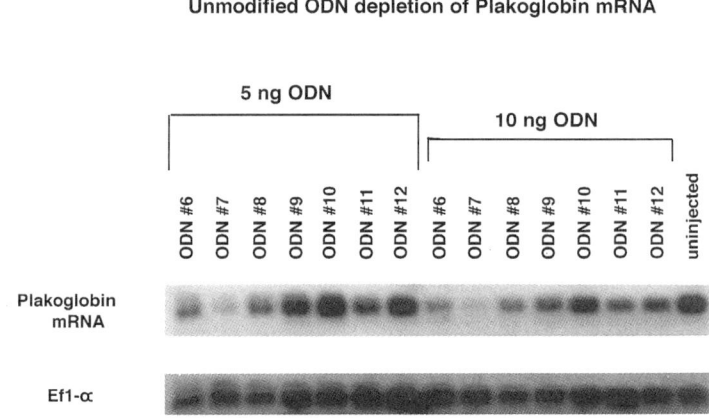

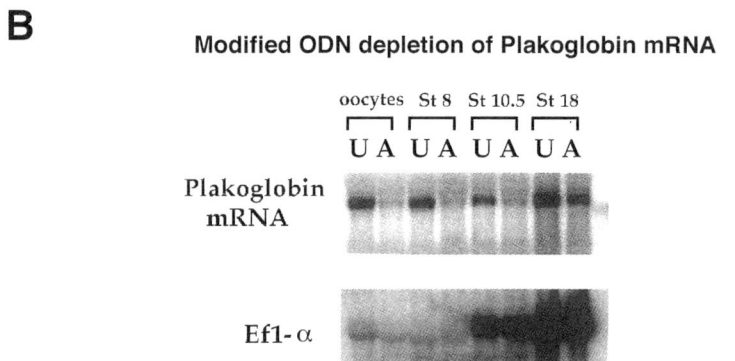

Figure 18-4 Oocytes in (A) have been depleted with one of seven unmodified ODNs, antisense to plakoglobin mRNA, in both 5- and 10-ng doses. Compare with levels of plakoglobin in oocytes in (B) with 2 ng of modified ODN injected. Ef1-α is provided as a loading control.

from the increased stability of phosphorothioate ODNs (Dagle et al. 1990; Woolf et al. 1990).

Other alternatives to achieve nontoxic depletions also exist. An alternative to phosphorothioate modification which may ameliorate the toxicity of the modified ODNs is to synthesize ODNs with 5′- and 3′-methoxyethyl phosphoroamidate linkages (Dagle et al. 1990; Heasman et al. 1992); however, we are unaware of any such ODNs that are commercially available at the time of publication. Another alternative to the use of a high-dose phosphorothioate-modified ODN is to utilize two (or theoretically more) ODNs in combination. It has been shown that the combination of oligodeoxynucleotides provides a superior depletion to any one alone, even at a higher dose of ODN (Morgan et al. 1993). Finally, HPLC purification also reduces toxic effects.

The Host Transfer Technique

Once a suitable ODN has been identified, the effect of depletion of the mRNA should be tested in developing embryos. To observe the effects of the ODN, oocytes depleted of the target mRNA are fertilized by the use of the host transfer technique (see Figure 18-2 for a schematic representation), as detailed below.

1. Selecting an Ovary The major determinant in the success of this technique is the quality of the oocytes injected and transferred. Unfortunately, there is no sure test of oocyte quality. Therefore, we routinely anesthetize several donor frogs consecutively and remove small pieces of ovary in a sterile fashion, via an abdominal incision (for further detail see "Transferring the Oocytes into an Egg-Laying Female" below). Frogs are selected to donate ovary based on two criteria. First, it has been our experience that frogs that do not look healthy and plump do not yield "good" ovary, so we pick healthy, mature females to operate on. Second, frogs which have been injected with human chorionic gonadotropin (hCG) in the previous 10 weeks are not used. We have had success with frogs purchased from Nasco, Xenopus1, and Xenopus Express.

The ovary itself is not used if there are very few full-grown oocytes, oocytes with small cavitations in the animal hemisphere, or many oocytes that are flaccid or wrinkled. Oocytes with conspicuous equatorial bands often fertilize well (but not always), although these bands may not be obvious on palely pigmented oocytes (Heasman et al. 1991). One test that may be used to determine the suitability of an ovary is to defolliculate a small number of oocytes and stimulate them to mature using a final concentration of 1 μM progesterone in OCM. This test is useful since maturation is a necessary step for fertilization, and oocytes which do not mature well (less than 80%) will not fertilize well. Care should be taken to defolliculate only the largest subset of healthy oocytes in order to ensure the best chance of fertilization, and uniformity of embryonic response to ODN.

2. Removal and Subdivision of the Ovary into Separate Dishes Having chosen a suitable ovary, enough for the experiment is withdrawn from the anesthetized frog via the abdominal incision. The number of "good" oocytes per lobe of ovary varies greatly, but two or three lobes will often yield about 500 oocytes. Ovary is washed in OCM to remove blood, and is then cut into small pieces (generally about 1 × 2 cm) and cultured in OCM at 18°C. Care is taken not to overcrowd the dishes with ovary, as the tissue is highly metabolically active (six or seven pieces per 90-mm dish). In these conditions, ovary can be maintained in culture for 2–4 days. Before subdividing ovary, however, the anesthetized frog should be sutured (for further detail see "Transferring the Oocytes into an Egg-Laying Female" below) and put into a clean bucket containing tap water and 3 ml/10 gallons of Amquel (Kordon).

3. Defolliculation and Injection To study the function of the test mRNA, large numbers of defolliculated oocytes are required (approximately 500 for a large experiment). We defolliculate oocytes manually using jeweler's forceps (#4, American Surgical Instruments Corporation), as collagenase-treated oocytes do not fertilize. In an attempt to keep the media as sterile as possible, we wipe down the watchmaker's forceps with 70% ethanol before use. Also, in a variation from previously published technique (Heasman et al. 1991), we incubate the defolliculated oocytes in dishes of OCM which have not been previously agarose coated. After defolliculation, modified ODNs are diluted with sterile filtered distilled water to a concentration less than 0.5 ng/nl. Often, ODNs are diluted to 0.2 ng/nl and 0.4 ng/nl, and are spun at 4°C and 20,000 rpm for 10 minutes. Oocytes are injected into the equatorial zones of oocytes in OCM, as superficially within the oocyte as possible in batches of 70–100. Initially, a typical experiment includes batches of 70–100 uninjected controls or 4 ng of sense ODN only, as well as 1, 2, 3, and 4 ng of modified ODN. We limit ourselves to five batches of oocytes per experiment due to the limitation of vital dyes (Appendix B). After initial successful underexpression experiments have been completed, a dose is found which causes an effective depletion. Then, that dose and one other are usually injected so that a phenotype may be observed even if there is variability in the response of the oocyte to ODN. Oocytes are all incubated for at least 24 hours at 18°C in order to allow the ODNs to decay.

4. Progesterone Treatment of Oocytes and Injection of Frogs with Human Chorionic Gonadotropin (hCG) Oocytes are prepared for fertilization by adding a final concentration of 1 µM progesterone (to the OCM and then culturing them at 18°C for 12 hours. Host frogs are stimulated with human chorionic gonadotropin (1000 U, Sigma) so that they will begin to lay eggs 8–12 hours later at room temperature. Typically progesterone treatment and frog hCG injection are carried out at 9–10 p.m. for a 9 a.m. transfer the following day.

5. Freezing Some Oocytes for RNA Analysis and Staining the Other Oocytes with Vital Dyes Before transferring the oocytes to a host frog, a small number of oocytes are frozen for Northern analysis and the rest are colored with vital dyes (see Appendix B). Dyes are added directly to the OCM containing the oocytes. We use 100 μl of red, blue, or brown dyes in 8 ml of OCM. Green is obtained by adding 100 μl each of blue and brown, and purple (mauve) by mixing 100 μl each of blue and red. Oocytes are colored for approximately 15 minutes on a rocking platform, and then washed in a large dish of OCM.

6. Transferring the Oocytes into an Egg-Laying Female A host frog that has just begun to lay eggs is selected for oocyte transfer. The newly laid eggs are examined to check that the host is not laying eggs that are in heavily jellied strings, or that are clearly dead. The host is then anesthetized using MS222 (3-aminobenzoic acid ethyl ester methanesulfonate salt, Sigma). As soon as the frog is no longer responsive to touch on the bottom of the jaw, a small incision is made through the skin in the lower half of the anterior abdominal wall, to one side of the midline. The connective tissue and muscle layer under the skin are also opened with one cut, using small scissors, thus making an incision into the coelomic cavity. The incision is made large enough to allow the entrance of the end of a Pasteur pipette, and about half the size of the incision which is used to remove ovary. The sterile Pasteur pipette is cut to a diameter slightly larger than a full-grown oocyte, using a diamond pencil, and flamed to produce smooth surfaces. Then the pipette is used to transfer all of the experimental oocytes into the coelomic cavity.

It is important in any operation on the frog to hold one side of the incision in the forceps while the oocytes are introduced with the other hand. If the forceps hold the edge of the incision a small distance above the frog, valuable time will not be wasted looking for the incision and valuable oocytes will not be lost as they are expelled out of the coelomic cavity. Care is also taken not to introduce air, or large excesses of OCM.

The oocytes can be injected into any part of the coelomic cavity, as they will be moved into the oviducts by the ciliary action of the coelomic epithelium. We have introduced as many as 700 oocytes into one host frog. The layers of the body wall are then sutured (45 cm of 4-0 black braided silk, C3, cutting 3/8 circle, 13 mm, Ethicon) using two or three stitches to sew together the muscle and overlying fascia, and two or three stitches to sew the skin, taking care not to sew skin and muscle or skin and fascia together. The sutures are applied using a hemostat. The frog is then returned to clean water and recovers from the anesthetic within 10–20 minutes. Care should be taken not to overexpose frogs to anesthetic, i.e., to carry out the transfer of oocytes as soon as movement has ceased, and to use sterile technique as much as possible when operating on the frog.

Fertilization and Development

1. Fertilization of Embryos Three hours after the oocytes are transferred to the host frog, the frog is squeezed to lay eggs at 20-30 minute intervals and fertilized with a sperm suspension, as per a normal in vitro fertilization (Holwill et al. 1987). The percentage of experimental oocytes that are laid by the female varies from one experiment to the next, but as long as the female does not stop laying eggs most of the dyed oocytes can be recovered. Of these only a portion of them will fertilize, and a 20-40% rate of fertilization is usual for transferred and uninjected oocytes. Usually a female will give three or four good fertilizations per experiment.

2. Dejellification of Embryos Embryos are dejellied in 2% cysteine, pH 7.8, in 0.1× MMR (Appendix A) after the first cell cycle. Briefly, they are gently swirled until the embryos lie close together instead of being held apart by the jelly coats. Then embryos are washed three times in a large excess volume of 0.1× MMR. Both host embryos and transferred embryos are then allowed to develop in 0.1× MMR.

As it is easier to distinguish the five colors from each other early in development rather than later, the embryos are subdivided into separate dishes at this time and are monitored carefully through development to observe differences between the control and experimental embryos. In addition, embryos may be fixed and frozen for further analysis at later stages.

Rescue Experiments

Although controls already exist within an underexpression experiment, the specificity of the ODN's effect(s) must be rigorously demonstrated. Injection of sense ODN alone, or injection of sense and antisense ODN together, may reveal whether the effect of an ODN is limited to its RNA-depleting abilities or due to nonspecific ODN toxicity, but it is insufficient to illustrate that the ODN's specific action is limited to the mRNA of interest. The only way to demonstrate convincingly that the ODN's effects are due to underexpression of the message of interest and not to the depletion of another mRNA, is to demonstrate that synthetic mRNA from the gene of interest can substitute for wild-type mRNA, i.e. the way to show specificity is to deplete the endogenous message and rescue the effect(s) of depletion by the addition of synthetic message of that gene.

1. RNA Synthesis The first step prior to a successful rescue experiment is to obtain mRNA which translates well in vivo. mRNAs transcribed from many bacterial plasmid vectors do not translate well in *Xenopus* oocytes. However, it has been shown that the 5′- and 3′-β-globin UTRs are sufficient to ensure the translation of transcripts in the *Xenopus* oocyte when attached to the coding region of another gene (Krieg and Melton 1987). Several bacterial plasmid vectors which contain the globin UTRs are available, of which we routinely use pSP64T.

2. RNA Overexpression in Host Transfer   Next, a dose–response experiment is undertaken to determine whether overexpression of synthetic mRNA in oocytes induces a phenotype. Injection of four batches of 75–100 oocytes per batch should be performed as discussed previously. We often use doses spread over a wide range (e.g., 50 pg, 200 pg, 1 ng, and 3 ng — any dose over 5 ng is very likely to be toxic) to get an idea of the highest dose that does not cause embryonic abnormalities. It is important to know (at least approximately) what is the largest dose of injected RNA that does not perturb normal embryonic development, so that that dose may be injected into the embryo to attempt a rescue.

An alternative to injection of mRNA into oocytes is injection of rescuing mRNA into embryos. Briefly, it is similar to injection into oocytes except that injection into embryos should be carried out in isotonic media (1× MMR) containing Ficoll (2–3%). Most embryonic injections take place at either the two- or four-cell stage, although they may take place at later stages as well.

3. Performing the Rescue Experiment   The rescue experiment involves injecting oocytes with ODNs, incubating the oocytes for 24 hours, injecting the oocytes with RNA, maturing them, and then transferring them to a new host frog. On the first day, ODN is injected into a large batch of oocytes, and even though a "good" dose has been found, both for ODN and RNA, we usually inject at least two doses of ODN since oocyte response to ODN may be variable.

Each batch of ODN-injected oocytes must be big enough (150–200 oocytes) to inject half of the oocytes with RNA. After ODN injection, incubation of the oocytes for 24 hours is necessary to allow virtually all of the ODN to degrade (Figure 18-1; Raats et al. 1997). The following RNA injection should then initiate protein expression and reveal the nature of ODN-induced abnormality. There need be no time lapse between RNA injection and transfer, but we usually allow the oocytes to incubate again overnight before transferring them.

Conclusions

This method is difficult because it is impossible to predict with certainty which batches of oocytes will fertilize and which host frog will faithfully fertilize those oocytes. Nevertheless, much information can be obtained even from relatively small numbers of fertilized eggs (10–20 eggs), and it has been our experience that once a characteristic phenotype is observed, it is highly reproducible from one experiment to the next. Other techniques which study gene function in vertebrates, such as knockouts in *Mus musculus* or the insertion of DNA into *Xenopus* embryos (Kroll and Amaya 1996), provide information about zygotic vertebrate gene function, while the technique described here is useful for maternal gene products. Despite the challenging nature of these experiments, they allow the direct study of the role of maternal genes in *Xenopus laevis*.

ACKNOWLEDGMENTS We are grateful to the NIH (grant number 1 RO1 HD33002) and the Institute of Human Genetics and the Harrison Fund for financing this work. We thank M. Kofron for the contribution of plakoglobin data and figures to this work, and A. Spagnuolo for data pertaining to α-catenin.

References

Baker, C., Holland, D., Edge, M., and Colman, A. (1990). Effects of oligo sequence and chemistry on the efficiency of oligodeoxyribonucleotide-mediated mRNA cleavage. Nucleic Acids Res. 18(12):3537–3543.

Brun, R. (1975). Oocyte maturation in vitro: contribution of the oviduct to total maturation in *Xenopus laevis*. Experientia 31(11):1275–1276.

Dagle, J. M., Walder, J. A., and Weeks, D. L. (1990). Targeted degradation of mRNA in *Xenopus* oocytes and embryos directed by modified oligonucleotides: studies of An2 and cyclin in embryogenesis. Nucleic Acids Res. 18(16):4751–4757.

Dagle, J. M., Weeks, D. L., and Walder, J. A. (1991). Pathways of degradation and mechanism of action of antisense of oligonucleotides in *Xenopus laevis* embryos. Antisense Res. Dev. 1(1):11–20.

Dash, P., Lotan, I., Knapp, M., Kandel, E. R., and Goelet, P. (1987). Selective elimination of mRNAs in vivo: complementary oligodeoxynucleotides promote RNA degradation by an RNase H-like activity. Proc. Natl. Acad. Sci. USA 84(22):7896–7900.

Heasman, J., Holwill, S., and Wylie, C. C. (1991). Fertilization of cultured *Xenopus* oocytes and use in studies of maternally inherited molecules. Methods Cell Biol. 36:213–230.

Heasman, J., Torpey, N., and Wylie, C. (1992). The role of intermediate filaments in early *Xenopus* development studied by antisense depletion of maternal mRNA. Development (Suppl.):119–125.

Heasman, J., Crawford, A., Goldstone, K., Garner-Hamrick, P., Gumbiner, B., McCrea, P., Kintner, C., Noro, C. Y., and Wylie, C. (1994). Overexpression of cadherins and underexpression of beta-catenin inhibit dorsal mesoderm induction in early *Xenopus* embryos. Cell 79(5):791–803.

Holwill, S., Heasman, J., Crawley, C. R., and Wylie, C. C. (1987). Axis germ line deficiencies caused by u-v irradiation of *Xenopus* oocytes cultured *in vivo*. Development 100:735–743.

Kofron, M., Spagnuolo, A., Klymkowsky, M., Wylie, C., and Heasman, J. (1997). The roles of maternal alpha-catenin and plakoglobin in the early *Xenopus* embryo. Development 124(8):1553–1560.

Krieg, P. A., and Melton, D. A. (1987). In vitro RNA synthesis with SP6 RNA polymerase. Methods Enzymol. 155:397–415.

Kroll, K. L., and Amaya, E. (1996). Transgenic *Xenopus* embryos from sperm nuclear transplantations reveal FGF signalling requirements during gastrulation. Development 122(10):3173–3183.

Morgan, R., Edge, M., and Colman, A. (1993). A more efficient and specific strategy in the ablation of mRNA in *Xenopus laevis* using mixtures of antisense oligos. Nucleic Acids Res. 21(19):4615–4620.

Raats, J. M. H., Gell, D., Vickers, L., Heasman, J., and Wylie, C. (1997). Modified mRNA rescue of maternal CK1/8 mRNA depletion in *Xenopus* oocytes. Antisense and Nucleic Acid Drug Development 7:263–277.

Shuttleworth, J., and Colman, A. (1988). Antisense oligonucleotide-directed cleavage of mRNA in *Xenopus* oocytes and eggs. EMBO J. 7(2):427–434.

Shuttleworth, J., Matthews, G., Dale, L., Baker, C., and Colman, A. (1988). Antisense oligodeoxyribonucleotide-directed cleavage of maternal mRNA in *Xenopus* oocytes and embryos. Gene 72(1–2):267–275.

Subtelny, S., and Bradt, C. (1961). Transplantations of blastula nuclei into activated eggs from the body cavity and from the uterus of *Rana pipiens* II. Development of the recipient body cavity eggs. Dev. Biol. 3:96–114.

Weeks, D. L., Walder, J. A., and Dagle, J. M. (1991). Cyclin B mRNA depletion only transiently inhibits the *Xenopus* embryonic cell cycle. Development 111(4):1173–1178.

Woolf, T. M., Jennings, C. G., Rebagliati, M., and Melton, D. A. (1990). The stability, toxicity and effectiveness of unmodified and phosphorothioate antisense oligodeoxynucleotides in *Xenopus* oocytes and embryos. Nucleic Acids Res. 18(7):1763–1769.

Appendix

A: Solutions for Maintenance and Manipulation of *Xenopus* Oocytes and Embryos

OCM (Oocyte Culture Media)
240 ml Liebovitz L-15 medium (Sigma)
160 ml Sterile doubled-distilled water
0.16 g Bovin serum albumin (BSA; Sigma)
2 ml Glutamine from 200 mM stock (Sigma) (Make fresh weekly.)
40 µl of 10 mg/ml Gentamicin
Adjust pH to 7.6–7.8.
Make fresh daily.

10× MMR (Marc's Modified Ringer's Solution)*
1 M NaCl
20 mM KCl
20 mM $CaCl_2$
10 mM $MgCl_2$
150 mM HEPES*
pH 7.6
Store at 4° Celsius.

---

*This recipe is actually a modification of MMR. We use 3× the normal HEPES concentration.

B: Vital Dyes

When making up vital dyes, be aware that different batches of dyes will stain oocytes differently. Therefore, it is necessary to visually inspect oocytes and to determine how much vital dye is required to stain the oocytes in each of the five colors in the 15-minute incubation used to dye the oocytes. Although we have attempted to make more than five colors using combinations of vital dyes, only five are easily distinguishable:

Blue — 0.1% Nile blue A (Sigma)
Red — 0.25% Neutral red (Sigma)
Brown — 2% Bismarck brown (Sigma)
Green — (100 µl blue + 100 µl brown)
Mauve — (100 µl blue + 100 µl red)

Red, blue, and brown dyes are dissolved in sterile distilled water and spun down after shaking for 0.5 hours. The supernatant is aliquotted as 1-ml aliquots in 1.5-ml microfuge tubes. The pellet is discarded.

Dyes are added directly to the OCM containing the oocytes. We use 100 µl of red, blue, or brown dyes in 8 ml of OCM. Green is obtained by adding 100 µl each of blue and brown dyes, and purple (mauve) by mixing 100 µl each of blue and red dyes. Oocytes are colored for approximately 15 minutes on a rocking platform, and then washed in a large dish of OCM.

19

Testing the Cell Fate Commitment of Single Blastomeres in *Xenopus laevis*

Sally A. Moody

Cell lineage studies, begun at the turn of the century in marine invertebrates and in the past two decades in several vertebrates, reveal what kinds of tissues descend from a single cell or defined regions of embryos. These studies demonstrate in many species (except mammals) that a regional, and sometimes cell-by-cell, pattern of the body plan is represented in the embryo before, or shortly after, fertilization of the egg. Defining precisely from which cells the various tissues and organs arise has been invaluable in elucidating the maternal control of body organization and in understanding how a flat sheet or ball of cells becomes transformed into a complex array of tubes, organs and muscle masses. However, unless the fate of a cell is completely determined by the inheritance of maternal factors, the fate maps of early embryonic cells only inform us of the developmental path taken by the cell under normal, intact embryo conditions. Such studies cannot describe the full array of developmental potential of a cell. The fate expressed by a cell, i.e., the different tissue types that descend from it, is usually influenced by a number of factors, which may include maternal determinant molecules, cell–cell interactions, growth factor signals, and position within a morphogen gradient. Therefore, to test the developmental potential of a cell, it is necessary to allow the cell to develop under novel experimental conditions.

There are at least three ways to test the fate commitment of a single embryonic cell, or blastomere. The cell may be transplanted to a novel region of the embryo to test what descendants are produced when the cell develops in a different embryonic environment. The cell may be left in situ, but manipulated in ways that prevent it from communicating with its normal neighbors. Alternatively, the cell may be cultured as an explant to test what cell types it can produce in the absence of direct contact with neighboring cells or their secreted factors. Additional information can be obtained by deleting the cell from the embryo to test whether its presence is necessary for the normal development of the remaining cells, and whether its former neighbors reconstitute its fate.

Different vertebrate embryos have different degrees of experimental accessibility, making them more or less amenable to these tests of developmental potential. In mammals, the early blastomeres can be marked with tracers (e.g., Pedersen et al. 1986), but the resulting lineages cannot be followed past implantation stages due to dilution of the tracer. Single cells have not been transplanted successfully, nor easily cultured, primarily due to the need for the protective membranes and follicular cells. Single-cell deletions have been developed, especially for the purpose of human in vitro fertilization biopsy, but these manipulations have not revealed which cells reconstitute pattern since lineage labeling does not survive the entire length of the embryos' prolonged gestation. Disruption of cell–cell communication in situ, however, can be elegantly achieved with transgenic approaches. In avians (and presumably reptiles, although I am not aware of any such studies), the fertilized egg remains within the mother's oviduct during the earliest stages, and is not accessible for manipulation until primitive streak stages when the eggs are laid (Patten 1929). Zebra fish are amenable to most of the necessary manipulations. However, the small size of the eggs, the cytoplasmic bridges between early blastomeres and the yolk mass (Kimmel and Law 1985a,b), and the indeterminate cleavages (Kimmel and Warga 1987) make single-blastomere approaches difficult. Amphibians, however, are ideal for testing the developmental potential of early embryonic cells. They have very large embryos (1–2 mm in diameter), whose cells are large, relatively easy to manipulate, and contain their own nutritive store in the form of intracellular yolk platelets. During the cleavage stages, blastomeres adhere primarily via calcium-dependent molecules, which provides an easy way to isolate single cells. Many amphibian species develop very rapidly and do so without the G-phases of mitosis, preventing the dilution of lineage dyes as development proceeds. Therefore, amphibians have been favored subjects for single-cell analyses of fate and cell–cell interactions.

Xenopus laevis became a popular model for amphibian development because it is easy to rear in the laboratory, can be induced to breed at any time of the year, and their embryos develop the basic tissues and organs within 2 days after fertilization. Unlike most other amphibian species, which are seasonal breeders, *Xenopus* can lay eggs simply in response to the injection of chorionic gonadotropin. Early *Xenopus* blastomeres are very amenable to tests of cell fate determination for many reasons, the foremost being that the identity of single cells can be recognized consistently across experiments by virtue of the early establishment of the embryonic axes, the asymmetric pigmentation pattern, and the relatively regular cleavage patterns (Figure 19-1). Furthermore, because zygotic transcription does not begin until shortly before the onset of gastrulation, many of the tests of cell fate determination can be analyzed without concern for regulators of transcription.

The cardinal axes (animal–vegetal; dorsal–ventral; right–left) of *Xenopus* embryos can all be identified soon after fertilization. The animal–vegetal axis is established during oogenesis and demarcates the pregastrula anterior–posterior axis. The animal–vegetal axis in the mature oocyte consists of a gradient of molecules (Hausen and Riebesell 1991), some of which act later in the establishment of the primary tissues. The most notable landmarks of this axis are that (1) only the animal hemisphere is darkly pigmented, and (2) the embryos orient with the vegetal hemisphere facing the center of gravity after fertilization, due to a gradient distribution of different sizes of yolk platelets. The dorsal–ventral axis is established at fertilization. Entry of the sperm nucleus initiates a cytoplasmic rotation that results in the dorsal axial tissue forming at approximately 180° from the sperm entry point (Vincent et al. 1986; Vincent and Gerhart 1987). This cytoplasmic rotation and concomitant contraction of the pigmented cortex toward the ventral side results in dorsal blastomeres usually having a lighter pattern of pigmentation than ventral blastomeres (Figure 19-1). The right–left axis is identifiable by the formation of the first cleavage furrow, which approximates the midsagittal plane (Klein 1987).

The outcome of having these early indicators of the cardinal axes is that specific regions, and at the early cleavage stages this means specific cells, can be reproducibly identified across a large population of embryos. In addition, it is possible for the experimenter to choose embryos that cleave in virtually identical patterns. Although the patterns of cleavage furrows are not invariant from one embryo to the next during the morula period (2-cells to 512-cells), as they are in several invertebrates, there are a few patterns that predominate (e.g., Figure 19-1). Using embryos with this stereotypic cleavage pattern minimizes experimental interspecimen variation (e.g., Moody 1987a,b; Moody and Kline 1990).

Normal Fate Maps and Lineage Tracing

The first step in testing fate determination and commitment is to establish the normal fate of a cell in its natural embryonic environment. Not only should all the major tissue descendants be identified, but population variance, minor descendants, and spatial distribution may enable one to test the effects of altering the location or gene expression of individual cells. Although a fate map needs to be simplified in order to provide a broad understanding of general trends, the subtle details and variations may become very important in examinations of particular cell types. There are several detailed fate maps available for the cleavage stages of *Xenopus* (Hirose and Jacobson 1979; Jacobson and Hirose 1981; Jacobson 1983; Masho and Kubota 1986; Dale and Slack 1987; Moody 1987a,b; Takasaki 1987; Masho 1988; Moody and Kline 1990). Nonetheless, because there can be differences in how laboratories identify the cardinal axes (see Masho 1990), the quality of the eggs, the genetic background of frog colonies, the diffusion of different tracer molecules, the experience of the experimenter in identifying individual cells, etc., it is worthwhile for experimenters to create their own small series of control embryos. Not only can subtle variations be tested under one's own laboratory conditions, but also one can check whether one is identifying the same cells as in the published work. This will give you confidence that you are consistently targeting the desired cells.

A molecule must meet several requirements to be a successful lineage tracer. The molecule must be nontoxic and nonreactive so that it does not change the recipient cell's developmental fate. This requirement is met for horseradish peroxidase (HRP) and fluorescent dextrans (Weisblat et al. 1978; Jacobson 1985; Stent and Weisblat 1985). Also, maps made with these different tracers are virtually identical and coincide with the results of classical vital dye maps of late blastulae and gastrulae (Bauer et al. 1994; Vodicka and Gerhart 1995). The tracer molecule also must be small enough to diffuse quickly through the injected cell before it divides, so that all of the descendants will be evenly and completely labeled (Figure 19-2). However, the tracer must be large enough not to pass through the numerous gap junctions between adjacent blastomeres (Guthrie et al. 1988). Horseradish peroxidase is about 44,000 MW and rapidly diffuses; cells fixed 5 minutes after injection are completely and evenly filled with the reaction product, whereas neighboring cells are completely unlabeled (Figure 19-3; see also Weisblat et al. 1978; Moody and Kline 1990). The fluorescent dextrans can be purchased in a variety of molecular weights, but the 10,000 MW and 40,000 MW versions are the most commonly used because they diffuse rapidly throughout the injected cell. With both of these markers, one sometimes observes a heavily labeled clone and a very lightly labeled clone. It seems that the heavily labeled clone is derived from the injected cell and the lightly labeled clone is derived from an adjacent sister cell that received some label via a cytoplasmic bridge, which had not yet closed during cytokinesis. This labeling pattern can be avoided if blastomeres are injected late in the cell cycle when these bridges should no longer be patent (Moody 1987a).

Two mRNAs that encode for tracer molecules, β-galactosidase β-GAL) and green fluorescent protein (GFP; Chalfie et al. 1994), are also useful. Both proteins are derived from nonvertebrates (β-GAL from bacteria and GFP from jellyfish), can be distinguished from endogenous vertebrate proteins, and have no known deleterious effects on developing vertebrate cells. However, diffusion is a particularly important issue when using these mRNAs as lineage tracers. Comparison of the complexity of clones derived from a cell labeled with HRP or dextran with those derived from the same cell injected with an mRNA tracer reveals that the mRNA clones often constitute only a subset. The clone is smaller and may not include all of the tissues. For example, we observe that GFP-labeled cells are not in the posterior tissues normally derived from 32-cell animal tier blastomeres (K. B. Moore, personal observations). This suggests that the mRNA does not diffuse evenly throughout the blastomere prior to the next cell division, either because it is too large and highly charged to diffuse freely or because it is rapidly loaded onto polysomes and anchored in place. Although this characteristic renders mRNAs less useful as straightforward lineage markers, they are ideal as markers to be used in combination with exogenous mRNAs. By mixing β-GAL or GFP mRNA with the test mRNA, it is likely that the tracer and the test mRNAs will diffuse to the same extent. Thus, when the tissue is analyzed at a later point in development, the tracer protein will identify those cells expressing the exogenous protein (Vize et al. 1991). However, the tracer mRNA and the test mRNA may not be distributed and/or expressed equally in all descendants of the injected blastomere. We have found, for example, that only a subset of GFP-labeled cells expresses FGF protein derived from an injected mRNA, as detected by immunocytochemistry (K. B. Moore, personal observations). Thus, if one needs to know exactly which cells express the exogenous protein, it is better to include an internal tag in the test mRNA (e.g., Witta et al. 1995).

A tracer molecule should be easily detectable by simple histological procedures. Enzymes, such as HRP and β-GAL, are detected by fixing them in place within cells, and providing them with a substrate that upon being altered by enzymatic action becomes an insoluble, colored precipitate. This precipitate therefore indicates the cellular location of the enzyme, which was either directly injected or encoded for by the injected mRNA. An advantage of this technique is that the preparations, either whole mounts (Figure 19-4) or tissue sections, are essentially permanent and can be referenced for years. In addition, they can be made clear in organic solvents for improved three-dimensional visualization of the staining pattern. An advantage of fluorescent tracers is that they can be viewed in the live embryo, although one needs to be very careful not to damage the fluorescent cells by photoablation (Shankland 1984). In addition, fluorescent tracers can be combined with other fluorescent methods, such as UV-excitable nuclear markers, fluorescent streptavidin to detect biotin-labeled compounds, and the immunofluorescent detection of cell type-specific proteins (Figure 19-5). One can either visualize the two (or more) labels with separate filter sets, or simultaneously visualize the labels with the new dual- and triple-pass fluorescence filter cube sets (from Omega or Chromatec), which can be fitted to any microscope. We typically combine Texas red-dextran as a lineage tracer (rather than the rhodamine conjugate) with fluorescein or GFP-labeled cell markers because their emission wavelengths do not overlap with these filter

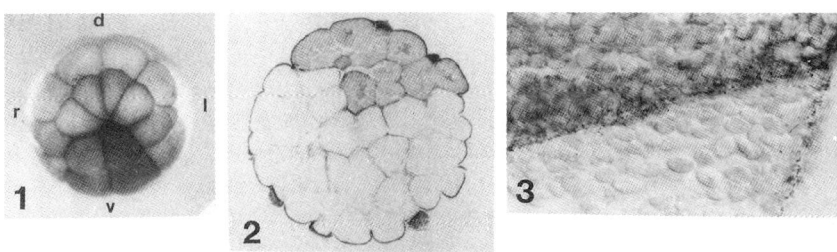

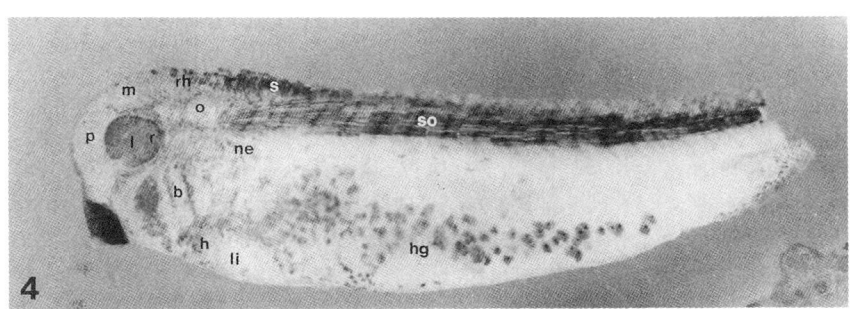

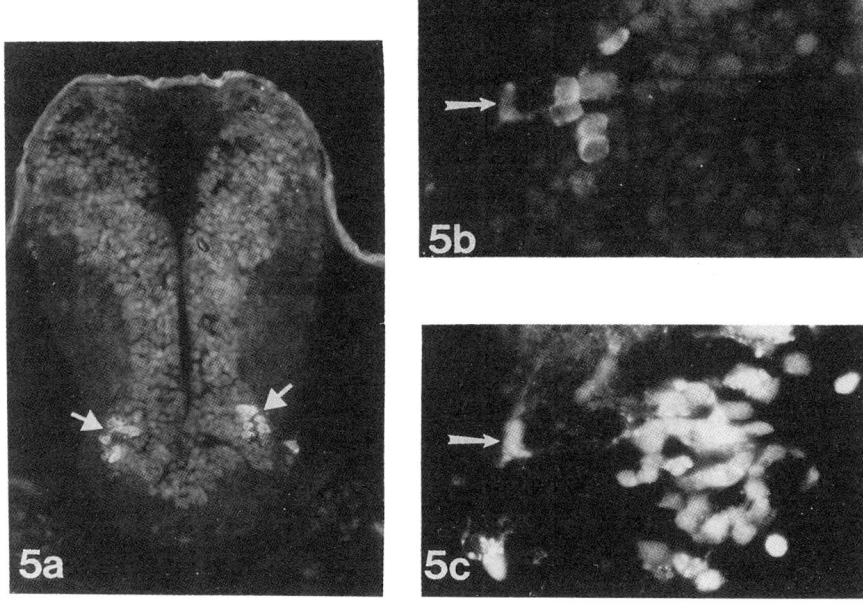

sets. This allows the clear and confident identification of double-labeled cells (Figure 19-5).

To perform lineage tracing, one needs to deliver the tracer molecule inside the blastomere without damaging that cell. For specifics on how to microinject cells, see Kay (1991). To perform this delicate procedure, a microinjection apparatus is required. There are several kinds available that range in complexity, but the most important feature to be considered is the typical injection volumes that you will use. Injection volumes into oocytes, fertilized eggs, and two-cell embryos can be as large as 10 nl (or sometimes larger), but for any older stages it is wise to keep injection volumes to around 1 nl per blastomere. Although some laboratories report injecting 4–10 nl of mRNA into each blastomere of a two-cell embryo, these volumes can be toxic if using HRP or dextran tracer molecules. Also, at stages later than two cells, large volume injections of either mRNAs or

Figure 19-1 Animal-pole view of a 32-cell embryo demonstrating the regular cleavage pattern on the dorsal side (d) used for fate-mapping studies. The dorsal animal blastomeres are more lightly pigmented than the ventral (v) animal blastomeres. The embryo is facing the camera so left (l) and right (r) are in mirror image to the viewer. (Prepared by Dr. Sen Huang.)

Figure 19-2 A section through the equatorial region of a 128-cell embryo that received an HRP injection into one animal blastomere at the 16-cell stage. The dark cells to the top of the figure contain the histochemical reaction product, and thus are the descendants of the injected blastomere. All other blastomeres are completely devoid of label.

Figure 19-3 The cellular interface between two blastomeres fixed 5 minutes after injection with HRP. The histochemical reaction product in the top cell has diffused all the way to the cell membrane, and does not enter the adjacent cell (bottom). The oval subcellular particles are yolk platelets.

Figure 19-4 A stage 33–34 embryo had one blastomere injected with HRP at the 16-cell stage. Descendants of the injected blastomere contain the dark reaction product. Anatomical structures: b, branchial arch; h, heart; hg, hindgut; l, lens; li, liver; m, mesencephalon; ne, nephric tubules; o, otic vesicle; p, prosencephalon; r, retina, rh, rhombencephalon; s, spinal cord; so, somites.

Figure 19-5 A combination of fluorescent dextral lineage-labeling and immunofluorescent detection of specific cell phenotypes allows specific lineage identification at the single-cell level. (a) A transverse section of the diencephalon of a stage 45 tadpole demonstrating bilateral clusters of dopamine neurons in the lateral hypothalamus (arrows). (b) Higher magnification of an adjacent tissue section illustrating dopmaine neurons identified by tyrosine hydroxylase primary antibodies and fluorescein secondary antibodies. (c) Same field as in (b), viewed for Texas red optics. Labeled cells are descendants of one 32-cell blastomere. The arrow points to a neuron that is double labeled. Thus, it is a dopamine neuron that has descended from the injected blastomere. (Prepared by Dr. Sen Huang.)

tracers are not well tolerated by the cells and can result in artifactual fate changes. We found that injecting > 10 nl of tracer into some 16-cell blastomeres can have significant effects on cell fate (e.g., drive epidermal lineages into brain), and thus should be avoided (Hainski and Moody 1992).

The concentration of the tracer to be injected is also an important consideration. One wants the label to be bright enough to be easily detected, but not so concentrated that it inhibits normal development. We use 1% fluorescein-dextran, 0.5% Texas red-dextran, 1% rhodamine-dextran, 5% HRP, and 300–500pg per cell of β-GAL mRNA or GFP mRNA. Fluorescent dextrans are dissolved in 0.2 M KCl (pH 6.8), HRP in sterile water, and mRNAs in sterile, RNase-free TE. Many laboratories microinject embryos in a 3–5% solution of Ficoll to prevent leakage of the tracer through the puncture wound. *Do not use Ficoll* if later you need to remove the vitelline membrane (see below). Ficoll withdraws the perivitellar fluid, causing the vitelline membrane to lie so tightly against the blastomeres that its manual removal is virtually impossible.

Although these tracer molecules are not cytotoxic, per se, when injected at too high a concentration or volume they can inhibit normal development. At the extreme, the injected cell will immediately stop dividing. This artifact is obvious when, shortly after injection, there are one or two large cells in a field of smaller ones. Mitosis also may be blocked at a later point in development; in this case, there will be larger-than-normal labeled cells within the embryo. Another indicator of toxicity is correctly sized, labeled cells that are spherical, rather than differentiated, in shape. This can happen to the entire clone or to only a subset of the clone. Often, these cells will move to the correct regions of the embryo, but never differentiate. Another sign of damage is the accumulation of labeled cells in the spaces within the embryo, i.e., the central canal and ventricles of the nervous system and the lumen of the gut, liver, and heart. These damaged cells probably dissociated from the rest of the embryo during gastrulation movements. If any of these signs of damage occur, discard the embryos and inject less tracer solution in the next experiment.

Embryos injected with enzyme tracers can be raised under normal laboratory conditions, whereas embryos injected with fluorescent tracers should be raised in the dark, as a precaution against photoablation of the labeled clone or photobleaching of the label. These embryos can be raised in an incubator or on the bench beneath a darkened plastic cover, such as the top of a slide box. They reportedly grow faster under red safelights (Wetts et al. 1989). Enzyme tracers can be visualized after fixation with a simple histochemical reaction (HRP, Moody 1987a; β-GAL Vize et al. 1991). Fluorescent dextrans can be immediately viewed with epifluorescence or laser-confocal microscopy, either in fixed or living embryos. However, excitation of the fluorophore releases free radicals that can damage living cells (Shankland 1984). Therefore, it is better to view living embryos under low-light conditions or very briefly. Detectable levels of GFP can be observed 1–2 hours after injection of the mRNA, and precautions should also be taken here to avoid photoablation of living cells. Fluorescent lineage tracers can be extremely hardy if the specimens are stored in the dark and refrigerated or frozen. For example, we have viewed 1-year-old dextran-labeled preparations stored at −80°C with no detectable diminution of the signal. Be aware, however, that most organic solvents used to store or clear tissues/embryos prepared with a histochemical reaction (HRP, β-GAL, alkaline phosphatase) will destroy fluorescence.

Experimental Tests of Developmental Potential and Fate Commitment

The developmental potential of a cell, or group of cells, is the limit of the types of tissues that can descend from that cell. The fertilized egg is totipotent because it has the potential to make all tissues of the embryo. The process of development gradually restricts the potential of the egg's descendants causing them to become unipotent at their terminal differentiation. The purpose of the experimental approaches described below is to discern when, and by what mechanisms, these restrictions occur. In theory, a cell's developmental potential can be assayed by raising it under defined conditions. However, the entire repertoire of possible environments cannot be reasonably tested to identify the cell's full potential. Accordingly, it is more practical to test a cell's commitment by comparing its normal repertoire of descendants with those expressed after experimental manipulation.

The term "commitment" refers to whether a cell's fate is fixed (committed) regardless of the surrounding cellular environment, or is still in a state that can be influenced by external factors (uncommitted). A cell that is committed to its fate will express that fate regardless of whether it is grown in other regions of the embryo or in culture. An intermediate term is also commonly used: if a cell does not express a fixed fate, but neither is the cell totipotent, it is said to be "biased." For example, after experimentation a cell may express some, but not all, of its original fate. In this chapter, I will discuss four ways to manipulate blastomeres in order to gain information about their state of commitment. A first approach is to delete the cell of interest. This will not reveal the state of commitment of the deleted cell, of course, but will test whether its presence is necessary for the normal development of the remaining cells, and whether they can change fate to reconstitute the entire embryo. Three manipulations directly test fate commitment of the test blastomeres. The cell may be cultured as an explant in defined medium, transplanted to a novel region of the embryo, or manipulated in situ in a way that prevents it from communicating with its normal neighbors.

For all of these methods, the following tools and solutions are recommended:

- 60-mm plastic Petri dishes, coated with about 1-mm thickness of 2% agarose, dissolved in either distilled water or culture medium.
- At least two finely sharpened forceps (Dumont #5, biologie), autoclaved or sterilized with ethanol. I prefer to have four forceps prepared in case one is damaged during the experiment.
- A good-quality stereomicroscope with at least 50× magnification capability.
- Culture medium; either Steinberg's solution, Marc's Modified Ringer's solution (MMR), or Modified Barth's Solution (MBS) can be used. Recipes can be found in Peng (1991). The medium should be filter-sterilized and can be supplemented with 0.1% gentamicin.
- Sterile 6″ glass Pasteur pipettes.
- Hair loops.

Deletion of a Blastomere

The rationale for deleting a cell or group of cells is to determine whether that cell is necessary for the continued development of the rest of the embryo or for a subset of its structures. In embryos in which fate is strictly determined by the subcellular localization of molecules, deletion experiments have revealed that particular blastomeres are necessary for certain primary tissue development (see Davidson 1990). In embryos in which cell–cell interactions are of primary importance in fate determination, single-cell deletions may not produce embryos with missing parts because the remaining cells may regulate their production of descendants and reconstitute the whole embryo. For example, in *Xenopus* there are nine blastomeres in the 32-cell embryo that contribute descendants to each retina (Huang and Moody 1993). One of these blastomeres (the nomenclature is either D1.1.1, according to Jacobson and Hirose 1981; or A1, according to Nakamura and Kishiyama 1971) gives rise to 50% of the retina, yet when it is deleted, the remaining blastomeres change fate so that a full-sized retina usually develops (Huang and Moody 1993). Likewise, lineage analysis showed that dorsal axial tissues are restored by both contralateral and ventral blastomeres after deletions of eight-cell stage blastomeres (Yuge and Yamana 1989). From these experiments, one can conclude that the presence of the deleted cell is not necessary at the time of the deletion, and that the remaining cells somehow are made aware of the deletion and alter the production of their own descendants to compensate for the single-cell loss. However, regulation does not always occur in response to single-cell deletions in *Xenopus*. For example, when dorsal blastomeres of the eight-cell embryo (Kageura and Yamana 1984; Yamana and Kageura 1987) or dorsal-animal blastomeres of the 16-cell embryo (Gallagher et al. 1991) are deleted, axial defects result. Thus, the time of the deletion, the cell that is deleted, and the final tissue that is monitored may determine whether one will observe regulation versus loss of structure. With this kind of an assay, the fate commitment of the deleted cell is not being tested, although it is inferred. Rather, the plasticity in fate of the remaining cells is under direct test.

In order to manipulate the cells, the protective coverings must be removed. The first sets of membranes are the extensive jelly coats. After fertilization, these can be removed chemically with either cysteine or dithiothreitol, as described in detail elsewhere (Moody 1987a; Peng 1991). All of the embryos can be processed at the same time in large volumes of solution. They need to be thoroughly washed over a period of about 10 minutes to make sure that the chemicals, which are toxic to further development, are completely removed. Survival appears to be improved if the embryos are transferred to a second or third set of Petri dishes after being washed free of the dejellying solutions, and are subsequently raised at low density. The more difficult membrane to remove is the vitelline membrane. This is a very tough, clear, thin membrane that tightly adheres to the oocyte, and lifts away, ever so slightly, from the cell surface after fertilization. By tilting the embryos at an angle under the stereomicroscope, one can see the membrane separated by a clear space above the surface of the embryo. This perivitellar space is largest at the animal pole. Although there are protocols for removing the vitelline membrane chemically from large numbers of embryos at once, these methods are not recommended for the types of manipulations described in this chapter. These protocols involve enzymatic digestion, which is difficult to control and usually damages the embryo. The healthiest embryos will be obtained when the vitelline membrane is manually removed from each embryo. This procedure requires some practice and manual dexterity, and should be mastered before embarking upon any of the following manipulations.

Since embryos removed from the vitelline membrane stick extremely well to Petri dish plastic, to avoid tearing and damaging the embryos, all manipulations and culture are carried out on the smooth, slippery, inert surface of an agarose bed. It is helpful to melt small depressions in the surface of the agarose. These will hold the embryos in place a little better during the operation, and also help them to maintain a rounded shape after the vitelline membrane has been removed. Normal embryos are firmly held together in a spherical shape by the vitelline membrane and perivitellar fluid (Figure Figure 19-6a). When the membrane is removed, embryos relax into a flattened spheroid, much like a nearly deflated soccer ball. This flattened shape is accentuated when they are grown on the flat surface of a Petri dish (Figure 19-6b). Although flattened embryos can develop normally, gastrulation movements are not always completed properly. This physical distortion may induce artifacts in the interpretation of your experiments because it may affect with whom the cells interact and where they reside during gastrulation. We have found that allowing the embryos to maintain a more spherical shape by culturing them in shallow depressions in the agarose (Figure 19-6c) helps reduce the number of nonspecific gastrulation defects.

To make the depressions, heat the tip of a glass Pasteur pipette in a flame until it melts into a small ball (about 2 mm in diameter). While the tip is still hot, touch it to the surface of the gelled agarose. Only a shallow depression

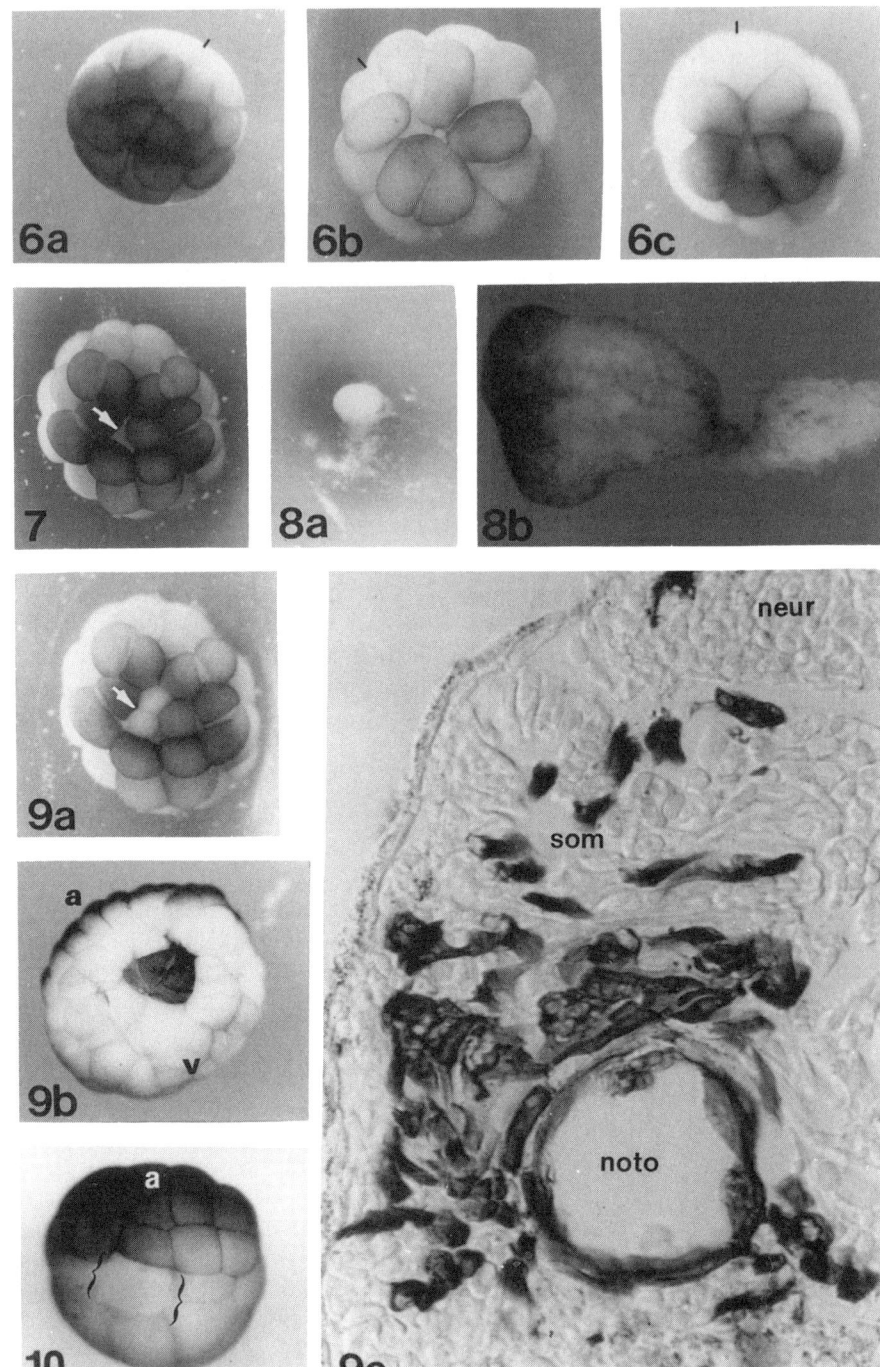

Cell Fate Commitment of Single Blastomeres in *Xenopus* 367

is necessary (about 0.5–1 mm deep and 1.5–2.0 mm in diameter) to support the embryo. Usually, the pipette needs to be reheated between making each well. Ten to twenty wells can be made in each 60-mm dish.

Blastomere deletions are performed in solutions containing a lowered concentration of cations in order to lessen the strength of the cell–cell adhesions. It is not necessary to use a calcium- and magnesium-free solution for deletion experiments, because (1) a cell can be pulled free with a little force from the forceps, and (2) a cation-free solution makes blastomeres very leaky and fragile and therefore too easily damaged or killed. After fertilization and removal of the jelly coat (see above), the embryos should be stored in 50%-strength culture medium. Some laboratories swear by MBS, some

Figure 19-6 Animal-pole views of cleavage-stage embryos cultured on a bed of agarose. (a) This embryo is still within the vitelline membrane and retains a spherical shape. (b) This embryo has been removed from the vitelline membrane and is sitting on a flat surface. It has flattened and the cells are flaccid and spread out. (c) This embryo also has been removed from the vitelline membrane, but is sitting in a depression melted into the agarose. Thus, it is able to better retain its normal shape. In each figure, the tick marks indicate the dorsal midline.

Figure 19-7 A 32-cell embryo in which a single ventral animal blastomere (V1.1.1) has been removed. The arrow points to the clean gap that is left after the entire cell and all debris has been removed.

Figure 19-8 Blastomere explant culture. (a) A single 32-cell blastomere has been removed from the donor without damage. It can either be placed in culture or transplanted to a host embryo. The surrounding debris is from its neighbors that were destroyed during the dissection. (b) An explant derived from four D1.1 blastomeres has elongated after 2-day culture in simple salt medium. (Prepared by Dr. Alexandra Hainski.)

Figure 19-9 Steps in transplantation of blastomeres. (a) A dorsal animal blastomere was transplanted into the gap created by deleting a ventral animal blastomere (the same embryo as in Figure 19-7). The transplanted cell (arrow) has divided once. (b) The dark patch in the center of the embryo is the four descendants of an animal-pole blastomere that was transplanted into a vegetal marginal position. In 1 hour, it has completely healed into place, and divided twice. (c) Transverse section of an embryo that received an HRP-labeled dorsal animal transplant into the ventral vegetal region of the host. Host cells are unstained, and descendants of the transplanted blastomere can be identified by their dark reaction product. This experiment shows that the transplant can autonomously differentiate into notochord (noto), neural tissue (neur), and muscle (som). Note that the cellular identity of the descendants of the transplanted blastomere can be clearly identified histologically. (Prepared by Dr. Betty Gallagher.)

Figure 19-10 A 32-cell embryo in which a filter barrier (brackets) was placed between two marginal-zone blastomeres. The placement of these barriers can be quite precise. (Prepared by Dr. Daniel Bauer.)

by MMR, and my laboratory uses Steinberg's solution; these are all very similar in composition, but the Steinberg's solution has the lowest ionic concentrations. When the embryos reach the appropriate stage for the deletion, transfer them to the agarose-coated Petri dishes and place one in each of the depressions. This dish should contain either 50% Steinberg's solution or 0.1× MMR (or MBS).

If the lineage labeling is to be done with blastomeres younger or the same stage as the deleted blastomere, microinject them prior to removing the vitelline membrane and let the injection site completely heal before transferring them to this dish. Do *not* do these injections in a Ficoll solution because the vitelline membrane will be very difficult to remove. If the lineage labeling is to be done with blastomeres older than the deleted blastomere, microinject them after the operation in that same culture dish. Because the support of the vitelline membrane is missing, the injection should be done with a very fine-tipped micropipette (about 10 μm) and a small volume of tracer (no more than 1 nl) to prevent leakage through the puncture wound.

However frustrating manual removal of the vitelline membrane is, with persistence it becomes an easy chore. It seems to work best if you use two different forceps: one sharpened to a very fine, pointed tip, and one sharpened with a slightly squared tip. With the squared-tip forceps in your subdominant hand, grasp the vitelline membrane and hold on to it; with the fine-tipped forceps in your dominant hand, grab the vitelline membrane next to the first forceps without stabbing the underlying embryo. Peel the membrane away by separating the two forceps. It works best if a large sheet is torn away; if just a little hole is made, the pressure inside the membrane will force cells of the embryo out through the hole, which nearly always causes the embryo to tear and break open. It is easiest and safest to grab the vitelline membrane at the animal pole where the perivitellar space is the largest. However, it also is useful to grab the vitelline membrane over the cell you wish to delete. Since you are going to remove the cell, it does not matter if you damage it in the process of removing the vitelline membrane. After a large tear has been made in the vitelline membrane, either the remaining membrane can be peeled away, or the embryo turned over to slide out through the tear.

Once the embryo has been removed from the vitelline membrane, it is very soft and squishy and *fragile*. It is very easy to poke it with the instruments and cause the cells to leak cytoplasm. To avoid such damage, the embryos can be pushed about and oriented with blunt forceps or with a hair loop. To perform the deletion, orient the embryo with the cell to be deleted (the victim) facing the optical field. Holding the squared forceps open, gently place them over the embryo using the subdominant hand to stabilize the embryo and hold it in place. Using the fine-tipped forceps, grab the victim right in the middle of the cell and gently pull. This works very well for 16-cell-stage and older blastomeres. Use the holding forceps to prevent neighboring cells from pulling free with the victim. In some batches of eggs, the cell to be deleted will simply lift out of the embryo cleanly. In others, you will just rip the cell open and cytoplasm will start leaking everywhere. This latter situation is not a problem: just pick out the chunks of the destroyed cell with both forceps until all the remnants of the damaged cell are completely removed. Again, by holding down the embryo with one forceps, you can prevent the tugging on the pieces of the damaged cell from also removing neighboring cells. The resulting embryo should have a single cell neatly removed (Figure 19-7), with no leaking neighbors or oozing debris that could compromise the rest of development (McClendon 1910). We have found that 50% Steinberg's solution and $0.1\times$ MMR (or MBS) have a low enough concentration of cations to help the cells let go of one another, but have a high enough cation concentration to allow the cells to maintain their membrane integrity. However, if cells fall apart without tugging on them, increase the cation concentration and if they are impossible to separate, lower it. The optimal ionic concentration needed for clean deletions can vary between batches of eggs.

For removal of cells at stages earlier than 16-cells, one needs to actually dissect the cells (Kageura and Yamana 1984; Yamana and Kageura 1987; Yuge and Yamana 1989). They are much too large and fragile to be tugged on without damaging their neighbors, and frequently they are directly connected to their neighbors by cytoplasmic bridges from ongoing cytokinesis. After incubating the embryos in the lowered cation medium for at least 10 minutes, the forceps can be used to gently separate the blastomeres. This can be accomplished by placing the tips of closed forceps into the cleavage furrow between cells and slowly allowing the forceps to open, pushing the adjacent cells apart. Some experimenters prefer to tease the cells apart with glass needles or hair loops. Once the cell surfaces are no longer adherent, the cytoplasmic bridges can be snipped with the forceps. These dissections are most successful if done very late in the cell cycle when the cytoplasmic bridges are small and the cells are more rounded. There is a high mortality rate with this early operation, so be prepared to perform large numbers in order to get 10–20 successful cases.

After each dissection, it is helpful to clean the tips of the forceps by jabbing them into the agarose bed. This will remove any adherent bits of vitelline membrane and the very sticky, yolk-filled cytoplasm. If the forceps are extremely dirty, they should be gently brushed along dampened labwipes, and then dipped in 70% ethanol to sterilize them. Do not dip dirty forceps into the ethanol, as this will cause the yolk to dry as a hard crust onto the tips. After removing cells from about 5–10 embryos, use a sterile Pasteur pipette to suck up the debris and remove it from the dish. These dying bits of cells release proteases, which theoretically could damage the operated embryos. In addition, they provide nutrients to unwanted bacteria.

The embryos heal quickly, and the gaping hole created in the embryo should close completely within 15–30 minutes, depending upon at which stage the deletion was made. Single cells can be injected with lineage tracer at this time, but because the support of the vitelline membrane is missing, the tip of the micropipette should be very fine (about 10 μm) and the volume of the injected material no larger than 1 nl. After all the embryos in the dish have healed, they can be transferred to a clean dish, each into their own depression, or the solution can be cleaned up by removing half of the old, dirty solution, and replacing it with fresh. If the embryos are not healing rapidly, try adding a higher percentage cation solution; increasing the cation concentration promotes healing. However, embryos removed from the vitelline membrane have great difficulty proceeding through gastrulation if the solutions are greater than 0.5×. Therefore, if cations are raised after the operation to promote healing, lower them again prior to the predicted time of gastrulation.

The embryos should be analyzed for gross morphological defects as they reach the desired stages of development. This is generally when the organ systems have all formed, 2–4 days after fertilization. If embryos do not heal after the deletion, nonspecific gastrulation defects may occur. Therefore, it is worthwhile to divide the embryos into groups ("healed perfectly" versus "not-so-perfect") as they develop so as to correlate later defects with problems with the operation. Also, it is a good idea to have control embryos that developed without the vitelline membranes to check for effects of the manipulation and/or the culture medium. Missing structures can be detected in histological sections or whole-mount embryos stained with cell-type-specific antibodies or RNA probes, and alterations in the lineages of neighboring cells can be identified by mapping the location of the descendants of the tracer-injected blastomeres.

Explant Culture of a Single Blastomere

One test of a blastomere's fate is to remove it from all other cellular interactions. Because *Xenopus* blastomeres contain large amounts of intracellular yolk stores, they can survive in culture with no nutritional supplements. Furthermore, one can test a cell's requirements for exogenous growth factors in normal-fate expression by culturing them with or without such factors in defined media. To eliminate the role of the production of, and binding to, the extracellular matrix, blastomeres can be cultured on an agarose bed (described below). Alternatively, to test the influence of various extracellular substrates on differentiation potential, the cells can be cultured in similar wells that previously were coated with these molecules (Godsave and Slack 1989; 1991).

Blastomere explant cultures require the same tools and solutions as those described above. Agarose-coated 60-mm dishes without depressions should be prepared for the dissections. In addition, wells in a 96-well culture plate should be coated with about two drops of molten agarose. These small wells reduce the required volume of culture media and prevent blastomeres from falling together into one large mass. The agarose forms a steep meniscus when it solidifies, obviating the necessity to melt depressions into the bed. Single blastomeres or small numbers (up to four) of the same blastomere will be cultured in one well. We have found that blastomeres cultured singly, or in small groups, survive best if cultured in Normal Amphibian's medium (NAM; Messenger and Warner 1979). This solution is poorly buffered, and thus cannot be stored; we always make it fresh on the morning of the experiment and change the solution with freshly made medium on each day of culture. Each well should be filled with 500–800 µl of sterile NAM solution. Others have successfully used 50% L-15 medium (Gibco) (Kageura and Yamana 1984; Gallagher et al. 1991).

To prepare the blastomeres for culture, fertilize, dejelly, and wash embryos as described above. The vitelline membrane should be removed as described above, but instead of grabbing the membrane over the cell of interest, grab it anywhere *but* over the cell of interest. In this case, one wants to avoid any damage to the explanted cell, but the rest of the embryo is dispensable. If holes are poked in the embryo during the removal of the vitelline membrane it will not matter, as long as the explanted cell is undamaged. To dissect out the desired cell, avoid touching that cell with the forceps, and, instead, pull away all of the neighboring cells. For this operation, it is useful to have two pairs of finely sharpened forceps, and to perform the dissection in 50–70% Steinberg's or 0.3–0.5× MMR (or MBS). The explant blastomere can be stabilized by holding onto one neighbor cell with the forceps in the subdominant hand, and removing all other neighbors with the forceps in the dominant hand. The neighbor cell used as a handle can then be picked away with both forceps (Figure 19-8a). Quickly place the cell in culture, using a 6″ sterile Pasteur pipette. We avoid using plastic transfer pipettes because the blastomeres stick extremely well to plastic. Transfer must be done gently with very little sucking on the cell, or it will tear apart in the shear forces of the pipette. Also, avoid air bubbles and make sure the tip of the pipette is under the surface of the medium in the culture well before releasing the blastomere; many a cell has been torn apart at the air–water interface. The wells of the 96-well plates are very narrow and may be difficult for the beginner to maneuver within; if your culture medium is not expensive (due to rare growth factors), culture blastomeres in 48-well or 24-well dishes until you are well practiced.

The blastomeres will naturally fall to the bottom of the agar-coated well and form an aggregate. We have had trouble successfully culturing single blastomeres older than the 16-cell stage with this method, but younger blastomeres (Hainski and Moody 1996) or small groups (2–6 cells) of the same blastomere (Gallagher et al. 1991) grow quite well. If a midline cell is to be cultured, it can be dissected free of the embryos in bilateral pairs. These explants can survive quite well, even if some of the cells added to the well explode or otherwise disintegrate. The next day, the turbid mass at the bottom of the dish should be explored with a hair loop to determine whether there is a buried, healthy mass of cells. If so, these can be transferred to a new well containing fresh culture medium. Alternatively, single blastomeres of the 32-cell embryo can be successfully cultured on an extracellular matrix substrate (Godsave and Slack 1989; 1991).

If it is desired to culture the blastomeres in growth factor, antibodies, or other reagents, it is useful to make up the additive at a 2× strength. Transfer the dissected blastomeres into a culture well containing 500 µl of NAM, and when all of the dissections are completed, add 500 µl of NAM containing the 2× additive. In this way, all cells will begin treatment synchronously. Be sure to transfer very little fluid from the dissection dish with the blastomeres into the culture well, so that the growth factors are not diluted.

Explants can be cultured for 36–48 hours with this method. They can be collected and fixed for whole-mount immunocytochemistry, histology, or in situ hybridization using standard techniques, or they can be homogenized and extracted for gene expression assays. In some cases, their external morphological appearance may indicate what tissues are expressed. Just like blastula animal cap explants, blastomere explants can remain rounded if expressing ventral mesodermal derivatives, or elongate moderately if expressing dorsal mesodermal derivatives (Figure 19-8b; Gallagher et al. 1991). However, we have found that morphology is only a rough estimator of tissue differentiation. It is best to test the explant with markers for tissue-specific differentiation. For example, we found that dorsal mesodermal markers can be expressed in explants that do not exhibit identifiable elongation (Hainski and Moody 1996).

Transplantation of Blastomeres to a Novel Location in a Host Embryo

A very powerful technique for testing whether a cell's location in the embryo determines its fate is to lineage-label the cell and then transplant it to a novel location in an unlabeled host. This technique has shown that some dorsal midline blastomeres signal others to form the primary embryonic axis (Gimlich and Gerhart 1984; Gimlich 1986) and that others autonomously express a dorsal axial fate by the 16- and 32-cell stages (Takasaki and Konishi 1989; Gallagher et al. 1991). Transplantation also has been used to demonstrate whether all blastomeres are competent to give rise to retinal tissue (Huang and Moody 1993). The basic information that is necessary is the normal fate of the cell to be transplanted and the normal fate of the cell in the location to which the cell will be transplanted. To perform this test, one needs embryos that comply with a regular cleavage pattern, a method for labeling the transplanted cell, and a method for specifically identifying the phenotype of the cells descended from the transplanted cell.

As mentioned in the fate map section, it is important to know not only what others have reported in their fate maps, especially regarding variance, reproducibility, and minor constituents to a particular clone, but also at what level one can discriminate between the clonal composition of the two different blastomeres. If the cell to be transplanted gives rise to tissue X in 20% of embryos and the cell that normally occupies that site gives rises to tissue X in 40% of embryos, one might not detect a statistically significant difference between the two fates after the experimental manipulation. There may be differences that are too subtle for your assay to detect. Likewise, it may be necessary to record the entire phenotypic composition of a clone, rather than just the tissue of interest to your study, because fate changes may be global and compensatory. Thus, it is well worth creating a series of control embryos in your own laboratory with which to compare your experimental results. This is especially true if using methods or tracers that differ from those used in the original reports.

To perform single-cell transplantation, there must be a pool of donors and a pool of hosts that are at the same stage of development. After dejellying the eggs, and picking through them for cells whose first cleavage furrow bisects the grey crescent, it is useful to separate them into dishes according to cleavage stage. It is easiest to transplant cells no younger than the 16-cell stage. In order to have enough time to make all the necessary preparations and for the lineage label to diffuse throughout the injected cell, we typically divide the embryos into two pools very early and lineage-label half of them. It is desirable to have all cells of the donor labeled, so these injections can be performed quickly at the two- or four-cell stages. One nanoliter of fluorescent dextran (1% solution), HRP (5% solution), or 500 pg of GFP or β-GAL mRNA per cell is adequate. After labeling with either fluorescent dextrans or GFP mRNA, one should raise the embryos in the dark to prevent any photoablation of the labeled cells or photobleaching of the label.

After the donors have been labeled, the hosts should be transferred to agar-coated Petri dishes in which depressions have been made. I prefer to make the depressions in a circle about two-thirds of the distance from the center of the dish, making about 10 depressions. With this number, I do not loose track of which embryos are donors and which are hosts, and they are not overcrowded during further development. These dishes should contain 50% Steinberg's (or 0.1× MMR or MBS) to facilitate dissecting the cells free from their embryos. The host embryos must come from the pool that was not lineage-labeled. One host embryo should be placed in each depression and the vitelline membrane removed, as described for the deletion experiments. Place an equal number of donors in the center of the dish, so they cannot fall into extra depressions and be mistaken for unlabeled hosts. Starting at one point in the circle of hosts, delete the appropriate cell in the host, as described for the deletion experiments (Figure 19-7). If the embryo is seriously damaged in the attempt, go on to the next host; wasted time can result in all the donors developing beyond the stage of interest before any hosts can be properly prepared. Immediately place a donor embryo next to the operated host, remove its vitelline membrane, and proceed to remove the desired cell as described in the explant culture section

(Figure 19-8a). Before trying to transport the isolated donor blastomere into the host embryo, make sure the wound in the host is large enough to accommodate it. Healing occurs very rapidly and the transplant cell may not fit into the available hole. If brute force is tried and the transplant is stuffed into the wound, it will break apart or begin to leak cytoplasm. Instead, place the tines of the closed forceps into the wound, and, as they open, allow them to stretch open the hole gently. Next, using either closed forceps or a hair loop, gently float the transplant cell over the host embryo and let it drop into the hole. It should be pushed gently deeper into the embryo for better healing (Figure 19-9a), but the less the cell is touched the more likely it will survive. Although one might be able to do the transfer using a glass pipette, I find the shear forces within the pipette to be too damaging. When one embryo is done, move on to the next pair. When all the embryos in the dish are completed, remove the debris with a Pasteur pipette, as described above.

These embryos take longer to heal than the deletion embryos. Try not to move the dish until there are sure signs of integration of the transplant, after about an hour (Figure 19-9b). If more than one dish of transplants needs to be made, it is best to move to another microscope. If the transplanted cell is leaking cytoplasm, that debris should be removed gently with forceps or a hair loop. Be aware that leaking cells may be leaking their lineage label as well as their cytoplasm, so it is best to raise these embryos separately from the ones that underwent a "perfect" transplantation. Allow the embryos to heal for several hours undisturbed. Once the transplant is healed, the embryos should be transferred to depressions in fresh dishes filled with 50% culture medium. A short stint in a higher cation concentration may promote healing, but embryos will disintegrate in concentrations higher than 50% as they proceed through gastrulation without a vitelline membrane. Embryos should be raised to the appropriate stage of development as determined by the tissues to be analyzed, fixed, and processed by methods appropriate to the lineage label utilized.

This is not a simple procedure, and the mortality is quite high. After a lot of practice, a 20–30% survival rate is outstanding. When fluorescent tracers are used, we always check the embryos for a signal with the epifluorescence microscope using the 10× objective before histological processing. If there is no lineage label by which you can distinguish transplanted from host descendants, there is no point in performing the histology or immunocytochemistry. At the cellular level, this technique can clearly distinguish the labeled descendants of the transplanted blastomere from the unlabeled host cells (Figure 19-9c).

Interfering with Intercellular Communication In Situ

Culture of single cells or small groups of blastomeres removes them from the intercellular communications in the intact embryo. But, it also removes them from long-range and short-range diffusible signals and whatever signaling pathways set up the positional values in the embryo. This can be avoided by directly blocking cell-to-cell adhesion in the embryo itself. We have used three different methods for this kind of perturbation: injection of antibodies, dissociation of the embryo within the vitelline membrane, and placement of temporary physical barriers between cells (Bauer et al. 1996).

Injection of Antibodies In short, I do not recommend this approach. There are some published records of success in blocking protein function in *Xenopus* blastomeres with this technique, but in our hands the likelihood of technical artifacts is too high. Some guidelines need to be followed if choosing to try this approach. The antibody must be known to block protein function. There is no point injecting precious antibodies unless there is clear evidence that they will have the required effect. Also, the access of an injected antibody to the correct subcellular compartment must be determined. Immunoglobulins are huge molecules that do not diffuse readily; will they ever reach the protein of interest? The length of time that the antibody can function within a blastomere must be known. Since this is difficult to determine, it is best to use this technique for studies in which you wish to block protein function at the time of the injection. Finally, there are many components in serum that can have nonspecific detrimental effects on a cell. Therefore, preimmune serum injection controls must be performed, preferably with serum from the same animal in which the antibodies were made. We have found it difficult to satisfy all of these criteria, and perform an interpretable experiment with this approach. A more useful alternative that is in common use is to construct a mutant protein that can function as a dominant-negative to the protein of interest, or to find an antagonist protein, and then to inject the mRNA encoding these proteins into the blastomere of interest (Vize et al. 1991; Paul et al. 1995).

Dissociation It is simple to dissociate cleavage-stage blastomeres within the vitelline membrane by disrupting the calcium-dependent adhesion between cells via incubation in calcium- and magnesium-free medium supplemented with a calcium chelator. The vitelline membrane is permeable to ions, and thus is not a barrier to the free exchange of cations between the embryonic cells and the culture medium. An advantage is that by keeping an embryo within the vitelline membrane, the blastomeres remain in their original positions during the dissociation period. Also, one can temporally control when cells are dissociated. We have incubated embryos for several 1-hour periods prior to gastrulation to define the time period during which a putative interaction occurs (Bauer et al. 1996). Readhesion between cells is accomplished simply by transferring the embryos to normal, cation-containing medium. A disadvantage is that cell–cell interactions are disrupted throughout the embryo, rather than between specific cells.

Incubation in calcium- and magnesium-free media can make the cells very fragile and their membranes leaky. We have experimented with several media and feel that we can keep the embryos healthiest if incubated in calcium- and magnesium-free modified Stern's solution (CMF-MSS; Nakatsuji and Johnson 1982). The embryos are transferred from normal medium into CMF-MSS for 5 minutes in order to reduce the amount of cations carried into the medium containing the chelator. The embryos are then incubated for 30 minutes in CMF-MSS containing 20 mM sodium citrate as a calcium chelator, and then transferred into CMF-MSS without citrate for the remaining dissociation period. We minimized the incubation in citrate because it can be toxic to the embryos. Following the dissociation period, the embryos are incubated in complete modified Stern's solution (MSS; Nakatsuji and Johnson 1982), supplemented with 5% bovine serum albumin (BSA) for 2–4 hours, and then transferred to 50% Steinberg's solution. Injection of a lineage tracer into the blastomere of interest can be done before, during, or after the dissociation incubation, but one should avoid transferring dissociated embryos by pipette so that the loose cells do not shear or rearrange.

To test the effectiveness of the dissociation, open up the vitelline membrane at the end of the dissociation period in the CMF-MSS. We found that the blastomeres of embryos treated for 1 hour between the 8- and 32-cell stages fell apart completely when the vitelline membrane was opened. Those treated beginning at the 128-cell stage could be easily separated with forceps, and, once a gap was made between the surface cells, the entire embryo dissociated. However, the surface cells of those treated beginning at the 512-cell stage were difficult to separate, although the internal cells had dissociated. However, a 2-hour dissociation period worked quite well for these later embryos (30 minutes in citrate, 90 minutes in CMF-MSS; Bauer et al. 1996).

To test whether gap junction communication is affected during dissociation, a low-molecular-weight marker can be injected into individual blastomeres. Although Lucifer yellow can be used for this purpose (Warner et al. 1984; Nagajski et al. 1989), we found that it can close gap junctions, giving a false negative result, if too much is injected into the cell. Based on studies of neuronal gap junction coupling showing that biotin derivatives more accurately demonstrate patent gap junctions, we have injected biocytin (373 D, Sigma), a fixable, uncharged derivative of biotin, into individual blastomeres of dissociated embryos. Embryos are dissociated for 50 minutes beginning at the eight-cell stage. Then, while still in the CMF-MSS, a single blastomere of the 32-cell embryo is injected with 1 nl of a mixture of 0.5% Texas red-labeled dextran amine (TRDA, 10,000 kDa, Molecular Probes) and 1% biocytin. The TRDA is used to distinguish blastomeres that are still connected via cytoplasmic bridges that persist between blastomeres following division from those coupled only by gap junctions. After 10 more minutes of dissociation in CMF-MSS, the embryos are fixed in 4% paraformaldehyde, frozen, sectioned in a cryostat, and the slides are incubated overnight at room temperature with 1% fluorescein-labeled streptavidin (Calbiochem). Because biocytin passes through gap junctions and labeled dextran does not, the originally injected cell and any sister cell connected to it by a cytoplasmic bridge will be doubly labeled with TRDA and fluorescein-tagged biocytin. Cells that are coupled to the originally injected blastomere only through gap junctions will be labeled only with fluorescein-tagged biocytin. We found that a 1-hour dissociation did not change the percentage of blastomeres that are coupled by gap junctions (Bauer et al. 1996).

Placing a Barrier between Blastomeres In an attempt to interfere with cell-cell interactions between individual blastomeres, we borrowed the idea of implanting a permeable, noninteractive synthetic filter. These nontoxic filters had been used in many studies of tissue interactions and inductions (e.g., Grunz and Tacke 1986), and had been shown to prevent cellular contact but to allow diffusion of molecules. We had hoped to use these to distinguish between signaling via diffusible molecules versus membrane-bound molecules, but this proved impossible, as discussed below.

Nuclepore filters (0.1 µm pores; Nuclepore Corp., Pleasanton, CA) are used as barriers between individual blastomeres. Wearing gloves and using sterile scissors, these are cut into 0.25 mm^2 pieces, washed in 95% ethanol to sterilize them, washed in sterile distilled water, and dried on sterile filter paper in a Petri dish. Just before use, the filter squares are soaked in a small volume of sterile culture medium to rehydrate them. After preparing the embryos and injecting the test blastomere with lineage tracer, the embryo is transferred to a depression in an agarose-coated Petri dish containing 25–50% Steinberg's solution (or 0.1× MMR or MBS). The vitelline membrane is removed, and the injected blastomere is carefully separated from the neighboring blastomere by teasing apart the adjacent cell surfaces with sharpened forceps. The cation concentration of this solution may need to be adjusted from one batch of eggs to the next. The goal is to lower the cations enough so that the cell membranes separate easily, without tearing, but not so much that the cells leak cytoplasm or that other blastomeres dissociate. This optimal concentration seems to vary between batches of eggs, and needs to be titrated at each experiment.

Embryos removed from the vitelline membrane will not survive in this lowered cation solution, but, instead, will completely dissociate. Therefore, after the blastomeres are separated from one another, the embryo is transferred to a new agarose-coated dish containing a higher cation medium. It is very easy to damage the embryos in this transfer because the cells are so loosely adherent. Any standard medium that promotes healing will be useful for this step; we use NAM (Messenger and Warner 1979). After transferring the embryo into a depression and orienting it so that the gap between the blastomeres is accessible, the filter square should be picked up with forceps and placed in the gap. If the wound has already begun to heal, it may be necessary to reopen the hole with the forceps. We have found it helpful to make filter squares of several different shapes and sizes, and to have them already rehydrated in the transfer dish before you start the operations. Then, as different-sized gaps are created, you can pick through the filter squares for one that fits snuggly. It is very easy to damage the embryos during this manipulation. They are fragile and you need to poke and prod them a lot to get the filter positioned properly. When the barrier is in place (Figure 19-10), it must be deep enough to prevent the cells from readhering (i.e., it must extend into the blastocoel), but it can tear underlying cells if shoved in too far.

The barrier can stay in place for a variable amount of time, but it must be removed prior to the onset of gastrulation movements so that it does not physically impede these critical morphogenetic events. For our experiments, we wanted the barrier in place for an hour (Bauer et al. 1996). We found that the Nuclepore material repels the cells, so not only do they not adhere to it, but also the gap actually widens up to 200 µm during the incubation. Thus, even though the filters contain pores, it is not feasible to test for the presence of diffusible signals over such a distance. One also needs to monitor these embryos frequently because the barriers can slide out of the widened gap and will need to be refitted into the hole. Because the blastomeres do not adhere to the filter, the squares can easily be removed at the end of the experimental period. After the barriers have been removed, transfer the embryos to 50% culture medium in an agarose depression to allow gastrulation to proceed normally.

It is very important to perform several controls because these embryos are extensively manipulated. Some embryos should be passed through the different media, their vitelline membranes removed, and two blastomeres separated, but no barrier inserted. Some embryos should be passed through the different media, and their vitelline membranes opened, but their blastomeres not separated. Some embryos should be passed through the different media with no other manipulations. These different controls will help to distinguish between the effects of the operations and the culture media versus the blockade of intercellular communication between the blastomeres.

Summary

Lineage tracers have made it possible to detect at the cellular level the fate of embryonic cells. However, to understand the mechanisms by which these fates are established, one must expose cells to novel experimental environments. In *Xenopus*, this can be done precisely because blastomeres are large enough to manipulate as single cells and are specifically identifiable from one embryo to the next. In this chapter, several different techniques for testing the fate commitment of blastomeres have been described. In combination with lineage tracing, immunological identification of specific cell phenotypes, and exogenous gene expression techniques, these experimental approaches provide powerful tools for elucidating the molecular and cellular mechanisms of cell fate commitment.

References

Bauer, D. V., Huang, S., and Moody, S. A. (1994). The cleavage stage origin of Spemann's Organizer: analysis of the movements of blastomere clones before and during gastrulation in Xenopus. Development 120:1179–1189.

Bauer, D. V., Best, D. W. Hainski, A. M., and Moody, S. A. (1996). A contact-dependent animal-to-vegetal signal biases neural lineages during Xenopus cleavage stages. Dev. Biol. 178:217–228.

Chalfie, M., Tu, Y., Euskirchen, G., Ward, W. W., and Prasher, D. C. (1994). Green fluorescent protein as a marker for gene expression. Science 263:802–805.

Dale, L., and Slack, J. M. W. (1987). Fate map of the 32-cell stage of *Xenopus laevis*. Development 100:279–295.

Davidson, E. H. (1990). How embryos work: a comparative view of diverse modes of cell fate specification. Development 108:365–389.

Gallagher, B. C., Hainski, A. M., and Moody, S. A. (1991). Autonomous differentiation of dorsal axial structures from an animal cap cleavage stage blastomere in *Xenopus*. Development 112:1103–1114.

Gimlich, R. L. (1986). Acquisition of developmental autonomy in the equatorial region of the *Xenopus* embryo. Dev. Biol. 115:340–352.

Gimlich, R. L., and Gerhart, J. C. (1984). Early cellular interactions promote embryonic axis formation in *Xenopus laevis*. Dev. Biol. 104:117–130.

Godsave, S., and Slack, J. M. W. (1989). Clonal analysis of mesoderm induction in *Xenopus laevis*. Dev. Biol. 134:486–490.

Godsave, S., and Slack, J. M. W. (1991). Single cell analysis of mesoderm formation in the *Xenopus* embryo. Development 111:523–530.

Grunz, H., and Tackle, L. (1986). The inducing capacity of the presumptive endoderm of *Xenopus laevis* studied by transfilter experiments. Roux's Arch. Dev. Biol. 195:467–473.

Guthrie, S., Turin, L., and Warner, A. E. (1988). Patterns of junctional communication during development of the early amphibian embryo. Development 103:769–783.

Hainski, A. M., and Moody, S. A. (1992). *Xenopus* maternal RNAs from a dorsal animal blastomere induce a secondary axis in host embryos. Development 116:347–355.

Hainski, A. M., and Moody, S. A. (1996). Activin-like signal activates dorsal-specific maternal RNA between 8- and 16-cell stages of *Xenopus*. Dev. Genet. 19:210–221.

Hausen, P., and Riebesell, M. (1991). *The Early Development of Xenopus laevis*. New York, Springer-Verlag.

Hirose, G., and Jacobson, M. (1979). Clonal organization of the central nervous system of the frog. I. Clones stemming from individual blastomeres of the 16-cell and earlier stages. Dev. Biol. 71:191–202.

Huang, S., and Moody, S. A. (1993). The retinal fate of *Xenopus* cleavage stage progenitors is dependent upon blastomere position and competence: studies of normal and regulated clones. J. Neurosci. 13:3183–3210.

Jacobson, M. (1983). Clonal organization of the central nervous system of the frog. III. Clones stemming from individual blastomeres of the 128-, 256-, and 512-cell stages. Dev. Biol. 71:191–202.

Jacobson, M. (1985). Clonal analysis and cell lineages of the vertebrate nervous system. Ann. Rev. Neurosci. 8:71–102.

Jacobson, M., and Hirose, G. (1981). Clonal organization of the central nervous system of the frog. II. Clones stemming from individual blastomeres of the 32- and 64-cell stages. J. Neurosci. 1:271–284.

Kageura, H., and Yamana, K. (1984). Pattern regulation in defect embryos of *Xenopus laevis*. Dev. Biol. 101:410–415.

Kay, B. K. (1991). Injection of oocytes and embryos. Methods in cell biology: *Xenopus laevis*: Practical Uses in Cell and Molecular Biology 36:663–669.

Klein, S. L. (1987). The first cleavage furrow demarcates the dorsal-ventral axis in *Xenopus* embryos. Dev. Biol. 120:299–304.

Kimmel, C. B., and Law, R. D. (1985a). Cell lineage of zebrafish blastomeres I. Cleavage pattern and cytoplasmic bridges between cells. Dev. Biol. 108:78–85.

Kimmel, C. B., and Law, R. D. (1985b). Cell lineage of zebrafish blastomeres III. Clonal analyses of the blastula and gastrula stages. Dev. Biol. 108:94–101.

Kimmel, C. B., and Warga, R. M. (1987). Indeterminant cell lineage of the zebrafish embryo. Dev. Biol. 124:269–280.

Masho, R. (1988). Fates of animal-dorsal blastomeres of eight-cell stage *Xenopus* embryos vary according to the specific patterns of the third cleavage plane. Dev. Growth Differ. 30:347–359.

Masho, R. (1990). Close correlation between the first cleavage plane and the body axis in early *Xenopus* embryos. Dev. Growth Differ. 32:57–64.

Masho, R., and Kubota, H. Y. (1986). Developmental fates of blastomeres of eight-cell stage *Xenopus laevis* embryos. Dev. Growth Differ. 28:113–123.

McClendon, J. F. (1910). The development of isolated blastomeres of the frog's egg. Am. J. Anat. 10:425–430.

Messenger, A. E. and Warner, A. E. (1979). The function of the sodium pump during the differentiation of amphibian embryonic neurones. J. Physiol. 292:85–105.

Moody, S. A. (1987a). Fates of the blastomeres of the 16-cell stage *Xenopus* embryo. Dev. Biol. 119:560–578.

Moody, S. A. (1987b). Fates of the blastomeres of the 32-cell stage *Xenopus* embryo. Dev. Biol. 122:300–319.

Moody, S. A., and Kline, M. J. (1990). Segregation of fate during cleavage of frog (*Xenopus laevis*) blastomeres. Anat. Embryol. 182:347–362.

Nagajski, D. J., Guthrie, S. C., Ford, C. C., and Warner, A. E. (1989). The correlation between patterns of dye transfer through gap junctions and future developmental fate in *Xenopus*: the consequences of u.v. irradiation and lithium treatment. Development 105:47–752.

Nakamura, O., and Kishiyama, K. (1971). Prospective fates of blastomeres at the 32-cell stage of *Xenopus laevis* embryos. Proc. Japan Acad. 47:407–412.

Nakatsuji, N., and Johnson, K. E. (1982). Cell locomotion in vitro by *Xenopus laevis* gastrula mesodermal cells. Cell Motil. 2:149–161.

Patten, B. M. (1929). *The Early Embryology of the Chick*, Philadelphia, P. Blakiston's Son.
Paul, D. L., Yu, K., Bruzzone, R., Gimlich, R. L., and Goodenough, D. A. (1995). Expression of a dominant negative inhibitor of intercellular communication in early *Xenopus* embryo causes delamination and extrusion of cells. Development 121:371–381.
Pedersen, R. A., Wu, K., and Balakier, H. (1986). Origin of the inner cell mass in mouse embryos: cell lineage analysis by microinjection. Dev. Biol. 117:581–595.
Peng, H. B. (1991). Solutions and protocols. Methods in Cell Biology: *Xenopus laevis*: Practical Uses in Cell and Molecular Biology 36:657–662.
Shankland, M. (1984). Positional determination of supernumerary blast cell death in the leech embryo. Nature 307:541–543.
Stent, G. S., and Weisblat, D. A. (1985). Cell lineage in the development of invertebrate nervous systems. Ann. Rev. Neurosci. 8:45–70.
Takasaki, H. (1987). Fates and roles of the presumptive organizer region in the 32-cell embryo in normal development of *Xenopus laevis*. Dev. Growth Differ. 29:141–152.
Takasaki, H., and Konishi, H. (1989). Dorsal blastomeres in the equatorial region of the 32-cell *Xenopus* embryo autonomously produce progeny committed to the organizer. Dev. Growth Differ. 31:147–156.
Vincent, J. P., and Gerhart, J. C. (1987). Subcortical rotation in *Xenopus* eggs: an early step in embryonic axis specification. Dev. Biol. 123:526–539.
Vincent, J. P., Oster, G. F., and Gerhart, J. C. (1986). Kinematics of gray crescent formation in *Xenopus* eggs: the displacement of subcortical cytoplasm relative to the egg surface. Dev. Biol. 113:484–500.
Vize, P. D., Melton, D. A., Hemmati-Brivanlou, A., and Harland, R. M. (1991). Assays for gene function in developing *Xenopus* embryos. Methods Cell Biol. 36:367–387.
Vodicka, M. A., and Gerhart, G. C. (1995). Blastomere derivation and domains of gene expression in the Spemann Organizer of *Xenopus laevis*. Development 121:3505–3518.
Warner, A. E., Guthrie, S. C., and Gilula, N. B. (1984). Antibodies to gap junctional protein selectively disrupt junctional communication in the early amphibian embryo. Nature 311:127–131.
Weisblat, D. A., Sawyer, R. T., and Stent, G. S. (1978). Cell lineage analysis by intracellular injection of a tracer enzyme. Science 202:1295–1298.
Wetts, R., Serbedzija, G. N., and Fraser, S. E. (1989). Cell lineage analysis reveals multipotent precursors in the ciliary margin of the frog retina. Dev. Biol. 136:254–263.
Witta, S. E., Agarwal, V. R., and Sato, S. M. (1995). *XlPOU2*, a noggin-inducible gene, has direct neuralizing activity. Development 121:721–730.
Yamana, K., and Kageura, H. (1987). A reexamination of the "regulative development" of amphibian embryos. Cell Differ. 20:3–10.
Yuge, M., and Yaman, K. (1989). Regulation of the dorsal axial structures in cell-deficient embryos of *Xenopus laevis*. Dev. Growth Differ. 31:315–324.

Part III

METHODS FOR STUDYING *DROSOPHILA* OOCYTES AND EMBRYOS

20

Eggshell Assembly in *Drosophila*

Gail L. Waring

The *Drosophila* eggshell is a specialized extracellular matrix formed between the oocyte and overlying follicle cells during late oogenesis. The eggshell is composed of three morphologically distinct proteinaceous layers: an oocyte proximal vitelline membrane layer produced during egg chamber stages 8–10; a crystalline inner chorionic layer (ICL), and an outer tripartite endochorion layer, both of which are produced during stages 12–14. The protective eggshell plays an essential role in fertilization and respiration (Margaritas 1985) and it has been postulated that the vitelline membrane layer may also serve as a repository for positional information that is utilized during early embryonic development (Stein et al. 1991; Hong and Hashimoto 1995). Several putative eggshell structural genes have been identified based on their selective expression by the follicle cells during the period of eggshell deposition. The deduced translation products encoded by these genes are consistent with the general profile of eggshell proteins in terms of size and amino acid composition. A high glycine, alanine, proline, and tyrosine content is typical of most eggshell proteins and several vitelline membrane proteins possess a highly conserved 38 amino acid "tag" (reviewed in Spradling 1993).

A central issue concerning the eggshell, as well as other complex macromolecular structures, is how the proteins assemble into such precise structures. Recent studies indicate that assembly entails far more regulation at the post-translational level than was previously envisioned (Nogueron and Waring 1995; Pascucci et al. 1996). Several eggshell gene products are cleaved in the vitelline membrane layer in a temporally regulated fashion. There is also a growing list of proteins which are synthesized during the early stages of eggshell production that become transiently associated with the vitelline membrane (Pascucci et al. 1996; M. I. Nogueron and G. L. Waring, unpublished data). Most of the retained proteins become incorporated into the assembling chorionic layers during late oogenesis; however, at least one proteolytic derivative is taken up by the oocyte. The timing and kinetics of release of these proteins from the vitelline membrane layer varies. Hence, for at least some eggshell proteins, their availability for assembly is dictated by transient intermolecular associations, as well as by their time of synthesis. Resolution of eggshell assembly at the molecular level has been greatly facilitated by the acquisition of specific antibodies. In this chapter I describe the use of bacterial expression vectors to produce specific eggshell protein antiserum, variations in Western blotting parameters used to optimize the signals for specific eggshell proteins, and evolving immunocytochemical methods used to localize eggshell components in the assembling and mature eggshell by electron microscopy.

Production of Specific Antiserum

Several eggshell genes have been cloned and sequenced (reviewed in Spradling 1993). Selected open reading frames from many of these genes have been cloned into pATH bacterial expression vectors and the resultant fusion proteins were expressed at high levels (Nogueron and Waring 1995; Pascucci et al. 1996) using standard protocols (Koerner et al. 1991). The time course of induced expression was established for each fusion protein; in general, the amount of fusion protein was maximal after 5–6 hours of induction. Bacterial lysates were prepared according to Sambrook et al. (1989) and novel gene products in the expected size ranges were detected by SDS–PAGE (Laemmli 1970). Fusion proteins were purified by preparative gel electrophoresis. Although the amount of protein recovered from the bacterial cultures varied, for most constructs we routinely recovered 1–2 mg of "purified" fusion protein per 50-ml of culture using the following protocol.

1. Bacterial cells from 50—200 ml of induced cultures were pelleted and resuspended in 1/20 the original volume of SDS gel loading buffer, as described by Sambrook et al. (1989). Proteins were resolved by preparative SDS–PAGE (usually 10% gels).

2. To recover unstained fusion protein from the preparative gels, guide lanes were cut from both sides of the gels and stained with Coomassie brilliant blue. To facilitate alignment of the guide lanes with the unstained portions of the gels, the guide lanes were notched in the region of the desired fusion proteins. Bands of interest were excised from the unstained gel and stored at $-20°C$. To ensure that

the desired region had been excised, the remaining gel was stained with Coomassie blue. Gel bands were diced, placed in an Elutrap chamber (Schleicher and Schuell) and electroeluted in buffer containing 81 mM glycine, 13 mM Tris, 0.02% SDS, and 0.05% PEG 8000 at 200 V for 5 hours. PEG was included in the buffer to minimize adsorption of fusion protein to the walls of the Elutrap vessel. The eluants were lyophilized, resuspended in PBS (100 mM phosphate buffer, pH 7.0, 150 mM NaCl) at a concentration of 100–200 µg/ml and stored at $-20°C$.

3. For primary immunizations, 1 ml of fusion protein in PBS (100–250 µg/ml) was emulsified in an equal volume of Freund's complete adjuvant and injected subcutaneously into New Zealand White rabbits at multiple sites on the back. After 1 month, the rabbits were reinjected at multiple subcutaneous sites with 100–150 µg of fusion protein emulsified in an equal volume of incomplete adjuvant. A second boost, 3 or more weeks later, was usually sufficient to obtain high-titer serum. Serum was collected by ear bleeding and processed as described in Harlow and Lane (1988). In several instances, unpurified serum was used directly for our analyses. When necessary, antiserum was purified by affinity chromatography using 0.5-ml columns of Affi-Gel 10 (Bio-Rad Laboratories) to which approximately 1 mg of electrophoretically purified fusion protein was attached according to the manufacturer's instructions. Following inactivation of complement, serum was passed over the column four times. Unbound antibodies were collected in two column volumes of 1M NaCl. Bound antibodies were eluted with three column volumes of 50 mM glycine, pH 2.5, and collected in Eppendorf tubes. Eluted fractions were neutralized with 0.5 M Tris-HCl, pH 8.5, and bovine serum albumin (BSA) was added to a final concentration of 1%. Eluted fractions were dialyzed overnight against phosphate-buffered saline (100 mM phosphate, pH 7.4, 150 mM NaCl). Dialyzed fractions were analyzed by Western blots (see below), aliquoted, and stored at $-20°C$.

Detection of Soluble Eggshell Antigens by Immunoblotting

Western blot analysis of proteins extracted from egg chambers at different developmental stages has proven useful in several respects: (1) evaluating the specificity of antiserum, (2) establishing when specific eggshell proteins become cross-linked into insoluble protein complexes, and (3) revealing unsuspected stage-specific post-translational processing events (Nogueron and Waring 1995; Pascucci et al. 1996).

The RNA and protein synthesis profiles have been established for several eggshell proteins; therefore, the accumulation profiles of specific eggshell antigens can be anticipated. For example, the s18 chorion protein is synthesized during stages 13 and 14. (Note: eggshell proteins are named based on their apparent size on SDS gels. The "s" prefix signifies "shell"; the "sV" prefix indicates the vitelline membrane layer of the eggshell.) An antiserum specific for s18 should detect an 18-kDa SDS-soluble antigen only in stage 13 and 14 egg chambers (Figure 20-1). Eggshell proteins become covalently cross-linked into an insoluble matrix during the later stages of oogenesis. While the timing varies, by late stage 14, eggshell proteins are no longer

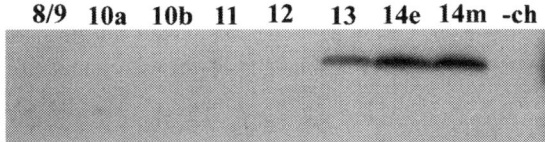

Figure 20-1 Stage-specific accumulation of eggshell antigens during oogenesis. SDS-soluble protein from 10 egg chambers at the developmental stages indicated at the top of the figure were resolved on a 16% gel, blotted overnight to PVDF membrane in an E-C transfer system and incubated with s18 antiserum, diluted 1:1000. A signal in the 18-kDa size range was observed in stage 13 and early or mid stage 14 egg chambers. No reactive species were detected in egg chambers from which the chorion layers had been manually removed (−ch). The sizes of the reactive species were estimated based on their migration relative to proteins of known molecular weight run in parallel. (Adapted from Figure 1; Pascucci et al. 1996).

soluble in SDS buffers. Insolubilization during stage 14 therefore serves as an additional criterion for identifying eggshell antigens. The chorion layers can be manually removed from stage 14 egg chambers by gently rolling the egg chambers over double-stick tape. Antigens which associate only with the chorion layers will be missing from the remaining vitelline-membrane-bound oocyte fraction (Figure 20-1). Staged egg chambers from female sterile mutants in which specific eggshell proteins are missing are the most definitive controls for antiserum specificity. However, to date, protein null mutants have been identified for only three eggshell genes: *sV23* (Savant and Waring 1989), *s36* (Digan et al. 1979), and defective chorion-1 (*dec-1*) (Bauer and Waring 1987; Nogueron and Waring 1995).

Preparation of Blot
For most eggshell proteins tested, standard blotting protocols using either nitrocellulose or PVDF membranes have given satisfactory and comparable results.

1. Two to three days prior to dissection, feed bottled flies (1–2 days post-eclosion) wet yeast paste (brewer's yeast).
2. Etherize well-fed females and remove the ovaries in a small dish containing insect Ringer's (150 mM NaCl; 5 mM KCl, 1.5 mM $CaCl_2$, 10 mM Tricine, pH 6.85) using a stereoscopic microscope. Place one pair of #5 watchmaker's forceps between the abdomen and the thorax; place another pair at the tip of the abdomen. After a gentle pull, white ovarian tissue containing egg chambers at different developmental stages should become apparent. Tease the egg chambers apart using watchmaker's forceps and separate the egg chambers according to stage, following the morphological criteria outlined in Mahowald and Kambysellis (1980).
3. Transfer the staged egg chambers to 1.5-ml Eppendorf tubes containing insect Ringer's. To transfer, draw a few microliters of Ringer's into a 50- or 100 μl capillary pipette, take up the pooled egg chambers into the pipette and dispense its contents into an appropriately marked Eppendorf tube.

4. Pellet the egg chambers by brief centrifugation (15 seconds in a microfuge), withdraw the Ringer's and resuspend the egg chambers in Laemmli sample buffer (50 mM Tris-HCl, pH 6.8; 100 mM DTT; 2% SDS; 10% glycerol, and 0.1% brom phenol blue) at a concentration of 1 or 2 egg chambers per microliter.
5. Boil the extracts for 5 minutes and pellet the SDS-insoluble material by brief centrifugation at 14,000 rpm. For abundant eggshell proteins, extract derived from 10–20 egg chambers per lane is sufficient; for minor eggshell components, such as fc177 and its derivatives, 100 egg chambers per lane is advised.
6. Fractionate egg chamber proteins by SDS–PAGE (0.75-mm gels). Vitelline membrane proteins (10–20 kDa size range) were resolved on 16% gels; 12% gels were used to separate eggshell proteins in the 20–130 kDa size range; 7.5% gels were used for the large *dec-1* derivatives (45–200 kDa).
7. In general, fractionated proteins were electroblotted to nitrocellulose or PVDF membranes using standard protocols (Harlow and Lane 1988; and NEN instruction manual, respectively). After rinsing the gels with deionized water, gels were equilibrated for 15 minutes or less in transfer buffer (360 mM glycine; 50 mM Tris; 0.1% SDS; and 20% methanol for the *dec-1* proteins; 190 mM glycine, 25 mM Tris, 20% methanol for all the other eggshell proteins). Proteins ranging in size from 10 to 200 kDa were successfully transferred using either an E-C transfer system (0.2 A for 16 hours at 4°C) or a BioRad transblot apparatus equipped with plate electrodes (35 V, 45–60 minutes at 10–15°C). Overall the results obtained with the nitrocellulose and PVDF membranes were comparable. A notable exception was the *dec-1* fc177 antigen and its derivatives. The fc177 antigen is a minor *dec-1* translation product that is processed via a 120-kDa intermediate into a stable 85-kDa C-terminal derivative (s85; Nogueron and Waring 1995). The 177- and 120-kDa forms were observed on both membranes; however, a s85 signal was consistently observed only on nitrocellulose (Figure 20-2). Furthermore, the intensity of the fc177 signal was enhanced significantly when pre-equilibration of the gel in transfer buffer was minimized and the concentration of methanol in the transfer buffer was reduced from 20% to 10%. These observations underscore the need to vary blotting parameters when one is dealing with unknown antigens.
8. To reduce the loss of protein during subsequent incubation and washing steps, blots were routinely air-dried prior to staining. Air-dried blots have been stored for up to 2 weeks in air-sealed bags at 4°C without effect.

Detection of Eggshell Antigens on Protein Blots
1. Dried blots were stained with 0.1% India ink (Higgin's Inc.) to monitor the transfer of proteins in general. Dry PVDF blots were wetted in 100% methanol for 30 seconds to 1 minute and rinsed twice (5 minutes each) in Tris-buffered saline (TBS: 50 mM Tris, pH 6.5, 150 mM NaCl) containing 0.4% Tween 20 (w/v), prior to staining. Nitrocellulose blots were wetted in water. The PVDF blots were stained for 15–30 minutes in 0.1% India ink in TBS, pH 6.5, containing 0.2% Tween 20, while nitrocellulose blots were stained in 0.1% India ink in phosphate-buffered saline (100 mM phosphate, pH 7.4, 150 mM NaCl) containing 0.3% Tween 20. The mildly acidic pH was critical for satisfactory staining of the PVDF membranes. Excess ink was removed by two 5-minute

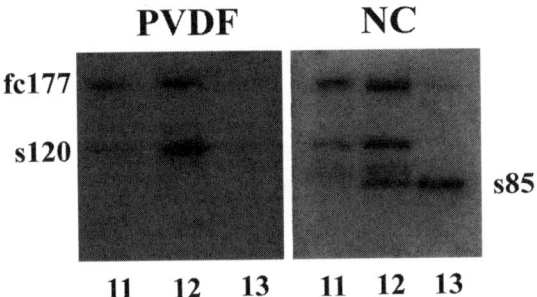

Figure 20-2 Membrane dependence of s85 Western blot signal. SDS-soluble proteins from 80-100 egg chambers at the developmental stages shown below the figure were resolved on 7.5% gels, electroblotted to either PVDF or nitrocellulose (NC) membranes, and reacted with a 1:150 dilution of affinity-purified serum directed against a fusion protein containing 200 amino acids unique to fc177 and its C-terminal derivatives. fc177, the primary translation product, is detected on both membranes in stage 11 and stage 12 egg chambers. s120, a processed C-terminal intermediate, is also observed on both membranes. s85, the mature C-terminal proteolytic derivative of fc177, is evident in stages 12 and 13 on the NC membrane but is not detected on the PVDF membrane. (Adapted in part from Figure 6, Nogueron and Waring 1995.)

rinses: TBS, pH 6.5, for the PVDF membrane; PBS, pH 7.4, containing 0.2% Tween 20 for the nitrocellulose membranes.

2. Nonspecific sites were blocked on the stained blots by incubating with PBS, pH 7.2–7.3 containing 5% nonfat powdered milk and 0.1% Tween 20 for a minimum of 3 hours at room temperature under gentle but continuous agitation. Prior to incubation with the antiserum, blots were washed twice with PBS, pH 7.4.

3. Washed blots were placed in sealed bags with antiserum diluted in PBS, pH 7.2–7.3, containing 3% BSA. In some cases, 0.1% Tween 20 was added to reduce the nonspecific background signals. Since binding of some antibodies may be inhibited by detergents such as Tween 20, it is useful to check this parameter when handling uncharacterized serum. Approximately 10 ml of diluted antiserum was used per 100 cm^2 of blot. Useful dilutions ranged from 1:1000 to 1:10,000. Incubations were carried out on a rocking platform overnight at room temperature.

4. After removing the primary antibodies, the blots were rinsed in several changes of PBS, pH 7.2. To visualize the antigen–antibody complexes $[^{125}I]$-labeled protein A (New England Nuclear) was used as a secondary reagent. For a 10 × 14 cm blot, 0.6 µCi of labeled protein A, diluted in 15 ml of buffer (PBS, pH 7.2, containing 3% BSA), was added to a bag containing the blot. After sealing, the blot was incubated at room temperature for 1–2 hours on a rocking platform.

5. Radiolabeled blots were rinsed as in step 4 and mounted on a piece of used X-ray film covered with wax paper. Saran wrap was placed over the blot and the blot was exposed to X-ray film at −70°C with the aid of an intensifying screen.

Using this detection method, we have analyzed 13 eggshell antigens to date: two vitelline membrane proteins (sV17 and sV23); two chorion proteins (s36 and s18) and nine *dec-1* derivatives (fc106, s80, s25, s20, s60, fc125, s95, fc177, and s85). With the exception of s25, these conditions gave a good signal-to-noise ratio for all of the other eggshell antigens. It was determined empirically that detection of s25 was greatly enhanced if the pH of all of the solutions was reduced to 6.8-7.0 (Figure 20-3).

Immunolocalization

Following its completion during late stage 10, the vitelline membrane appears as a uniform layer, 2-3 µm in thickness. As the egg chamber matures and the oocyte grows, the vitelline layer thins to 0.5-1 µm at stage 14. The tripartite endochorion, at 1 µm thickness can be resolved from the vitelline membrane at the light microscope level. To distinguish the ICL from the endochorion floor and to resolve molecular asymmetries within the vitelline membrane and endochorion layers, the resolving capacity of the electron microscope (EM) is needed. The need for resolution at the EM level has been underscored by recent studies which have revealed molecular asymmetries within the vitelline membrane layer during its assembly at stage 10, and later when eggshell components, such as the *dec-1* derivatives, redistribute (M. I. Nogueron and G. L. Waring, unpublished).

In this section, I describe fixation, embedding, and staining protocols which we have used with success. Fixation conditions are critical for preservation of both morphology and antigenicity. The reactivity of antigens within the vitelline membrane and endochorion layers appears to be differentially affected by fixation conditions. Using 2% formaldehyde and 1% glutaraldehyde, we retained adequate eggshell morphology and antigenicity to localize definitively two abundant vitelline membrane proteins, sV17 and sV23, and two abundant chorion proteins, s36 and s18 (Figure 20-4). Differences in the relationship between fixation and immunodetection of antigens within the vitelline membrane and chorion layers became apparent in studies on the less abundant *dec-1* derivatives. Biochemical fractionation and immunogold labeling studies both indicate that the s60 *dec-1* derivative is present in both layers. In lightly fixed tissue (4% formaldehyde, 1% glutaraldehyde for 60 minutes at room temperature), eggshell ultrastructure is compromised and there is light but specific labeling of the endochorion layer (Figure 20-5A). With longer fixation times (overnight at 4°C), eggshell morphology is restored, and while there is limited labeling of the endochorion (Figure 20-5B), the majority of gold particles are within the vitelline membrane (Figure 20-5C). More rigorous fixation (4% formaldehyde, 2.5% glutaraldehyde, overnight at 4°C) eliminates the endochorion signal, leaving the vitelline membrane signal seemingly unaffected (not shown). While detection of s60 in the endochorion layer was problematic with fixation overnight, other endochorion components, such as s25, s36, and s18, were

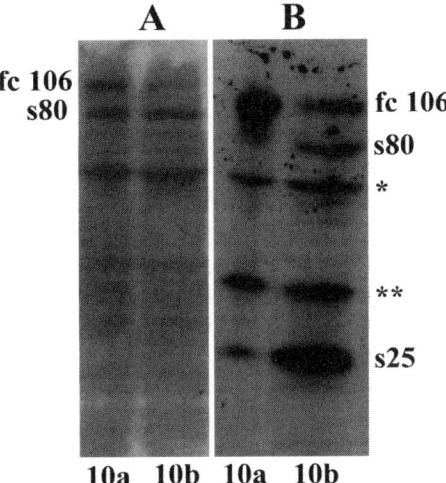

Figure 20-3 pH dependence of the s25 *dec-1* signal. Proteins extracted from early (10a) or late (10b) stage 10 egg chambers were resolved on 12% gels, electroblotted to PVDF membrane and reacted with an antiserum directed against the N-terminal region of the *dec-1* proteins (Nfc106; Nogueron and Waring 1995). In (A), all of the incubations and washes were done with solutions at standard pHs (pH 7.2–7.4); in (B), the pH of all of the solutions was decreased to 6.8–7.0. The proprotein fc 106 and its cleaved C-terminal derivative, s80, were detected using either of these pH conditions. A *dec-1*-related processing intermediate (**) and the 25-kDa N-terminal derivative (s25) were observed only with the more acidic pH conditions (B). The *dec-1* relatedness of all bands was verified by their absence in extracts from *dec-1* protein null mutants. A non-*dec-1*-related band in the 67-kDa size range was prominent on both blots (*).

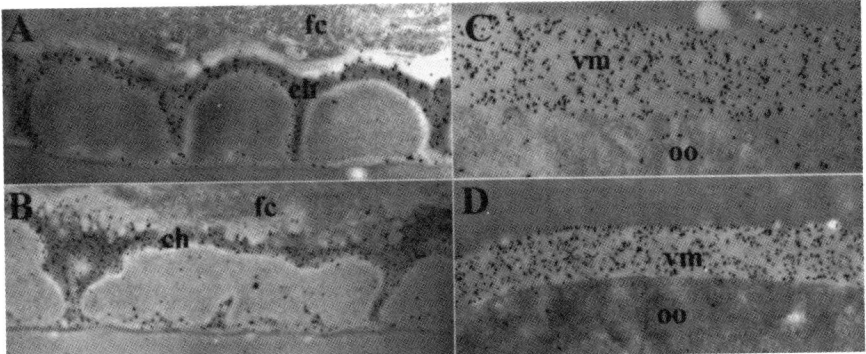

Figure 20-4 Immunolocalization of abundant eggshell antigens. Immunogold labeling of (A and B) the chorion and (C and D) the vitelline membrane layers of a stage 14 egg chamber with s18, s36, sV23, and sV17 antisera, respectively. fc, follicle cell; ch, chorion; vm, vitelline membrane, and oo, oocyte.

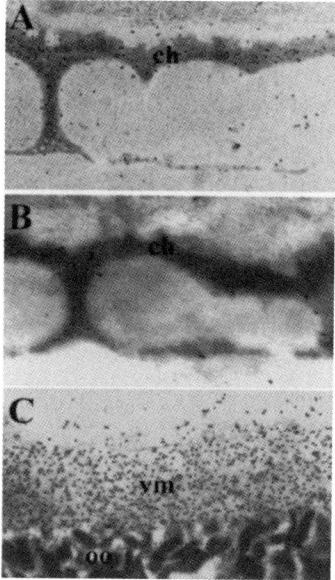

Figure 20-5 Effect of fixation conditions on immunogold labeling of the chorion and vitelline membrane. Thin sections showing the immunoreactivity of the chorion and vitelline membrane layers of a stage 14 egg chamber reacted with a *dec-1* antiserum (Cfc106; Nogueron and Waring 1995) under different fixation conditions. In stage 14 egg chambers, the most abundant *dec-1* product recognized by the Cfc106 serum is s60, the mature C-terminal derivative of fc106. (A) Labeling of the endochorion layer (ch) over background is seen when the tissue is subjected to light fixation (60 minutes, room temperature). Tissue fixed for a more extended period (overnight at 4°C) shows specific, but limited labeling of (B) the endochorion but prominent labeling of (C) the vitelline membrane layer. ch, chorion; vm, Vitelline membrane; oo, oocyte.

readily detected under these conditions. Like s60, the s25 chorion signal was eliminated when the tissue was fixed with a higher concentration of glutaraldehyde (2.5%). These results emphasize the need to vary fixation conditions with every eggshell antigen tested, in order to minimize the risk of false negatives, especially within the endochorion layer. Antigens in both the vitelline membrane and endochorion layers were detected using the following protocol.

Preparation of the Specimen
1. Remove ovaries from rapidly laying females as outlined in the previous section.
2. Transfer the ovaries with forceps to a small screw-cap cryovial containing Ringer's solution.
3. After the ovaries sink to the bottom of the vial, remove the Ringer's with a Pasteur pipette and replace it with cold fixative (1% glutaraldehyde and 4%

formaldehyde in 0.06 M sodium phosphate buffer, pH 7.4). Fix overnight at 4°C.
4. Rinse the tissue four times at 4°C (30 minutes each) in 0.06 M phosphate buffer, pH 7.4, containing 0.2 M sucrose.
5. Dehydrate the ovaries by passing them through the following alcohol dilution series, using a volume that is at least 10× greater than the sample:

30% ethanol	60 minutes	4°C
50% ethanol	60 minutes	−20°C
70% ethanol	60 minutes	−20°C (tissue can be held overnight at this stage)
95% ethanol	60 minutes	−20°C
100% ethanol	60 minutes	−20°C
100% ethanol	overnight	−20°C

6. Infiltrate the tissue with Lowicryl-K4M resin (Electron Microscopy Sciences) according to the following schedule:

35% resin in 100% ethanol	−20°C	4 hours
50% resin in 100% ethanol	−20°C	overnight
67% resin in 100% ethanol	−20°C	4 hours
100% resin	−20°C	overnight
100% resin	−20°C	1 hour

7. Using a prechilled 100 µl capillary pipette, transfer the ovaries from the cryovial to a dissecting dish, on ice, filled with Lowicryl that has been prechilled to −20°C. Allow the ovaries to sink to the bottom of the dish. Using a chilled capillary pipette, immediately transfer one ovary at a time to BEEM capsules half-filled with cold Lowicryl placed in a UVC2 ultraviolet Cryo Chamber (Ted Pella, Inc.) at −25°C. Let each ovary sink to the bottom of the BEEM capsule; if necessary, orient the ovaries with a wooden applicator stick so that the longitudinal axis of the ovary is parallel to the face of the BEEM capsule. Do not introduce bubbles in the process. Fill the capsule with cold resin.
8. Polymerize in the UVC2 Cyro chamber under ultraviolet light for 24 hours at −25°C. As the dry ice is consumed, allow the temperature within the chamber to rise to room temperature and continue polymerization for an additional 1–3 days. Polymerized blocks can be stored over desiccant at room temperature prior to sectioning.

Immunostaining
1. Collect 60–90-nm thin sections on formvar-coated (Dykstra 1993) nickel grids for immunostaining. Use nonmagnetic forceps to handle the nickel grids. To facilitate staging of the egg chambers within the ovarian sections, also collect 1 µm-thick sections on glass slides and stain with 0.1% toluidine blue in 1% sodium borate.
2. To minimize nonspecific binding, float the grids section-side down on 50–60 µl drops of blocking solution (4% BSA in 0.1 M phosphate, pH 7.4) for 60 minutes at room temperature. All solutions used for immunostaining were prepared with Milli-Q deionized water and filtered through 0.22-µm Millipore disposable acrodisks immediately prior to their use.

3. Following incubation in blocking buffer, remove excess liquid by gently touching the edge of the grid to a piece of filter paper. Transfer the grids to drops of antiserum diluted in buffer containing 1% BSA, 500 mM NaCl, 0.1% Tween 20, and 100 mM phosphate, pH 7.4 (PBS-A). To minimize evaporation and the accumulation of dust particles, place the drops on a piece of Parafilm and cover with a large Petri dish. Centrifuge the diluted primary and secondary antiserum for 5 minutes at 5000 rpm immediately prior to use. Incubate for 1.5–2.5 hours at room temperature.
4. After incubation, remove unbound antibody by washing the grids in six drops (5 minutes each) of 0.5% BSA in 100 mM phosphate, pH 7.4.
5. To detect bound antibody, incubate the reacted grids with commercially supplied (Amersham Life Sciences) goat antirabbit IgG coupled to 15- or 20-nm colloidal gold particles diluted 1:30 in PBS-A, pH 7.4, for 1.5–2 hours.
6. Remove unbound secondary antibody by washing as in step 4.
7. Postfix the antibody complexes by a brief incubation in 2% glutaraldehyde (2 minutes). Wash the fixed grids in two drops of 0.5% BSA/PBS, pH 7.4 (5 minutes each) and two drops of deionized water (<2 minutes each).
8. Dry the grids and store at room temperature.
9. Stain postfixed grids in alcoholic uranyl acetate for 15–30 seconds. To prepare methanolic uranyl acetate add 1.25 g of uranyl acetate to 25 ml of 50% methanol in a light-proof container. Float the grid, specimen-side down, on a drop of methanolic uranyl acetate placed on a small sheet of Parafilm. Rinse the grids in a slow stream of Milli-Q deionized water. Remove excess liquid and dry. Stain in a drop of lead citrate (0.02 g of lead citrate, 0.1 ml of 10 N NaOH, 10 ml of water) for 15–30 seconds. Rinse immediately and thoroughly as above; dry on Whatman filter paper. To avoid lead dirt on sections, it is important that the lead citrate have minimal contact with air. Keep the Parafilm containing the lead citrate drop covered with a Petri dish lid whenever possible and minimize air contact during manipulations. After drying, the grids should be ready for examination.

The specificity of the antiserum in situ often differs from that inferred by Western blot analysis. If available, the most informative controls are sections from protein null mutants. Stage specificity is also a useful parameter.

References

Bauer, B. J., and Waring, G. L. (1987). 7C female sterile mutants fail to accumulate early eggshell proteins necessary for later chorion morphogenesis in Drosophila. Dev. Biol. 121:349–358.

Digan, M. E., Spradling, A. C., Waring, G. L., and Mahowald, A. P. (1979). The genetic analysis of chorion morphogenesis in *Drosophila melanogaster*. In *Eucaryotic Gene Regulation ICN-UCLA Symposium*, Axel, R., et al., eds. New York, Academic Press, pp. 171–181.

Dykstra, M. J. (1993). *Biological Electron Microscopy. Theory, Techniques and Troubleshooting.* New York, Plenum Press, pp. 130–133.

Harlow, E., and Lane, D. (1988). *Antibodies: A Laboratory Manual.* Cold Spring Harbor, NY, Cold Spring Harbor Laboratory Press, pp. 471–510.

Hong, C. C., and Hashimoto, C. (1995). An unusual mosaic protein with a protease domain, encoded by the nudel gene, is involved in defining embryonic dorsoventral polarity in Drosophila. Cell. 82:785–794.

Koerner, T. J., Hill, J. E., Meyers, A. M., and Tzagoloff, A. (1991). High-expression vectors with multiple cloning sites for construction of trpE-fusion genes: pATH vectors. Methods Enzymol. 194:477–490.

Laemmli, U. K. (1970). Cleavage of structural proteins during assembly of the head of bacteriophage T4. Nature (London) 227:680–685.

Mahowald, A. P., and Kambysellis, M. P. (1980). Oogenesis. In *Genetics and Biology of Drosophila*, Ashburner, M., and Wright, T. R. F., eds. London, Academic Press, pp. 141–224.

Margaritis, L. H. (1985). Structure and physiology of the eggshell. In *Comprehensive Insect Physiology, Biochemistry, and Pharmacology*, Vol. 1, Kerkurt, G. A., and Gilbert, L. I., eds. New York, Pergamon, pp. 153–173.

Nogueron, M. I., and Waring, G. L. (1995). Regulated processing of *dec-1* eggshell proteins in Drosophila. Dev. Biol. 172:272–279.

Pascucci, T., Perrino, J., Mahowald, A. P., and Waring, G. L. (1996). Eggshell assembly in Drosophila: processing and localization of vitelline membrane and chorion proteins. Dev. Biol. 177:590–598.

Sambrook, J., Fritsch, E. F., and Maniatis, T. (1989). *Molecular Cloning: A Laboratory Manual*, Vol. 3, Cold Spring Harbor, NY, Cold Spring Harbor Laboratory Press, pp. 18.40–18.41.

Savant, S. S., and Waring, G. L. (1989). Molecular analysis and rescue of a vitelline membrane mutant in *Drosophila melanogaster*. Dev. Biol. 135:43–52.

Spradling, A. C. (1993). Developmental genetics of oogenesis. In *The Development of Drosophila melanogaster*, Vol. 1, Bates, M., and Martinez Arias, A., eds. Cold Spring Harbor, NY, Cold Spring Harbor Laboratory Press, pp. 1–70.

Stein, D., Roth, S., Vogelsang, E., and Nusslein-Volhard, C. (1991). The polarity of the dorsoventral axis in the Drosophila embryo is defined by an extracellular signal. Cell 65:725–735.

21

Cell Biological and Genetic Methods for Studying *Drosophila* Oogenesis

Denise J. Montell

Drosophila oogenesis has emerged as a model system for the study of many fundamental biological problems including stem cell biology, germ cell fate specification, localization of morphogenetic determinants, cell migration, and dynamic remodeling of the actin cytoskeleton, to name a few. In this chapter, many of the experimental methods used to study *Drosophila* oogenesis are described. A detailed description of oogenesis appears elsewhere (King 1970; Spradling 1993). Additional methods can be found in other chapters of this book, as well as in other publications (Verheyen and Cooley, 1994). Basic *Drosophila* biology is described in Ashburner (1989) and techniques in fly husbandry and genetics can be found in Grigiatti et al. (1986). General information concerning *Drosophila* genes and stocks can be found on the World Wide Web (http://flybase.bio.indiana.edu:82/). Fly stocks can be obtained from the *Drosophila* stock center at the University of Indiana in Bloomington, and, in many cases, stocks can be ordered over the Web.

Tools and Equipment Needed
Stereo microscope
Compound microscope equipped with Nomarski optics and fluorescence
Depression slides (VWR 48339-009)
Dumont style No. 5 forceps (Charles Nusinov & Sons, Baltimore, MD)
Transfer pipettes (e.g., PGC Scientific, Gaithersburg, MD 313-001)

Basic Skills

Dissecting Ovaries

Prior to dissection, place female flies on fresh food supplemented with baker's yeast (either a few grains of dry yeast or, even better, a paste of wet yeast), for 24-36 hours at room temperature. Prepare Ringer's solution (EBR), or insect cell culture medium (Schneider's or Grace's) supplemented with 10% serum. You will also need a dissecting microscope, a depression slide, a pair of fine-tip forceps, and either an etherizer with ether in it or access to carbon dioxide to anesthetize the flies.

Dissecting the pair of ovaries from a well-fed *Drosophila* female is a simple matter because of the large size of the tissue relative to the rest of the fly. First anesthetize the fly with carbon dioxide or ether. Grasp one wing of the fly, using a forceps, and hold the fly to the bottom of a depression slide so that the fly is immersed in buffer or culture medium. Gently hold the fly at the level of the thorax using a second forceps (in your left hand, if you are right-handed). Using the right forceps, grasp the end of the abdomen and pull. Usually, the ovary pair will emerge from the broken cuticle intact and still connected to the gut. Simply tease the ovary pair away from any other tissues and then remove the carcass and other tissues from the medium. A dissected ovary pair is shown in Figure 21-1.

The ovaries can readily be further dissected into ovarioles by sliding the forceps in between the ovarioles (the strings of egg chambers that run the length of the ovary), along their length, and pulling them apart at the base, adjacent to the most mature egg chambers. When dissected in this way, the egg chambers will remain encased in a sheath which can slow penetration of antibodies or other probes. Therefore, it is sometimes desirable to dissect ovaries into individual egg chambers devoid of sheath. One way to do this is, starting with a whole ovary, to grasp the base of the ovary with the left forceps and the tip of an ovariole with the other forceps. Then, slowly pull on the tip. A string of egg chambers will emerge out of the sheath, still connected to each other. One can reliably obtain stage 9 and younger egg chambers using this technique. Except for the damage inflicted by the forceps at the tip, the egg chambers are completely viable and can be further cultured (see last section for culturing methods). It is very difficult to dissect unfixed ovaries further into individual egg chambers without damaging them to some extent. Stage 10 egg chambers can be liberated from the ovary simply by using a forceps to stroke the ovary from the bottom toward the top.

The hand-plucking method is reasonable if only small numbers of egg chambers are required. However, if large numbers of sheath-free egg chambers are required, it is less tedious to employ collagenase treatment to dissociate whole ovaries. After dissecting the ovaries, the medium is replaced with fresh medium to which 10 mg/ml collagenase A has been added. As

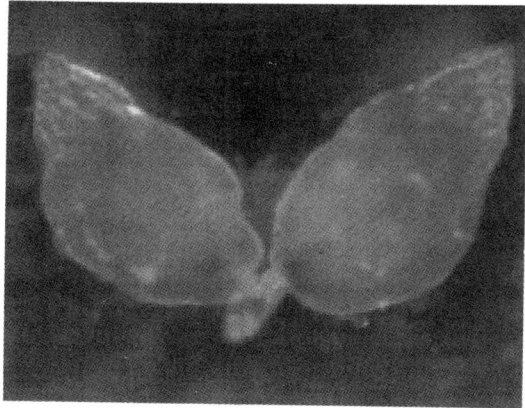

Figure 21-1 Bright-field micrograph of a dissected ovary pair.

soon as the solution is added, begin pipetting the ovaries up and down using a 200-μl micropipetter. As the enzymatic digestion of the sheath proceeds, the pipetting aids in separating the ovarioles and egg chambers so that a reasonably uniform treatment takes place. As soon as many individual egg chambers are available, transfer them to a microfuge tube and allow them to settle to the bottom of the tube. Keep the tube on ice to prevent further collagenase digestion. Remove the medium and rinse the egg chambers several times to wash away the collagenase. The egg chambers should then be fixed and washed.

Basic Fixation Techniques

Note: Formaldehyde and gluteraldehyde are noxious and mutagenic. Wear protective clothing and gloves, perform all fixations in a fume hood, and consult with your safety office concerning proper disposal of fixatives.

Formaldehyde Fixative Combine 37% formaldehyde, buffer B, and deionized water in a 1:1:4 ratio. Remove excess buffer from dissected ovaries or egg chambers and add 100 μl of fixative. Incubate for 10 minutes at room temperature. Remove fixative and discard. Wash three times for 15 minutes each with PBT.

Gluteraldehyde Fixative Dilute buffer B with deionized water in a 1:5 ratio. Make a 0.2% gluteraldehyde solution using diluted buffer B and 8% gluteraldehyde. Remove excess buffer from dissected ovaries or egg chambers. Add 100 μl of fixative and incubate for 10 minutes at room temperature. Remove fixative and discard. Wash ovaries or egg chambers in PBT three times for 15 minutes each.

After fixing egg chambers, it may be useful to transfer them back to the dissecting dish to separate them by stage, unless you are working with whole ovaries. If your interest is in egg chambers younger than stage 10, it is wise to separate the older egg chambers away from the young ones and then transfer only the stages of interest back to the microfuge tube for further processing. The reason is that the large older egg chambers are more visible and it is common to lose the younger ones upon subsequent washings if the older ones are present.

Staging Egg Chambers

Most of the 14 stages of oogenesis described by King (1970) can be distinguished on the basis of a few, easy-to-recognize features, such as the size of the egg chamber, the size of the oocyte, the proportion of the egg chamber occupied by the oocyte, and the shapes and positions of follicle cells. A brief description of oogenesis will aid in describing how to stage egg chambers.

Drosophila oogenesis begins at the tip of the ovary, in a structure called the germarium (Figures 21-2 and 21-3). At the very tip of the germarium reside two to three germ-line stem cells which divide asymmetrically to produce a new stem cell and a committed precursor cell known as a cystoblast. The cystoblast divides four times, each time with incomplete cytokinesis, yielding 16 interconnected germ-line cells. A monolayer of approximately 14 somatic cells then envelops the germ-line cluster. At first, this collection of cells is shaped like a lens; however, at the posterior end of the germarium, the structure reorganizes into a sphere, at which point it is a stage 1 egg chamber, in which the oocyte is clearly recognizable at the posterior-most position in the germ-line cluster. This egg chamber then pinches off from the germarium. During stages 2 through 7, the egg chamber increases in size and DNA content as the cells endoreplicate (Figures 21-2 and 21-3). In stage 8, yolk can first be observed in the oocyte. Stage 9 is characterized by rearrangement of the initially uniform, cuboidal follicle cell monolayer (Figure 21-2A). Most of the follicle cells become columnar and stack up in the posterior of the egg chamber, in contact with the oocyte. The remaining cells become squamous and stretch out to cover the entire nurse cell cluster. A group of six to ten cells, known as border cells, delaminates from the follicle epithelium and migrates through the center of the nurse cell cluster until it reaches the border between the nurse cells and oocyte.

By stage 10, the yolk-filled oocyte fills one-half of the volume of the egg chamber and the most anterior of the columnar follicle cells begin an inward, or centripetal, migration to cover the anterior end of the oocyte (Figure 21-2B). In stage 11 of oogenesis, the oocyte is larger than the entire nurse cell cluster and a cage of actin filaments polymerizes around each of the nurse cell nuclei. Then, the nurse cells squeeze all of their cytoplasm into the oocyte, through the ring canals. In stage 12, little is left of the nurse cells other than the nuclei (Figure 21-3C) and, on the dorsal side, small snorkel-

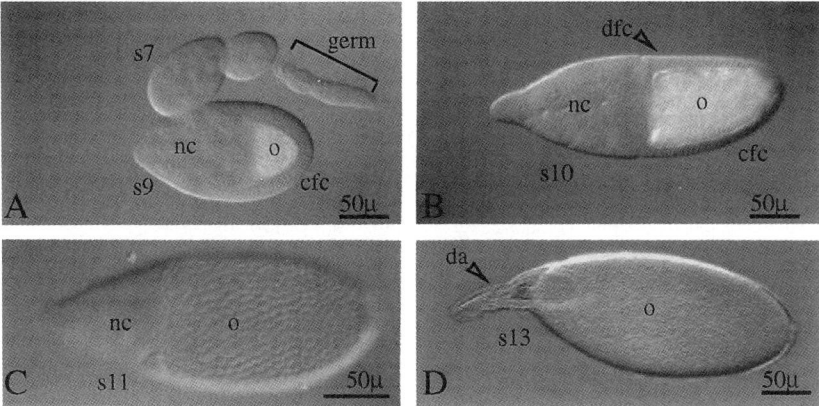

Figure 21-2 Nomarski optics images of egg chambers of various stages. (A) A dissected ovariole containing the germarium (germ) through stage 9 (s9) egg chambers. The stage 7 (s7) egg chamber is characterized by its size, oval shape, and lack of accumulation of yolk (the yolk appears white). The small oocyte is not clearly visible at this magnification. Stage 9 (s9) is characterized by reorganization of the follicle cells. The follicle cells adjacent to the oocyte take on a columnar morphology, whereas more anterior follicle cells become squamous and are therefore too thin to see. (B) A stage 10 egg chamber. Note that the oocyte occupies one-half of the volume of the entire egg chamber at this stage. (C) A stage 11 egg chamber in which the nurse cell contents have been partially transferred to the oocyte. This surface view shows the hexagonal array of the columnar follicle cells which cover the ooctye. (D) A stage 13 egg chamber in which the dorsal appendages are obvious and little remains of the nurse cells. In the stage 9–13 egg chambers, anterior is to the left and dorsal is up. In the germarium through stage 7 egg chambers, anterior is to the right. nc, Nurse cells; o, oocyte; cfc, columnar follicle cells; dfc, dorsal follicle cells; da, dorsal appendages.

like respiratory appendages, called the dorsal appendages, can be seen. In stage 13 egg chambers, tiny remnants of the nurse cells persist while the dorsal appendages become longer and more prominent (Figure 21-3D). In stage 14 egg chambers, the nurse cell remnants are completely gone and the oocyte and eggshell are completely mature.

Identifying Cell Types

As mentioned above, the various cell types of the ovary can be identified based on their relative sizes and positions. The central cells are the germ-line cells. The nurse cells can be distinguished from the oocyte because the nurse cells are polyploid whereas the oocyte is not. Thus, nurse cell nuclei are larger and stain more intensely with DNA stains than the oocyte (Figure 21-3A). Furthermore, the oocyte is the most posterior of the germ-line cells.

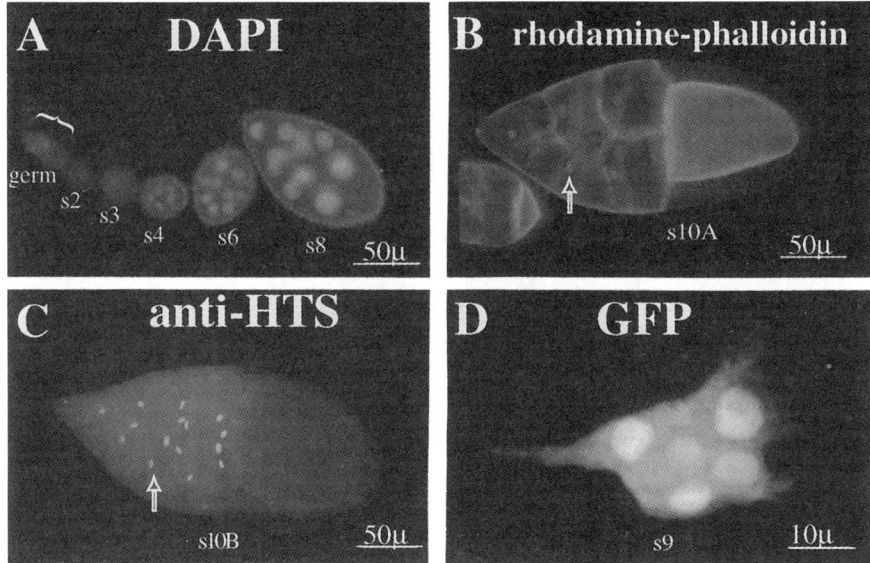

Figure 21-3 Egg chambers stained by a variety of techniques. (A) Fluorescence micrograph of a whole-mount dissected ovariole containing the germarium through stage 8, stained with DAPI to label all nuclei. The nurse cell DNA content is much greater than that of the surrounding follicle cells, which is, in turn, much larger than that of the oocyte. The early stages are out of focus. (B) Fluorescence micrograph of a whole-mount stage 10A egg chamber stained with rhodamine-conjugated phalloidin to label filamentous actin. Cortical actin in the nurse-cell–oocyte complex is clearly seen, as are the ring canals which are intercellular bridges through which cytoplasm flows from the nurse cells to the oocyte. The arrow indicates one ring canal. (C) Fluorescence micrograph of a whole-mount stage 10B egg chamber labeled with a monoclonal antibody against the HTS protein (Robinson et al. 1994) and a fluorescein-coupled secondary antibody. The ring canals are specifically stained. The arrow indicates one ring canal. (D) Confocal micrograph of border cells migrating from left to right during stage 9 of oogenesis. The cells are expressing GFP via the GAL4/UAS system. Some GFP partitions into the nucleus. The rest fills the cytoplasm revealing the fibroblast-like morphology of the cells, which is not visible with other staining methods. In all panels, anterior is to the left.

The follicle cells are the smaller, outer cells of the egg chamber. They are also polyploid, although less so than the nurse cells. Stalk cells are specialized follicle cells that separate and connect neighboring egg chambers.

Mounting Egg Chambers on Microscope Slides

To view egg chambers in the microscope, place four small dabs of vacuum grease or petroleum jelly where the corners of the coverslip will be. The dabs are usually delivered from a 3-ml syringe which has been loaded with the grease or jelly. Pipette about 20 μl of medium with egg chambers onto the slide. Place a coverslip on top and press down just enough so that the egg chambers do not move when you gently touch the coverslip. This makes the chambers flat enough for viewing without crushing them. If there is too much liquid under the coverslip, then wick some of it away by placing a Kimwipe at the edge of the coverslip. Watch under the dissecting scope to see when the egg chambers stop moving. If there is too little liquid under the coverslip, pipette a small drop at the edge of the coverslip and it will be drawn underneath. The amount of liquid is critical for getting good images.

Staining Egg Chambers

Labeling Nuclei

DAPI Staining Staining with DAPI is simple, reliable, and produces beautiful results (Figure 21-3A). The stain is also quite robust in that samples retain their fluorescence for several days when kept in the dark. To stain ovaries with DAPI, dissect ovaries in EBR (or Schneider's or Grace's medium). If desired, dissect further into individual egg chambers. Fix in formaldehyde fix for 10 minutes. Rinse three times in PBT. Incubate egg chambers in 0.5 μg per milliliter of DAPI for 1 hour or more at room temperature. In this and all subsequent steps, wrap the tube in aluminum foil to prevent bleaching. The length of incubation required depends on whether whole ovaries, ovarioles, or individual egg chambers are being stained. The thicker the tissue, the longer it takes for the DAPI to penetrate evenly to all the nuclei. After staining, rinse three times in PBT for 5 minutes each. Then, equilibrate egg chambers in 50% glycerol/50% PBS with an antibleaching additive. Home-made antibleaching reagent can be made by dissolving *p*-phenylenediamine at a concentration of 0.1% in the mounting medium. This substance is toxic and will discolor plastic surfaces it contacts, so we prefer to purchase mounting medium with antibleaching reagent in it (Vectashield).

SPIF Staining Most confocal microscopes are not equipped to detect DAPI fluorescence. Therefore, for confocal microscopy, an alternative fluorescent DNA label has been used (Edwards and Kiehart 1996). The exact chemical nature of the compound is unknown; however, the staining method is as simple as DAPI staining.

First prepare the stain as follows. To a solution of 4 mM Na_2CO_3 (pH 9.0) and 90% glycerol in a 15-ml conical tube, add *p*-phenylenediamine (1,4-diamino-benzene; Sigma) to a concentration of 1 mg/ml and mix briefly to disperse the flakes. Sonicate on ice until the *p*-phenylenediamine solution turns bright yellow, which should occur within 30–60 minutes. Sonicate with the microprobe $\frac{1}{4}''$ into the solution and set the power setting to generate a rapid roil. After sonication, leave the solution overnight at room temperature. Then store in aliquots at $-80°C$. (Note that the *p*-phenylenediamine is a carcinogen and so gloves should be worn).

Prior to mounting, be sure to remove all liquid from the specimen. Then add approximately 30 µl of SPIF and mount the sample on a slide, cover with a 22 × 22-mm coverslip and seal with nail polish. For best results, allow 1 hour before viewing the specimen using either 488-nm or 514-nm light. Although emission is strongest in the fluorescein channel, there is some spillover into the rhodamine channel.

BrdU Labeling   To label cells that are progressing through the cell cycle, BrdU incorporation has been empoloyed (Margolis and Spradling 1995; Lilly and Spradling 1996). To label ovaries by following BrdU incorporation, dissect ovaries in Grace's medium. Incubate in 1 µg/ml bromodeoxyuridine (BrdU) diluted in Grace's medium for 1 hour. Rinse ovaries in EBR to remove excess BrdU. Fix for 15 minutes in buffer B, 37% formaldehyde, and deionized H_2O (1:1:4). Rinse ovaries twice in PBT. Wash twice for 15 minutes each in PBT. Denature for 30 minutes in 2 M HCl. Neutralize for 2 minutes in 100 mM borax. Wash ovaries twice for 15 minutes each in PBT. Incubate overnight with a 1/20 dilution of mouse anti-BrdU at 4°C. Rinse three times and wash three times. Add secondary antibody of your choosing. Incubate for 2 hours at room temperature. Rinse three times and wash three times as described above. Equilibrate stained egg chambers in 50% glycerol/50% PBS/0.1% PPD or in Vectashield.

Rhodamine-Phalloidin Staining

Filamentous actin can be specifically labeled using rhodamine-conjugated phalloidin (Figure 21-3B). Fix egg chambers with formaldehyde. Gluteraldehyde fixation is not compatible with phalloidin staining. Rinse egg chambers three times with PBT for 5 minutes each. Rhodamine-conjugated phalloidin arrives dissolved in methanol. Phalloidin is a poison and should be handled with care; wear gloves. Pipette the desired amount of rhodamine–phalloidin into a microfuge tube and evaporate the methanol solvent. If only 1 or 2 µl is to be used, this can be done by leaving the tube open in a fume hood for a few minutes. If a larger volume is in use, then evaporate the solvent in a speed vac. Resuspend phalloidin in 200 µl of PBT for each microliter of phalloidin that was dried down.

Incubate egg chambers in PBT/phalloidin for 30 minutes to overnight. In this and all subsequent steps, wrap the tube in aluminum foil to prevent bleaching of the rhodamine. The time necessary to obtain good staining depends upon the degree of dissection of the egg chambers. If the egg chambers have been plucked out of the sheath, 30 minutes is sufficient. If the sheath remains, an overnight incubation with agitation is recommended. Note that the sheath will stain with phalloidin, so if the sheath is left on, it will be necessary to view the egg chambers using confocal microscopy in order to see the follicle cell and nurse cell staining. Also note that DAPI (or SPIF) and phalloidin staining can be achieved at the same time, in one incubation. After the desired incubation, wash egg chambers three times for a total of 30 minutes. Equilibrate egg chambers in Vectashield and view using a fluorescence microscope.

Antibody Staining

Dissect and fix egg chambers as described for phalloidin staining. In general, the more thoroughly dissected the egg chambers are, the better the antibody staining results will be. Lowering the formaldehyde concentration to 2–4% may be necessary for some antibody–antigen combinations. Incubate egg chambers in primary antibody (typically a 1:200 to 1:500 dilution of a polyclonal antiserum or 1:1 to 1:10 dilution of a monoclonal antibody, just as for other tissues). Affinity-purified polyclonals typically work better than unpurified antibodies. For details on affinity purification, please see Harlow and Lane (1988). Incubation in primary antibody can be carried out for a few hours at room temperature or overnight a 4°C. Even better penetration of the antibody into the tissue might be obtained by overnight incubation at 25°C with agitation.

Remove as much of the primary antibody solution as possible using a transfer pipette. Rinse (i.e., add buffer and allow egg chambers to settle; no incubation time necessary) egg chambers three times with PBT. Wash egg chambers three times for 20–30 minutes each. Add secondary antibody. Fluorescently labeled secondary antibodies produce superior results to enzyme-linked secondaries, probably because the enzyme-linked antibodies are larger and do not penetrate the egg chambers as well. When using fluorescently labeled antibodies, keep the tube wrapped in foil at all times during incubation, as well as during subsequent washings, to prevent bleaching. Incubate for 2 hours at room temperature. Rinse three times and wash three times in PBT. If desired, egg chambers can also be labeled with phalloidin and/or DAPI at this point. Equilibrate stained egg chambers in mounting medium (50% glycerol/50% PBS or Vectashield). Mount egg chambers on a microscope slide and view using a fluorescence microscope or a confocal microscope. An example of a stage 10 egg chamber stained with a monoclonal antibody to the hu-li tai shao protein (HTS; Yue and Spradling 1992; Robinson et al. 1994) and a fluorescein-conjugated goat antimouse secondary antibody is shown in Figure 21-3C.

Green Fluorescent Protein

The complete morphology of egg chamber cells can be visualized when the green fluorescent protein (GFP) is expressed in the ovary. For example, the fibroblast-like morphology of the border cells can be appreciated better when the cells express GFP (Figure 21-3D). A *UAS-GFP* transgene has been described (Yeh et al. 1995) and a number of transgenic lines expressing GAL4 in various subsets of follicle cells in the *Drosophila* ovary have been generated (Manseau et al. 1997). Therefore, it is now possible to express GFP in these subpopulations by crossing these two strains together.

Staining Ovaries for β-GAL Activity

A large variety of enhancer trap lines exists (see next section for methods used to generate enhancer trap lines), which express a nuclear β-GAL fusion protein in specific subsets of ovarian cells (Figure 21-4). Staining ovaries for β-GAL activity is one of the simplest and most reliable staining methods available. To carry out staining for β-GAL activity, dissect ovaries in EBR or Schneider's + serum. If you are only dissecting a few ovaries and the ovaries will be fixed within 10 minutes of dissection, then EBR is satisfactory. If you are dissecting hundreds of lines (e.g., when conducting an enhancer trap screen), then dissection in Schneider's supplemented with serum is highly recommended. In this way, the ovaries will be kept alive for up to 2 hours prior to fixation. Ovaries can be kept in a 96-well microtiter dish on ice during the dissections.

Following dissection, fix ovaries in either formaldehyde or gluteraldehyde fix for 10 minutes. Longer fixation will reduce staining intensity. Formaldehyde fix produces superior egg chamber morphology; however, gluteraldehyde fix results in stronger staining. While the ovaries are fixing, warm an aliquot of staining solution to 37°C. Staining solution can be made up and stored at 4°C for about 1 year; however, the X-GAL must be added to the prewarmed solution immediately prior to use. Remove fixative and wash the ovaries once with PBT. For each 100 µl of staining solution, add 2.5 µl of 8% X-GAL dissolved in dimethylsulfoxide or dimethylformamide (this solution can be stored at −20°C in the dark for about 6 months and is more stable in dimethylformamide than in dimethylsulfoxide). Add 25–100 µl of staining solution + X-GAL to fixed ovaries. If one or two ovary pairs are being stained, 50 µl is adequate. If a large number of ovaries is to be stained in one tube, the volume should be increased and some kind of mechanical agitation should be applied to ensure uniform staining. Incubate samples at 37°C for as long as necessary for the blue precipitate to become apparent. Different enhancer trap lines vary widely in the length of time required. For example, staining can be observed in PZ1310 (*slbo^1*) homozygotes after only 10 minutes (Montell et al. 1992), whereas other lines require overnight staining.

Genetic Methods

There are several different genetic approaches one can take to analyze the effects of mutations on *Drosophila* oogenesis. Oogenesis has been primarily

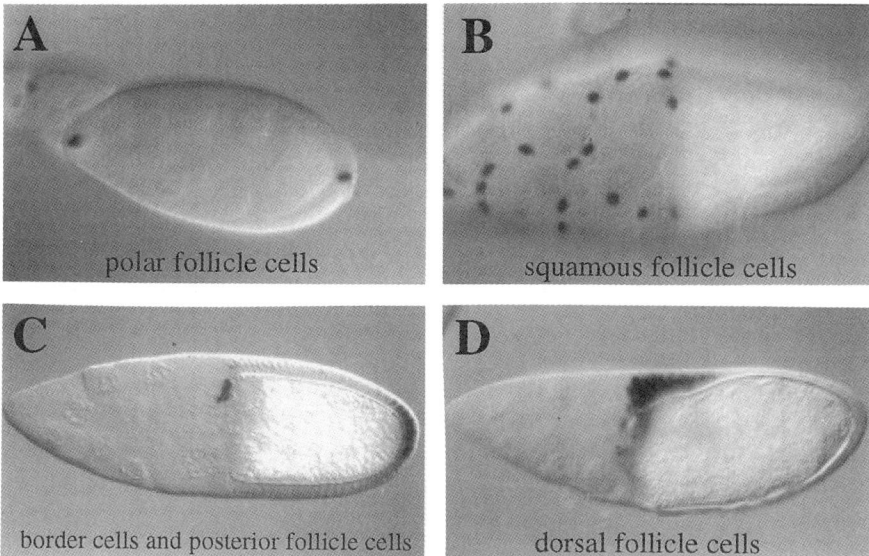

Figure 21-4 β-GAL activity staining in egg chambers from four PZ enhancer trap lines. Nomarski optics images of stained egg chambers. In all cases, anterior is to the left and dorsal, if known, is up. The staining is nuclear because β-GAL is expressed as a fusion protein which contains a nuclear localization sequence. (A) Stage 8 egg chamber from line PZ80, an insertion into the *fasciclin III* locus (D. J. Montell and A. C. Spradling, unpublished results) in which two cells at each end of the egg chamber are labeled; (B) Line PZ311, an insertion into an unknown gene (D. J. Montell and A. C. Spradling, unpublished results) expresses β-GAL in the squamous nurse-cell-associated follicle cells; (C) Staining of border cells (anterior cluster) and posterior follicle cells from enhancer trap line PZ0617, an insertion into the *torso-like* locus (Savant-Bhonsale and Montell 1993); (D) Stage 11 egg chamber from enhancer trap line PZ413 (Lee and Montell 1997) showing staining of the dorsal and anterior follicle cells that will give rise to the dorsal appendages.

studied by identifying and characterizing female sterile mutations (Schupbach and Wieschaus 1991). Female sterility can result from complete loss of function of a gene required only for oogenesis, or from partial loss of function of a gene required for other developmental events as well. However, not every locus with a function in oogenesis can be mutated to a female sterile phenotype. The major limitation of studying oogenesis is that it is challenging to study the functions of genes that are also required for earlier developmental events. The following methods have been developed and provide a variety of helpful tools in the study of oogenesis, although in most cases they are not specific to the study of oogenesis.

Enhancer Trapping

Enhancer trapping is a variation of insertional mutagenesis, in which the transposable P-element that is used as the mutagen contains a reporter gene, usually the *lacZ* gene, under the control of a weak, constitutive promoter. The element is mobilized by crossing the flies to a strain which expresses the transposase enzyme. Then, the transposase gene is crossed away and stocks are made, each with the P-element in a new location. If the P-element inserts into or near a gene, the reporter gene within the P-element often comes under the control of neighboring enhancers. Therefore, by staining tissues for β-GAL activity, one can infer the pattern of expression of a nearby gene. This approach has been used widely to identify new genes expressed in interesting patterns, throughout *Drosophila* development. Enhancer trapping is also useful in the analysis of oogenesis, particularly in recognizing functional subsets of follicle cells (see Figure 21-4). The crosses used in a typical enhancer trap experiment are shown in Figure 21-5. Several different enhancer trap vectors have been used successfully (O'Kane and Gehring 1987; Bellen et al. 1989; Bier et al. 1989; Ghysen and O'Kane 1989; Karpen and Spradling 1992; Török et al. 1993), which employ different eye color markers and different plasmids for cloning purposes.

One of the advantages of enhancer trapping over chemical mutagenesis is that an indication of the expression pattern of a gene is available along with the phenotype, prior to cloning the gene. In addition, a molecular tag is inserted into the gene, which facilitates cloning. However, the reporter gene is not always expressed in exactly the same pattern as a nearby gene. Often the enhancer trap expression pattern reports only part of the actual gene expression pattern. Less commonly, the enhancer trap may reflect a combination of expression patterns of more than one gene. Sometimes, the pattern of expression is exactly like that of the neighboring gene but the level of expression indicated by the enhancer trap is not accurate. Furthermore, the P-elements tend to insert into introns and flanking regions, so sometimes the transcribed sequences are actually located tens of kilobases away from the P-element insertion site. Even with these limitations, however, enhancer trap expression patterns have served as useful cell-type markers and have been helpful in identifying new genes of interest.

If an enhancer trap insertion is to be used to clone the corresponding gene, it is advisable to select enhancer traps that cause a mutant phenotype, in addition to exhibiting an interesting expression pattern. This is because the mutant phenotype indicates that the gene has a discernible function and, second, that the P-element is usually (but not always) in closer proximity to transcribed sequences when a phenotype is apparent. Prior to attempting to clone the gene, it is important to confirm that there is only one P-element in the stock and to confirm that the phenotype of interest is actually caused by the P-element insertion, rather than by some other aberration on the chromosome.

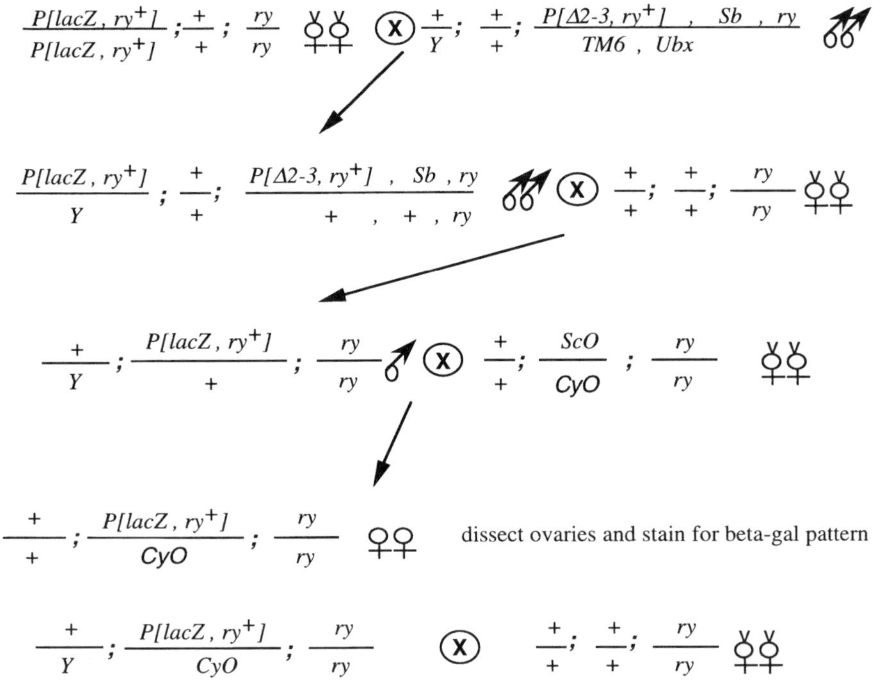

$$\frac{P[lacZ, ry^+]}{P[lacZ, ry^+]} ; \frac{+}{+} ; \frac{ry}{ry} \; \female\female \quad \text{(X)} \quad \frac{+}{Y} ; \frac{+}{+} ; \frac{P[\Delta 2\text{-}3, ry^+] \;,\; Sb \;,\; ry}{TM6 \;,\; Ubx} \; \male\male$$

$$\frac{P[lacZ, ry^+]}{Y} ; \frac{+}{+} ; \frac{P[\Delta 2\text{-}3, ry^+] \;,\; Sb \;,\; ry}{+ \;,\; + \;,\; ry} \; \male\male \quad \text{(X)} \quad \frac{+}{+} ; \frac{+}{+} ; \frac{ry}{ry} \; \female\female$$

$$\frac{+}{Y} ; \frac{P[lacZ, ry^+]}{+} ; \frac{ry}{ry} \; \male \quad \text{(X)} \quad \frac{+}{+} ; \frac{ScO}{CyO} ; \frac{ry}{ry} \; \female\female$$

$$\frac{+}{+} ; \frac{P[lacZ, ry^+]}{CyO} ; \frac{ry}{ry} \; \female\female \quad \text{dissect ovaries and stain for beta-gal pattern}$$

$$\frac{+}{Y} ; \frac{P[lacZ, ry^+]}{CyO} ; \frac{ry}{ry} \quad \text{(X)} \quad \frac{+}{+} ; \frac{+}{+} ; \frac{ry}{ry} \; \female\female$$

If all ry$^+$ flies are Cy$^+$ and all CyO flies are ry$^-$, then the P-element is inserted on the second chromosome

If Cy flies are a mixture of ry and ry$^+$, then the P-element is not on the second and is probably on the third.

Figure 21-5 Crosses carried out in an enhancer trap mutagenesis screen. P[*lacZ,ry*$^+$] indicates the enhancer trap P-element described in Karpen and Spradling (1992), which contains the gene encoding a transposase/β-GAL protein with a nuclear localization signal, as well as the ry+ selectable eye color marker. P[*Δ2-3,ry*$^+$] is a P-element encoding functional transposase but with damaged P-element ends such that it cannot itself transpose (Robertson et al. 1988). The same chromosome is marked with the dominant bristle marker Sb. Please see Lindsley and Zimm (1992) for details of balancer stocks such as TM6 and CyO and other genetic markers such as ScO. The basics of carrying out fly crosses are described in Grigiatti et al. (1986). Semicolons are used to separate loci on different chromosomes whereas commas separate loci on the same chromosome. Progeny of the third cross can be dissected and the ovaries stained for β-GAL activity to determine the expression pattern of the new insertion. Although the insertion is shown on the second chromosome, further crosses are actually required to determine onto which autosome the element has inserted. To analyze the recessive phenotype of new insertions, balanced males and virgin females are crossed and progeny lacking the balancer are analyzed phenotypically. If no progeny lacking the balancer are obtained, then the chromosome with the insertion carries a lethal mutation which may or may not be due to the P-element.

There are three ways to clone the DNA flanking an enhancer trap insertion: plasmid rescue, PCR, and making a library from the P-element-containing line. In all cases, it is necessary first to prepare genomic DNA from adult flies. This is done by collecting up to 40 anesthetized flies into an Eppendorf tube and freezing them to immobilize them. Then, add 10 µl of homogenization buffer, at room temperature, for every fly, up to 400 µl. Crush the flies as much as possible using a pestle that fits in the microfuge tube. Add an equal volume of DNA extraction buffer and vortex briefly. Incubate at 65°C for 30 minutes. Spin in a microfuge for 1 minute at top speed to pellet the debris. Remove the supernatant and transfer to a fresh tube. Add 190 µl of 5 M potassium acetate and mix. Incubate on ice for 15 minutes (longer is okay). Centrifuge for 5 minutes. Collect 750 µl of supernatant and transfer to a fresh tube. Add 750 µl of absolute ethanol and mix gently. Let sit at room temperature for 5 minutes. Centrifuge for 5 minutes. Discard supernatant. Add 500 µl of 70% ethanol to the pellet and remove (this rinse is to remove residual salt). Dry the pellet in a speed vac for 5 minutes. Resuspend in 360 µl of TE and pipette up and down until the DNA is dissolved. Add 40 µl of 3 M sodium acetate and vortex briefly. Add 1 ml of absolute ethanol and mix gently until the DNA precipitates. Spin in a microfuge for 5 minutes. Remove the supernatant and discard. Dry the pellet in a speed vac for 5 minutes. Resuspend the DNA in 100 µl of TE, pipetting up and down to be sure that the DNA is dissolved.

To clone flanking DNA by plasmid rescue, prepare genomic DNA from flies and assume 0.3 µg of DNA per fly. Digest 1 µg (~3 fly equivalents or f.e.) of DNA in a 30 µl reaction mix with one or more restriction enzymes that cut once in the P-element. Phenol-extract and ethanol-precipitate the DNA. Dissolve DNA in 60 µl of TE (1 f.e./30 µl, assuming 70% recovery). Use 20 µl (0.67 f.e.) in a 200-µl ligation reaction at 15°C overnight. Phenol-extract and ethanol-precipitate the DNA. Resuspend in 5 µl of 0.1× TE and use 1 µl (~0.5 f.e.) to transform *Escherichia coli*. A transformation efficiency of at least 10^7 colonies per microgram of DNA is necessary in order to get colonies. Electroporation is the transformation procedure of choice because one almost always recovers colonies using this method. Standard calcium chloride preparation of competent cells is successful sometimes, but not as reliably. Spread the transformation on standard L broth plates with ampicillin (100 µg/ml), or kanamycin (50 µg/ml), depending on the enhancer trap vector in use.

To amplify flanking DNA sequences using the polymerase chain reaction (PCR), digest 1 f.e. of DNA with a restriction enzyme that cuts within the P-element just inside the 5′-P-end. For the PZ element, appropriate enzymes are EcoR1, Hpa II, and Sau3a. Phenol extract and ethanol-precipitate the DNA. Ligate DNA in a 100 µl reaction, using 1 µl of T4 ligase overnight (15°C). Ethanol-precipitate DNA (phenol-extraction is optional). Resuspend in 10 µl of TE. (*Optional*) Digest 7 µl of DNA with Ase1 in a 10 µl reaction at 37°C for 1 hour or longer. This enzyme cuts in between the two primer sites, linearizing the template. Some fragments seem to amplify better as linear templates, others as circles. Use 5 µl in the subsequent PCR reaction containing the following:

5 µl DNA
1.5 µl 10 mM each of dACGT
25 µl 2× PCR buffer (4 mM $MgCl_2$; 50 mM KCl; 200 µg/ml gelatin; 10 mM Tris-HCl, pH 8.4)
2 µl 0.1 mg/ml P32 primer (5′-GTA TAC TTC GGT AAG CTT CGG CTA T-3′)
2 µl 0.1 mg/ml P561 primer (5′-CGA AT GCG TCG TTT AGA GCA GCA G-3′)
13.5 µl H_2O
1 µl Taq 1 polymerase

Add 25 µl paraffin oil and spin briefly in a microfuge in a cold-room. Denature the sample by heating above 94°C for 5 minutes. Then, carry out 30 cycles of 1 minute at 94°C, 1 minute at 72°C, and 1 minute at 55°C. Amplified fragments can be cloned as EcoR1–HindIII fragments because the HindIII site is within the P32 primer and the EcoR1 site is a P-element sequence included in the amplified fragment. Alternatively, PCR fragments can be labeled and used directly to probe wild-type genomic libraries, according to standard procedures (Sambrook et al. 1989).

If neither of the above methods is successful, as a last resort, one can make a genomic library from the P-element containing DNA and then probe it with the 5′-P-end. The simplest way to do this is to digest genomic DNA to completion with EcoR1 and clone it into a prepared vector such as λZAP (Stratagene, La Jolla, CA), according to the manufacturer's instructions.

Mosaic Analysis

Another approach to learning the function of a gene in oogenesis is to employ mosaic analysis. Using this experimental approach, patches of homozygous mutant cells can be analyzed in an otherwise heterozygous animal. Adult flies that are heterozygous for a mutation of interest (m), which can even be a recessive lethal mutation, are treated with an agent that induces mitotic recombination. (Recombination is normally restricted to meiosis.) Mitotic recombination can produce one daughter cell that is homozygous for m and one daughter cell that is homozygous for the wild-type allele, as shown in Figure 21-6. Subsequent cell divisions produce a clone of cells homozygous for m. Elegant methods for producing and detecting mosaic clones in germ-line and follicle cells of the *Drosophila* ovary have been established.

Mitotic recombination can be induced either by treating organisms with X-rays or γ-rays, or by taking advantage of specific DNA sequences (FRTs) and the enzyme that recognizes them (FLP) to generate site-specific mitotic

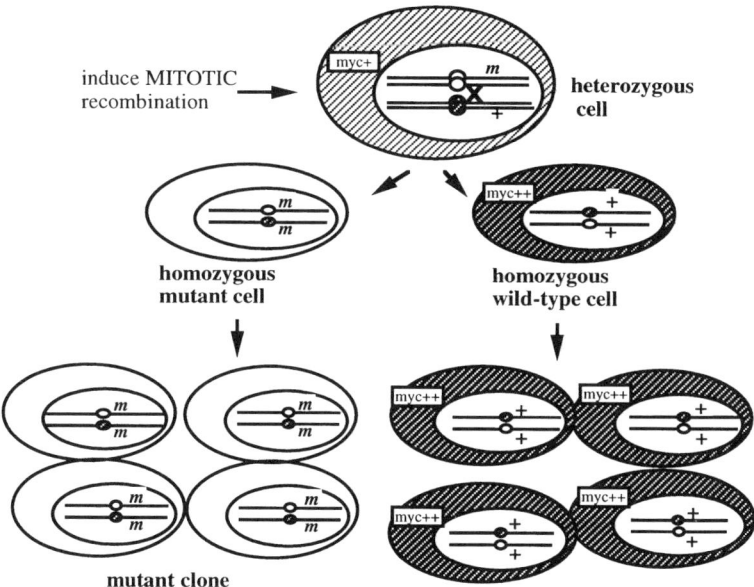

Figure 21-6 Schematic diagram of clones produced by mitotic recombination. At the top, a cell is shown which is heterozygous for the mutation to be analyzed (m) and a wild-type allele ($+$). Since the organism is heterozygous, all cells start out this way. This cell is shown after DNA replication has occurred in preparation for cell division and the sister chromatids are shown right next to each other. The two centromeres are colored differently in order to facilitate following the products of recombination. Following segregation of the chromsomes, each cell gets one of each type of centromere, one daughter cell acquires two mutant alleles and the other daughter cell acquires two wild-type alleles. Following further cell divisions, a clone of mutant cells and a clone of wild-type cells result. If the chromosome arm with the wild-type allele also contains a gene that codes for a protein with a myc-epitope tag, then the heterozygous cell will stain with an antibody against c-myc (will be myc$^+$, indicated by the lighter hatching) and the homozygous wild-type cell will stain twice as intensely (will be myc^{++}, indicated by the dark hatching). However, the homozygous mutant clone will be myc$^-$ (no hatching). See text for details.

recombination (Golic and Lindquist 1989; Xu and Harrison 1994). The frequency of X-ray-induced mitotic recombination is quite low, whereas very high frequencies can be achieved with the FLP/FRT system. In either case, it is important to have a simple and reliable way of detecting the homozygous mutant clones, so that any phenotype that is observed can be correlated with the genotype of the cells.

Analyzing Germ-line Clones Using the Dominant Female Sterile Technique The "dominant female sterile" technique is an elegant genetic technique for creating and marking germ-line clones and is outlined in Figure 21-7. The

A

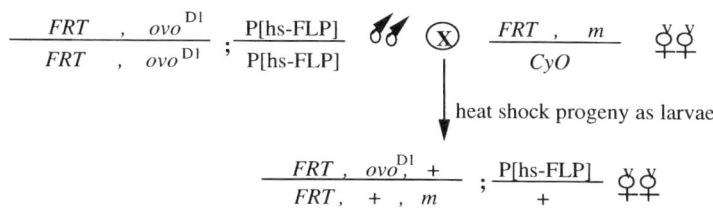

B

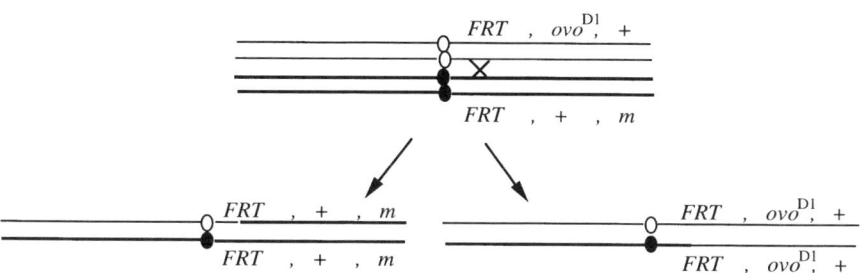

Figure 21-7 Making germ-line clones using the dominant female sterile technique. (A) Cross necessary to generate the genotype which can be used to make clones. Progeny should be heat shocked as larvae in order to induce the FLP enzyme which catalyzes recombination specifically at the FRT DNA sequences. The dominant female sterile mutation ovo^{D1} arrests oogenesis if it is present in the germ line either in one or two copies. Therefore, the only eggs that can be laid are those that have resulted from egg chambers that lack the ovo^{D1} mutation because of a mitotic recombination event. Therefore, all mature eggs are also homozygous for the mutation of interest (m) and the phenotype can be analyzed. (B) Schematic diagram of the chromosomes undergoing mitotic recombination and the products of that event.

rationale is to start with female flies that are transheterozygous for a mutation of interest (m) and a dominant female sterile mutation, usually ovo^{D1}, which causes an early arrest of oogenesis. If mitotic recombination takes place between the chromosome carrying ovo^{D1} and the chromosome carrying m, then mature egg chambers can result only if the germ-line has lost the ovo^{D1} mutation and is therefore homozygous for m. The only exception is if the recombination event takes place between ovo^{D1} and m when m is proximal to ovo. In this case, egg chambers wild-type for both loci can result. Thus, an added advantage of the FLP/FRT system is that the site of recombination is fixed; therefore, selection of an FRT proximal to both mutations avoids this ambiguity entirely and all mature egg chambers can be assumed to be homozygous for m. It is important to note that the dominant female sterility of ovo^{D1} is autonomous to the germ line cells; thus, egg chambers will only develop if the germ-line has lost the ovo^{D1} allele. The genotype of the follicle cells cannot be discerned in this experiment.

The dominant female sterile technique as originally described was actually used to determine the germ-line requirements for all X chromosome lethal mutations (Perrimon et al. 1986). However, the technique had two limitations: it was limited to the X chromosome, where the endogenous *ovo* locus resides; and gene functions could only be analyzed in the germ line. The first limitation has been overcome by cloning the *ovo* locus from the ovo^{D1} mutant and making transgenic fly lines that have insertions of the ovo^{D1} mutant transgene on each of the autosomes (Chou et al. 1993; Mevel-Ninio et al. 1994). Therefore, the dominant female sterile approach can now be used to make clones for mutations on any one of the chromosome arms, including the tiny fourth chromosome. In addition, this approach has been made more efficient by placing FRT sites at the base of each of the chromosome arms containing the ovo^{D1} transgene. By inducing expression of the FLP site-specific recombinase, high frequencies of mitotic recombination can be induced (Chou and Perrimon 1992, 1996).

To make mosaic clones using the FLP/DFS technique, first recombine your mutation of interest onto a chromosome with an FRT on the same chromosome arm (these stocks are available from N. Perrimon, Harvard University). Then cross these flies to a stock with the FRT and ovo^{D1} on the same chromsome arm, as well as a heat-inducible FLP transgene on another chromosome (*hs-FLP*). To induce recombination, heat shock the progeny of this cross during the third instar larval stage (5–6 days after egg laying). When the adults emerge, collect the females of the appropriate genotype with males of any genotype, feed them yeast for a day or two, and then collect embryos and analyze them for phenotypic effects. This procedure is outlined in Figure 21-7.

Analyzing Follicle Cell Clones To identify and characterize mitotic clones in the follicle cells, several approaches are available, though none is quite as elegant as the dominant female sterile approach. X-ray treatment and the FLP/FRT system are both suitable for generating mosaic clones in follicle cells, just as for the germ line. The difference is that there is no known dominant female sterile mutation, autonomous to the follicle cells, that prevents egg chamber development. However, several other ways to mark follicle cell clones have been developed.

To analyze the somatic maternal effect of a particular mutation on embryonic development, a mutation that causes eggshell defects, such as the *fragile chorion* (*fch*) locus on the third chromosome, can serve as a marker for the mutant clones, as shown in Figure 21-8. This approach was used to analyze whether relatively small mutant clones in the follicle cells were sufficient to cause a *torso-like* (*tsl*) phenotype in the embryo

A
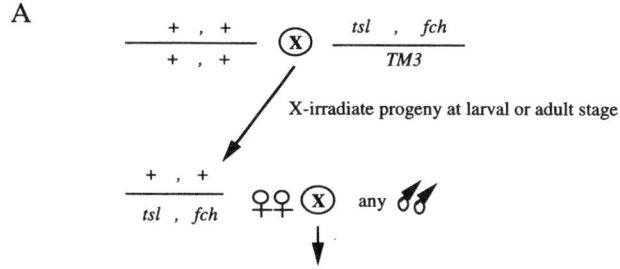

collect embryos and analyze for chorion phenotype and for maternal effect

B
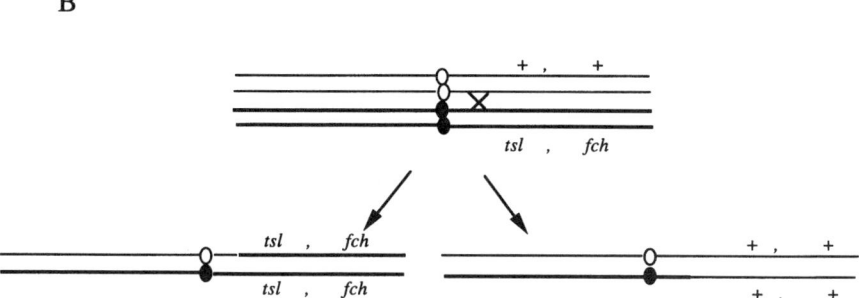

Figure 21-8 Making clones in follicle cells using irradiation. (A) Cross necessary to produce the genotype in which mosaic clones can be generated. In this example, the *fragile chorion* (*fch*) mutation is used to mark the mosaic clones. Since the *fch* mutation is on the same chromosome arm as the *torso-like* (*tsl*) mutation, when mitotic recombination takes place the mutant clones will be mutant for both loci. By examining the embryos for patches of fragile chorion and then correlating that with their cuticle phenotype, Stevens et al. (1990) were able to ascertain that small clones of follicle cells at either the anterior or posterior end of the egg chamber produced local disruptions of terminal pattern formation. Progeny of the cross should be irradiated with 1250 rads, using a calibrated source of X-rays or γ-rays, at the third instar larval stage. Irradiation at the adult stage can also produce clones, whose size and frequency will depend on the time interval between irradiation and analysis. (B) Schematic diagram of the recombination event and its products.

(Stevens et al. 1990). In this case, the *fch* marker mutation was recombined onto the same chromosome arm as the *tsl* mutation. Then, these flies were crossed to wild-type flies and the progeny were X-irradiated (1250 rads), at either the larval or adult stage, and eggs were collected from the treated females and allowed to develop. These embryos were then analyzed one by one for the phenotypic effects of *fch* and patterning defects associated with *tsl*. This approach was successful; however, there were two severe limitations to it. First, the low frequency of mosaic clones induced by X-rays meant that over 10,000 embryos had to be examined in order to find seven mosaic clones. Furthermore, scoring the *fch* marker restricts this analysis to maternal effects on embryonic development and precludes analyzing earlier events in oogenesis. The advantage of this approach is that it is simple: all one needs are two fly stocks and a calibrated source of X-rays or γ-rays.

Alternatively, if one wishes to analyze the effects of a particular mutation on earlier events in oogenesis, then one can use an immunocytochemical marker for the follicle cells. For example, if the chromosome arm that is wild-type for m (i.e., m^+) contains a transgene that drives the expression of a myc-epitope-tagged protein, then, following recombination, the homozygous mutant cells will fail to label with an anti-myc monoclonal antibody, whereas the heterozygous and homozygous wild-type cells will label, as illustrated in Figures 21-6 and 21-9 (Xu and Harrison 1994). After setting up the appropriate crosses, flies should be heat shocked once to induce the expression of FLP. Then the flies should be put on fresh food every day for 1–5 days to allow stem cells that might have undergone recombination to produce egg chambers. The shorter the interval between heat shock and analysis, the smaller and more frequent the clones will be. Conversley, longer periods will produce fewer, larger clones. The flies must be put on fresh food every day because if they are not, they will stop laying eggs and oogenesis will not continue at the normal pace, preventing the mutant cells from producing clones. When using myc-staining to mark clones, it is necessary to heat shock the flies a second time, just prior to ovary dissection, in order to induce expression of the myc-epitope-tagged protein. Best results are achieved by heat shocking for 40 minutes to 1 hour and then allowing 1 hour at 25°C prior to dissection.

A disadvantage of the myc-staining method for marking clones is that, since the clones fail to stain, one must have very reliable and uniform staining of many egg chambers to be confident that an absence of staining is indicative of a true clone rather than a technical artifact. One way to be sure is to double-label the egg chambers with a second antibody that is expected to label all of the cells of interest, regardless of genotype. Then, look for cells that label for this marker but fail to label for myc. Another alternative would be to use a transgene expressing GFP or a GFP fusion protein under the control of a ubiquitous promoter to mark clones. The advantage of GFP compared to myc is that one avoids the added steps of a staining procedure; however, the FRT, hs-myc stocks already exist and are available from Tian Xu (Yale University School of Medicine) and Gerry Rubin (University of California, Berkeley).

Another method for marking clones gets around this problem by creating clones that express β-GAL. Harrison and Perrimon (1993) created a transgene that contains a tubulin promoter followed by an FRT sequence. A second transgene, inserted at the identical position in the genome, has an FRT sequence upstream of β-GAL coding sequences. When recombination takes place between the two FRT sites, the tubulin promoter is brought into proximity to the *lacZ* gene and expression of β-GAL results. Thus, only recombined chromosomes express *lacZ*, as diagrammed in Figure 21-10. This method was employed to investigate cell lineages and stem cell beha-

Drosophila Oogenesis: Cell Biological and Genetic Study Methods

A

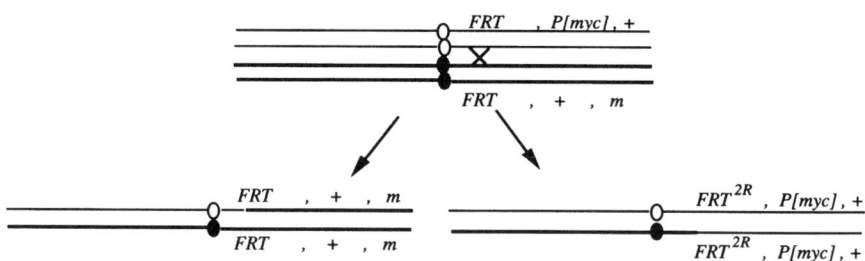

B

Figure 21-9 Making clones using a myc-tagged transgene as the marker. (A) Cross necessary to generate the genotype for inducing clones. P[*myc*] indicates a transgene on a P-element that expresses a heat-inducible fusion protein with a myc-epitope tag. FRT indicates the FRT sequence, proximal to both the marker and the mutation of interest (*m*) on the same chromosome arm. P[*hs-FLP*] indicates a transgene on another chromosome which encodes the FLP recombinase under the control of the hsp70 heat-inducible promoter. Progeny of the cross should be heat shocked at the third instar larval stage. (B) Schematic diagram of the recombination event and its products.

vior in the ovary (Margolis and Spradling 1995), i.e., for following which groups of cells result from a common precursor cell. In principle, the method can also be used to analyze the effects of mutations. By starting with a mutation of interest distal to the P[*FRT-lacZ*] transgene, recombination can generate homozygous mutant clones that stain for β-GAL. However, it is important to note that one-half of the stained clones will be heterozygous, not homozygous for the mutation of interest. So while all homozygous mutant clones should express β-GAL, some stained clones are expected not to display a mutant phenotype. This introduces an uncertainty that must be kept in mind. If, for example, the mutation of interest impedes further cell division or interferes with cell survival, then one could end up with all of the β-GAL-expressing cells being heterozygous and one might erroneously conclude that the mutation did not have a dramatic effect.

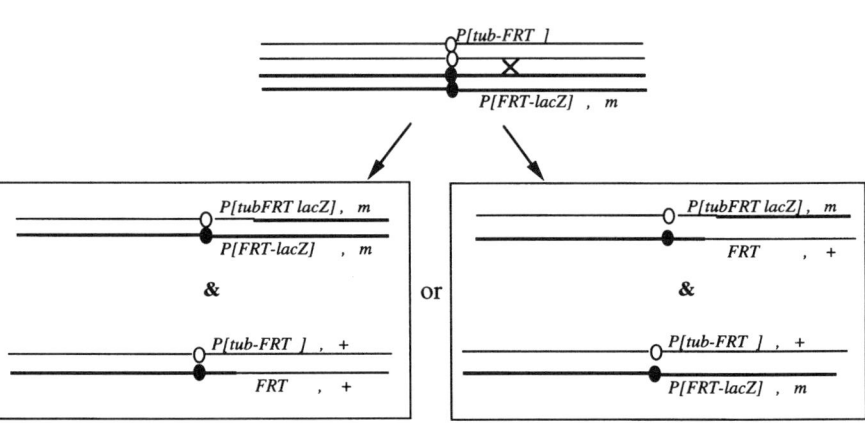

Figure 21-10 Making clones that express β-GAL. (A) Cross necessary to generate the genotype for inducing clones. P[*tub-FRT*] indicates a P-element containing a tubulin promoter upstream of an FRT. P[*hs-FLP*] indicates a transgene on another chromosome which encodes the FLP recombinase under the control of the hsp70 heat-inducible promoter. P[*FRT-lacZ*] indicates a P-element inserted at the identical position as P[*tub-FRT*], but which contains a promoterless *lacZ* gene downstream of the FRT sequence. Recombination between the two FRTs results in a functional *lacZ* gene and expression of β-GAL. (B) Schematic of the recombination event and its products. Note that there are two equally likely ways for the recombinant chromosomes to segregate. In one case, the cells expressing β-GAL are also homozygous for *m*; but in the other case, the cells expressing β-GAL are only heterozygous for *m*. Thus, β-GAL is not an unambiguous marker for the mutant clones.

Transgenic Lines Expressing Dominant Negative Constructs

A third method for analyzing gene function in the ovary is to express dominant-negative or constitutively active forms of particular proteins, an approach that has been used to study the roles of 21-kDa G proteins in oogenesis (Lee et al. 1996; Murphy and Montell 1996). This is accomplished by making transgenic fly lines expressing the construct of interest under the control of inducible and/or ovarian specific enhancers and promoters, using standard P-element-mediated germ-line transformation. A list of useful promoters for this approach is found in Table 21-1.

To induce expression of transgenes from heat-inducible promoters, female flies of the desired genotype should be placed on fresh food supplemented with baker's yeast for 24–36 hours at room temperature before treatment. Transfer females to fresh food without yeast. Prepare a 37°C water bath. Place fly vials in a test tube rack (those designed to hold 50-ml centrifuge tubes are the right size). Secure the vials by wrapping a large rubber band around them. This will prevent them from floating in the water bath. Put the rack in the water bath, taking care that the water level reaches the bottom of the cotton plug in the vial. Incubate for 10–60 minutes, depending on the degree of induction desired. In general, constitutively active proteins seem to be quite toxic and shorter heat shocks produce more interpretable phenotypes. By contrast, dominant-negative constructs must be expressed at levels higher than the endogenous protein to be effective therefore, 1 hour heat shocks are typically employed. Heat shocks for longer than 1 hour will produce heat shock artifacts and will result in poor viability. The water bath temperature is critical: if it is too high, the flies will not survive; if it is too low, poor induction will result. So be sure to use an accurate thermometer, or, better yet, several thermometers and heat up the water bath well in advance to avoid temperature fluctuations during the heat shock.

Isolation of Nuclei from *Drosophila* Ovaries

Methods for biochemical fractionation of ovaries are not as fully developed as those for cell biological and genetic approaches. However, nuclei can be purified, e.g., in preparation for fluorescence-activated sorting. Treat all glassware and plasticware with bovine serum albumin (BSA) to prevent sticking of egg chambers and nuclei. Dissect ovaries in EBR. Treat with 5 mg/ml collagenase in EBR for 15 minutes. After incubation is complete, rinse ovaries with EBR to remove residual collagenase. Homogenize ovaries in nuclear isolation buffer (15 mM Tris-HCl, pH 7.4, 60 mM KCl, 15 mm NaCl, 250 mM sucrose, 1 mM EDTA, 0.1 mM EGTA, 0.15 mM spermine, 0.5 mM spermidine) and 1.5% NP40, using a 2-ml Kontes dounce with the B pestle. Filter the homogenate 2× through a 100 μm Nitex mesh. Place the homogenate on a sucrose step gradient of 2.5 M, 1.6 M, and 0.8 M sucrose and centrifuge for 20 minutes at 20,000g. (Adding a small amount of DAPI to the homogenate before placing it on the sucrose gradient allows one to follow the pelleted nuclei using a UV light source.) Remove the sucrose gradient and resuspend the pellet in nuclear isolation buffer. Add DAPI to a concentration of 0.5 μg/ml. Keep nuclei on ice before sorting.

Table 21-1 Promoters that drive gene expression in *Drosophila* oogenesis

Gene/vector name	Expression pattern in the ovary	Expression elsewhere	References
Nanos	Nurse cells from stage 6	None	Gavis and Lehmann 1992
torso-like	Anterior and posterior follicle cells from stage 2	Embryonic expression	Savant-Bhonsale and Montell 1993
otu/pcog	Germ-line cells from stem cell (expression level variable)	None	Robinson and Cooley 1997
hsp26,K10/pGerm	High-level, constitutive, germ-line expression not heat inducible	None	Serano et al. 1994
hsp70/pCaSpeR-hs	Low-level leaky expression; inducible expression in germ line through stage 9; in follicle cells, throughout oogenesis	Low-level leaky expression; high-level inducible expression in most tissues	Murphy and Montell 1996
GAL4 enhancer traps	A variety of patterns	?	Manseau et al. 1997

Culturing Egg Chambers

In Vitro Culture

In vitro culture of *Drosophila* egg chambers can be useful in some circumstances. While egg chambers can be kept alive in culture for short periods, it is unfortunate that no medium has been defined that will support full egg chamber development at most stages. Schneider's medium supplemented with 10% fetal calf serum is suitable to keep all stages of egg chambers alive for 1–2 two hours at room temperature. However, prolonged incubation will result in death of the cells and a failure of the egg chamber to develop. Therefore, in vitro culture is typically only used in order to carry out manipulations such as BrdU labeling (see above) or laser ablations of particular cells (Montell et al. 1991; Lin and Spradling 1993; Lilly and Spradling 1996). To obtain further development, the egg chambers must then be injected into a host female and cultured in vivo (see below).

Stage 10 and older egg chambers can be cultured to maturity in vitro. The percentage of egg chambers developing normally increases with increasing age of the starting egg chamber, such that only 50% of stage 10A egg chambers develop normally in vitro, whereas 70% of stage 10B egg chambers develop normally.

In Vivo Culture

Development of all stages of egg chambers can occur normally following short-term in vitro culture, if egg chambers are subsequently injected into the abdomen of a host female and cultured in vivo (Gutzeit and Koppa 1982; Montell et al. 1991; Lin and Spradling 1993). It is convenient to use ovo^{D1} females as the host in such experiments because this dominant female sterile mutation prevents the host from making mature egg chambers of its own. Therefore, there is more space in the abdomen into which one can inject egg chambers, and when one dissects the host to recover the injected egg chambers following culture, the only mature egg chambers in an ovo^{D1} female are those that were injected.

The success of injections is critically dependent on the injection needle. The needle can be pulled out to a fine point manually over a Bunsen burner flame. Then use a forcep to break the tip of the needle at the desired thickness so that the opening is just big enough to accommodate the egg chamber but no larger than necessary. Try to break the needle such that there is a bevel at the tip. Attach rubber tubing to a needle holder and attach a 50-ml syringe to the other end of the tubing. Pulling on the syringe will suck air or liquid into the needle. Pushing on the syringe plunger will then expel the air or liquid. One can attach this apparatus to a ringstand with clamps if desired. The basic egg chamber injection procedure is as described below.

Etherize ovo^{D1} females for 5 minutes in a standard etherizer with approximately 50 µl of anhydrous ether added to the cotton. Place a piece of double-sided tape on a microscope slide. Line up the host females on the tape on their backs, sticking their wings down. If anesthetized properly, the females will remain asleep for approximately 10 minutes. Dissect the egg chambers to be injected, into ovarioles, using the plucking method. Draw some culture medium up and down in the needle a few times to coat the needle with protein and prevent sticking of the egg chambers inside the needle. Draw up an ovariole into the needle, keeping the ovariole near the tip. Puncture the abdomen of the host female (with the bevel of the needle facing upward) and inject the ovariole into the abdomen. If successful, immediately transfer the female into a fresh vial with a small piece of Kimwipe inside. Place the female on the Kimwipe, so it does not get stuck in the food. Culture injected flies at 18°C, room temperature, or 25°C for the desired period of time. Then, dissect open the female and recover the developed egg chamber.

ACKNOWLEDGEMENTS   I thank Allan Spradling for my introduction to *Drosophila* oogenesis and the techniques used to study it. I thank Janice Crawford, Dan Kiehart, Mary Lilly, and Lynn Cooley for providing updated protocols for inclusion in this chapter. Thanks also to Lynn Cooley for providing the anti-HTS antibody. Thanks to Celeste Berg, Gary Karpen, Dennis McKearin, and members of my lab for valuable insights into these techniques.

References

Ashburner, M. (1989). *Drosophila, a Laboratory Handbook*. Cold Spring Harbor, NY, Cold Spring Harbor Laboratory Press.

Bellen, H. J., O'Kane, C. J., Wilson, C., Grossniklaus, U., Pearson, R. K., and Gehring, W. J. (1989). P-element mediated enhancer detection: a versatile method to study development in Drosophila. Genes Dev. 3:1288–1300.

Bier, E., Vaessin, H., Shepherd, S., Lee, K., McCall, K., Barbel, S., Jan, L. Y., and Jan, Y. N. (1989). Searcing for pattern and mutation in the *Drosophila* genome with a *P-lacZ* vector. Genes Dev. 3:1288–1300.

Chou, T. B., and Perrimon, N. (1992). Use of a yeast site-specific recombinase to produce female germline chimeras in Drosophila. Genetics 131:643–653.

Chou, T. B. and Perrimon, N. (1996). The autosomal FLP-DFS technique for generating germline mosaics in *Drosophila melanogaster*. Genetics 144:1673–1679.

Chou, T. B., Noll, E., and Perrimon, N. (1993). Autosomal P[ovoD1] dominant female sterile insertions in Drosophila and their use in generating germ-line chimeras. Development 119:1359–1369.

Edwards, K. A., and Kiehart, D. P. (1996). Drosophila nonmuscle myosin II has multiple essential roles in imaginal disc and egg chamber morphogenesis. Development 122:1499–1511.

Gavis, E. R., and Lehmann, R. (1992). Localization of nanos RNA controls embyonic polarity. Cell 71:301–313.

Ghysen, A., and O'Kane, C. J. (1989). Neural enhancer-like elements as specific cell markers in Drosophila. Development 105:35–52.

Golic, K. G., and Lindquist, S. (1989). The FLP recombinase of yeast catalyzes site-specific recombination in the Drosophila genome. Cell 59:499–509.

Grigiatti, T., Hafen, E., Johnston, P., Jowett, T., Kidwell, M. G., Lawrence, P. A., Levine, M., Morata, G., Nusslein-Volhard, C., Pardue, M. L., Pirrotta, V., Roberts, D. B., Santamaria, P., Spradling, A. C., Wieschaus, E., and Wilcox, M. (1986). *Drosophila A Practical Approach*. Oxford, IRL Press.

Gutzeit, H. O., and Koppa, R. (1982). Time-lapse film analysis of cytoplasmic streaming during late oogenesis of Drosophila. JEEM 67:101–111.

Harlow, E. and Lane, D. (1988). *Antibodies A Laboratory Manual*, Cold Spring Harbor, NY, Cold Spring Harbor Laboratory Press.

Harrison, D. A., and Perrimon, N. (1993). Simple and efficient generation of marked clones in *Drosophila*. Current Biol. 3:424–433.

Karpen, G., and Spradling, A. C. (1992). Analysis of subtelomeric heterochromatin in a Drosophila minichromosome Dp1187 by single P-element insertional mutagenesis. Genetics 132:737–753.

King, R. C. (1970). *Ovarian Development in Drosophila melanogaster*. New York, Academic Press.

Lee, T., and Montell, D. J. (1997). Multiple Ras signals pattern the *Drosophila* ovarian follicle cells. Dev. Biol. 185:25–33.

Lee, T., Feig, L., and Montell, D. J. (1996). Two distinct roles for Ras in a developmentally regulated cell migration. Development 122:409–418.

Lilly, M. A. and Spradling, A. C. (1996). The Drosophila endocycle is controlled by Cyclin E and lacks a checkpoint ensuring S-phase completion. Genes Dev. 10:2514–2526.

Lin, H., and Spradling, A. C. (1993). Germline stem cell division and egg chamber development in transplanted *Drosophila* germaria. Dev. Biol. 159(1):140–152.

Lindsley, D. L. and Zimm, G. G. (1992). *The Genome of Drosophila melanogaster*. San Diego, Academic Press.

Manseau, L., Baradaran, A., Brower, D., Budhu, A., Elefant, F., Phan, H., Philp, A. V., Yang, M., Glover, D., Kaiser, D., Plater, K., and Selleck, S. (1997). GAL4 Enhancer traps expressed in the embryo, larval brain, imaginal discs, and ovary of Drosophila. Dev. Dyn. 209:310–322.

Margolis, J., and Spradling, A. C. (1995). Identification and behavior of epithelial stem cells in the Drosophila ovary. Development 121:3797–3707.

Mevel-Ninio, M., Guenal, I., and Limbourg-Bouchon, B. (1994). Production of dominant female sterility in *Drosophila melanogaster* by insertion of the ovoD1 allele on autosomes: use of transformed strains to generate germline mosaics. Mech. Dev. 45:155–162.

Montell, D. J., Keshishian, H., and Spradling, A.C. (1991). Laser ablation studies of the role of the Drosophila oocyte nucleus in pattern formation. Science 254:290–293.

Montell, D. J., Rqrth, P., and Spradling, A. C. (1992). *slow border cells*, A locus required for a developmentally regulated cell migration during oogenesis, encodes Drosophila C/EBP. Cell 71:51–62.

Murphy, A. M., and Montell, D. J. (1996). Cell type-specific roles for Cdc42, Rac, and RhoL in *Drosophila* oogenesis. J. Cell Biol. 133:617–630.

O'Kane, C. J., and Gehring, W. J. (1987). Detection in situ of genomic regulatory elements in Drosophila. Proc. Natl. Acad. Sci. USA 84:9123–9127.

Perrimon, N., Mohler, D., Engstrom, L., and Mahowald, A. P. (1986). X-linked female sterile loci in *Drosophila melanogaster*. Genetics 113:695–712.

Robertson, H. M., Preston, C. R., Phillis, R. W., Johnson-Schlitz, D. M., Benz, W. K., and Engels, W. R. (1988). A stable genomic source of P element transposase in *Drosophila melanogaster*. Genetics 118:461–470.

Robinson, D. N., and Cooley, L. (1997). Examination of the function of two kelch proteins generated by stop codon suppression. Development 124:1405–1417.

Robinson, D. N., Cant, K., and Cooley, L. (1994). Morphogenesis of *Drosophila* ovarian ring canals. Development 20:2015–2025.

Sambrook, J., Fritsch, E. F., and Maniatis, T. (1989). *Molecular Cloning: A Laboratory Manual*. Cold Spring Harbor, NY, Cold Spring Harbor Laboratory Press.

Savant-Bhonsale, S., and Montell, D. J. (1993). *torso-like* encodes the localized determinant of Drosophila terminal pattern formation. Genes Dev. 7:2548–2555.

Schupbach, T., and Wieschaus, E. (1991). Female sterile mutations on the second chromosome of *Drosophila melanogaster*. II. mutations blocking oogenesis or altering egg morphology. Genetics 129:1119–1136.

Serano, T. L., Cheung, H. K., Frank, L. H., and Cohen, R. S. (1994). P element transformation vectors for studying *Drosophila melanogaster* oogenesis and early embryogenesis. Gene 138:181–186.

Simon, J. A., Sutton, C. A., Lobell, R. B., Glaser, R. L., and Lis, J. T. (1985). Determinants of heat shock-induced chromosome puffing. Cell 40:305–817.

Spradling, A. C. (1993). Developmental genetics of oogenesis. In *The Development of Drosophila melanogaster*. Bate, M., and Martinez-Arias, A., eds. Cold Spring Harbor, NY, Cold Spring Harbor Laboratory Press, pp. 1–70.

Stevens, L. M., Frohnhoffer, H. G., Klingler, M., and Nusslein-Volhard, C. (1990). Localized requirement for *torso-like* expression in follicle cells for development of terminal anlagen of the Drosophila embryo. Nature 346:660–663.

Török, T., Tick, G., Alvarado, M., and Kiss, I. (1993). *P-lacW* insertional mutagenesis on the second chromosome of *Drosophila melanogaster*: isolation of lethals with different overgrowth phenotypes. Genetics 135:71–80.

Verheyen, E., and Cooley, L. (1994). Looking at oogenesis. Methods Cell Biol. 44:545–561.

Xu, T., and Harrison, S. D. (1994). Mosaic analysis using FLP recombinase. Methods Cell Biol. 44:655–681.

Yeh, E., Gustafson, K., and Boulianne, G. L. (1995). Green fluorescent protein as a vital marker and reporter of gene expression in Drosophila. Proc. Natl. Acad. Sci. USA 92:7036–7040.

Yue, L., and Spradling, A. (1992). *hu-li tai shao*, a gene required for ring canal formation during Drosophila oogenesis, encodes a homolog of adducin. Genes Dev. 6:2443–2454.

Appendix: Reagents and Buffers

Anti-BrdU antibody (Becton Dickinson)
Anti-myc antibody (Santa Cruz Biotechnology, Santa Cruz, CA)
Fluoroscein goat antimouse secondary antibody (Vector Labs, Burlingame, CA)
Collagenase A (Boehringer Mannheim)
4′, γ-diamidino-2-phenylindole (DAPI, Sigma, St. Louis, MO)
Dimethylformamide or dimethylsulfoxide (Baker, Phillipsburg, NJ)
Etherizer (Carolina Biologicals, Burlington, NC)
Fetal calf serum, heat inactivated (Hyclone, Logan, UT)
Formaldehyde (Baker, 37% solution)
Gluteraldehyde (Polysciences, 8% solution)
Grace's medium (Gibco BRL, Grand Island, NY)
Rhodamine-conjugated phalloidin (Molecular Probes, Eugene, OR)
Schneider's medium (Sigma)
Triton X-100 detergent (Sigma)
Vectashield mounting medium (Vector Labs)
X-GAL (5-bromo-4-chloro-3-indolyl-β-D-galactopyranoside, Sigma, St. Louis, MO)

Buffer B
100 mM KH_2PO_4/K_2HPO_4 (pH 6.8)
450 mM KCl
150 mM NaCl
20 mM $MgCl_2 \cdot 6H_2O$

DNA Extraction Buffer
0.1 M NaCl
0.2 M Sucrose

0.1 M Tris-HCl (pH 9.1)
50 mM EDTA
0.5% SDS

Modified Ringer's solution (EBR, Am. Nat. 70:218-225)
10 mM HEPES buffer (pH 6.9)
130 mM NaCl
4.7 mM KCl,
1.9 mM $CaCl_2$

Phosphate-Buffered Saline
20 g/liter NaCl
5 g/liter KCl
5 g/liter KH_2PO_4
27.8 g/liter $Na_2HPO_4 \cdot 2H_2O$

PBT (PBS + 0.1% Triton X-100)

β-Galactosidase (β-GAL) staining solution (Simon et al. 1985)
10 mM $NaH_2PO_4 \cdot H_2/Na_2HPO_4 \cdot 2H_2O$ (pH 7.2)
150 mM NaCl
1.0 mM $MgCl_2 \cdot 6H_2O$,
3.1 mM $K_4[Fe^{II}(CN)_6]$, 3.1 mM $K_3[Fe^{III}(CN)_6]$,
0.3% Triton X-100

22

Techniques for Analyzing Protein and RNA Distribution in *Drosophila* Ovaries and Embryos at Structural and Ultrastructural Resolution

Satoru Kobayashi
Reiko Amikura
Akira Nakamura
Paul Lasko

The fruit fly *Drosophila* is a premier model organism for studying the developmental genetics of oogenesis and embryonic patterning. A key factor in the rapid progress made by researchers using *Drosophila* has been the development of techniques for determining the spatial localization of RNAs and proteins with extremely high precision, high reproducibility, and relative ease. Methods based on in situ hybridization or antibody staining of whole organisms are facilitated in *Drosophila* by the small size of the ovaries and embryos and their permeability to macromolecular probes through most developmental stages. This chapter will provide a set of protocols for determining the localization of RNAs and proteins at the structural and ultrastructural levels in *Drosophila* ovaries and embryos.

The first method developed for determining the localization of low-abundance mRNAs in *Drosophila* embryos involved in situ hybridization to paraffin- or cryostat-sectioned material using radiolabeled DNA or RNA probes (Akam 1983; Hafen et al. 1983, Ingham et al. 1985). While this method proved extremely useful, it has today been almost entirely superseded by the nonradioactive whole-mount in situ hybridization technique using digoxigenin-labeled probes, first developed by Tautz and Pfeifle (1989). The advantages of the nonradioactive method are many. First, results are obtained much faster as the detection method takes no incubation time (for fluorescent detection) or minutes (for alkaline phosphatase detection) rather than the days or weeks required for autoradiographic exposure. Second, the precision of the nonradioactive method is far greater, as it is not compromised by the track lengths of the decaying β-particles traveling through the photographic emulsion, nor by the thickness or graininess of the emulsion. Finally, whole-mount methods, unlike sectioning, allow visualization of all three spatial dimensions in a single sample, and obviate the need for the most laborious aspect of the earlier protocol; namely, preparing high-quality tissue sections. However, for applications requiring precise quantitation of signal, autoradiographic methods remain best.

The whole-mount in situ hybridization method was quickly adapted to *Drosophila* ovaries (Suter et al. 1989, Ephrussi et al. 1991). However, one disadvantage of this technique is that whole stage 11–14 oocytes are poorly permeable to probes and the antibody molecules required to detect them, and therefore are not easily and reproducibly stained. In situ hybridization to thin sections remains the surest way of obtaining information about RNA localization in these developmental stages. Such techniques using digoxigenin-labeled probes on cryostat-sectioned material (Schären-Wiemers and Gerfin-Moser 1993) are commonly used with vertebrate embryos, but have not been published for *Drosophila*. In situ hybridizations using digoxigenin-labeled probes on paraffin-sectioned *Drosophila* embryos have been reported (Kobayashi et al. 1995); this protocol is reproduced here.

Antibody staining protocols in use today are based upon those developed for *Drosophila* embryos by Mitchison and Sedat (1983) and for *Drosophila* ovaries by Brower et al. (1981). A similar permeability problem exists with antibody staining of stage 11–14 oocytes as for in situ hybridizations; to circumvent this, cryostat-cut 10-μm sections can be stained (Hay et al. 1988). A freeze-substitution protocol (Hay et al. 1988; Lasko and Ashburner 1990) has no major advantage over antibody staining of undisrupted tissue in solution (St Johnston et al. 1991; Suter and Steward 1991), and requires larger volumes of antiserum, as well as more work.

Whole-Mount In Situ Hybridization Using Digoxigenin-Labeled RNA Probes

IA. Probe Labeling
1. Subclone the fragment of interest in an in vitro transcription vector such as pBluescript (Stratagene) or pGEM (Promega).
2. Linearize the plasmid DNA (usually 1 μg or more) containing the insert by restriction enzymes that produce 5′-cohesive or blunt ends, or amplify the insert by PCR using primers complementary to the T7 and T3 (or SP6, depending on the vector) promoters. Prepare DNA for the sense-strand negative control at the same time. The plasmid or PCR fragment should be purified by phenol–chloroform extraction and ethanol-precipitation. If you

use a PCR-amplified fragment as the template, carry out the ethanol-precipitation using ammonium acetate.
3. Add the following reagents, in the order listed, to a reaction tube at *room temperature*. (Note: We *do not* treat our water with DEPC; trace amounts of DEPC inhibit RNA polymerase activity.)

   2.0 μl of 10× transcription buffer (Boehringer Mannheim) *or* 4.0 μl of 5× buffer (Gibco BRL)

   2.0 μl of 0.1 M DTT (*only* if the 5× Gibco BRL buffer is used, which lacks DTT)

   Autoclaved deionized and distilled water (to make a final volume of 20.0 μl)

   2.0 μl of DIG NTP labeling mixture (Boehringer Mannheim)

   ≥1.0 μg of linearized DNA

   2.0 μl of RNA polymerase (T7 or T3 or SP6, depending on the promoter)

   1.0 μl of RNase inhibitor [such as RNasin (Promega)]

   Total 20.0 μl

4. Incubate at 37°C for 1–2 hours.

   (Using 1 μg of linearized DNA as a template, more than 5 μg of DIG-labeled RNA should be synthesized, which is enough for about 50 in situ hybridization experiments. The above reaction can easily be scaled down by one-half if convenient.)

5. Add 2 μl of RNase-free DNase (Boehringer Mannheim), and 1 μl of yeast tRNA (10 mg/μl), and incubate at 37°C for 15 minutes. Add 30 μl of double-distilled water and go directly to step 8. Alternatively, some workers carry out the following two steps:

6. Ethanol-precipitate. *Do not* extract with phenol–chloroform!!

7. Dissolve the precipitate in 50 μl of DEPC-treated double-distilled water.

8. To reduce the probe length, add 50 μl of alkaline solution (80 mM $NaHCO_3$, 120 mM Na_2CO_3), and incubate at 60°C for t minutes [t (minutes) is calculated using the following formula: $t = (L_0 - L_f)/(0.11 \times L_0 \times L_f)$; L_0 = initial length of transcript (kb); L_f = desired probe length (usually 0.1–0.15 kb)]. t is approximately 60 minutes for most transcript lengths.

9. Neutralize the solution by adding 100 μl of 300 mM sodium acetate (pH 5.2), and then ethanol-precipitate.

10. Dissolve the precipitate in 20–50 μl of hybridization solution (50% formamide, 5× SSC).

IB. Estimation of Probe Concentration

1. Dilute the labeled control RNA (DIG Kit, vial 5) at 100, 400, and 1600-fold. Spot 1 μl of each dilution onto a positively charged nylon membrane (for example Boehringer Mannheim #1 209 272).

2. Dilute the synthesized probe at 100, 400, and 1600-fold. Spot 1 μl of each dilution onto the same membrane.

3. Dry the membrane with an electric hair dryer, or UV cross-link in a Stratalinker.

4. Wet the membrane in buffer 1 [100 mM Tris-Cl (pH 7.5), and 150 mM NaCl].

5. Wash the membrane in buffer 2 [0.5% blocking reagent (Boehringer Mannheim, included in the "DIG Detection Kit") in buffer 1] for 15–30 minutes at room temperature.
6. Incubate the membrane in buffer 1 containing the AP-conjugated anti-DIG antibody (Boehringer Mannheim; 1:5000 dilution) at room temperature for 20 minutes.
7. Wash twice with buffer 1 (15 minutes each).
8. Wash the membrane in buffer 3 [100 mM Tris-Cl (pH 9.5), 150 mM NaCl, and 50 mM $MgCl_2$] for 1 minute.
9. Incubate the membrane with freshly prepared color-substrate solution [45 µl of NBT (75 mg/ml stock in 70% (v/v) dimethyl formamide] and 35 µl of X-phosphate (50 mg/ml stock of the ditoluidine salt in dimethyl formamide) in 10 ml of buffer 3 for 15 minutes in the dark.
10. Estimate the concentration of the probe by comparing the color intensities of the probes with those of control. [The starting concentration of the undiluted control labeled-RNA (vial 5) is 100 ng/µl.]

IC1. Preparation and Fixation of Embryos

1. Collect the embryos, wash with water, and then dechorionate for 30 seconds in a 6% solution of sodium hypochlorite (VWR, reagent grade, use undiluted). Wash well with water. Do not allow dechorionated embryos to become dry.
2. Transfer the embryos into a glass tube containing 4 ml of PP [4% paraformaldehyde in PBS (130 mM NaCl, 7 mM Na_2HPO_4, 3 mM NaH_2PO_4)] and 16 ml of heptane. For small-scale embryo collections, a 2.0-ml polypropylene microcentrifuge tube is appropriate, in which case the above volumes are reduced by a factor of 10.
3. Rotate the tube on a rotating wheel for 20 minutes. (*Note:* Gentle agitation of the tissue throughout the fixation and washing steps greatly improves signal uniformity. 2.0-ml Eppendorf Safe-Lock tubes fit tightly into a 40-place autoclavable microtube rack manufactured by Treff AG (Degersheim, Switzerland, fax +41 71 371 2943; distributed by Diamed Lab Supplies, Mississauga, ON, Canada, fax +1 416 625 6280). We attach such a rack to a Cole Parmer rotator device ("Roto-Torque," model 7637-01), insert the incubation tubes in the rack, and rotate slowly (speed setting 4) throughout fixation and washing. The particular rotator used is small enough to fit within incubators and refrigerators, for carrying out steps at various temperatures.)
4. Remove as much of the lower phase as possible and about half of the upper phase, then add 10 ml (1 ml if small-scale prep) of 100% methanol. Shake vigorously for 20 seconds.
5. Transfer the embryos that sink to the bottom to a new tube. Wash in several changes of methanol. *Do not* try to recover floating embryos; they are not devitellinized or properly fixed and they will not work in hybridization experiments.

At this stage, the embryos can be stored at $-20°C$ for several months or even years. Also, embryos fixed by this procedure can be used for immunohistochemistry (see below).

6. If large-scale fixation was used, transfer appropriate amounts of embryos (50 μl maximum) to 2.0-ml microcentrifuge tubes. Refix the embryos and rehydrate by passage through a series of washes consisting of various ratios of ME [50 mM EGTA (pH 8.0) in 90% MeOH] and PP. The first step is for 5 minutes in 7:3 ME/PP, the second for 5 minutes in 1:1 ME/PP, the third for 5 minutes in 3:7 ME/PP, and the last step for 20 minutes in PP alone. As rehydration proceeds, the embryos will tend to adhere to the wall of the tube.
7. Wash the embryos in PBS for 10 minutes, then three times each for 5 minutes in PBT (0.1% Tween 20 in PBS).
8. Incubate for 3 minutes in 50 μg/ml proteinase K in PBT at room temperature (23°C). This is a *critical* step for ensuring that the embryos are permeable to the probe and to the antibodies used for its detection. While this incubation time and enzyme concentration is a good starting point, it may need to be modified for particular batches of proteinase K, and for particular probes. Also, late-stage embryos (later than stage 14) require longer proteinase K treatment to give acceptable levels of signal. If proteinase K treatment is insufficient, signal will be weak or nonexistent; if treatment is excessive, the embryos will fall apart in the subsequent steps. For a new probe or a new lot of proteinase K, it is best to divide the embryos and try three or four different treatment conditions, carrying them in parallel through the subsequent steps.
9. Stop the proteinase K treatment by washing the embryos twice (1 minute each) with 2 mg/ml glycine in PBT (freshly prepared).
10. Wash the embryos twice for 5 minutes each in PBT, refix for 20 minutes with PP, and wash three times for 10 minutes each in PBT.

IC2. Preparation and Fixation of Ovaries
1. Dissect ovaries from well-fed and well-mated 2–3 day old females in PBS, and fix ovaries with 4% paraformaldehyde in PBS + 1% DMSO for 20 minutes. Dechorionation with bleach will destroy the ovaries, and heptane/methanol permeabilization is not necessary. Teasing apart the individual ovarioles before fixation will give more uniform staining results as macromolecular reagents are not always permeable to the interior of a tightly packed cluster of ovarioles. However, complete separation of ovarioles will result in more tissue loss during subsequent manipulations. We partly tease ovaries apart at this stage, leaving the ovarioles attached at the smaller (early-stage) end.
2. Wash ovaries with PBS for 5 minutes, followed by two washes in PBT for 5 minutes each.
3. Carry out proteinase K treatment as in step 8 above, but for 20 minutes with 50 μg/ml of proteinase K at room temperature. Again, this is a *critical* step and the precise conditions may have to be determined empirically. If signal is only apparent in early-stage egg chambers (< stage 6), proteinase K treatment is probably too short; conversely, overly harsh proteinase K treatment results in a loss of signal from early-stage egg chambers. Often, it is impossible to obtain optimum staining for all stages of oogenesis from a single proteinase K treatment, and different sets of ovaries digested under various conditions have to be processed in parallel.
4. Stop the proteinase K treatment, wash, refix, and wash again as in steps 9 and 10 of Protocol IC1.

To improve detection efficiency in late-stage oocytes (> stage 10), ovaries can be incubated in 90% MeOH + 10% DMSO for 1 hour at −20°C after step 4 (Ephrussi et al. 1991). Ovaries are then rehydrated by three washes in PBT for 5 minutes each. In our hands, this additional step improves signal in late-stage oocytes but reduces signal (sometimes to invisibility) in early stages. Also, for analysis of late-stage oocytes, diffusion washing after hybridization (Protocol IE2) is superior to washing by RNase treatment (Protocol IE1).

ID. Hybridization
1. Prepare the hybridization solution [HS; 50% formamide, 5× SSC (autoclaved), 50 µg/ml heparin, 0.1% Tween 20, and 100 µg/ml heat-denatured salmon sperm DNA].
2. Wash the embryos or ovaries for 10 minutes in 1:1 HS/PBT, and then for 10 minutes in HS alone.
3. Change the HS, and prehybridize for 30–60 minutes (or more) at 45–60°C. We prehybridize in Parafilm-sealed microcentrifuge tubes in a water bath in a floating rack.

The embryos or ovaries can be stored at −20° C for up to a few months after this step.

4. Divide the embryos or ovaries into 2.0-ml microcentrifuge tubes. We hybridize in a volume of 50–100 µl in 2.0-ml Eppendorf Safe-Lock tubes. Heat-denature probes with 10–20 µg of yeast tRNA at 85–95°C for 5 minutes, then chill on ice. For hybridization, add heat-denatured probe at a final concentration of 100 ng/50 µl (for RNase treatment, Protocol IE1) or 20 ng/50 µl (for diffusion wash, Protocol IE2). Hybridize overnight (at least 12 hours at 56°C) without shaking. We hybridize in Parafilm-sealed microcentrifuge tubes in a water bath in a floating rack. If diffusion washing is chosen for the subsequent step, probe concentration is critical, and may need to be empirically determined each time.

After this step carry out one of the three washing methods presented below.

IE1. Washing by RNase Treatment
The major advantage of this washing method is that, even if too high a probe concentration is used for the hybridization, background staining is usually low because RNase treatment degrades most of the unhybridized probe. This method is particularly useful in experiments involving hybridizations with many different probes, for which optimizing the concentration of each is laborious.

1. Add 50 µl of MAB (100 mM maleic acid, pH 7.5, 150 mM NaCl, and 0.1% Tween 20), and let stand for 5 minutes. PBT may be used as an alternative to MAB in this and the subsequent steps; MAB may slightly improve the signal-to-noise ratio in the final color development.
2. Add 100 µl of MAB, and let stand for 5 minutes. Then fill the tube with MAB, and wash the embryos at room temperature for 10 minutes.
3. Incubate the embryos or ovaries with 20 µg/ml of RNase A in MAB at 37°C for 30 minutes. Some workers also add RNase T1 at a concentration of 125 U/ml; we observe no difference when this is done.

4. Wash at least five times for 1 minute each in MAB. Proceed to Protocol IF.

IE2. Washing by Diffusion
1. After hybridization, add washing HS (wHS; 5× SSC, 50% formamide, 0.1% Tween 20) and incubate at the same temperature as the hybridization for 15 minutes. Repeat at least four times, turning the tube upside down occasionally.
2. Incubate at room temperature in a wHS/PBT series (3:2, 1:1, 2:3) for 5–10 minutes each.
3. Wash five times for 5–10 minutes each with PBT. Proceed to Protocol IF.

IE3. Washing by Electrophoresis
This washing method (Kobayashi, et al. 1994) is quick and gives a signal-to-noise ratio superior to that of the other two methods, particularly in later stage embryos. It is probably not appropriate for ovaries because of the larger size of that tissue, unless all individual late-stage egg chambers are dissected apart.

1. Prepare a 2% agarose minigel in Tris-acetate buffer [4.84 g Tris base, 1.14 ml of glacial acetic acid, 2 ml of 0.5 M EDTA (pH 8.0)] in 1 liter of water; concentrated stocks of up to 50× can be prepared. Insert a comb of a thickness of at least 1 mm, and with enough teeth to make a sample well for each hybridization being processed. When the gel has solidified, remove the comb and mount it in a standard horizontal electrophoresis tank.
2. Rinse the embryos in PBT for 1 minute, then transfer the embryos from each hybridization into a sample well using a Pipetman (Gilson) or similar device. The extreme end has to be cut off of many brands of 200-μl micropipetter tips for the embryos to flow through easily.
3. Carry out electrophoresis at 100 V for 20–60 minutes.
4. Remove the embryos from the wells using a micropipetter and transfer the contents of each well into a 2.0-ml microcentrifuge tube.
5. Wash embryos in PBT for 10 minutes at room temperature. Proceed to Protocol IF.

IF. Detection
1. Incubate the embryos or ovaries with 20% normal sheep serum (previously heat-inactivated at 56°C for 30 minutes) in MAB for 1 hour. (Do not exceed 1.5 hours.)
2. Incubate the embryos or ovaries for 1 hour at room temperature in 100–500 μl of anti-DIG antibody (Fab' fragment, Boehringer Mannheim, freshly diluted 1:3000 in MAB). It is usually not necessary to preabsorb anti-DIG antibody with fixed embryos.
3. Wash the embryos or ovaries four times *or more* for >20 minutes each in MAB, then twice for 5 minutes each in buffer 3 (see Protocol IB, step 8) supplemented with 0.1% Tween 20. Addition of levamisole is not necessary.
4. Start color development by incubating the embryos or ovaries in buffer 3 containing 2.25 μl of NBT [75 mg/ml stock in 70% (v/v) dimethyl formamide] and 1.75 μl X-phosphate [50 mg/ml 5-bromo-4-chloro-3-indolyl phosphate (toluidine salt) in dimethyl formamide] per milliliter. Transfer the embryos or ovaries, along with the staining solution, to a suitable vessel, such as a 12-well

plastic tissue culture plate or a depression slide, to observe the progress of the color reaction under a dissecting microscope. For embryos, the color should develop fully within 30 minutes, and usually the reaction is complete within only 5 minutes. Staining times longer than 40–60 minutes result in high levels of background, and the signal-to-noise ratio is enhanced if staining times are kept to a minimum. Ovaries usually show less background staining, and sometimes take over 1 hour of staining for optimal signal.
5. Stop the reaction by washing the embryos or ovaries with PBT or MAB at least three times for 5 minutes each.
6. Dehydrate the embryos or ovaries in an ethanol series (once in 70% ethanol for 3 minutes, and 3 times in 100% ethanol for 3 minutes each).
7. Remove as much ethanol as easily possible, and resuspend the embryos or ovaries in Histoclear (National Diagnostics; nontoxic and smells nice) or xylene (toxic and smells bad).
8. Mount the embryos or ovaries on clean microscope slides in Eukitt or Permount.

To double-stain the embryos or ovaries with antibody after in situ hybridization, block them by resuspending in 1% BSA in PBT after step 5 of Protocol IF. Then carry out the primary antibody incubation (Protocols II, see below). For antibody staining of tissue that has already undergone in situ hybridization, more dilute concentrations of both primary and secondary antibodies are usually required than for a single antibody staining. This is probably because the proteinase K treatment improves the accessibility to the antibodies to their epitopes.

IG. Troubleshooting High Background Staining

The most usual difficulty encountered in whole-mount in situ hybridization experiments is unacceptably high levels of background staining. Background can be reduced by (1) altering the conditions of the proteinase K treatment, (2) preabsorbing the anti-DIG antibody against ovaries or embryos, (3) increasing the hybridization temperature, (4) reducing the probe concentration, or (5) reducing the amounts of NBT and X-phosphate by a factor of 2–10, while maintaining the same ratio.

In Situ Hybridization to Paraffin-Sectioned Tissue Using Digoxigenin-Labeled RNA Probes

IIA. Preparation of Siliconized Microscope Slides
1. Place microscope slides in metal staining trays (such as VWR #25445-011).
2. Soak the slides for 30 minutes in a hot detergent solution.
3. Rinse the slides extensively in tap water.
4. Autoclave at 120°C for 20 minutes.
5. Rinse the slides in distilled water.
6. Bake the slides at 180°C for 3 hour.
7. Place the slides in a solution of 2% 3-aminopropyltriethoxysilane (Sigma) in acetone for 10 seconds.
8. Wash the slides briefly in acetone.
9. Wash the slides with autoclaved distilled water.

10. Air-dry.

Slides siliconized in this manner can be stored in closed boxes at room temperature for at least 6 months.

IIB1. Fixation, Dehydration, and Embedding of *Drosophila* Ovaries

1. Dissect ovaries in PBS and wash briefly in PBS.
2. Transfer the ovaries to 4% paraformaldehyde in PBS and fix for 2 hours at room temperature.
3. Dehydrate in an ethanol series: 70% ethanol for 15 minutes, 90% ethanol for 15 minutes, 100% ethanol (two changes, 15 minutes each), and xylene (two changes, 15 minutes each). Histoclear (National Diagnostics) can be substituted for xylene in this and the subsequent steps. If tissue is not *absolutely transparent* at this stage, return to 100% ethanol and repeat the subsequent steps.
4. Process the ovaries for embedding: xylene/Paraplast Plus (1:1) for 30 minutes, Paraplast Plus (four changes, 2 hours total). Paraplast Plus (Oxford Labware) is available from VWR, Fisher, and other major lab suppliers. These manipulations are carried out at 58–60°C; do not overheat!
5. Transfer ovaries in Paraplast to embedding molds [e.g., Fisher #15-182-501A or -505A or Simport Plastics #M475-1 (Beleoil, QC, Canada, fax 1 514 464 3394] and embed on ice.
6. Cut sections (7–10 μm thick) on a microtome and collect on siliconized slides (prepared as in Protocol IIA).

IIB2. Fixation, Dehydration, and Embedding of *Drosophila* Embryos

1. Collect, dechorionate, devitellinize, and fix embryos as in Protocol IC1, steps 1–5.
2. Dehydrate the embryos in three changes of 100% ethanol (15 minutes each).
3. Clear the embryos in three changes of xylene (15 minutes each). Histoclear (National Diagnostics) can be substituted for xylene in this and the subsequent steps. If tissue is not *absolutely transparent* at this stage, return to 100% ethanol and repeat the subsequent steps.
4. Embed and section as in Protocol IIB1, steps 4–6.

IIC. Pretreatment of Slides and Hybridization

Note: Depending on the number of slides being processed, various vessels are appropriate for steps 1–8. Slides can be processed in pairs (back-to-back) in 50-ml disposable plastic conical tubes. Larger numbers can be processed in Coplin jars (such as VWR 25460-000) or in racks in rectangular staining dishes (such as VWR 25445-007). If glassware is chosen, it should be sterilized before use.

1. Remove wax from the sections by passing the slides through two changes of xylene (10 minutes each).
2. Rehydrate the sections in an ethanol series (100%, 90%, 70%, for 10 minutes each), then wash in PBT for 1 minute.
3. Fix the sections in 4% paraformaldehyde in PBS (freshly prepared) for 15 minutes.
4. Wash in PBS for 15 minutes.

5. Wash in 0.1 M triethanolamine (diluted from a 2 M stock solution, adjusted to pH 8.0 with HCl, and sterilized by autoclaving) for 1 minute.
6. Acetylate the tissue by placing in 0.25% acetyl anhydride in 0.1 M triethanolamine for 10 minutes. This solution must be prepared *immediately* before use, as acetic anhydride is extremely unstable in aqueous conditions.
7. Wash in PBS for 5 minutes.
8. Dehydrate in an ethanol series (70%, 80%, 90%, 100%, 100%, for 1 minute each).
9. Allow slides to air-dry.
10. Prehybridize in HS (see Protocol ID, step 1) at 47°C for 1 hour in a moist chamber.
11. Hybridize in HS containing 1–5 µg/ml DIG-labeled RNA probe for at least 12 hours at 47°C in a moist chamber. Be sure that the probe solution covers all the sections, and cover each slide with a sheet of Parafilm, or a glass coverslip, to reduce evaporation. Moisten the chamber with 50% formamide, 5× SSC, not with water. Seal the entire chamber with Parafilm, again to reduce evaporation.

IID. Washing and Detection
1. Wash the sections four times (20 minutes each) in 50% formamide, 5× SSC, at 47°C.
2. Wash twice (30 minutes each) in TNE (10 mM Tris-HCl, pH 7.8, 500 mM NaCl, 1 mM EDTA) at 37°C.
3. Incubate the sections in TNE containing 20 µg/ml of RNase A at 37°C for 20 minutes.
4. Wash three times (20 minutes each) in 2× SSC at 47°C.
5. Incubate the sections in 100–300 µl of anti-DIG antibody (Fab' fragment, Boehringer Mannheim, freshly diluted 1:1000 in PBS). It is usually not necessary to preabsorb anti-DIG antibody with fixed embryos.
6. Wash four times (20 minutes each) in PBS at room temperature.
7. Wash three times (5 minutes each) with buffer 3.
8. Apply 1 ml of color-substrate solution (see Protocol IB, step 9) to the sections.
9. Allow color to develop in the dark. Color can develop within a few minutes, or take as long as 24 hours. Background staining is usually very low with this method, so long color reactions are possible. Monitor the progress of color development periodically under a dissecting microscope.
10. Stop color development by washing the slides in PBS.
11. Dehydrate and mount as in Protocol IF, steps 6–8.

Antibody Staining of Ovaries and Embryos

IIIA. Antibody Reactions
1. Fix embryos as described above in Protocol IC1, steps 1–5. For antibody staining of ovaries, dissect in PBS and fix for 15 minutes in 4% paraformaldehyde in PBS.
2. Rehydrate embryos from methanol by washing (5 minutes each) in 70% methanol in PBS, 50% methanol in PBS, and 30% methanol in PBS.

3. After fixation of ovaries (step 1) or rehydration of embryos (step 2), wash three times (2 minutes each) in PBST (PBS + 0.2% Tween 20), then once (2 minutes) in PBSBT (PBST + 1% bovine serum albumin + 0.2% Triton X-100).
4. Block nonspecific protein–protein binding by incubating ovaries or embryos in four changes of PBSBT (total time 4 hours). Attaching the reaction tubes to a rotating device for this and subsequent steps is helpful; we use the apparatus described in Protocol IC1, step 3, above.
5. Incubate ovaries or embryos with an appropriate dilution of primary antiserum in PBSBT overnight at 4°C.
6. Rinse three times (1 minute each) with PBST. Higher concentrations of Tween 20 (up to 1%) in the PBST used for steps 6–7 and 9–10 can be useful for reducing nonspecific staining.
7. Wash four times (20 minutes each) with PBST.
8. Incubate with conjugated secondary antibody at an appropriate dilution in PBSBT overnight at 4°C.
9. Rinse three times (1 minute each) with PBST.
10. Wash four times (20 minutes each) with PBST.

IIIB. Preparation of Sectioned Ovaries for Antibody Staining
1. At least 1 day before the experiment, prepare subbed microscope slides (Ashburner 1989). Dip clean microscope slides into a solution of 1% gelatin and 0.1% chromium potassium sulfate ($Cr_2[SO_4]_3K_2SO_4 \cdot 24H_2O$; prepare solution without boiling at 60°C). Allow the subbed slides to air-dry overnight standing on end on paper towels. Subbed slides can be stored indefinitely in closed slide boxes; subbing improves the adhesion of tissue sections to the slide.
2. Dissect ovaries in PBS. When 5–10 pairs of ovaries are dissected, cover the end of a cryostat chuck with a bead of O.C.T. embedding fluid (Tissue-Tek, available from VWR), transfer the ovaries in a cluster into the center of the bead, and freeze rapidly on dry ice. When the embedding fluid has completely frozen, transfer the chuck to a cryostat chamber at −18°C, and allow its temperature to equilibrate for 1 hour.
3. Cut 10-μm sections, picking them up individually with a room-temperature subbed slide. Depending on how scattered the ovaries are in the embedding fluid, 3–12 or more sections can be placed on an individual slide. Smaller volumes of reagents can be used in subsequent steps if the sections are positioned toward one end of the microscope slide.
4. When sufficient sections have been transferred onto a slide, place the slide in a glass Coplin jar containing freshly made 4% paraformaldehyde in PBS. Fix for 10 minutes, then transfer the slide into another Coplin jar containing PBSBT. If multiple slides are being prepared at a single sitting, the fixative can be reused and the slides can accumulate in the PBSBT.
5. Transfer all the slides into a fresh PBSBT solution and block for at least 2 hours.
6. Carry out antibody staining as in Protocol IIA above, starting from step 5. We use glass Coplin jars which hold five slides each for the primary and secondary antibody reactions; if sections are placed toward one end of the

slides, a volume of 15–20 ml in these jars is required to cover the tissue. Washing steps can be carried out using slide racks in larger volume dishes if convenient. Do *not* agitate the slides throughout the antibody staining and washing as constant agitation will cause the sections to fall off the slides.

IIIC. Detection

For secondary antibodies conjugated to a *fluorochrome*, no further processing is necessary. Tissue should not be dehydrated, and is mounted on microscope slides in an aqueous-based mountant such as 50–80% glycerol in PBS or Vectashield (Vector Laboratories); the latter contains an anti-photobleaching agent which is particularly important for confocal microscopy.

For *biotinylated* secondary antibodies, signal can be enhanced by treatment with horseradish peroxidase (HRP)-conjugated avidin, and detected using the DAB/hydrogen peroxide reaction. Signal enhancement is usually carried out with the Vectastain ABC or ABC Elite kits (Vector Laboratories). The following protocol is used for enhancement and detection. Note: All buffers must be prepared with *deionized and distilled water (resistivity > 15 MΩ/cm)*; the DAB reaction is very sensitive to water quality.

1. During the washing step (step 10 of Protocol IIIA), prepare the ABC reaction mix by adding 10 μl of solution A and 10 μl of solution B from the Vectastain kit to 1 ml of PBST. Mix and let stand at room temperature for 30 minutes.
2. After the final wash is removed from the tissue, add 400–1000 μl of ABC solution. Incubate for 30 minutes on the rotator and for another 30 minutes on the bench at room temperature.
3. Rinse three times (1 minute each) with PBST.
4. Wash three times (10 minutes each) with PBST.
5. Prepare staining solution. *Important:* Make this solution fresh each time. Add 0.5 ml of a 4 mg/ml DAB solution (made in *deionized and distilled water*) to 4.5 ml of PBS (made in *deionized and distilled water*). Add 5 μl of 8% $NiCl_2 \cdot 6H_2O$. (The presence of nickel ions changes the color of the precipitate from brown to a more easily photographed grey-black.)
6. Prepare hydrogen peroxide solution by diluting a concentrated stock (30%) by 1:100. We use Fisher ACS grade 30% hydrogen peroxide.
7. Incubate tissue in staining solution for 5 minutes. Transfer ovaries and embryos, along with the staining solution, to a vessel which allows the progress of the staining to be observed (as in Protocol IF, step 4).
8. Add 8 μl of diluted hydrogen peroxide for ovaries (16 μl for embryos) and very gently swirl the vessel to mix. DAB staining is usually very rapid (often within seconds), so be sure not to overstain.

For *HRP*-conjugated secondary antibodies, follow only steps 5–8 in Protocol IIIC.

For *alkaline phosphatase*-conjugated secondary antibodies, follow the staining procedure for in situ hybridizations (Protocol IF, steps 1–8). For all detection methods, permanent mounted preparations can be made as in Protocol IF, steps 6–8.

For *tissue sections on slides*, monitor HRP or alkaline phosphatase reactions by periodically observing a slide under a dissecting microscope. Do not observe for so long that the slide dries out!

In Situ Hybridizations at the Ultrastructural Level

We present details of two in situ hybridization methods for electron microscopy (EM) applicable to *Drosophila* ovaries and embryos. The first is a preembedding in situ hybridization method (Pomeroy et al. 1991), which has been used to show the presence of mitochondrial large ribosomal RNA (mtlrRNA) on *Drosophila* polar granules (Kobayashi et al. 1993), and to detect *fushi tarazu* mRNA at the apical portion of blastodermal cells (Amikura et al. 1993). The advantages of this method over post embedding protocols are that hybridization, washing, and detection take place on whole tissue in solution and therefore extensive manipulation of EM grids is not necessary, and the intensity of the hybridization signal can be previewed under a light microscope before carrying out EM sectioning.

The second is a postembedding in situ hybridization method, which was used to show the localization of *Polar granule component* (*Pgc*) RNA on polar granules (Figure 22-1a; Nakamura et al. 1996). While postembedding is more laborious, its advantage is that it resolves all problems of permeability of the probes and antibodies into tissues or embryos. Figure 22-1b shows mtlrRNA in mitochondria, which is undetectable by preembedding because of the impermeability of mitochondrial membranes. This method has also been used in *Drosophila* ovaries (Binder et al. 1986).

IVA. A Pre-embedding In Situ Hybridization Method for *Drosophila* Embryos

1. Collect and dechorionate embryos with sodium hypochlorite as in IC1, step 1.
2. Soak the dechorionated embryos in heptane saturated with 25% glutaraldehyde for 1 minute.
3. Fix the embryos in 4% paraformaldehyde/0.5% gluataraldehyde in 0.1 M cacodylate buffer (pH 7.4) for 30 minutes. In this fixative, remove vitelline membranes manually with a pair of tungsten needles according to the method of Zalokar and Erk (1977). To prepare 0.1 M cacodylate buffer (pH 7.4), add 4.28 g of sodium cacodylate to 75 ml of double-distilled water, and adjust the pH with 0.1 N HCl.
4. Wash the embryos with 0.1 M cacodylate buffer (pH 7.4), twice for 5 minutes each.
5. Dehydrate the embryos in 70%, 80%, and 90% ethanol for 10 minutes each.
6. Wash the embryos in two changes of 100% methanol, for at least 5 minutes each. Embryos can be stored in methanol for several days at −20°C at this stage.
7. Carry out steps 6–10, Protocol IC1, for rehydration, refixation, proteinase K treatment, and additional refixation.
8. Carry out steps 1–3, Protocol ID, for prehybridization. Prehybridize for 1 hour at 45°C.

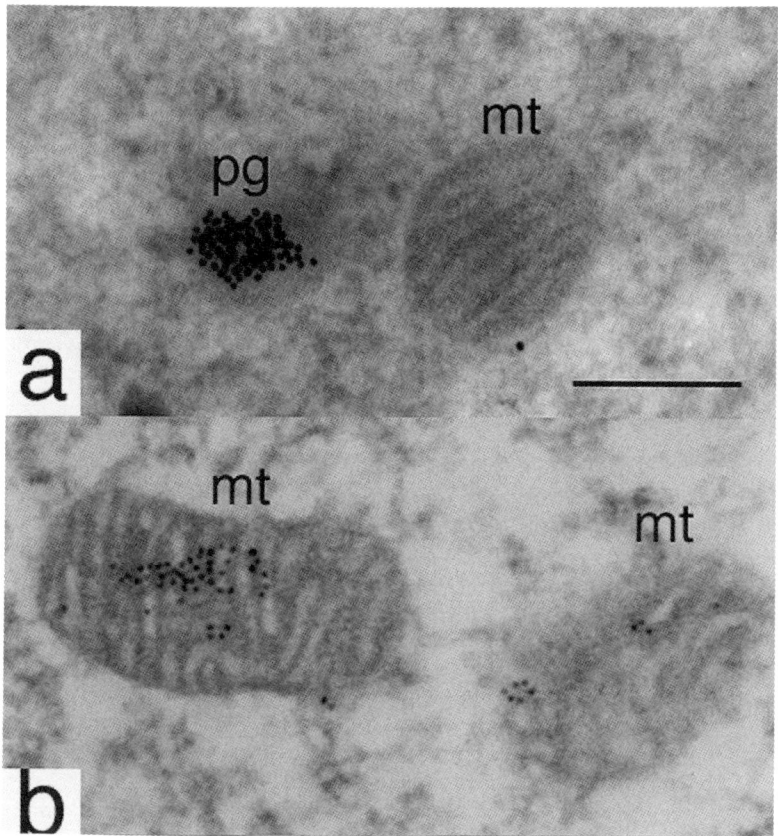

Figure 22-1 Electron micrographs of in situ hybridizations on an early *Drosophila* embryo using the postembedding Protocol IVB. pg, polar granule; mt, mitochondrion. (a) A probe for Pgc RNA intensely labels polar granules. (b) Using this technique, mtlrRNA is detected within the mitochondria. Scale bar: 0.2 µm.

9. Hybridize as in Protocol ID, step 4, with a DNA probe overnight at 45°C. To make DNA probes, PCR-amplified DNA is labeled with digoxigenin for 20 hours using the Digoxigenin-High Prime Kit (Boehringer Mannheim). Ethanol-precipitate, then resuspend the labeled DNA in deionized and distilled water, and estimate its concentration as in Protocol IB, using control DIG-labeled DNA (included in DIG Nucleic Acid Detection Kit, Boehringer Mannheim). Adjust the concentration of the labeled probe to 50 µg/ml before use. Probes can be stored for several months at −20°C. For hybridization, 0.5 ml of denatured labeled probe was added to 50 µl of HS (see Protocol ID, step 1). The probe in HS was denatured at 70°C for 5 minutes and immediately cooled in ice water containing NaCl. We hybridize approximately 50 embryos at a time in 50 µl of HS.

10. Wash the embryos in HS for 20 minutes at room temperature.
11. Wash in varying ratios of HS/PBT (4:1, 3:2, 2:3 and 1:4) for 20 minutes each.
12. Wash the embryos in PBT, twice for 20 minutes each.
13. Incubate the embryos in 200 µl of PBT containing 1 µl of 1-nm-gold-conjugated anti-digoxigenin antibody (Biocell, UK) for 1 hour at room temperature.
14. Wash in PBT, four times for 20 minutes each.
15. Wash in cacodylate buffer (pH 7.4), twice for 10 minutes each.
16. Refix in 2.5% glutaraldehyde in cacodylate buffer (pH 7.4) for 30 minutes.
17. Wash in double-distilled water containing 0.1% Tween-20 (DWT), three times for 5 minutes each.
18. Enhance the signal using the Silver Enhancement Kit (Biocell, UK) for 3 minutes.
19. Wash in DWT, three times for 5 minutes each.

At this stage, the success of the hybridization and the intensity of the signal can be checked under a dissecting microscope. If all is satisfactory, proceed with the following steps to prepare the sample for electron microscopy.

20. Fix with 1% OsO_4 in 0.1 M cacodylate buffer (pH 7.4) for 60 minutes.
21. Rinse the embryos in cacodylate buffer once for 5 minutes.
22. Dehydrate the embryos in an ethanol series (70%, 80%, 90%, and 95%) for 15 minutes each, followed by two changes of 100% ethanol (30 minutes each).
23. Embed the embryos in Spurr resin (TAAB, Inc. also available from Sigma), according to the method of Spurr (1969) outlined below. Incubate the embryos in acetone for 5 minutes, in QY-1 (*n*-butyl glycidyl ether; Nisshin EM, Tokyo; also available from Sigma) for 1 hour, in QY-1/Spurr resin (1:1) for 1 hour, and in three changes of Spurr resin (twice for 1 hour each, followed by overnight). We use small dishes of aluminum foil (2-cm diameter) as containers for Spurr resin, and we transfer the embryos from one container to the next with a disposable yellow pipette tip. Finally, the embryos are placed in a silicon-rubber mold filled with neat (100%) Spurr resin. Allow the resin to polymerize overnight at 65°C.
24. Cut thin sections (80–90 nm thick) and collect on copper grids (150-gauge mesh).
25. Stain sections with 2% uranyl acetate in water for 30 minutes.
26. Wash grids in double-distilled water, six times for 2 minutes each.
27. Stain sections with lead citrate solution (Reynolds 1963) for 2 minutes. Prepare lead citrate solution as follows: dissolve 1.33 g of lead nitrate and 1.76 g of sodium citrate in 30 ml of double-distilled water. Stir for 30 minutes, then add 8 ml of 1 N NaOH and bring to 50 ml with double-distilled water. Final pH should be 12.0 ±0.1.
28. Wash grids with double-distilled water once for 2 minutes.

Note: For steps 25–28, a small number of grids can be processed by floating them on a drop of double-distilled water or staining solution. Larger numbers can be processed with an apparatus described by Toyoda et al. (1991).

29. Observe under the electron microscope.

IVB. A Postembedding In Situ Hybridization Method for Electron Microscopy

1. Collect and process embryos as in Protocol IVA, steps 1–4.
2. Rinse the embryos with PBT, three times for 5 minutes each.
3. Treat the embryos with proteinase K (Merck), 50 µg/ml, for 3 minutes at 23°C. Proteinase K treatment is not always necessary.
4. Wash the embryos in 2 mg/ml of glycine in PBT for 30 seconds.
5. Wash the embryos in PBT, three times for 5 minutes each.
6. Fix the embryos in 4% paraformaldehyde, 0.5% glutaraldehyde in PBT for 20 minutes.
7. Wash the embryos in PBT, three times for 10 minutes each.
8. Wash the embryos with TBST (100 mM Tris-HCl, pH 7.5, 150 mM NaCl, 0.1% Tween 20) once for 5 minutes.
9. Incubate in 0.2 M ammonium chloride for 10 minutes.
10. Wash the embryos with TBST, three times for 5 minutes each.
11. Dehydrate the embryos in an ethanol series as in Protocol IVA, step 22, at 4°C.
12. Embed using an automatic freeze-substitution system (Reichert) according to the method of Robertson et al. (1992). The following incubations are required: 100% acetone at 4°C for 15 minutes; 100% acetone at −20°C, two changes for 1 hour each; acetone/Lowicryl HM20 resin (3:1) at −50°C for 1 hour; acetone/Lowicryl HM20 resin (1:1) at −50°C for 1 hour; acetone/Lowicryl HM20 resin (1:3) at −50°C for 1 hour; neat (100%) Lowicryl HM20 resin, at −50°C, three changes, two for 1 hour each, the third overnight. To make Lowicryl HM20 resin (Polysciences, Inc.), mix 2.8 g of cross-linker D with 17.02 g of monomer E, then add 0.1 g of initiator C.
13. Transfer the embryos into gelatin capsules filled with neat resin (Lowicryl HM20) and polymerize by UV irradiation at −50°C for 48 hours.
14. Increase the temperature slowly up to room temperature (take at least 12 hours to do this) and then remove the capsules.
15. Cut thin sections (80–90 nm thick) and collect on nickel grids (150-gauge mesh).
16. Hybridize using DNA probes to the sections on the grids, in a 50-µl drop of HS containing 1 µl of DNA probe at 45°C for 5 hours in a moist chamber. Prepare probes as in Protocol IVA, step 9.
17. Transfer the grids through the following series of washes (float on a 50-µl drop for each wash): HS, six times for 2 minutes each; HS/TBST (1:1), six times for 2 minutes each; TBST, six times for 2 minutes each.
18. Float grids on a 50-µl drop of TBST containing 1 µl of 10-nm-gold-conjugated anti-digoxigenin antibody (Biocell, UK) for 1 hour at room temperature.
19. Float grids on a drop of TBST, six times for 2 minutes each.
20. Fix sections by floating the grids in 0.5% glutaraldehyde in 0.1 M cacodylate buffer (pH 7.4) for 10 minutes.
21. Float grids on a drop of double-distilled water, six times for 2 minutes each.
22. Stain and observe as in steps 25–29, Protocol IVA.

Antibody Staining at the Ultrastructural Level

We also present a immunohistochemical method to detect proteins in *Drosophila* embryos at ultrastructural resolution. Vasa protein is a component of polar granules and nuage particles in *Drosophila* (Hay et al. 1988; Liang et al. 1994). Figure 22-2 shows the localization of Vasa protein on polar granules detected by this method.

1. Process embryos as in Protocol IVB, steps 1–2 and 9–14. *Do not* treat with proteinase K (i.e., steps 3–8 of Protocol IVB).
2. Cut thin sections (80–90 nm thick) and collect on Formvar-coated nickel slot grids.
3. Float grids on a drop of 0.5% BSA in TBST on a sheet of Parafilm for 5 minutes.
4. Float grids on a drop of a solution of a primary antibody diluted in TBST (1:50 for the affinity-purified anti-Vasa polyclonal antibody used for Figure 22-2) at room temperature for 1 hour in a moist chamber.
5. Rinse the grids on a drop of TBST, six times for 5 minutes each.
6. Float grids on 50 ml of TBST containing secondary antibody [15-nm-gold-conjugated antirabbit IgG (Biocell, UK) 1:50 dilution] for 1 hour at room temperature in a moist chamber.
7. Rinse the grids on a drop of TBST, six times for 5 minutes each.

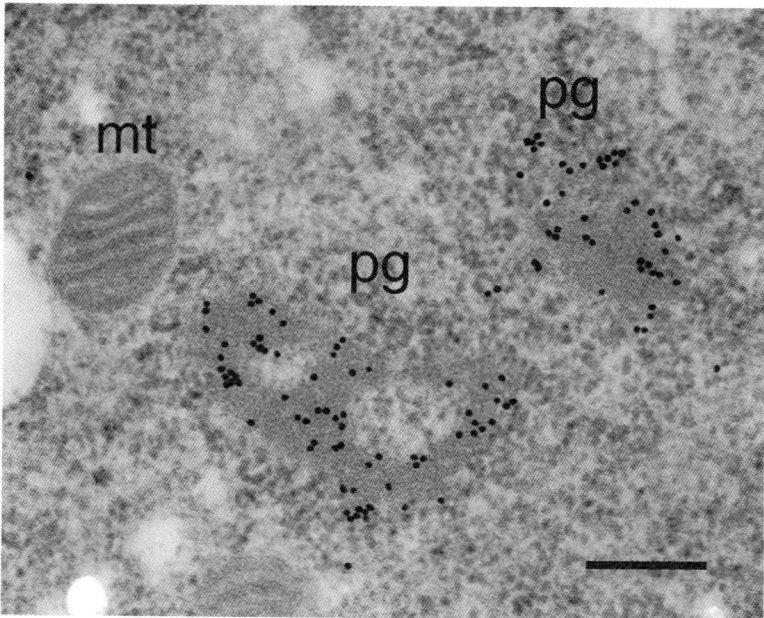

Figure 22-2 Electron micrograph which reveals the ultrastructural localization of Vasa protein in the pole plasm of an early *Drosophila* embryo. pg, polar granule; mt, mitochondrion. Many colloidal gold particles are detected on the polar granules. Scale bar: 0.2 µm.

8. Postfix in 0.5% glutaraldehyde in 0.1 M cacodylate buffer (pH 7.4) for 10 minutes at room temperature.
9. Rinse the grids on a drop of double-distilled water, six times for 5 minutes each.
10. Stain and observe as in Protocol IVA, steps 25–28.

ACKNOWLEDGMENTS We are grateful to Sylvia Styhler and Beat Suter for sharing protocols and for critical reading of the manuscript. S. K. and R. A. are supported by a Grant-in-aid from the Ministry of Education, Sports and Culture, Japan, by the Tsukuba Advanced Research Alliance Project, and by the Research Project for Future Program from the Japan Society for the Promotion of Science (JSPS). A. N. was supported by a JSPS postdoctoral fellowship, and is presently a Medical Research Council of Canada postdoctoral fellow. Research in P. L.'s laboratory is supported by operating grants from the National Cancer Institute of Canada with funds from the Canadian Cancer Society, and from the Medical Research Council of Canada, the Natural Science and Engineering Research Council of Canada, and the Fonds pour la formation de chercheurs et l'aide à la recherche du Québec. P. L. is a Research Scientist of the National Cancer Institute of Canada.

References

[Note: Morel (1993) and Newman and Hobot (1993) are good general reference monographs for novice electron microscopists.]

Akam, M. E. (1983). The location of *Ultrabithorax* transcripts in *Drosophila* tissue sections. EMBO J. 2:2075–2084.
Amikura, R., Kobayashi, S., Endo, K., and Okada, M. (1993). Nonradioactive *in situ* hybridization methods for *Drosophila* embryos detecting signals by immunogold-silver or immunoperoxidase method for electron microscopy. Dev. Growth Differ. 35:617–623.
Ashburner, M. (1989). *Drosophila: A Laboratory Manual*. Cold Spring Harbor, NY, Cold Spring Harbor Laboratory Press, p. 358.
Binder, M., Tourmente, S., Roth, J., Renaud, S., and Gehring, W. J. (1986). *In situ* hybridization at the electron microscope level: localization of transcripts on ultrathin sections of Lowicryl K4M-embedded tissue using biotinylated probes and protein A-gold complexes J. Cell Biol. 102:1646–1653.
Brower, D. L., Smith, R. J., and Wilcox, M. (1981). Differentiation within the gonads of *Drosophila* revealed by immunofluorescence. J. Embryol. Exp. Morphol. 63:233–242.
Ephrussi, A., Dickinson, L. K., and Lehmann, R. (1991). oskar Organizes the germ plasm and directs localization of the posterior determinant nanos. Cell 66:37–50.
Hafen, E., Levine, M., Garber, R. L., and Gehring, W. J. (1983). An improved *in situ* hybridization method for the detection of cellular RNAs in *Drosophila* tissue sections and its application for localizing transcripts of the homeotic *Antennapedia* gene complex. EMBO J. 2:617–623.

Hay, B., Ackerman, L., Barbel, S., Jan, L. Y., and Jan, Y. N. (1988). Identification of a component of *Drosophila* polar granules. Development 103:625–640.

Ingham, P. W., Howard, K. R., and Ish-Horowicz, D. (1985). Transcription pattern of the *Drosophila* segmentation gene *hairy*. Nature 318:439–445.

Kobayashi, S., Amikura, R., and Okada, M. (1993). Presence of mitochondrial lrRNA outside mitochondria in germ plasm of *Drosophila melanogaster*. Science 260:1521–1524.

Kobayashi, S., Saito, H., and Okada, M. (1994). A simplified and efficient method for *in situ* hybridization to whole *Drosophila* embryos, using electrophoresis for removing non-hybridized probes. Dev. Growth Diff. 36:629–632.

Kobayashi, S., Amikura, R., Nakamura, A., Saito, H. and Okada, M. (1995). Mislocalization of *oskar* product in the anterior pole results in ectopic localization of mitochondrial large ribosomal RNA in *Drosophila* embryos. Dev. Biol. 169:384–386.

Lasko, P. F., and Ashburner, M. (1990). Posterior localization of vasa protein correlates with, but is not sufficient for, pole cell development. Genes Dev. 4:905–921.

Liang, L., Diehl-Jones, W., and Lasko, P. (1994). Localization of vasa protein to the *Drosophila* pole plasm is independent of its RNA-binding and helicase activites. Development 120:1201–1211.

Mitchison, T. J., and Sedat, J. (1983). Localization of antigenic determinants in whole *Drosophila* embryos. Dev. Biol. 99:261–264.

Morel, G. (1993). *Hybridization Techniques for Electron Microscopy*. Boca Raton, FL, CRC Press.

Nakamura, A., Amikura, R., Mukai, M., Kobayashi, S., and Lasko, P. F. (1996). Requirement for a non-coding RNA component of *Drosophila* polar granules for germ cell establishment. Science 274:2075–2079.

Newman, G. R., and Hobot, J. A. (1993). *Resin Microscopy and On-Section Immunocyto-chemistry*. Heidelberg, Germany, Springer-Verlag.

Pomeroy, M. E., Lawrence, J. B., Singer, R. H., and Billings-Gagliardi, S. (1991). Distribution of myosin heavy chain mRNA in embryonic muscle tissue visualized by ultrastructural *in situ* hybridization. Dev. Biol. 143:58–67.

Reynolds, E. S. (1963). The use of lead citrate at high pH as an electron opaque stain in the electron microscope. J. Cell Biol. 17:208–212.

Robertson, D., Monaghan, P., Clarke, C., and Atherton, A. J. (1992). An appraisal of low-temperature embedding by progressive lowering of temperature into Lowicryl HM20 for immunocytochemical studies. J. Microscopy 168:85–100.

Schären-Wiemers, N., and Gerfin-Moser, A. (1993). A single protocol to detect transcripts of various types and expression levels in neural tissue and cultured cells: *in situ* hybridization using digoxigenin-labeled cRNA probes. Histochemistry 100:431–440.

Spurr, A. R. (1969). A low-viscosity epoxy resin embedding medium for electron microscopy. J. Ultrastruct. Res. 23:31–43.

St Johnston, D., Beuchle, D., and Nüsslein-Volhard, C. (1991) *staufen*, A gene required to localize maternal RNAs in the *Drosophila* egg. Cell 66:51–63.

Suter, B., and Steward, R. (1991). Requirement for phosphorylation and localization of the *Bicaudal-D* protein in *Drosophila* oocyte differentiation. Cell 67:917–926.

Suter, B., Romberg, L., and Steward, R. (1989). *Bicaudal-D*, a *Drosophila* gene involved in developmental asymmetry: localized transcript accumulation in ovaries and sequence similarity to myosin heavy chain tail domains. Genes Dev. 3:1957–1968.

Tautz, D., and Pfeifle, C. (1989). A non-radioactive *in situ* hybridization method for the localization of specific RNAs in *Drosophila* embryos reveals translational control of the segmentation gene *hunchback*. Chromosoma 98:81–85.

Toyoda, M., Kita, S., Furiya, K., and Osamura, Y. (1991). Characterization of AL amyloid protein identified by immunoelectron microscopy: a simple method using the protein A-gold technique. J. Histochem. 39:239–242.

Zalokar, M., and Erk, I. (1977). Phase-partition fixation and staining of *Drosophila* eggs. Stain Technol. 52:89–95.

23

Analysis of RNA Distribution during *Drosophila* Oogenesis Using Fluorescent In Situ Hybridization

Jolanta B. Glotzer
Anne Ephrussi

The technique of RNA hybridization in situ is a valuable tool for investigating the mechanisms and patterns of gene expression within a cell, allowing one to study the spatiotemporal distribution of messenger RNAs (mRNAs). The technique relies on hybridization between specific mRNA molecules expressed by the cell and complementary RNA or DNA probes which are visualized by a variety of methods. In this chapter, we will first briefly discuss the RNA hybridization in situ protocols that have been used in the past to investigate the distribution of mRNAs during oogenesis in *Drosophila melanogaster*. We will next describe a novel RNA hybridization in situ protocol that we have developed, which utilizes fluorescently labeled RNA probes (Glotzer et al. 1997), and we will discuss in detail its advantages and applications.

The continuous identification of genes that are required for embryonic development of the fruit fly, and the discovery that mRNAs encoded by many of these genes are localized to discrete regions within the oocyte, has led to the realization that localization of mRNA, in some cases, is an important step in the establishment of cell polarity (for a review, see Curtis et al. 1995; St Johnston 1995). The asymmetric distribution of mRNAs in the *Drosophila* egg, and, consequently, of the encoded protein products, triggers the cascade of events leading to the development of the embryo and of the adult fly. Localization of *bicoid (bcd)* mRNA to the anterior pole of the egg and of *nanos (nos)* mRNA to the posterior pole controls, respectively, the development of the anterior embryonic structures, including the head and thorax, and of the posterior structures, including the abdomen (Berleth et al. 1988; St Johnston et al. 1989; Wang and Lehmann 1991). Localization of *gurken (grk)* mRNA to the anterodorsal region of the oocyte induces dorsal cell fates, and ultimately determines the dorsoventral polarity of the embryo (Neuman-Silberberg and Schüpbach 1993). Hierarchies of mRNA localization also exist, in that localization of some mRNAs may require the prior localization and translation of others. One example is that of *oskar (osk)* mRNA, whose localization and translation at the posterior pole of the oocyte are required for subsequent localization of *nos* mRNA (Ephrussi et al. 1991; Wang et al. 1994).

Oogenesis in *Drosophila* consists of 14 stages, defined according to distinct morphological criteria (for a review of oogenesis, see Spradling 1993). Development and maturation of the oocyte occurs in a cluster of cells known as an egg chamber. The egg chamber is composed of a syncytium of 15 germ-line-derived nurse cells and a single oocyte, surrounded by a layer of somatic follicle cells (Figure 23-1A). The nurse cells and the oocyte are connected by cytoplasmic bridges or ring canals that allow the intercellular trafficking of molecules and organelles. The oocyte is arrested in meiosis and is transcriptionally inactive during most of oogenesis. In contrast, the nurse cells are transcriptionally active and produce mRNAs and proteins that are transferred to the developing oocyte and subsequently distributed to distinct locations within this cell.

The distributions of many mRNAs expressed during *Drosophila* oogenesis have been examined in detail. The in situ hybridization techniques that have been used most commonly include hybridization to sections or to whole-mount preparations (Frigerio et al. 1986; Stephenson et al. 1988; Ephrussi et al. 1991). To allow their visualization, probes have been labeled by incorporation either of radioactive nucleotides or of digoxigenin-linked nucleotides that are subsequently detected with antidigoxigenin antibodies. These antibodies can be coupled to an enzyme, often alkaline phosphatase. The presence and location of the enzyme within the egg chamber is then detected by the appearance of a visible precipitate, the result of the enzymatic conversion of a provided substrate. The major drawback of these methods is that they are time consuming due to the number of steps involved, each of which requires controls (e.g., follicle cells contain endogenous alkaline phosphatase activity, which occasionally increases the background staining; hence, careful control of the background is necessary: the secondary antibodies must be preadsorbed against fixed egg chambers to prevent nonspecific cross-reactivity, etc.).

The recent development of the fluorescent in situ hybridization technique (FISH), which has been successfully used to study mRNA expression in cultured cells (reviewed by Heng et al. 1997), has encouraged researchers to employ fluorescent methods in studies of RNA distribution in *Drosophila*.

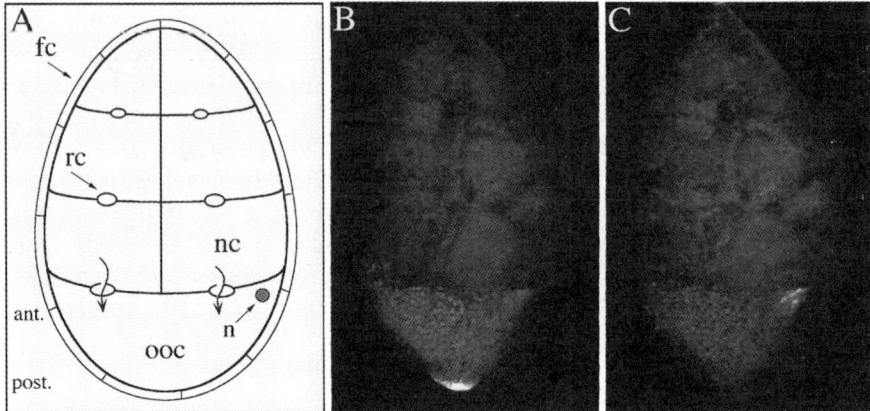

Figure 23-1 Double-label whole-mount RNA in situ hybridization of an egg chamber at stage 9 showing the relative distributions of *osk* and *grk* RNAs. (A) Schematic organization of the egg chamber: nurse cells (nc) are at the top and oocyte (ooc) is at the bottom; the oocyte nucleus (n) is indicated. The nurse cells and the oocyte are connected by cytoplasmic bridges or ring canals (rc) that allow the intercellular movement of cytoplasm (arrows). The egg chamber is surrounded by follicle cells (fc). The anterior (ant.) and the posterior (post.) of the oocyte are indicated. (B) Confocal image of a stage 9 egg chamber showing *osk* RNA localized to the posterior pole of the oocyte. *osk* RNA probe was labeled with fluorescein. (C) Confocal image of the same egg chamber showing *grk* RNA localized to the anterodorsal corner of the oocyte, near the nucleus. *grk* RNA probe was labeled with rhodamine.

For example, probes containing fluorescein-coupled nucleotides that were subsequently detected by a combination of antifluorescein primary and FITC-conjugated secondary antibodies have been used to study RNA expression in *Drosophila* embryos (Hughes et al. 1996).

To study mRNA distribution during oogenesis in *Drosophila*, we have developed a fluorescent RNA hybridization in situ technique involving antisense RNA molecules covalently linked to a fluorochrome (Glotzer et al. 1997). Synthesis of the fluorescent probe is a two-step process. In the first step, antisense RNA is synthesized by transcription in vitro. During the transcription reaction, an amino-allyl analog of UTP is incorporated uniformly throughout the transcript, resulting in RNA containing free amino groups. In the second step, the amino groups of the RNA are covalently linked to succinimidyl esters of fluorochromes, resulting in fluorescent RNA.

The method we describe allows visualization of the hybridized probe immediately after the hybridization, which significantly shortens the procedure by reducing the number of steps involved. In addition, this method has a lower risk of nonspecific cross-reactivity. The only control required is for the specificity of the RNA–RNA hybridization, and this is easily revealed using the corresponding sense probe. The hybridized probe does not diffuse; hence, the resolution of the image is quite sharp. In addition, the method is much less expensive than other fluorescent RNA labeling procedures. Finally, fluorescence allows visualization by confocal microscopy and, therefore, resolution of internal signals. This allows optical sectioning in different planes and reconstruction of sections to create a three-dimensional image of the sample. The method has many applications. In the paragraphs below, we list a few that we have adapted for our own research.

1. Analysis of the Relative Spatial Distribution of Two or More Distinct mRNA Species within the Same Cell
   Double- and potentially multilabeling in situ hybridization allows one to assess the distribution of different mRNAs within the same cell, in our case in the oocyte. This can provide valuable information when working with mutants that are not fully penetrant, and therefore present phenotypes that vary from egg chamber to egg chamber, and when working with mutants that affect the level of expression of one but not all mRNAs of interest. In addition, with this method it is possible to include, in the hybridization, specific probes that can serve as internal controls, for normalization.
   In Figures 23-1B and C we show the relative distributions of two mRNAs, *osk* and *grk*, in the same egg chamber at stage 9. The *osk*-specific probe was labeled with fluorescein and the *grk*-specific probe was labeled with rhodamine. *osk* RNA is localized to the posterior pole of the oocyte, whereas *grk* RNA is localized to the anterior of the oocyte, in the vicinity of the oocyte nucleus.
2. Analysis of the Three-Dimensional Distribution of mRNA within a Cell
   Analysis of the three-dimensional distribution of mRNA within a cell is especially interesting in the case of RNAs that are localized to distinct subcellular regions. Visualization of the localization sites of the RNA on its way, between the site of its synthesis, the nucleus, and the site of its final accumulation, often at the cell periphery, may provide valuable information about the mechanisms underlying the localization processes.
   Figure 23-2 shows a three-dimensional reconstruction of *osk* RNA localization in a stage 8 egg chamber. *osk* RNA was detected with a rhodamine-labeled RNA probe and the images were acquired with a confocal microscope. The images shown here represent stacks of images superimposed to create the three-dimensional image. Two interesting observations can be made from analyzing these images. First, within the nurse cells, where *osk* RNA is synthesized, the cytoplasmic population of the RNA appears granular. These granules may represent transport particles on their way to the oocyte. Second, within the oocyte, *osk* RNA accumulates at the border between the nurse cells and the oocyte. In this three-dimensional reconstruction, the anterior accumulation is visualized as a ring. This anterior accumulation of *osk* RNA is transient and represents a distinct step in the localization process (Ephrussi et al. 1991; Kim-Ha et al. 1991). It precedes the final localization of *osk* mRNA at the posterior pole of the oocyte that is seen later in oogenesis (compare with Figure 23-1).

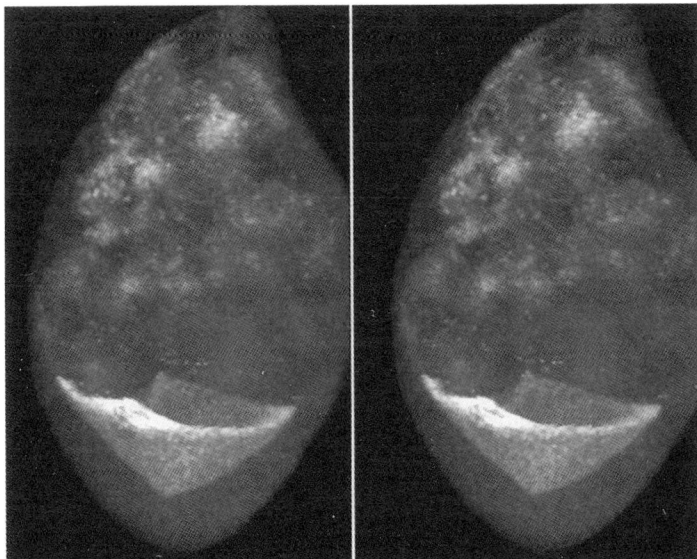

Figure 23-2 Tridimensional reconstitution of *osk* RNA distribution in an egg chamber at stage 8. *osk* RNA is associated with the oocyte cortex and is specifically enriched at the oocyte anterior, where its distribution resembles a ring. In addition, *osk* RNA is detected in the nurse cells, where it appears granular. *osk* RNA probe is labeled with rhodamine. The images are projections of 15 optical sections acquired 2 μm apart using confocal microscopy. To obtain the best tridimensional view, it is recommended to use stereoscopic glasses. For orientation of the egg chamber, see Figure 23-1.

The cytoskeletal elements, including microtubules and actin filaments, have been implicated in the localization of *osk* and other mRNAs (for a review, see Cooley and Theurkauf 1994; St Johnston 1995; Glotzer and Ephrussi 1996). We believe that the fluorescent RNA in situ hybridization technique will be valuable in developing further protocols where the distributions of RNA and the cytoskeletal elements are visualized simultaneously and are reconstructed in three dimensions. This will further our understanding of the function that the cytoskeleton plays in the localization of mRNAs.

3. Quantitative Analysis of the Relative Amounts of mRNA Expressed and/or Localized during Different Developmental Stages of Oogenesis

Although hybridization in situ, a technique that studies the steady-state distribution of the RNA, cannot provide direct information about the dynamics of the localization processes — for instance, the pathway, directionality, and rate of movement of the RNA molecules — it can be valuable in answering other related questions. For example, when does the localization process start and how long does it continue? Fluorescent in situ hybridization allows one to reliably and reproducibly quantitate the amount of fluorescent probe in oocytes at different developmental stages, provided that the images are acquired with a camera that responds to light in a linear manner over the range of intensities present in the sample. The previously used procedures of RNA detection, which rely on a secondary, enzymatic, step for detection of the probe, are not easily applicable to the evaluation of relative levels of gene expression. Such hybridizations are difficult to control and compare, given the nonlinearity of enzymatic reactions.

Using fluorescent in situ hybridization, we have analyzed the relative accumulation of *osk* RNA at the posterior pole of the oocyte at two developmental stages of the oogenesis (Figure 23-3). Whole-mount egg chambers were hybridized to rhodamine-labeled *osk* RNA probe and the images of the oocytes were acquired with a cooled CCD camera. To quantitate the accumulation of fluorescence at the posterior pole, the images were acquired under identical conditions, in which the full dynamic range of fluorescence was registered by the camera (Figure 23-3). Image analysis was performed using the NIH Image program. In each image, two areas of identical size were chosen: first, covering the posteriorly localized fluorescence, and second, in the neighboring cytoplasm (background). The fluorescence in both areas was integrated and the background was subtracted from the fluorescence accumulated at the posterior pole. For each developmental stage (stage 9 and stage 10), the posteriorly accumulated fluorescence was averaged and the standard deviation was calculated. The average obtained for stage 9 was used as a base to normalize the data. These data show that the process of *osk* RNA localization to the posterior pole continues beyond stage 9, as nearly twice as much RNA is detected at the posterior pole of oocytes at stage 10 compared with oocytes at stage 9.

Methods

Protocol 1 describes the production of RNA containing free amino groups. This is the substrate to which succinimidyl esters of fluorochromes are coupled, generating the fluorescent RNA (described in Protocol 2).

Protocol 1: Template Preparation and Transcription In Vitro

1. Linearize 10 μg of plasmid, extract twice with phenol/chloroform and once with chloroform/isoamyl alcohol. Add NH_4OAc (final concentration approximately 1.7 M) and precipitate with 2 volumes of EtOH (prechilled at $-20°C$). Resuspend the DNA in 10 μl of DEPC-treated H_2O. Verify that the linearization was complete and the quality of the DNA by agarose gel electrophoresis.
2. Assemble the following components at room temperature (final concentrations are given in parentheses): in vitro transcription buffer (1×), $MgCl_2$ (14 mM), ATP, CTP, GTP (2.5 mM each ribonucleotide), amino-allyl-UTP and UTP (2.5 mM combined final concentration; amino-allyl-UTP/UTP ratio is typically 1:10 or 1:5), RNasin (50 U total), plasmid DNA (10 μg), RNA polymerase (40 U total), DEPC-treated H_2O (100 μl final volume of the reaction).
3. Incubate the reaction at 37°C for 1 hour; add a second aliquot of RNA polymerase (40 U) and incubate the reaction for 1 more hour at 37°C.

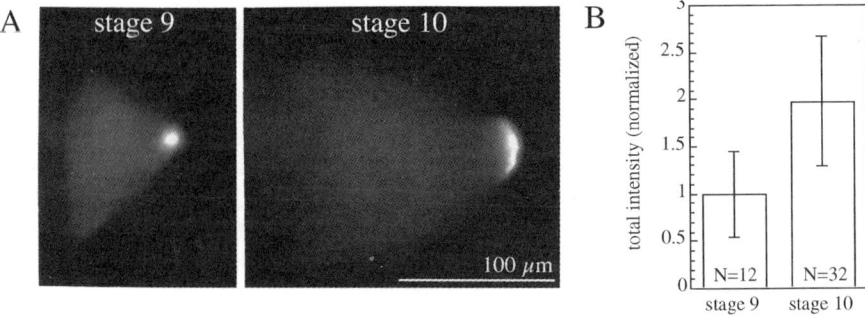

Figure 23-3 Quantitative analysis of *osk* RNA accumulation at the posterior pole of the oocyte at stages 9 and 10. *osk* RNA is detected by a whole-mount RNA hybridization in situ with an RNA probe labeled with rhodamine. (A) Examples of images of stage 9 and stage 10 egg chambers, respectively, acquired with a cooled CCD camera. (B) Comparison of the relative amounts of *osk* RNA accumulated at the posterior pole of the oocyte at stages 9 and 10. The columns represent a difference between the total fluorescence localized at the posterior pole and the fluorescence that is unlocalized and remains dispersed in the ooplasm (background). From these data, it appears that *osk* RNA continues to localize to the posterior pole of the oocyte beyond stage 9, as two times more fluorescence is detected at this location in oocytes at stage 10 than in oocytes at stage 9. Numbers of oocytes examined of each stage are indicated in parentheses.

4. Add RNase-free DNase (40 U) and incubate the reaction for 15–30 minutes at 37°C.
5. Extract the RNA twice with phenol/chloroform and once with chloroform/isoamyl alcohol.
6. Remove unincorporated nucleotides by gel filtration on a Sephadex G50 spin-column, following the manufacturer's instructions.
7. Add NH$_4$OAc (final concentration 2.5 M) and precipitate the RNA with 3 volumes of EtOH (prechilled at −20°C) overnight at −80°C.
8. Spin the RNA for 15 minutes at 14,000g; wash with 80% EtOH (prechilled at −20°C) and spin for 5 minutes at 14,000g.
9. Air-dry the RNA pellet in an open Eppendorf tube at 60°C in a heating block (do not overdry the pellet, as this could impede resuspension of the RNA).
10. Resuspend the RNA in 25 μl DEPC-treated H$_2$O.
11. Assess the quality and quantity of the RNA by agarose-gel electrophoresis and spectrophotometry. Up to 100 μg of RNA are obtained per reaction.

Protocol 2: Fluorescent Labeling of the RNA

We have successfully and reproducibly coupled amino-allyl modified RNAs with the following fluorochromes: 5-(and-6)-carboxy-X-rhodamine, succinimidyl ester; 6-(fluorescein-5-(and-6)-carboxamido)hexanoic acid, succinimidyl ester (SFX); and Rhodamine-green™-X, succinimidyl ester, hydrochloride (mixed isomers, see note below). Stock solutions of fluorochromes are made in dimethyl-formamide (DMF) at 100 mM. This RNA-labeling protocol was adapted from the protein-labeling protocol provided by Molecular Probes.

1. Assemble at room temperature the following components (final concentrations are given in parentheses): RNA (25 µg or more), bicarbonate buffer (pH 9, 0.15 mM), fluorochrome (5 mM), and DEPC-H$_2$O (final volume of the reaction 100 µl).

   Note
   If Rhodamine-green is used, 50 µl of DMF is added before the fluorochrome to facilitate its resuspension; in this case, the final volume of the reaction is 150 µl.

2. Incubate the reaction at room temperature for 1–3 hours in the dark.
3. Remove unincorporated fluorochromes by gel filtration on a Sephadex G50 spin-column.
4. Assess the quality and quantity of the RNA by agarose-gel electophoresis and spectrophotometry; the gel should not contain ethidium bromide. A fluorescent band corresponding to the labeled RNA should be visible with UV light from a hand-held apparatus. Measure the absorption spectrum of the labeled RNA at 260 nm (RNA) and the appropriate second wavelength (i.e., that characteristic of the fluorochrome used). Two peaks should be obtained: one corresponding to the RNA and the second corresponding to the fluorochrome, from which an estimate of the concentrations of the RNA and the fluorochrome, and the efficiency of labeling can be made. Typically, 10–30 fluorochromes are coupled per 1 kb of an RNA molecule.
5. Precipitate the RNA with NH$_4$OAc/EtOH overnight at $-80°C$, as in step 7 of Protocol 1.
6. Spin, wash, and air-dry the RNA pellet (see Protocol 1, steps 8–11). Resuspend the RNA in 100 µl of hybridization buffer (see Protocol 3, step 9).

Protocol 3: Hybridization In Situ
The whole-mount in situ hybridization protocol is that of Ephrussi et al. (1991), with the following modifications.

1. Anesthetize 2–3-day-old, well-fed female flies on a pad using CO$_2$. Using dissecting forceps, transfer the flies individually into a Petri dish containing PBS and 0.1% Tween 20 (PBT) at room temperature. Using a dissecting microscope and forceps, open the abdomen of each fly and dissect out the ovaries into the PBT. Using tungsten needles, tease apart the ovaries slightly (this facilitates the even access of solutions into the egg chambers located deep within the ovary). Transfer the ovaries into an Eppendorf tube containing fresh PBT using pipette tips, cut such that the ovaries pass through easily.
2. Fix the ovaries for 20 minutes. Fixation solution contains 1 ml of 4% paraformaldehyde in PBS, 100 µl of DMSO, and 100 µl of sodium hypochloride (household bleach). One-milliliter volumes of solutions are used in this and the following steps, and the incubations are performed at room temperature, with gentle rocking, unless indicated otherwise.
3. Wash the fixed ovaries in PBT four times for 5 minutes each.

4. Digest the ovaries with proteinase K at a concentration of 50 µg/ml in PBT for 45 minutes. It is recommended to perform a titration with different concentrations of proteinase K, as the activity of the enzyme varies from batch to batch.
5. Stop the digest by rinsing once in glycine (2 mg/ml in PBT), followed by two washes with PBT for 5 minutes each.
6. Postfix the ovaries in 4% paraformaldehyde/PBS for 20 minutes, followed by washing with PBT, three times for 5 minutes each.
7. Permeabilize the ovaries using MeOH (prechilled to −20°C) for 1 hour at −20°C.
8. Rehydrate the ovaries step-wise, by successive washes in PBT/MeOH (3:7, 1:1, 7:3), for 5 minutes each, followed by washes with PBT, three times for 5 minutes each. The step-rehydration helps preserve good morphology of the fixed egg chambers.
9. Rinse the ovaries in PBT/hybridization buffer (1:1), then prehybridize in hybridization buffer (HB) for 1 hour at 60°C in a water bath. HB contains 50% formamide, 5× SSC, 100 µg/ml tRNA, 50 µg/ml heparin, 0.1% Tween 20, pH 4.5, with citric acid.
10. Exchange the buffer with 50–100 µl of prewarmed HB containing 1–5 µl of fluorescently labeled RNA probe (titration is recommended). Incubate the hybridization reaction overnight at 60°C in a water bath.
11. Wash the hybridized egg chambers first in prewarmed HB, three times for 20 minutes each at 60°C, then once in prewarmed PBT/HB (1:1) for 20 minutes at 60°C, and finally four times in PBT, for 20 minutes each at room temperature.
12. Remove most of the PBT and replace with a few drops of the *n*-propyl gallate mounting medium (80% glycerol, 10 mM Tris, pH 8.5, 4% *n*-propyl gallate). Allow the ovaries to equilibrate for a few hours at room temperature or overnight at 4°C.
13. Cut off the end of a 1-ml plastic pipette tip (so that its diameter allows free movement of the ovaries), and, using a Pipetman, transfer the ovaries in a drop of mounting medium onto a glass slide. Under the dissecting microscope, carefully split the ovaries into single ovarioles using tungsten needles. Place a coverslip over them and seal with nail polish.

Microscopy and Image Analysis

The samples can be evaluated using a standard microscope equipped for epifluorescence. However, images of far better resolution can be obtained using a laser scanning confocal microscope equipped with filters to detect the fluorochrome of choice. We have used the EMBL compact confocal microscope, mounted on a Zeiss Axiovert equipped with 25× (0.8 NA) and 100× (1.3 NA) oil-immersion lenses. The images can be processed using the Adobe Photoshop Program and presented as gray-scale files (see Figures 23-1 and 23-2) or RGB files (an overlay of the two single files; not shown). If quantitative analysis of the levels of expression/accumulation of RNA is attempted, we recommend that the images be acquired using a CCD camera (see Figure 23-3). We used a Photometrics cooled CCD camera mounted on a Zeiss Axiovert controlled by software developed at the EMBL (Herr et al. 1993). Image analysis was performed on a Macintosh computer using the public domain NIH Image Program (developed at the U.S. National Institutes of Health and available on the Internet at http://rsb.info.nih.gov/nih-image/).

ACKNOWLEDGMENTS We thank Rainer Saffrich for help with CCD camera, Daniel St Johnston for providing the grk cDNA plasmid and for helpful comments on the in situ hybridization protocol, Elisa Izaurralde and Jim Deshler for suggestions on fluorescent RNA labeling, and David Drechsel and Michael Glotzer for advice on image analysis.

References

Berleth, T., Burri, M., Thoma, G., Bopp, D., Richstein, S., Frigerio, G., Noll, M., and Nüsslein-Volhard, C. (1988). The role of localization of *bicoid* RNA in organizing the anterior pattern of the *Drosophila* embryo. EMBO J. 7:1749–1756.

Cooley, L., and Theurkauf, W. E. (1994). Cytoskeletal functions during *Drosophila* oogenesis. Science 266:590–595.

Curtis, D., Lehmann, R., and Zamore, P. D. (1995). Translational regulation in development. Cell 81:171–178.

Ephrussi, A., Dickinson, L. K., and Lehmann, R. (1991). *oskar* Organizes the germ plasm and directs localization of the posterior determinant *nanos*. Cell 66:37–50.

Frigerio, D., Burri, M., Bopp, D., Baumgartner, S., and Noll, M. (1986). Structure of the segmentation gene *paired* and the *Drosophila* PRD gene set as part of a gene network. Cell 47:479–746.

Glotzer, J. B., and Ephrussi, A. (1996). mRNA localization and the cytoskeleton. Seminars Cell Dev. Biol. 7:357–365.

Glotzer, J. B., Saffrich, R., Glotzer, M., and Ephrussi, A. (1997). Cytoplasmic flows localize injected *oskar* RNA in *Drosophila* oocytes. Curr. Biol. 7:326–337.

Heng, H. H., Spyropoulos, B., and Moens, P. B. (1997). FISH technology in chromosome and genome research. Biosessays 19:75–84.

Herr, S., Bastian, T., Pepperkok, R., Boulin, C., and Ansorge, W. (1993). A fully automated image acquisition and analysis system for low light level fluorescence microscopy. Methods Mol. Cell. Biol. 4:164–170.

Hughes, S. C., Saulier-Le Drean, B., Livne-Bar, I., and Krause, H. M. (1996). Fluorescence in situ hybridization in whole-mount *Drosophila* embryos. Biotechniques 20:748–750.

Kim-Ha, J., Smith, J. L., and Macdonald, P. M. (1991). *oskar* mRNA is localized to the posterior pole of the *Drosophila* oocyte. Cell 66:23–35.

Neuman-Silberberg, F. S., and Schüpbach, T. (1993). The *Drosophila* dorsoventral patterning gene *gurken* produces a dorsally localized RNA and encodes a TGFα-like protein. Cell 75:165–174.

Spradling, A. C. (1993). Developmental genetics of oogenesis. In *The Development of Drosophila melanogaster*, Bate, M., and Martinez-Arias, A., eds. Cold Spring Harbor, NY, Cold Spring Harbor Laboratory Press, pp. 1–70.
St Johnston, D. (1995). The intracellular localisation of messenger RNAs. Cell 81:161–170.
St Johnston, D., Driever, W., Berleth, T., Richstein, S., and Nüsslein-Volhard, C. (1989). Multiple steps in the localization of *bicoid* RNA to the anterior pole of the *Drosophila* oocyte. Development (Suppl. 107):13–19.
Stephenson, E. C., Chao, Y.-C., and Fackenthal, J. D. (1988). Molecular analysis of the *swallow* gene of *Drosophila melanogaster*. Genes Dev. 2:1655–1665.
Wang, C., and Lehmann, R. (1991). *Nanos* is the localized posterior determinant in *Drosophila*. Cell 66:637–648.
Wang, C., Dickinson, L. K., and Lehmann, R. (1994). Genetics of nanos localization in Drosophila. Dev. Dyn. 199:103–115.

Appendix: Reagents

The ovaries shown in the figures were dissected from Oregon R (wild-type) flies.
RNAs used as probes in these experiments were transcribed from plasmids containing full-length *osk* and *grk* cDNAs (Ephrussi et al. 1991; Neuman-Silberberg and Schüpbach 1993).
The proteinase K stock is prepared in H_2O at 4 mg/ml.
Proteinase K, transcription buffer, NTPs, RNA polymerases, RNasin, RNase-free DNase, and Sephadex G50 spin-columns are from Boehringer Mannheim.
Amino-allyl UTP, dimethyl-formamide, *n*-propyl gallate, tRNA, and heparin are from Sigma.
Fluorochromes are from Molecular Probes.

24

Assessment of Poly(A) Tail Lengths on Specific mRNAs during Development

Fernando J. Sallés
Sidney Strickland

The generation of viable offspring is completely dependent upon maternally derived factors deposited within the oocyte. These maternal factors serve as a molecular road map guiding development during a period of transcriptional inactivity that occurs in early embryogenesis across a wide variety of species.

The requirement for these maternal factors is strikingly demonstrated in *Drosophila* development. In *Drosophila*, the oocyte grows over a 12-day period during which it accumulates cytosol and mRNAs generated primarily by its 15 conjoined nurse cells. After the nurse cells have transported most of their contents into the oocyte, the oocyte cell cycle arrests at metaphase I of meiosis until passage through the uterus, when fertilization occurs. The metaphase I arrest marks the end of the transcriptional contribution of the maternal genome to its offspring and marks the beginning of a period of transcriptional inactivity that remains in effect through the completion of both meiotic divisions, fertilization, and early embryonic cleavages until zygotic genome activation. Therefore, all developmental decisions that occur during this transcriptionally inactive period must be orchestrated by products, either proteins or mRNAs, stored in the cytoplasm of the oocyte. If deleterious mutations arise that affect either a factor's activity or its precise chronological activation, then development is compromised. Oocytes produced by females with such mutations give rise to abnormal, at times grossly deformed, embryos, as in the case of *bicoid*-deficient females whose embryos completely lack any head or thoracic structures.

These critical factors are produced during the oocyte growth phase that precedes meiotic arrest. Among them are a unique class of translationally regulated mRNAs (including *bicoid* mRNA described above). These mRNAs are collectively known as the stored or dormant maternal mRNAs (abbreviated here as dmRNAs). dmRNAs serve as a maternal care package for the female's future offspring. A primary form of dmRNA translational regulation is cytoplasmic lengthening of the poly(A) tail (cytoplasmic polyadenylation) and this has been extensively studied in *Xenopus*, in mouse, and more recently in *Drosophila* (for reviews, see Richter 1996; Wickens et al. 1996). A generalized regulatory scheme amalgamated from studies in all three systems shows that dmRNAs receive a long poly(A) tail upon transcription that is shortened during localization in the cytoplasm. The dmRNAs with short poly(A) tails are not translated. Upon the resumption of meiosis, and/or at various points during fertilization or early embryonic development, subsets of these dmRNAs are translationally activated by cytoplasmic polyadenylation, thereby providing chronological developmental cues in the absence of transcription. The timing and extent of a dmRNA's cytoplasmic polyadenylation and subsequent translation is encoded for in its 3'UTR and varies for different dmRNAs. Therefore, to aid in understanding early developmental decisions and the *cis*-acting RNA sequences that regulate translational control by cytoplasmic polyadenylation, it is necessary to have a quantitative assay to follow changes in dmRNA poly(A) tail length.

This chapter briefly reviews methods for determining specific mRNA poly(A) tail lengths and provides two detailed protocols for PCR-based assays that allow rapid (1-day) and sensitive (subnanogram) determination of polyadenylation.

Analysis of Polyadenylation: The Poly(A) Test (PAT) Assays

Several assays have been developed to study polyadenylation, the most commonly used being oligo(dT) and RNase H treatment of poly(A) or total RNA, followed by Northern analysis (Vournakis et al. 1975). A second method is RNase T1 protection using a radiolabeled RNA probe to the 3' end of the RNA of interest, followed by native gel analysis (Wilson and Treisman 1988). Both of these techniques require large quantities of RNA and multiple RNA manipulations. Additionally, they are labor intensive and do not accurately quantitate poly(A) tail length. A third method, RNA ligation–PCR (Sallie 1993), can be applied to the analysis of polyadenylation. This technique is very sensitive, given the use of PCR, and can readily quantitate poly(A) tail lengths, but it requires multiple RNA manipulations and relies on an inefficient single-stranded RNA ligation.

We have developed two PCR-based assays that have several advantages for poly(A) tail analysis. The techniques quantitatively determine poly(A) tail length, can be used on minute quantities (subnanogram) of total RNA with a minimum of RNA manipulations, and can be carried out in a single tube reaction in less than 1 day. The first assay, hereby designated RACE-PAT (Sallés et al. 1992; an application of the Rapid Amplification of cDNA Ends (RACE) protocol (Frohman et al. 1988), is simpler but less sensitive with respect to changes in poly(A) length. The second assay, hereby designated ligase-mediated PAT (LM-PAT) (Sallés and Strickland 1995), yields cleaner results especially in situations where poly(A) changes are more subtle. These assays have been used to study oogenesis and early development in mouse (RACE-PAT: Sallés et al. 1992; Temeles and Schulz 1997; LM-PAT: Sallés and Strickland 1995;), *Drosophila* (LM-PAT: Sallés et al. 1994; Gavis et al. 1996; Lieberfarb et al. 1996), *Caenorhabditis elegans* (LM-PAT: B. Goodwin, personal communication), and zebra fish (LM-PAT: M. O'Connell, personal communication). Additionally, the assays have been applied in nondevelopmental systems to study mRNA polyadenylation of specific mRNAs in somatic tissues (RACE PAT: Sheflin et al. 1996), cell culture lines (LM-PAT: Muckenthaler et al. 1997), and heteronuclear polyadenylation in mouse oocytes (RACE-PAT: Huarte et al. 1992) and in *Xenopus* (RACE-PAT: Rao et al. 1996).

RACE-PAT Theory

RACE-PAT relies on the ability of oligo(dT) to prime reverse transcription of an mRNA from any location along its poly(A) tail (Figure 24-1A, B). Therefore, for each polyadenylated mRNA, there will be heterogeneous cDNAs whose $5'$ ends depend on the site of oligo(dT) annealing. If the length of the poly(A) tail on a specific mRNA changes from one time point to another, then the cDNA heterogeneity also changes. The oligo(dT) primer used for reverse transcription is linked to a GC-rich anchor sequence [oligo(dT)-anchor] which serves to maintain cDNA heterogeneity during subsequent PCR amplification. Once generated, the PAT cDNAs allow the analysis of any mRNA within the population by PCR using the oligo(dT)-anchor in conjunction with a message-specific primer. The differences in $3'$ end heterogeneity between two samples for a given mRNA can be visualized by standard electrophoretic techniques (see Figure 24-2A). Additionally, PAT cDNAs are stable, allowing PCR analysis even after several years (unpublished results).

One can confirm proper amplification of the mRNA of interest by cleaving a fraction of the PCR products with a restriction endonuclease known to recognize a specific site in the sequence. This digest should generate a constant $5'$ portion and the heterogeneous $3'$ cDNA ends (see scheme, Figure 24-1B, D; Sallés et al. 1992). Restriction digest can also aid in allowing a more accurate quantitation of the size of the poly(A) tail by allowing the

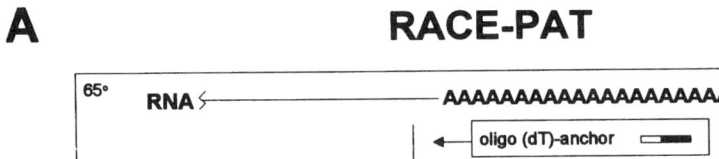
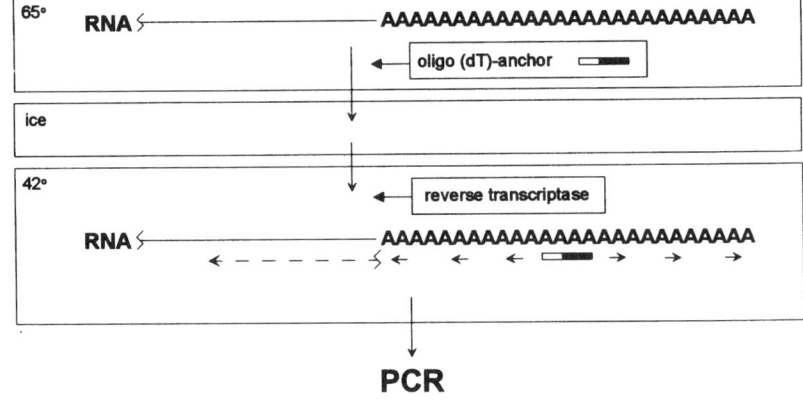
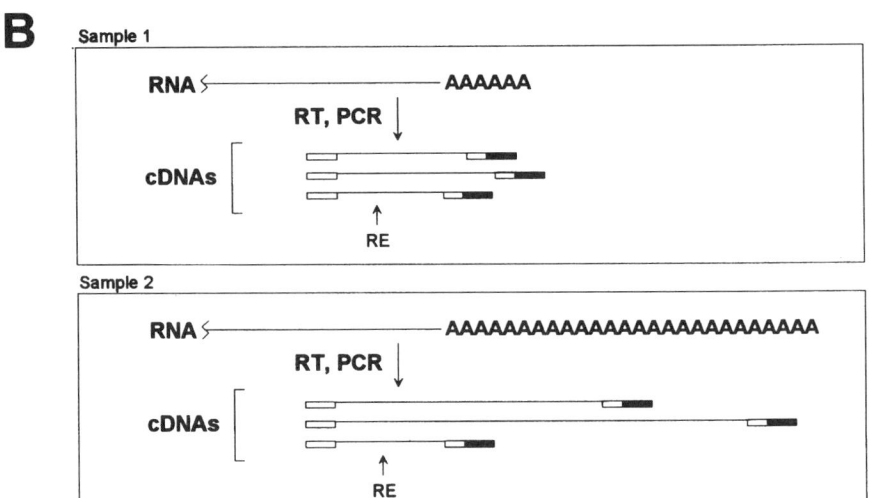

Figure 24-1 Schematic representation of the PAT assays. (A) RACE-PAT — refer to the section "RACE-PAT Theory" for details. Open/filled box, oligo(dT)-anchor primer; arrows flanking the oligo(dT)-anchor represent the range of possible annealing locations to prime reverse transcription; dashed line, first-strand cDNA. (B) RACE-PAT results — two samples (1 and 2) with differing poly(A) tail lengths on the same RNA. RNA is represented as a solid line, with the size of the poly(A) tail depicted by the stretch of letters A. Samples are reverse transcribed (RT) and subjected to PCR as shown in (A). Possible representative PAT PCR products are shown for each sample to illustrate heterogeneity [bracket, (cDNAs)]. Open box, message-specific primer; RE, unique (within amplified region) restriction endonuclease site used to control specific amplification. (C) LM-PAT — refer to the section "LM-PAT Theory" for details. Hatched box, phosphorylated

C LM-PAT

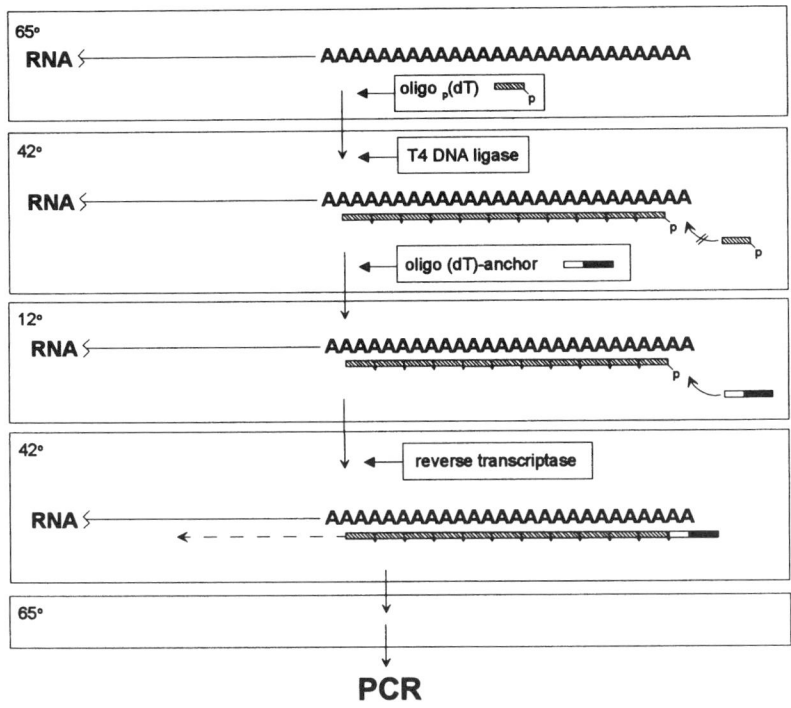

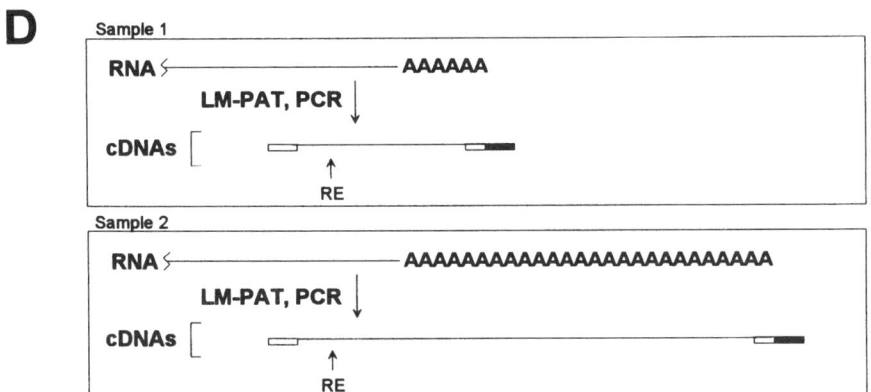

oligo(dT)$_{12-18}$. (D) LM-PAT results—two samples (1 and 2) with differing poly(A) tail lengths on the same RNA. RNA is represented as a solid line with the size of the poly(A) tail depicted by the stretch of letters A. Samples are subjected to LM-PAT cDNA synthesis and PCR amplified as shown in (C). Representative PAT PCR products are shown for each sample [bracket, (cDNAs)]. Open box, message-specific primer; RE, unique (within amplified region) restriction endonuclease site used to control specific amplification.

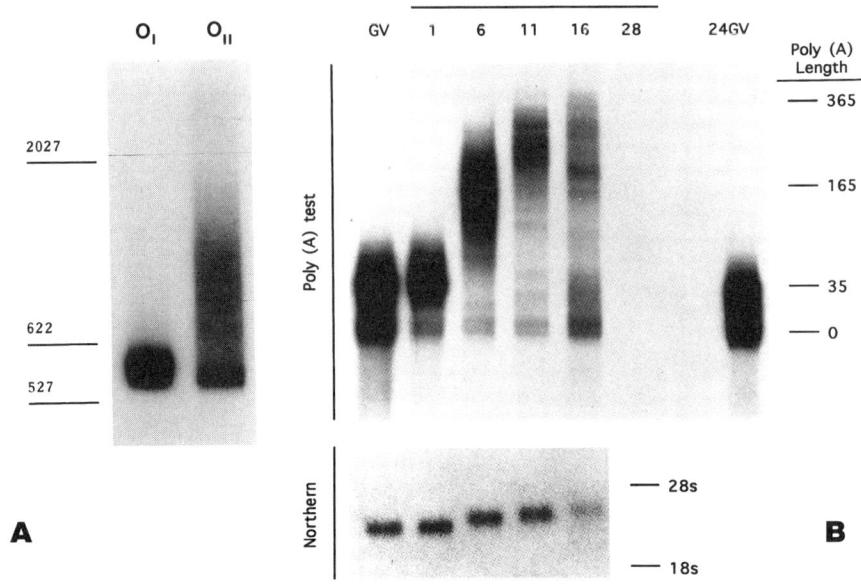

Figure 24-2 Comparative PAT analysis. Autoradiogram of ^{32}P-labeled PAT PCR products after 5% nondenaturing polyacrylamide gel electrophoresis. (A) RACE-PAT analysis of the mRNA for tissue plasminogen activator (tPA) in mouse oocytes (Reprinted with permission from Sallés et al. 1992). O_I — germinal vesicle (GV)-stage oocytes (arrested in meiosis I). tPA mRNA has a short poly(A) tail (approximately 30 nt; Huarte et al. 1987). O_{II} — fertilization-competent secondary oocytes having progressed through meiosis I [appoximately 16 hours after germinal vesicle breakdown (GVBD)]. tPA mRNA has a long poly(A) tail (approximately 400 nt). Migration of molecular-weight markers is indicated. PCR (27 cycles) was performed on one oocyte equivalent of RACE-PAT cDNAs derived from total RNA, and 5% (O_I) or 50% (O_{II}) of the products were analyzed by electrophoresis. The difference in loading amounts reflects the stability of the RNA (see Northern analysis below, GV vs. 16-hour GVBD). (B) LM-PAT (taken from Sallés and Strickland 1995) and Northern (Huarte et al. 1987) analysis of tPA mRNA in mouse oocytes. LM-PAT analysis [poly(A) test, top] of GV-stage oocytes [equivalent to O_I in (A)] and at various times after the resumption of meiosis I as evidenced by germinal vesicle breakdown (GVBD). 24GV — control GV-stage oocytes prevented from resuming meiosis for the duration of sample collection; tPA still has a short poly(A) tail. Poly (A) tail length indications are derived from markers. PCR (30 cycles) was performed on one oocyte equivalent of LM-PAT cDNAs derived from total RNA, and 25% (GV, 1 hour, 24GV) or 50% (6, 11, 16 hours) of the PCR products analyzed by electrophoresis. Estimates of the poly(A) tail length at the various time points correlate well with those derived from Northern analysis (Northern analysis; Huarte et al. 1987). Compare lanes O_I and O_{II} in RACE-PAT with GV and 16-hour GVBD in LM-PAT, respectively, to contrast the results of the two assays.

analysis of a smaller fragment of DNA. Additionally, If the mRNA contains alternate polyadenylation sites, they will appear as a double (or triple, etc.) profile (data not shown).

When using RACE-PAT, the heterogeneity seen in the PCR products is a consequence of oligo(dT)-anchor annealing along the entire poly(A) tail during the initial reverse transcription. This makes visualization of minor, but possibly significant, changes between samples difficult. In order to decrease this effect, we developed LM-PAT.

LM-PAT Theory

In LM-PAT, the oligo(dT)-anchor is targeted to the 3' end of the poly(A) tail (Figure 24-1C, D). First, the poly(A) tail is saturated with phosphorylated oligo(dT)$_{12-18}$ [p(dT)$_{12-18}$] at 42°C and the oligos ligated together using T4 DNA ligase creating a poly(dT) copy of the poly(A) tail. A short stretch of adenylate residues ($n \leq 10$) on the 3' and 5' ends of the poly(A) tail should remain unpaired due to unfavorable annealing at 42°C. A five-fold molar excess of the oligo(dT)-anchor is then added and the temperature lowered to 12°C. By lowering the temperature and increasing the oligo(dT) [p(dT)$_{12-18}$ and oligo(dT)-anchor] concentration, annealing of the oligo(dT)-anchor to the poly(A) tail termini is favored and ligation of the oligo(dT)-anchor occurs on the 3' end via the phosphorylated poly(dT) stretch. The temperature is then raised to 42°C and reverse transcriptase is added to synthesize cDNAs using the nascent poly(dT)-anchor as the primer. After reverse transcription, the reaction is heat-inactivated to eliminate any residual ligase activity. As with RACE-PAT, these cDNAs are very stable and the poly(A) state of any mRNA within the pool can be analyzed by PCR and electrophoresis, as described above, using a message-specific primer and the oligo(dT)-anchor primer (see Figure 24-2B).

Materials

RACE-PAT cDNAs

All of the following enzymes and solutions should be stored at −20°, except where noted.

DEPC (Diethyl pyrocarbonate)-treated dH$_2$O (Sambrook et al. 1989; stored at room temperature
Superscript II RNase H$^-$ reverse transcriptase (BRL; 200 U/μl).

[For RACE-PAT, it has been reported that results are improved if first-strand synthesis is carried out with AMV-RT vs. superscript II or unmodified MMLV-RT (Brooks et al. 1995)].

5× Superscript II RNase H$^-$ RT buffer
0.1 M DTT in DEPC-dH$_2$O

40 mM dNTP solution in DEPC-dH$_2$O (10 mM each dATP, dTTP, dCTP, dGTP)

An oligo(dT)-anchor primer (5'-GCG AGC TCC GCG GCC GCG T$_{12}$; 200 ng/µl for a 30-mer, in DEPC-dH$_2$O)

[Any oligo with a 3'-(dT) stretch ($n > 10$) and a 5'-GC-rich region that can serve as an anchor for PCR can substitute for this primer.]

Message-specific primer in dH$_2$O (see PCR section in the "Technical notes")
(*Optional*) RNase inhibitor (e.g., RNasin, 40 U/µl; Promega)

LM-PAT cDNAs
In addition to the reagents required for RACE-PAT, the following are needed.

T4 DNA ligase high concentration [10 weiss units/µl (USB)]

[Excess T4 DNA ligase is used to compensate for the enzyme's half-life during the initial 42°C incubation. A variety of enzyme preparations at different concentrations have been used successfully e.g., BRL (4–6 BRL units/µl ≈ 18–27 weiss U/µl); Fermentas (40 weiss U/µl; Muckenthaler et al. 1997); NEBL (400 cohesive end ligation units/µl ≈ 6 weiss U/µl; Promega (20 weiss U/µl); US Biochemical (10 weiss U/µl) (see "Acknowledgments").]

10 mM ATP in DEPC-dH$_2$O (required by the ligase)
Phosphorylated oligo(dT)$_{12-18}$ [p(dT)$_{12-18}$] (10 ng/µl in DEPC-dH$_2$O) (oligo phosphorylation is required by the ligase)

Methods

RACE-PAT cDNAs
1. Isolate total RNA (see "RNA Isolation" in the "Technical Notes") from cells or tissues and resuspend at the appropriate concentration(s) in DEPC-dH$_2$O. Place 5 µl in a sterile RNase-free microfuge tube.
2. Add 1 µl of oligo(dT)-anchor (200 ng) to each RNA sample.
3. Heat-denature at 65°C for 5–10 minutes and place on ice.
5. Add 14 µl of the following mixture and incubate at 42°C for 60 minutes. Make a master mix with enough reagents for all of the PAT reactions:
    4 µl 5× Superscript II H$^-$ RT buffer
    2 µl 0.1 M DTT
    1 µl 40 mM dNTP mixture
    6 µl DEPC-dH$_2$O
    1 µl Superscript II RT
  Addition of 1 µl of an RNase inhibitor is optional, reduce water by the same volume.
6. Carry out PCR (see below).

LM-PAT cDNAs
1. Isolate total RNA (see "RNA Isolation" in the "Technical Notes") from cells or tissues and resuspend at the appropriate concentration(s) in DEPC-dH$_2$O. Place 5 µl in a sterile RNase-free microfuge tube.

2. Add 2 µl of p(dT)$_{12-18}$ (20 ng) to each RNA sample.
3. Heat-denature at 65° for 5–10 minutes.
4. Transfer the tube to a 42°C water bath without an ice-quenching step. [The immediate transfer to 42°C theoretically aids in preventing p(dT)$_{12-18}$ annealing to the extreme ends of the poly(A) tail.]
5. Add 13 µl of the following *prewarmed* (42°C) mixture, mix by pipetting up and down, and incubate at 42°C for 30 minutes.
   Make a master mix with enough reagents for all of the PAT reactions:
      4 µl 5× Superscript II H$^-$ RT buffer
      2 µl 0.1 M DTT
      1 µl 40 mM dNTP mixture
      1 µl 10 mM ATP
      4 µl DEPC-dH$_2$O
      1 µl T4 DNA ligase (10 weiss U/ µl; see "Materials")
   Addition of 1 µl of an RNase inhibitor is optional, reduce water by the same volume.
6. At the end of the incubation, while at 42°C, add 1 µl of oligo(dT)-anchor [200 ng/µl, final concentration of oligo(dT)-anchor is in five-fold excess over p(dT)$_{12-18}$]. Vortex, quickly microfuge, and incubate at 12°C for 2 hours. [We recommend starting with 12°C, but other temperatures ranging from 0°C (ice) to 25°C have successfully been used (F. Sallés, unpublished data; A. Verroti, personal communication).]
7. Transfer the samples to 42°C for 2 minutes. [This short 42°C incubation before adding the RT should destabilize hybridization of the oligo(dT)-anchor to the 5' end of the poly(A) tail which would preferentially prime first-strand synthesis over the ligated poly(dT)-anchor.]
8. Add 1 µl of Superscript II RNase H$^-$ RT, vortex, and incubate at 42°C for 1 hour.
9. Heat-inactivate RT and ligase at 65°C 20 minutes.
10. Carry out PCR (see below).

PAT PCR

We refer you to molecular biology/PCR handbooks for the basics and discuss only those aspects relevant to this particular application (see Sambrook et al. 1989; Mullis et al. 1994).

1. Set up a standard 25–50 µl PCR reaction using 0.5–1 µl of each PAT cDNA. If necessary, the PAT cDNAs can be diluted.
2. Take an aliquot of the amplified products and digest them to completion with a restriction endonuclease to aid in analysis of the 3' ends (see PAT theory above and "Message-Specific Primer Choice" in the "Technical Notes").
3. Analyze the PCR products by gel electrophoresis (see "PAT Analysis" in the "Technical Notes").

Technical Notes

Both of these assays can be performed from RNA isolation to agarose gel analysis of PCR products in a single day. Should difficulties be encountered, controls should be performed to ensure that steps in the reaction are working correctly. These are routine controls and are addressed in many cloning manuals (e.g., Sambrook et al. 1989), but a few particularly relevant points are addressed below. The four major steps and control assays are (1) RNA isolation and integrity controls (RACE-PAT and LM-PAT—isolate large quantities of RNA and analyze by gel electrophoresis and UV shadowing); (2) reverse transcription (RACE-PAT and LM-PAT—radiolabeling first-strand synthesis); (3) ligation reaction (LM-PAT—incubate large overhang DNA markers with ligase); and (4) PCR amplification (RACE-PAT and LM-PAT—include positive controls).

PAT cDNA Synthesis

RNA Isolation We use the acid phenol method (Chomczynski and Sacchi 1987) without the addition of carrier RNA, even for small RNA preparations (< 10 ng). However, any RNA isolation technique leaving the RNA in dH_2O should be compatible with either of the PAT assays and, in our experience, the use of carrier tRNA does not interfere with either assay. For most instances, genomic DNA contamination is not a problem; however, in certain cases some background amplification is removed by prior DNase treatment of the RNA.

To DNase treat the sample, resuspend the RNA in RNase-free dH_2O with DNase buffer (10×: 500 mM Tris-HCl, pH 7.5; 10 mM $MgCl_2$; 1 mg/ml BSA) or an RNA polymerase buffer (commercially supplied RNA polymerase buffers work well), and add 1 µl of RNase-free DNase. The use of an RNase inhibitor is optional. Incubate at 37°C for 30 minutes. Phenol-extract and precipitate the RNA with 0.3 M NaOAC and 2.5 volumes of ethanol (Sambrook et al. 1989). Resuspend in dH_2O.

RNA Concentration The sensitivity of the reaction depends on several factors, including transcript abundance. Although smaller amounts can be used, a good starting point is between 3 ng and 5 ng of poly(A)-enriched RNA (Muckenthaler et al. 1997) or between 10 ng and 250 ng of total RNA. Should problems arise, it may be helpful to perform several PAT cDNA reactions for each sample at different RNA concentrations, e.g., 0.001, 0.01, 0.1, and/or 1µg total, and test each for optimal PCR results with your specific mRNA (B. Goodwin, personal communication). Altering RNA concentrations should eliminate the need to change any other reagent concentration.

RACE-PAT Theoretically, the amount of oligo(dT)-anchor present should be stoichiometric (one oligo/mRNA strand) since multiple priming would decrease cDNA length heterogeneity due to preferential use of the 5′-most annealed oligo. However, results show that even with a wide range of total RNA (approximately 4 ng–200 ng) and using 200 ng of oligo(dT)-anchor (a vast molar excess), reasonable length estimates can be derived (Figure 24-2 and unpublished data; Brooks et al. 1995; Temeles and Schultz 1997; L. Wu and J. Richter, personal communication).

LM-PAT A major variable in LM-PAT is the proper saturation of the poly(A) tail with p(dT)$_{12-18}$. If too much input RNA is used, complete saturation and/or annealing of the p(dT)$_{12-18}$ subunits will not occur and could lead to very poor amplification or a "laddering" of the products (see Figure 24-3). "Laddering" occurs when the oligo(dT)-adapter ligates internally to a partial poly(dT) stretch. The spacing between the "rungs" is the average size of the oligo p(dT) subunits. However, "laddering" usually does not interfere with the interpretation of the results.

PCR and Electrophoretic Analysis

Message-Specific Primer Choice General primer selection guides should be followed (see Sambrook et al. 1989; Mullis et al. 1994). Any PCR primer should theoretically work; however, certain primers never yield clean results and a primer to a different region should be tested. For maximal electrophoretic separation, the primer should be 100–400 nt 5' of the 3' end of the mRNA. In addition, it is convenient to have a unique restriction site (within the amplified region) approximately 50–100 bp from the 5' primer to test amplification specificity (see Figure 24-1B, D). It will simplify analysis if the restriction endonuclease site yields 5' fragments that are shorter than the smallest expected 3' fragments (see Sallés et al. 1992). If analyzing hnRNA, a primer recognizing an intronic region should be chosen and PCR amplification carried out with ^{32}P-end-labeled oligo(dT)-anchor followed by restriction digest to allow visualization of only the 3' end of the molecules (Huarte et al. 1992).

Cycling conditions One should use only the number of cycles necessary to properly visualize the results. Too many cycles can increase background bands. This number has to be determined empirically but 30–35 cycles is a good starting point for ethidium bromide-stained agarose gels, with fewer cycles necessary for the radiolabeled PCR analysis (see below).

PAT Analysis For screening and/or initial purposes, electrophoretic analysis can be conducted using ethidium-stained agarose gels. Agarose gel electrophoresis allows the visualization of even small changes (< 50 nt) in poly(A) tail length (Sallés and Strickland 1995) and is the easiest method. However, for greater resolution and quantitation of size, we recommend spiking the PCR reaction with a small amount of α^{32}P-dATP (any α-labeled dNTP can be used). After amplification, the samples are precipitated using

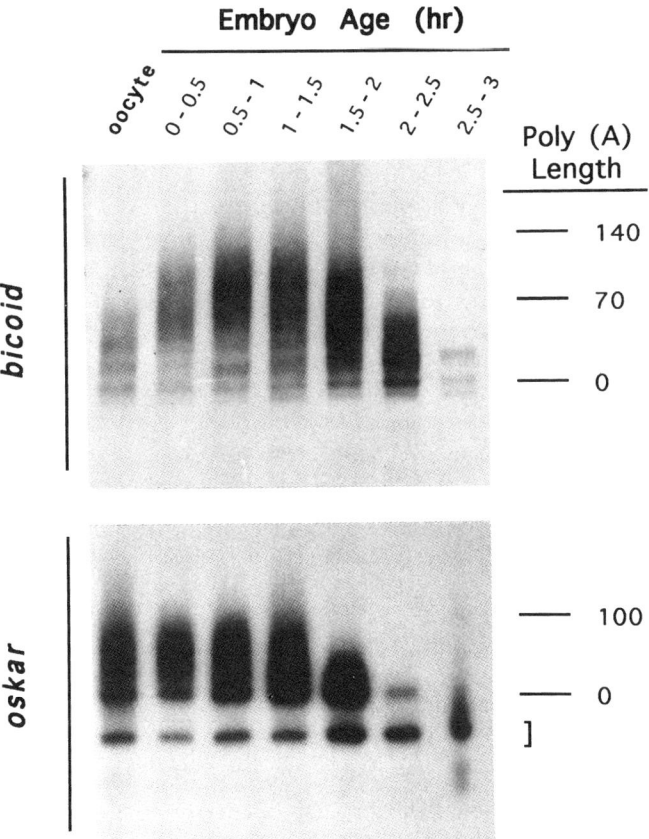

Figure 24-3 LM-PAT time courses for *bicoid* and *oskar* mRNAs. Autoradiogram of ^{32}P-labeled LM-PAT PCR products after 5% nondenaturing polyacrylamide gel electrophoresis. Reprinted with permission from Sallés et al. (1994). Copyright © 1994 American Association for the Advancement of Science. Total RNA was collected from ovaries (oocyte) or eggs at various intervals after laying and subjected to LM-PAT analysis. Both time courses are derived from the same LM-PAT cDNA preparations. Poly(A) tail length indications are derived from markers. The mRNA for *bicoid* undergoes cytoplasmic polyadenylation shortly after egg deposition and experiences tail shortening after 1.5 hours. The mRNA for *oskar* is not cytoplasmically polyadenylated during this window, but is deadenylated like *bicoid* after 1.5 hours. A "laddering" effect is clearly visible in the *bicoid* samples, especially in the oocyte and the first three embryo lanes.

sodium acetate and ethanol and resuspended in dH$_2$O, and a portion restriction digested (if desired) and/or electrophoresed on polyacrylamide gels (generally 4–6 % acrylamide). Nondenaturing and denaturing acrylamide gels have both been used effectively. Nondenaturing gels are generally more aesthetic; however, they can be trickier to electrophorese due to gel migration artifacts. Finally, for some RNAs, changes in polyadenylation alter their stability and hence concentration; therefore, when analyzing PCR products from two samples, it is sometimes necessary to adjust the amounts of each sample used for electrophoresis (see Figure 24-2A and B).

ACKNOWLEDGMENTS   The authors acknowledge the excellent technical assistance of M. Lieberfarb, as well as the excellent technical and editorial assistance of C. Wreden. The authors are indebted to, and graciously thank, the following individuals for sharing unpublished experiences and modifications to the original assays: T. Chu, B. Conne, M. Gallegos, E. Gavis, B. Goodwin, M. Hentze, J. Kimble, M. Lieberfarb, Z.-J. Liu, Y. Minami, S. Moody, M. Muckenthaler, M. O'Connell, J. Richter, J. Schisa, R. Schultz, A. Stutz, S. Sullivan, J.-D. Vassalli, A. Verrotti, P. Webster, M. Wickens, C. Wreden, and L. Wu. This work was supported by NIH GM 51584 to S.S.

References

Brooks, E. M., Sheflin, L. G., and Spaulding, S. W. (1995). Secondary structure in the 3' UTR of EGF and the choice of reverse transcriptases affect the detection of message diversity by RT-PCR. Biotechniques 19:806–815.

Chomczynski, P., and Sacchi, N. (1987). Single step method of RNA isolation by acid guanidinium thiocyanate phenol-chloroform extraction. Anal. Biochem. 162:156–159.

Frohman, M. A., Dush, M. K., and Martin, G. R. (1988). Rapid production of full-length cDNAs from rare transcripts: amplification using a single gene-specific oligonucleotide primer. Proc. Natl. Acad. Sci. USA 85:8998–9002.

Gavis, E. R., Lunsford, L., Evans Bergsten, S., and Lehmann, R. (1996). A conserved 90 nucleotide element mediates translational repression of nanos RNA. Development 122:2791–2800.

Huarte, J., Belin, D., Vassalli, A., Strickland, S., and Vassalli, J.-D. (1987). Meiotic maturation of mouse oocytes triggers the translation and polyadenylation of dormant tissue-type plasminogen activator mRNA. Genes Dev. 1:1201–1211.

Huarte, J., Stutz, A., O'Connell, M. L., Gubler, P., Belin, D., Darrow, A. L., Strickland, S., and Vassalli, J.-D. (1992). Transient translational silencing by reversible mRNA deadenylation. Cell 69:1021–1030.

Lieberfarb, M. E., Chu, T., Wreden, C., Theurkauf, W., Gergen, J. P., and Strickland, S. (1996). Mutations that perturb poly(A)-dependent maternal mRNA activation block the initiation of development. Development 122:579–588.

Muckenthaler, M., Gunkel, N., Stripecke, R., and Hentze, M. W. (1997). Regulated poly(A) tail shortening in somatic cells mediated by cap-proximal translational repressor proteins and ribosome association. RNA 3(9):983–995.

Mullis, K. B., Ferré, F., and Gibbs, R. A., eds. (1994). *PCR: The Polymerase Chain Reaction.* Boston: Birkäuser.

Rao, M. N., Chernokalskaya, E., and Schoenberg, D. R. (1996). Regulated nuclear polyadenylation of Xenopus albumin pre-mRNA. Nuc. Acids Res. 24:4078–4083.

Richter, J. D. (1996). Dynamics of poly(A) addition and removal during development. In *Translational Control*, Hersey, J. W. B., Mathews, M., and Sonenberg, N., eds. Cold Spring Harbor, NY, Cold Spring Harbor Laboratory Press, pp. 481–503.

Sallés, F. J., and Strickland, S. (1995). Rapid and sensitive analysis of mRNA polyadenylation states by PCR. PCR Methods Appl. 4:317–321.

Sallés, F. J., Darrow, A. L., O'Connell, M. L., and Strickland, S. (1992). Isolation of novel murine maternal mRNAs regulated by cytoplasmic polyadenylation. Genes Dev. 6:1202–1212.

Sallés, F. J., Lieberfarb, M. E., Wreden, C., Gergen, J. P., and Strickland, S. (1994). Coordinate initiation of Drosophila development by regulated polyadenylation of maternal messenger RNAs. Science 266:1996–1999.

Sallie, R. (1993). Characterization of the extreme 5' ends of RNA molecules by RNA Ligation-PCR. PCR Methods Appl. 3:54–56.

Sambrook, J., Fritsch, E. F., and Maniatis, T. (1989). *Molecular Cloning: A Laboratory Manual*, 2nd ed. Cold Spring Harbor, NY, Cold Spring Harbor Laboratory Press.

Sheflin, L. G., Brooks, E. M., Keegan, B. P., and Spaulding, S. W. (1996). Increased epidermal growth factor expression produced by testosterone in the submaxillary gland of female mice is accompanied by changes in poly-A tail length and periodicity. Endocrinology 137(5):2085–2092.

Temeles, G. L., and Schultz, R. M. (1997). Transient polyadenylation of a maternal mRNA following fertilization of mouse eggs. J. Reprod. Fert. 109(2):223–228.

Vournakis, J. N., Efstratiadis, A., and Kafatos, F. C. (1975). Electrophoretic patterns of deadenylated chorion and globin mRNAs. Proc. Natl. Acad. Sci. USA 72:2959–2963.

Wickens, M., Kimble, J., and Strickland, S. (1996). Translational control of developmental decisions. In *Translational Control*, Hersey, J. W. B., Mathews, M., and Sonenberg, N., eds. Cold Spring Harbor, NY, Cold Spring Harbor Laboratory Press, pp. 411–450.

Wilson, T., and Treisman, R. (1988). Removal of poly(A) and consequent degradation of c-fos mRNA facilitated by 3' AU-rich sequences. Nature 336:396–399.

25

Three-Dimensional, Time-Lapse Microscopy of *Drosophila* Embryogenesis

Jonathan Minden

Drosophila embryonic development can be divided into two distinct stages lasting 3 hours and 20 hours, respectively (Campos-Ortega and Hartenstein 1985). The first stage entails 13 rounds of rapid nuclear division in the absence of cytokinesis producing the syncytial blastoderm of approximately 6000 nuclei (Foe and Alberts 1983; Foe et al. 1993). The nuclear divisions appear as synchronous waves of mitosis emanating from the anterior and posterior poles of the egg. After the 13th nuclear division, the plasma membrane grows down around each nucleus to form individual cells yielding the cellular blastoderm. The first eight nuclear cycles occur deep within the cytoplasm of the embryo. During interphase 9 and 10, the nuclei migrate to the periphery of the embryo. A small number of nuclei that reach the posterior pole first form individual pole buds, which eventually form the pole cells—these are the germ cell progenitors. The remaining nuclei form an evenly spaced monolayer, with the exception of ~200 nuclei that remain in the yolk. The nuclei, with their centrosomes oriented toward the plasma membrane, are held close to the plasma membrane by an association of microtubules emanating from the centrosomes and membrane-associated actin microfilament caps located above each nucleus. The first 13 nuclear divisions are driven solely by maternally supplied transcripts and proteins. Zygotic transcription begins at a low level during nuclear cycle 11 and steadily increases. After three additional nuclear divisions, the process of cellularization begins immediately at the start of interphase 14. Cellularization lasts approximately 55 minutes and is divided into an initial slow phase followed by a rapid phase (Fullilove and Jacobson 1971, Schejter and Weischaus 1993). During the slow phase, cellularization furrows form between adjacent nuclei to form a hexagonal pattern. The coordinated fusion of small membrane vesicles provides the bulk membrane material needed for the cellularization furrows to grow down and around each nucleus (Loncar and Singer 1995). The nuclei become elongated perpendicular to the apical plasma membrane (toward the outside of the embryo) and their centrosomes remain apically oriented. The rapid phase of cellularization begins when the cellularization furrow grows past the bottom of the nucleus and the rate of addition of membrane vesicles is increased. Cellularization is completed by a cytokinesis-like event that purse-strings off the bottom of the cell. The resulting embryo has ~6000 columnar cells that are approximately 5 μm across and 30 μm long. The nucleus occupies the apical end of the cell.

When cellularization is completed, the second stage of embryonic development begins with the intricate movements of gastrulation (Costar et al. 1993; Leptin 1995). Most cells in the postcellularization embryo undergo two or three more mitoses; exceptions are neuronal cells, which divide many more times, and amnioserosa cells, which cease dividing. The pattern of mitosis is very different from the syncytial stage where synchronous waves of mitosis encompass the whole embryo. Instead, small patches of cells divide as distinct groups, or mitotic domains, in a bilaterally symmetric pattern (Foe 1989). Cells within the same mitotic domain behave similarly with respect to their division time and orientation, as well as cell shape changes and movement patterns. Four major morphogenetic movements occur during gastrulation, these are ventral furrow formation, posterior midgut invagination; cephalic furrow formation, and germ-band extension. The first three movements are driven by cell shape changes, while germ-band extension is driven by cell intercalation. Ventral furrow formation leads to the internalization of mesodermal precursors. Posterior midgut invagination leads to the formation of the posterior endoderm. (The anterior endoderm is formed slightly later in development by the anterior midgut invagination.) The cephalic furrow forms transiently—its role is not well understood. Germ-band extension occurs when tissue posterior to the cephalic furrow contracts ventrally, curls dorsally at the posterior tip, and extends anteriorly along the dorsal surface of the embryo, forming the U-shaped extended germ-band. After the extension of the germ-band, neurogenesis, myogenesis, and organogenesis occur to form the mature larva. All of these processes depend on the different cell types adopting specific shapes and contacts. These shapes and contacts are mediated by the association of cytoskeletal proteins with localized membrane proteins.

Development is a dynamic process that involves mitosis, morphogenesis, cell shape changes, gene expression, and apoptosis. In order to capture these events, a number of time-lapse microscopy techniques have been developed to track these cellular functions (Fullilove and Jacobson 1971; Foe and Alberts 1983; Campos-Ortega and Hartenstein 1985; Kellogg et al. 1988; Foe 1989; Hiraoka et al. 1989; Minden et al. 1989; Minden et al. 1990; Sullivan et al. 1990; Kam et al. 1991; Miller et al. 1991; Vincent and O'Farrell 1992; Costa et al. 1993; Foe et al. 1993; Schejter and Wieschaus 1993; Sullivan et al. 1993a,b; Swedlow et al. 1993; Cooley and Theurkauf 1994; Fogart et al. 1994; Kalpin et al. 1994; Mermall et al. 1994; Theurkauf 1994; Wang and Hazelrigg 1994; Field and Alberts 1995; Kallum et al. 1995; Leptin 1995; Loncar and Singer 1995; Mermall and Miller 1995; Oegema et al. 1995; Davis et al. 1996; Debec et al. 1996; Endow and Komma 1996; Matthies et al. 1996; Minden 1996; Paddy et al. 1996; Namba et al. 1997). These techniques were made possible by a number of advances in optical microscopy and fluorescent reagents over the last decade. They have provided numerous methods for following cell behavior in live embryos in three dimensions over long periods of time. Two major advances have come in the field of fluorescence microscopy to remove out-of-focus fluorescence light, which is a major source of image degradation: serial plane deconvolution, a numerical method pioneered by Agard and Sedat and colleagues (Agard 1984; Hiraoka et al. 1987), and laser confocal microscopy, an optical method developed by White and colleagues (Hyman and White 1987). While the confocal microscope provides high axial resolution, the laser light is often too intense for live imaging, particularly for long time-lapse recordings. The numerical approach allows one to make low-light level, time-lapse recordings without harming the embryo. The limitation of this approach is the computationally intensive postprocessing of images. Similarly, there has been a large increase in the range of fluorescent reagents used to tag macromolecules (such as proteins, lipids, and nucleic acids) and monitor various cellular parameters (such as pH, concentration of Ca^{2+}, membrane potential, and enzymatic activity) (Haugland 1996; Minden 1996). The most recent addition to the reagent list is green fluorescent protein (GFP), which is intrinsically fluorescent. This allows one to visualize GFP-expressing cells without injection of fluorescent dyes (Chalfie et al. 1994). In addition to the devices and reagents, the use of transgenic flies has provided a wide variety of reporter- and inducible-gene constructs for following and manipulating development in vivo.

In this chapter, I will review a number of techniques that we use to generate time-lapse, three-dimensional fluorescence microscopy recordings of live *Drosophila* embryos. All of these methods were pioneered in the laboratories of Drs. Bruce Alberts, Tim Mitchison, David Agard, and John Sedat. I have attempted to present the methods in great detail to provide a guide to performing time-lapse experiments. I will first describe how to collect, mount, and inject embryos. This is the most crucial part of time-lapse microscopy and requires a little practice to master. Next, I will describe the microscope system and time-lapse acquisition considerations.

There are a number of commercially available microscope systems that are easy to use. Before purchasing a system, it is recommended that one tests the system for sensitivity, speed, and versatility. The greatest concern should be that the injection and time-lapse experiment do not perturb embryonic development. Finally, I will describe the array of fluorescent reagents that we have used to monitor a variety of processes in the embryo. Many of these are commercially available or relatively easy to make, or can be obtained from individual laboratories. Since data obtained from all of these methods have been published elsewhere, I will make reference to the relevant publications and not show figures of the results.

Egg Collection and Mounting

Embryos are collected for injection and time-lapse analysis by placing flies that are grown on standard fly media at 25°C (Ashburner 1989) in plastic bottles that do not contain any food. Two bottles, each containing 50–100 flies, generate a sufficient number of embryos for injection experiments. A Kimwipe is placed in the bottle to absorb moisture and increase the surface area for the flies to walk. The bottle is capped with an egg collection plate made from a 35-mm plastic Petri dish filled with apple juice/agar.* The bottle is inverted and left in a quiet, dark room (preferably at 18°C or 25°C with high humidity) for 1–2 hours. The embryos are washed off the plate with a stream of egg wash** into a 150-ml beaker. To dechorionate the embryos, the volume of egg wash is brought to 40 ml and 40 ml of bleach is added. This solution is incubated for 90 seconds with occasional swirling. The embryos are then collected by filtering through a nested pair of stainless steel filters (85 mm, Bellco Biotechnology, Inc.): first, a coarse mesh (40 mesh) to remove flies and agar fragments and then a fine mesh (150 mesh backed by 20 mesh for stiffness) to collect the embryos. The fine filter is then washed three or four times with 100 ml each of distilled water and then egg wash. The embryos will become sticky in the distilled water. The Triton X-100 in the egg wash solution prevents this sticking. The washed embryos are transferred to a 60-mm glass Petri dish by inverting the filter and washing the embryos out with a stream of egg wash.

Time-lapse recording of embryos often requires that the embryos be viewed from a particular orientation. Generally, we record laterally or ventrally mounted embryos. The embryos in the glass Petri dish are examined under a dissecting stereo microscope with transmitted-light illumination to identify the embryos at the desired stage. We typically select embryos between nuclear cycles 10 and 13; for some applications, nuclear cycle 14, cellularizing, or postgastrulation embryos are collected. The desired

*Collection plates were made with apple juice/agar according to Ashburner (1989).
**Egg wash contains 0.7% M NaCl and 0.4% Triton X-100.

embryos are gently pushed around the bottom of the dish with a fine, but dull, stainless steel probe† to form an island of 10–30 embryos; unwanted embryos are pushed away from this island of embryos. The desired embryos are then gently sucked into a glass Pasteur pipette and deposited onto a piece of nylon mesh (1.5 × 1.5-cm, 3–130 mesh; Tetko, Inc.) on a paper towel. The embryos are washed with a few drops of distilled water and the nylon mesh is transferred to a block of 2.25% agar. The agar block is trapezoid-shaped to aid in viewing the embryos with transmitted-light illumination. The top surface, which is about 2 × 2-cm square, is narrower than the bottom. The embryos are very gently transferred from the mesh to one edge of the agar block. The embryos are aligned either parallel to the edge for side injections or perpendicular to the edge for end injections. Generally, there are one or two rows of 20–25 embryos each per coverglass. The embryos will be viewed through a coverglass. Therefore, to view the ventral surface, the ventral side of the embryo must be oriented away from the agar block. For a lateral view, the embryo is positioned laterally. The embryos are oriented by gentle turning with the probe.

Once oriented, the embryos are picked up off the agar block by adhering to a 22 × 50 mm, No. 1 coverglass that has one or two adhesive strip(s) painted across the middle of the coverglass along the short axis. The adhesive is made by dissolving the glue from double-sided tape (Scotch brand) with heptane. Typically, one foot of tape is packed into a 25-ml glass scintillation vial, filled with heptane, and rocked gently over night. The adhesive solution is clarified by centrifugation and stored in a tightly closed scintillation vial. The adhesive stripe is made by streaking 1.5 µl of glue solution across the coverglass; the heptane is allowed to evaporate for a minute or two before picking up the embryos by gently lowering the coverslip into contact with the embryos. Care must be taken to not crush or compress the embryos too much during any of the manipulation steps. Mishandled embryos develop with uneven cellularization or cytoplasmic leakage into the intervitelline space. Proper handling of the embryos up to this point should yield an 80–90% hatching rate. The hatching rate can be tested by immediately covering the embryos with halocarbon oil (700 Series, Halocarbon Products Corp.) and incubating the embryos for 24 hours at 25°C in a humid chamber.

Injection

In order to inject liquid into the embryo cytoplasm or into the intervitelline space, the embryos must be slightly desiccated. This is done by placing the coverglass with the mounted embryos in a dish that contains Drierite. We

---

†The probe is made from one-half of a pair of No. 5 Dumont forceps (Ted Pella, Inc.). The tip was rubbed dull on extra-fine, wet–dry sand paper.

use a 150-mm plastic Petri dish with a layer of Drierite. To determine the dryness of the embryos, the dish is placed under the dissecting microscope and light from the epi-illuminator is swept over the embryos. The surface of hydrated embryos is very reflective and smooth. Dehydrated embryos that are ready for injection appear very slightly dimpled and less reflective. At this point, the embryos are covered with halocarbon oil.

The glass needles used for injection are borosilicate capillaries (1.2 mm o.d., 0.94 mm i.d., and 10 cm long) with an internal filament to aid in loading. These are drawn to give a sealed tip that is about 3–4 mm long and 5–10 μm in diameter at the tip after breaking. A variety of devices can be used to pull needles. We use a horizontal needle puller (Model P-87, Sutter Instrument Co.). The needles are back-loaded with about 1.5 μl of solution delivered from a 10-μl glass syringe. To prevent the needle from clogging, the solution is generally spun for several minutes in a microcentrifuge prior to loading. Once the needle is loaded, it is placed in the needle holder and mounted in the micromanipulator (MO-203, Narashige). The tip of the needle is broken by gently running the tip into the side of a coverglass under halocarbon oil. Positive air pressure is applied to the needle through a 30-ml syringe to force out any air bubbles in the needle. The ideal tip should have a slight bevel to make penetration easier.

To inject the embryo, the needle is positioned at the side of the embryo, generally at the midpoint. To check that the needle is at the appropriate height, a slight negative air pressure is created by pulling the plunger out, not so far that the halocarbon oil is sucked up, and the needle is gently pushed against the embryo. This push should not be hard enough to puncture the embryo, but enough to deform the surface. If the needle is too high or low, it will glance off the side. A correctly positioned needle will slightly collapse the side of the embryo. Once the desired position is found, the needle is pulled back from the embryo and then plunged into it. The embryo will deform slightly and then the needle will penetrate. Care must be taken not to drive the needle entirely through the embryo. Once the needle is in the cytoplasm, positive air pressure is applied and a small plume of solution can be seen entering the cytoplasm, and the embryo will begin to swell. The needle is withdrawn as soon as the embryo appears to reach its full size. This is repeated down the row of embryos.

In some cases, one wants to inject into the intervitelline space. This is done by plunging the needle into the cytoplasm and carefully withdrawing the needle until the tip is just at the vitelline membrane. During the insertion of the needle, make sure that there is negative pressure inside the needle. This will prevent injection of liquid into the cytoplasm. Once the tip of the needle is in place, apply positive air pressure to the needle to inject the material into the intervitelline space; a small amount of material may leak into the cytoplasm. A good intervitelline injection will cause the plasma membrane to separate from the vitelline membrane.

After all the embryos are injected, one should suction off any cytoplasm or injection solution that may leak out of the embryo. Occasionally, there are air bubbles adhering to the embryos. These should be removed by gently pushing them away with the stainless steel probe.

Microscope System

Time-lapse recording of the injected embryos can be performed on a number of microscope systems. We have used both a wide-field microscope with a digital imaging camera and confocal microscopy. The digital camera imaging system is centered around a scientific-grade, cooled CCD camera (CH200, Photometrics with a KAF1400 CCD chip, Kodak) based on the design of Agard and Sedat (Swedlow et al. 1993). This is an extremely light-sensitive camera with a 1024 × 1024 pixel imaging area and 12-bit dynamic range. This provides 4096 gray-levels as compared with 256 gray-levels for video-rate CCD cameras and the photomultiplier tubes of most confocal microscopes. It has been our experience that the CCD camera microscope system is better for time-lapse microscopy than second-generation confocal microscopes, such as the BioRad MRC 600, because it requires less illumination and acquires images more rapidly. The intense laser light causes photobleaching of the fluorescent dyes, which produces free radicals that damage the embryo. The latest generation confocal microscopes and two-photon confocal microscopes may also be suitable for time-lapse microscopy, but we have not tested these systems.

The CCD microscope system has the following components: an inverted microscope (Olympus), a precision XYZ stage (Newport Corp.) fitted with high-precision, computerized stepper-motors (Nanomover, Melles Griot, Inc.), excitation light from a 100-W mercury lamp first passing through an electronic shutter and then a double filter wheel assembly that contains excitation band-pass filters (Chroma Technology) and neutral density filters (Newport Optical); the light is delivered to the microscope via a 1 mm diameter quartz fiber-optic cable; the light is then directed through the objective lens by a quadruple dichroic mirror (Chroma Technology); the emitted fluorescent light passes through a filter wheel assembly that contains emission band-pass filters (Chroma Technology); and finally the image is collected by the CCD camera. The filter wheels, camera, shutters, and stage are controlled by a UNIX work station (Personal Iris 4D35, Silicon Graphics, Inc.) through the VME bus. Software controlling time-lapse acquisition was written in-house. An integrated assembly of these components and software is commercially available from Applied Precision, Inc.

Data Acquisition

There are several parameters to consider for each time-lapse recording experiment; these are the frequency of image acquisition, the number of focal planes, the number of fluorescent signals to be recorded, and the number of embryos per session. The frequency of image acquisition depends of the process being monitored: e.g., chromosome movement during mitosis is quite rapid and requires the recording more than three image stacks per minute; while cell lineage tracing during development requires the recording of an image stack every 5–10 minutes.

The number of focal planes depends on the three-dimensional character of the structure being investigated: e.g., a nucleus is 5 µm in diameter and requires about 25× 0.2-µm optical sections to generate a good three-dimensional reconstruction, while tracking a cell in the whole embryo requires about 10× 5-µm sections to capture to depth the cell may travel through. The wide-field microscope system has limited depth of focus of about 50 µm. The confocal microscope can image more deeply and the two-photon confocal microscope can, in principle, optically section through the 200-µm depth of the embryo. For most time-lapse applications, recording 0.2-µm optical sections over 5 µm will take too long to capture very dynamic processes. Therefore, one must find a balance between the frequency of acquisition and optical sectioning in order to generate a data set that satisfies the experimental requirements. In some cases, we have had to make two different time-lapse recordings: one at high time density and low spatial density and another at high spatial density and low time density.

Many time-lapse microscopy experiments employ multiple fluorescent reagents. The main goal of multireagent experiments is to provide a relational marker; e.g., we have simultaneously examined cell shape and nuclear position during ventral furrow morphogenesis (Kam et al. 1991), chromosome and microtubule dynamics during mitosis (Sullivan et al. 1990), and cell death relative to gene expression and segment boundaries (T. M. Pazdera and J. S. Minden, personal communication). The main issue for multiple fluorescence experiments is that the fluorophores have sufficiently different spectral characteristics and that the optical filters discern between these reagents. Care should be taken to balance approximately the intensity of the different fluorophores at the level of injection. Further adjustments can be made with exposure time and neutral density filter selection.

Knowing the genotype of the embryo is a major issue when examining the effects of different mutations. The mutations most often studied are recessive and only one-quarter of the embryos are of the desired genotype. It is therefore advantageous to be able to determine the genotype of each embryo and to generate multiple recordings per session. Currently, the computer-controlled microscope stage is capable of cycling between 20 embryos per session. The position of the appropriate embryos is logged into the computer prior to the start of the time-lapse series. This allows one to obtain data from a reasonable number of embryos in order to draw statistically significant conclusions.

One major concern with digital time-lapse microscopy is the size of the data set. Each 1024 × 1024 image requires 2 Mbytes of memory, where each pixel is composed of two 8-bit bytes. We frequently make time-lapse recordings that occupy 2 Gbytes of memory. There are two methods to reduce the data load. One is to bin the pixel data in the CCD chip so that each pixel is composed of the sum of four adjacent pixels. This reduces a 1024 × 1024 image to a 512 × 512 image. This gives a four-fold reduction in data storage and only halves the image resolution. The other method is to reduce the image area. A 20× objective is capable of capturing the entire embryo, but only half of the image field is used. Acquiring a 512 × 256 two-fold binned image will capture the whole embryo. This combination of binning and image cropping reduces the data set by a factor of eight.. An important consideration when doing time-lapse microscopy is to arrange for large-scale data archiving. The archival system should be chosen to match the size of the largest anticipated data set. For long-term storage, we use 8-mm data tapes that hold up to 2 Gbytes of data.

Fluorescent Reagents and Probes

A large number of fluorescent reagents have been created to monitor a number of cellular and embryonic processes, including cytoskeletal and nuclear structures, mitosis, morphogenesis, cell death, gene expression, cell fate tracing, and cell fate manipulation. Three general categories of reagents will be described; these are fluorescent protein analogs, fluorescent indicators, and caged compounds.

Fluorescent Protein Analogs

The most commonly used compounds for studying *Drosophila* embryogenesis in vivo are fluorescent protein analogs. These are proteins that carry a fluorescent tag that allows one to follow the dynamic character of the labeled protein. The first application of fluorescent analogs in *Drosophila* embryos was to monitor cytoskeletal and chromosomal behavior in syncytial-stage embryos. These were tubulin, actin (Kellogg et al. 1988) and histones (Minden et al. 1989; Valdez-Perez and Minden 1995). The basic premise behind labeling of these proteins was to ensure that the covalent attachment of the fluorescent dye molecule does not affect the function of the protein. For actin and tubulin, conditions were found that allowed for repeated polymerization and depolymerization without significant loss of protein between cycles. Histone labeling was performed while the histones were bound as nucleosomes to double-stranded DNA–cellulose (Valdez-Perez and Minden 1995). To ensure the proper functioning of the label histones, only fractions that were eluted from the DNA at the expected salt concentration were used in injection experiments.

In addition to these three proteins, topoisomerase II (Swedlow et al. 1993), nuclear lamins (Paddy et al. 1996), cytoskeletal proteins (Field and Alberts 1995; Kallum et al. 1995; Oegema et al. 1995) and various antibodies

(Mermall et al. 1994; Mermall and Miller 1995) have been fluorescently tagged and injected into syncytial embryos. A rule of thumb for fluorescent labeling is to keep the labeling stoichiometry to less than one dye molecule per polypeptide. In most cases, the fluorescent dye was attached to the protein by forming an amide bond between lysine and an activated N-hydroxy succinimidyl ester derivative of the dye. The most commonly used dyes are carboxyfluorescein, tetramethyl rhodamine, and the cyanine dyes. We have found that concentrations of protein or dextran in excess of 10 mg/ml inhibit cellularization. Exogenously injected proteins are often toxic at high concentrations. One must carefully determine the safe concentration range for each new protein tested.

The advantage of proteins such as actin, tubulin, and histones is that they are abundant and readily available proteins. Other proteins may not be as easily prepared. Two routes for obtaining fluorescent protein analogs are to construct recombinant proteins that contain either an epitope tag, to aid in protein isolation, (Field and Alberts 1995; Kallum et al. 1995; Oegema et al. 1995) or a GFP fusion protein (Wang and Hazelrigg 1994; Davis et al. 1996; Endow and Komma 1996). The advantage of the GFP constructs is that stably transformed stocks can be generated, obviating the need for injection. The main drawbacks of GFP-tagged proteins are the time required for the GFP to undergo the internal cyclization required to form the fluorescent core, and the limited number of spectral choices. However, as more research is done on these proteins, brighter, faster, and other different-color fluorescent proteins will surely emerge.

Fluorescent Indicators

The second class of fluorescent reagents is fluorescent indicators. These can be fluorogenic enzyme substrates, benign volume markers, or environment markers. One of the main features of development is the spatial and temporal pattern of specific gene expression. Analysis of fixed embryos has shown that these patterns are very dynamic. A prominent tool for analyzing gene expression patterns is to monitor the expression of a reporter gene, such as *lacZ*, driven by the enhancer of the gene of interest. In order to gain a better appreciation of gene expression dynamics, we developed a substrate for the *lacZ* gene product, β-galactosidase (Minden 1996). This compound, RGPEG, is based on the red fluorogenic substrate resorufin-β-galactoside and bears a polyethylene glycol (PEG) tail that prevents the dye from exiting the cell through the plasma membrane or via gap junctions. RGPEG has been used to visualize the initial expression of a variety of enhancer-*lacZ* constructs. A major limitation of RGPEG is that the cleaved substrate is long lived and the β-galactosidase also has a long half-life, which does not permit one to detect the turning off of an enhancer. RGPEG is quite useful for genotyping embryos in vivo. This is done in conjunction with "blue" balancer chromosomes that carry *lacZ* under the control of a patterning gene promotor, such as *ftz*. Judging from the intensity of the fluorescent signal, one can determine the number of blue balancer chromosoms in the embryo. This allows one to correlate time-lapse phenotype with genotype.

One limitation of using a single fluorescent marker for cell lineage or cell death analysis is that there are no reference signals for where in the embryo the marked cell(s) resides. One simply sees fluorescent cells in a dark volume. To provide the reference signal, a membrane-impermeable fluorescent dye is injected into the intervitelline space (Warn and Magrath 1986; Kam et al. 1991). The fluorescent dye diffuses throughout the intervitelline space and provides a bas-relief image of the contours of the embryo surface. Segmental grooves and furrows appear as bright fluorescent valleys. Cells in mitosis swell up out of the epithelium and appear as small dark holes surrounded by fluorescent rings. These structures are easy to visualize and relate perfectly to the known morphological features. There are several types of fluorescent dyes that can be used for intervitelline marking: (1) fluorescently tagged polymers, such as dextran or polyethylene glycol (PEG), (2) calcein, which is a carboxylated form of fluorescein that can be injected directly into the cytoplasm and is actively pumped out of the embryo into the intervitelline space, and (3) RGPEG, the β-galactosidase substrate that bears a PEG tail. The first two provide a very good signal for the first few hours after injection. However, the dye compounds appear to be taken up into small vesicles within the cells and become punctate over time. Calcein has the advantage that it does not need to be injected directly into the intervitelline space. We noticed that a number of small fluorescent dye molecules, even those with short, M_r 600, PEG tails, are actively transported out of the cells into the intervitelline space. Very little is known about this phenomenon, but it probably depends on the function of an ABC transporter.

RGPEG that is injected into the intervitelline space gradually becomes more fluorescent over time, indicating that there is an endogenous β-galactosidase activity in the intervitelline space. This activity is not due to the known β-galactosidase since null mutations in this gene still show the intervitelline activity. Intervitelline injection of RGPEG has the advantage that the intervitelline signal overwhelms the signal caused by the vesicular uptake of dye. This produces a more stable and better defined intervitelline signal.

A large array of physiologically sensitive dyes exist, ranging from pH- and Ca^{2+}-sensitive dyes to those sensitive to mitochondrial function and cell death (Haugland 1996). Our experience has primarily been with the cell-death-sensitive dye, acridine orange (White et al. 1994; Namba et al. 1997). This weak basic dye is sequestered in a quenched state in the lysosomes of living cells. When a cell dies by apoptosis, the pH gradient across the lysosomal membrane is lost and the dye diffuses into the cytoplasm and quickly enters the nucleus, where it binds to the chromosomal DNA and becomes highly fluorescent. To visualize apoptotic cell death, a 0.25 mM solution of acridine orange in PBS is injected into the cytoplasm of syncytial-stage embryos (Namba et al. 1997). Time-lapse recordings are initiated at the end of stage 11, when the germ-band begins to retract and the first cell deaths appear. Cell death is a very dynamic process. Within 30–40 minutes of acridine orange entering the nucleus of an apoptotic cell, the dead cell corpse is engulfed by neighboring cells or scavenging macrophages.

Caged Compounds

Understanding development requires a complete description of cellular fates and the genes that specify and manifest these fates. The aforementioned fluorescent reagents allow one to monitor a wide variety of structures and functions during embryogenesis. In order to complete the description, one needs to be able to determine cell fates and perturb those fates in a controlled fashion. This level of control requires precise marking of single cells at specific times of development. In an effort to control the temporal and spatial expression of selected genes at the single-cell level for the purpose of fate mapping and ectopic gene expression, we devised a method for "caging" the activity of the potent transcriptional activator, GAL4-VP16, which is a chimeric protein bearing the DNA-binding domain of yeast GAL4 protein and the transcriptional activation domain of herpesvirus VP16 protein (Cambridge and Minden 1997). Caging refers to a form of photoreversible chemical modification. This allows one to inhibit the activity of a protein by in vitro modification. The inactivated protein is then injected into the embryo or cell, where it can be reactivated with a localized pulse of long-wavelength UV light.

The strategy for caging GAL4-VP16 was to modify lysine residues with an amine-reactive caging compound, 6-nitroveratrylchloroformate (NVOC-Cl). Several lysine residues in the binuclear Zn cluster of GAL4 are known to be essential for DNA binding as shown by mutational analysis. GAL4-VP16 DNA-binding activity was completely abolished after a 30-minute incubation with 2 mM NVOC-Cl under mildly basic conditions, as monitored by a gel mobility shift assay. More than 50% of the initial binding activity was recovered by irradiating the caged GAL4-VP16 with a low-intensity, long-wavelength (365 nm) UV lamp. These results clearly showed that it is possible to reversibly cage the DNA-binding activity of GAL4-VP16 in vitro.

To determine the appropriate concentration range of GAL4-VP16 for in vivo experiments, syncytial blastoderm-stage embryos were injected with unmodified protein. A transgenic fly strain carrying *UAS-lacZ* was used as a reporter of GAL4-VP16 activity (fly strain provided by N. Perrimon). The level of *lacZ* gene expression was monitored in vivo using RGPEG. Previous experience with *Drosophila* embryos showed that most injected molecules under 100 kDa distributed evenly throughout the cytoplasm. However, injected GAL4VP16 did not diffuse evenly throughout the embryo cytoplasm, but was localized to one-quarter to one-third of the egg volume around the injection site, as seen by staining with anti-GAL4 antibody and by *lacZ* expression. Anti-GAL4 antibody staining showed that caging of GAL4-VP16 did not perturb nuclear localization. GAL4-VP16 was usually injected at a concentration of 0.2 mg/ml or less. Concentrations of unmodified or caged GAL4-VP16 greater than 0.4 mg/ml caused developmental defects. This was most likely due to squelching effects, a phenomenon thought to be caused by the binding of general transcription factors to the acidic domain of excess free transcriptional activator (Gill and Ptashne 1988).

Caging of GAL4VP16 with 0.5 mM NVOC-Cl, which modified approximately eight of the fourteen GAL4-VP16 lysine residues, completely inhibited in vivo transcriptional activation. Curiously, this level of caging did not affect GAL4-VP16 DNA-binding activity in vitro. It is not known why the lower level of caging inhibited in vivo activity. Perhaps this level of modification inhibited GAL4-VP16 binding to chromatin, rather than to naked DNA, or interfered with specific protein–protein interactions. The inhibition of caged GAL4-VP16 could be reversed in vivo with 365-nm light from a standard 100-W mercury lamp shone through a microscope objective via the epifluorescence light path of an inverted microscope. Caged GAL4-VP16 could be photoactivated as early as nuclear cycle 11 and up to 6 hours after gastrulation at 25°C. Earlier and later activation has not yet been tested. The majority of the irradiated embryos developed into well-formed, normal stage 17 embryos.

Single-cell photoactivated gene expression was accomplished by narrowing the UV irradiation beam to approximately 5 mm, the diameter of a diploid *Drosophila* nucleus, by inserting a pinhole aperture at the field-stop position of the epifluorescence light path of the microscope. Single cells of *UAS-lacZ* embryos that had been injected with caged GAL4-VP16 were irradiated shortly after the onset of gastrulation. Irradiated embryos were incubated until they reached the desired developmental stage and were then fixed and stained with anti-β-galactosidase antibody (Bomze and Lopez 1994). To show that only single cells were being activated, amnioserosa cells, which do not divide after gastrulation, were irradiated at the start of gastrulation. These experiments showed that in more than 90% of cases, the activation was restricted to a single cell; the remainder had two activated cells. To show that photoactivated gene expression did not perturb cellular division or induced cell death, dorsal epidermis cells, which undergo three rounds of postblastoderm division, were irradiated at the start of gastrulation. Embryos were allowed to develop to stage 13, which is after the normal onset of programmed cell death. In general, we observed single clusters of eight marked cells and occasionally saw clusters of six or seven marked cells (data not shown). The caged GAL4-VP16 system is quite versatile. We have used it to drive the ectopic expression of a number of UAS constructs, such as *Ubx, cyclin E, ricin A, hid,* and *p35*.

It takes about 20–30 minutes after photoactivation in order to detect the presence of the caged GAL4-VP16-induced protein. To mark cells for visualization more rapidly, one can use caged fluorescein or rhodamine derivatives (Mitchison 1989). Vincent and O'Farrell (1992) used caged fluorescein that bore a membrane-impermeable PEG tail to mark clones of epidermal cells. This method has also been used for lineage tracing in vivo (Minden et al. 1990). Unfortunately, the limited photostability of the fluorescein only permits lineage tracing over short time periods for cells that do not migrate deeply into the interior of the embryo. This reagent is still a useful tool for short-term dynamic analysis of marked cells.

Image Processing

A major difference between wide-field microscopy and confocal microscopy is that the wide-field images require a significant amount of postacquisition image processing. There is wide variation in the types of processing steps required to generate the desired image quality. There are basically two levels of image processing: low-end image enhancement, and high-end processing that includes feature extraction, iterative deconvolution, and three-dimensional reconstruction. Low-end image processing is used to (1) balance, or normalize, the images in a time-lapse series to remove image-intensity differences caused by lamp flicker and reagent changes and (2) subtract background fluorescence, or glare. Each digital image is composed of an array of pixel values, where each pixel represents the intensity of light collected at that position in the image. The range of intensity depends on the detector. In our case, the intensity ranges between 0 and 4095. In a number of applications, the intensity of the fluorophore changes over time or between different fluorophores. Enzyme substrates and environment-sensitive probes are particularly prone to intensity changes. The software for the microscope system has an automatic exposure option, where the exposure time or neutral density filter is automatically adjusted to maintain the maximum pixel value within a certain range. In some instances, one wants to normalize all the images to the same intensity range. This is done by determining the minimum and maximum pixel values for each image and then scaling all intermediate values between preset values, such as 0 and 4095. In other applications where one wants to quantify the fluorescent intensity, particularly for enzyme substrates, the image intensity must be adjusted to compensate for exposure time and neutral density filter changes.

Wide-field images contain both in-focus and out-of-focus information. The deconvolution method, which will be described later, relies on closely spaced optical sections, usually less than 0.5 µm, to estimate the in- and out-of-focus contributions. Many of our applications that monitor cell death and cell lineage analysis are made at a much greater optical section spacing, on the order of 5 µm. This level of sectioning is not sufficient for three-dimensional deconvolution. To remove out-of-focus glare, we perform a local background subtraction operation, where a small box is scanned over the image, subtracting the minimum pixel value within the box from the pixel in the center of the box. This effectively removes diffuse signals. The size of the box is selected based on the average size of the object being analyzed. For example, when imaging nuclei, which are about 5 μm or 12 pixels when imaged with a 20× objective and binned, the box size is set to 15, i.e., slightly greater than the width of the nucleus. This increases the contrast of the image. To accentuate nuclei, the box size can be set to one-half the width of the nucleus. This produces a ring-like appearance for each nucleus.

These low-end operations produce images that are free of flicker and have improved contrast. To gain better contrast, definition, and feature recognition, more sophisticated image processing must be performed. Cell death patterns are complicated by the fact that dead cells are engulfed by neighboring cells and macrophages, both of which continue to fluoresce for some time after the dead cell is ingested. The macrophages present a particular problem because they are highly motile and continually scavenge dead cells; thus, they remain fluorescent for long periods. To discriminate between newly dying cells and macrophages, the time-lapse images are processed with a spot-detecting filter followed by a spot-tracking program (P. Janardhan and J. S. Minden, personal communication). This provides an automated method for determining the time and location of each cell death event in the embryo.

Another function that is very useful is to project a stack of optical sections into one image or a pair of tilted images to form stereo pairs. Before the images can be projected, the out-of-focus information must be removed. There are two deconvolution-based methods to remove this information. If the optical sections are spaced more than 0.5 μm apart, one can perform a two-dimensional deconvolution, which essentially performs a size-filtering function and removes objects that are larger or smaller than the particular structure of interest. Using nuclei as an example, one would set the deconvolution filters to remove objects greater than 10 μm and smaller than 2.5 μm. The two-dimensional deconvolved images can then be summed to generate a through-image projection or computationally tilted to generate a stereo pair of projected images.

The most precise reconstruction method is a full three-dimensional deconvolution, which requires closely spaced optical sections (Agard 1984; Hiraoka et al. 1987). In this method, one image plane in the stack is blurred by a function that approximates the point spread function of the objective lens used to collect the images. This blurred image is then subtracted from the neighboring images in the data stack and back-transformed to the original image. This process is reiterated until the deconvolved images form the best approximation of the original data. The advantage of this method is that all of the light data is used to form the deconvolved, three-dimensional image. In essence, the light is reassigned to its point of origin. This method is computationally intensive, but the results are often better than confocal microscopy. While this is probably the best reconstruction method, time-lapse data are not usually collected with less than 1-μm optical spacing, making it a seldom-used method for time-lapse analysis.

Conclusions and Future

Multicellular organisms develop by transforming a mass of undifferentiated cells into intricately organized groups of cells that coalesce to form functional structures. This chapter describes a number of methods that allow one to monitor processes such as cell division, chromosome dynamics, cytoskeletal dynamics, tissue morphogenesis, cell shape, gene expression, and cell death in living *Drosophila* embryos. The hope of these and future studies is to compile a complete atlas of the dynamic events that occur during embryogenesis.

ACKNOWLEDGMENTS   Much of the initial work on three-dimensional, time-lapse microscopy of *Drosophila* embryos was performed the labs of Drs. Alberts, Sedat, and Agard at UCSF. I would like to thank Tim Mitchison for the original work on caged fluorophores. This work was supported, in part, by scholarships from the Helen Hay Whitney Foundation and the Lucille P. Markey Charitable Trust, and a grant from the NIH, R01-HD/GM31642.

References

Agard, D. A. (1984). Optical sectioning microscopy: cellular architecture in three dimensions. Annu. Rev. Biophys. Bioeng. 13:191–219.
Ashburner, M. (1989). In *Drosophila: A Laboratory Manual*. Cold Spring Harbor, NY, Cold Spring Harbor Laboratory Press.
Bomze, H. M., and Lopez, A. J. (1994). Evolutionary conservation of the structure and expression of alternatively spliced ultrabithorax isoforms from *Drosophila*. Genetics 136:965–977.
Cambridge, S. B., and Minden, J. S. (1997). *Drosophila* mitotic domain boundaries as cell fate boundaries. Science, 277:825–828.
Campos-Ortega, J. A., and Hartenstein, V. (1985). *The Embryonic Development of Drosophila melanogaster*. New York, Springer-Verlag.
Chalfie, M., Tu, Y., Euskirchen, G., Ward, W. W., and Prasher, D. C. (1994). Green fluorescent protein as a marker for gene expression. Science 263:802–805.
Cooley, L., and Theurkauf, W. E. (1994). Cytoskeletal functions during *Drosophila* oogenesis. Science 266:590–596.
Costa, M., Sweeton, D., and Wieschaus, E. (1993). Gastrulation in *Drosophila*: cellular mechanisms of morphogenetic movements. In *The Development of Drosophila melanogaster*. Cold Spring Harbor, NY, Cold Spring Harbor Laboratory Press, pp. 425–465.

Davis, I., Girdham, C. H., and O'Farrell, P. H. (1996). A nuclear GFP that marks nuclei in living Drosophila embryos; maternal supply overcomes a delay in the appearance of zygotic fluorescence. Dev. Biol. 170:726–729.

Debec, A. Kalpin, R. F., Daily, D. R., McCallum, P. D., Rothwell, W. F., and Sullivan, W. (1996). Live analysis of free centrosomes in normal and aphidicolin-treated *Drosophila* embryos. J. Cell Biol. 134:103–115.

Endow, S. A., and Komma, D. J. (1996). Centrosome and spindle function of the *Drosophila* Ncd microtubule motor visualized in live embryos using Ncd-GFP fusion proteins. J. Cell Sci. 109:2429–2442.

Field, C. M., and Alberts, B. M. (1995). Anillin, a contractile ring protein that cycles from the nucleus to the cell cortex. J. Cell Biol. 131:165–178.

Foe, V. E. (1989). Mitotic domains reveal early commitment of cells in *Drosophila* embryos. J. Cell Biol. 102:1494–1509.

Foe, V. E., and Alberts, B. M. (1983). Studies of nuclear and cytoplasmic behaviour during the five mitotic cycles that precede gastrulation in *Drosophila* embryogenesis. J. Cell Sci. 61:31–70.

Foe, V. E., Odell, G. M., and Edgar, B. A. (1993). Mitosis and morphogenesis in the *Drosophila* embryo. In *The Development of Drosophila melanogaster*. Cold Spring Harbor, NY, Cold Spring Harbor Laboratory Press, pp. 149–300.

Fogart, P., Kalpin, R. F., and Sullivan, W. (1994). The *Drosophila* maternal-effect mutation *grapes* causes a metaphase arrest at nuclear cycle 13. Development 120:2131–2142.

Fullilove, S. L., and Jacobson, A. G. (1971). Nuclear elongation and cytokinesis in *Drosophila montana*. Dev. Biol. 26:560–577.

Gill, G., and Ptashne, M. (1988). Negative effect of the transcriptional activator *GAL4*. Nature 334:721–724.

Haugland, R. P., *Handbook of Fluorescent Probes and Research Chemicals*, Eugene, OR, Molecular Probes Inc.

Hiraoka, Y., Sedat, J. W., and Agard, D. A. (1987). The use of a charge-coupled device for quantitative optical microscopy of biological structures. Science 238:36–41.

Hiraoka, Y., Minden, J. S., Swedlow, J. R., Sedat, J. W., and Agard, D. A. (1989). focal points for chromosome condensation and decondensation revealed by Three-dimensional *in vivo* Time-lapse Microscopy. Nature 342:293–296.

Hyman, A. A., and White, J. G. (1987) Determination of division axes in the early embryogenesis of *Caenorhabditis elegans*. J. Cell Biol. 105:2123–2136.

Kallum, R., Raff, J. W., and Alberts, B. M. (1995). Heterochromatin protein 1 distribution during development and during the cell cycle in *Drosophila* embryos. J. Cell Sci. 108:1407–1418.

Kalpin, R. F., Daily, D. R. and Sullivan, W. (1994). Use of dextran beads for live analysis of the nuclear division and nuclear envelope breakdown/reformation cycles in the *Drosophila* embryo. BioTechniques 17:732–733.

Kam, Z., Minden, J. S., Agard, D. A., Sedat, J. W., and Leptin, M. (1991). *Drosophila* gastrulation: analysis of cell shape changes in living embryos by three-dimensional fluorescence microscopy. Development 112:365–370.

Kellogg, D. R., Mitchison, T. J., and Alberts, B. M. (1988). Behavior of microtubules and actin filaments in living *Drosophila* embryos. Development 103:675–686.

Leptin, M. (1995). *Drosophila* gastrulation: from pattern formation to morphogenesis. Annu. Rev. Cell Dev. Biol. 11:189–212.

Loncar, D., And Singer, S. J. (1995). Cell membrane formation during the cellularization of the syncytial blastoderm of *Drosophila*. Proc. Natl. Acad. Sci. USA 92:2199–2203.

Matthies, H. J., McDonald, H. B., Goldstein, L. S., and Theurkauf, W. E. (1996). Anastral meiotic spindle morphogenesis: role of the non-claret disjunctional kinesin-like protein. J. Cell Biol. 134:455–464.

Mermall, V., and Miller, K. G. (1995). The 95F unconventional myosin is required for proper organization of the *Drosophila* syncytial blastoderm. J. Cell Biol. 129:1575–1588.

Mermall, V., McNally, J. G., and Miller, K. G. (1994). Transport of cytoplasmic particles catalyzed by an unconventional myosin in living *Drosophila* embryos. Nature 369:560–562.

Miller, K. G., Field, C. M., Alberts, B. M., and Kellogg, D. R. (1991). Use of actin filament and microtubule affinity chromatography to identify proteins that bind to the cytoskeleton. Methods Enzymol. 196:303–319.

Minden, J. S. (1996). Synthesis of a new substrate for detection of *lacZ* gene expression in live *Drosophila* embryos. BioTechniques 20:122–129.

Minden, J. S., Agard, D. A., Sedat, J. W., and Alberts, B. M. (1989). Direct cell lineage analysis in *Drosophila melanogaster* by time-lapse, three-dimensional optical microscopy in living embryos. J. Cell Biol. 109:505–516.

Minden, J. S., Kam, Z., Agard, D. A., Sedat, J. W., and Alberts, B. M. (1990). Embryonic lineage analysis using three-dimensional time-lapse *in vivo* fluorescence microscopy. In *Bioimaging and Two-dimensional Spectroscopy*, Smith, I., ed. SPIE. Society of Photo-optical Instrumentation Engineers, Vol. 1205, Bellingham, WA, pp. 29–42.

Mitchison, T. J. (1989). Poleward microtubule flux in the mitotic spindle. J. Cell Biol. 109:637–652.

Namba, R., Pazdera, T. M., Cerrone, R., and Minden, J. S. (1997). *Drosophila* embryonic pattern repair: how embryos respond to *bicoid* RNA dosage alteration. Development 124:1393–1403.

Oegema, K., Whitfield, W. G. F., and Alberts, B. M. (1995). The cell cycle-dependent localization of the CP190 centrosomal protein is determined by the coordinate action of two separable domains. J. Cell Biol. 131:1261–1273.

Paddy, M. R., Saumweber, H., Agard, D. A., and Sedat J. W. (1996). Time-resolved, in vivo studies of mitotic spindle formaton and nuclear lamina breakdown in early *Drosophila* embryos. J. Cell Sci. 109:591–607.

Schejter, E. D., and Wieschaus, E. (1993). Functional elements of the cytoskeleton in the early *Drosophila* embryo. Annu. Rev. Cell Biol. 9:67–99.

Sullivan, W., Minden, J. S., and Alberts, B. M. (1990). *Daughterless-abo-like*, a *Drosophila* maternal-effect mutation that exhibits abnormal centrosome separation during the late blastoderm divisions. Development 110:311–323.

Sullivan, W., Fogarty, P., and Theurkauf, W. (1993a). Mutations affecting the cytoskeletal organization of syncytial *Drosophila* embryos. Development 118:1245–1254.

Sullivan, W., Daily, D. R., Fogarty, P., Yook, K. J., and Pimpinelli, S. (1993b). Delays in anaphase initiation occur in individual nuclei of the syncytial *Drosophila* embryo. Mol. Biol. Cell 4:885–896.

Swedlow, J. R., Sedat, J. W., and Agard, D. A. (1993). Multiple chromosomal populations of topoisomerase II detected in vivo by time-lapse, three-dimensional wide-field microscopy. Cell 73:97–108.

Theurkauf, W. E. (1994). Premature microtubule-dependent cytoplasmic streaming in *cappuccino* and *spire* mutant oocytes. Science 265:2093–2096.

Valdes-Perez, R. E., and Minden, J. S. (1995) *Drosophila melanogaster* syncytial nuclear divisions are patterned: time-lapse images, hypothesis, and computational evidence. J. Theor. Biol. 175:525–532.

Vincent, J.-P., and O'Farrell, P. H. (1992). The state of *engrailed* expression is not clonally transmitted during early *Drosophila* development. Cell 68:923–931.

Wang, S., and Hazelrigg, T. (1994). Implications for *bcd* mRNA localization from spatial distribution of *exu* protein in *Drosophila* oogenesis. Nature 369:400–403.

Warn, R. M., and Magrath, R. (1986). Observations by novel method of the surface changes during the syncytial blastoderm stage of the *Drosophila* embryo. Dev. Biol. 89:540–548.

White, K., Grether, M. E., Abrams, J. A., Young, L., Farrell, K., and Steller, H. (1994). Genetic control of programmed cell death in *Drosophila*. Science 264:677–683.

26

Anatomical Techniques for Analysis of Nervous System Development in the *Drosophila* Embryo

David Van Vactor
Casey Kopczynski

Although the embryonic nervous system of *Drosophila* has long been appreciated as a powerful system to study the specification of cell fate, it has more recently proven to be a useful model system to approach later events in neuronal development. The strong parallels between mechanisms that control axon guidance in *Drosophila* and vertebrate species suggest that *Drosophila* genetics will continue to provide important insights into the functions of conserved molecules in neural development (for a recent review, see Tessier-Lavigne and Goodman 1996). As the sophistication of questions concerning neural differentiation and morphogenesis increases, so does the need for improved methods of analysis. Anatomical techniques are especially important for understanding the mechanisms of neuronal connectivity; however, it is often difficult to learn new histochemical procedures solely from the methods described in the primary literature. This chapter is thus meant as a guide to the novice on rapid immunocytochemistry and tissue in situ hybridization.

The analysis of development at an anatomical level has often relied heavily on the availability of specific marker reagents. Within the nervous system, individual neurons display striking diversity in their characteristics, such as axonal morphology, revealing an underlying diversity in cellular identities (Goodman and Doe 1993). The resulting complexity of the developing nervous system means that specific molecular probes are necessary to distinguish particular cells or structures (e.g., axons) from their many neighbors. As developmental analysis proceeds toward molecular precision, the resolution of histochemical techniques becomes even more important.

In order to interpret molecular mechanisms, we need to know exactly what cells express candidate genes, and exactly what effect mutations in these genes have upon the development of neurons whose normal development has been characterized in detail. The methods described in this chapter allow juxtaposition of in situ hybridization patterns with antibody stains, while maintaining the tissue preservation necessary to identify specific cells and structures with Nomarski optics. These techniques permit the anatomical analysis at single-cell resolution that is so important for a thorough understanding of gene function.

In recent years, the combination of *Drosophila* genetics and high-throughput immunocytochemistry has allowed a number of direct anatomical screens for genes that control different aspects of neuronal development, from cell migration to axonal pathfinding and synaptic target recognition (e.g. Seeger et al. 1993; Van Vactor et al. 1993; Salzberg et al. 1994; Kolodziej et al. 1995, C. Korey and D. Van Vactor, in preparation). Similar adaptations of methods for high-quality tissue in situ hybridization of cDNA clones have allowed large-scale screens for genes with interesting patterns of embryonic gene expression (Kopczynski et al. 1996; C. C. Kopczynski et al., in preparation). Here, we describe simple methods which are easily adapted to high-throughput processing of embryonic samples. Since many other useful protocols have been recently compiled by other authors (e.g., Patel, 1994), this chapter is meant to compliment other resources with minimal redundancy.

Multiwell Plate Protocol for Immunocytochemistry

Most immunocytochemical methods for preparing *Drosophila* embryos under the light microscope are based on the two-phase fixation protocol designed by Zalokar and Erk (1977). Various modifications have provided refinements for staining embryos under different conditions (an excellent guide to embryonic histology has been compiled by Patel 1994). However, for large-scale projects, it has been necessary to develop efficient methods for high-throughput staining. The simple protocol below provides a method for staining hundreds of individual lines in a relatively short time. The main feature of this approach is the multiwell staining dish described in Figure 26-1. The staining dish, which can be manufactured from inexpensive materials, allows many embryo collection samples to be stained and processed simultaneously.

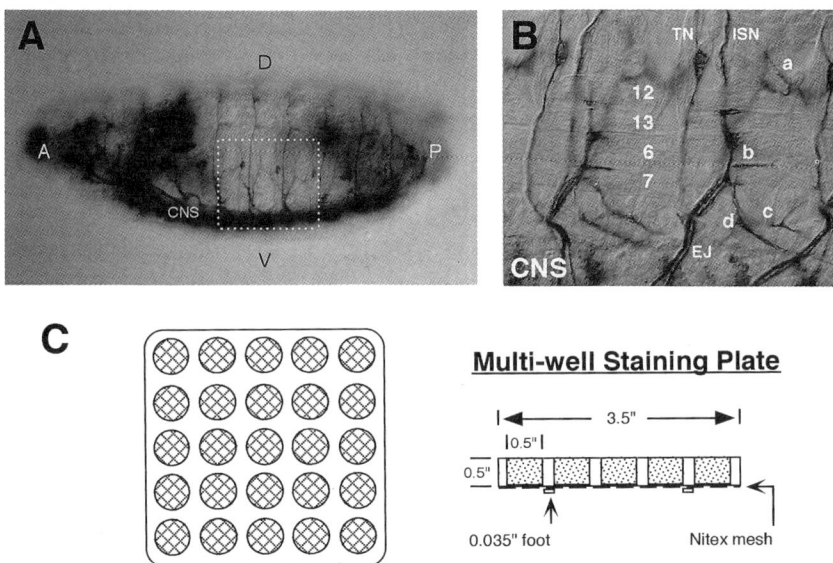

Figure 26-1 Staining with the multiwell plate. (A) An early stage 17 embryo is shown in whole mount, stained with the anti-Fasciclin II monoclonal antibody 1D4 (Van Vactor et al. 1993). The view is lateral (A, anterior; P, posterior; D, dorsal; V, ventral). The central nervous system (CNS) is out of focus. The segmentally repeated pattern of motor axons can be seen in the periphery. The ventral region outlined in the dotted box is shown in more detail in panel B. The length of the embryo is approximately 400 µm. (B) Two segments of a stage 17 embryo are shown in a fillet preparation, stained as in panel A. Fasciclin II is expressed by motor axons that emerge from the CNS at the exit junction (EJ); each branch (ISN, SNa, b, c, and d) innervates a distinct domain of body wall muscles. Each ventral longitudinal muscle (as visualized by Nomarski optics) is also labeled (7, 6, 13, and 12) TN, transverse nerve. (C) A schematic drawing of a 25-well version of the multiwell staining plate is shown (see text for details).

The Multiwell Plate

We use a simple staining plate constructed from a block of acrylic plastic (Figure 26-1C). The plate is fashioned to fit into a 100 × 100-mm square Petri dish. Nylon mesh (3-100/38XX Nitex nylon; Tetko, Inc.) is fused with bonding solvent to the base of the plate for solution access to the well contents. The dimension and number of wells can vary; however small-diameter wells have a limited capacity and overloaded wells do not stain uniformly. We find the 36-well, 25-well, and 16-well designs most useful, depending on the number of embryos in each sample. The placement of small plastic "feet" underneath the mesh surface is important to give solutions the ability to move freely between wells during incubations.

Basic Protocol

Embryo Collection

1. Place 100–200 flies in a collection chamber (egg-lay cup + grape juice agar plate); a small dab of yeast paste on the plate will attract the females to lay eggs. The flies should be prepared for egg laying by feeding for a day or two on wet yeast paste (the consistency of peanut butter); adults often peak in fecundity at 3 or 4 days after eclosion. Egg collections are made over a defined period (e.g., a 20-hour collection will contain a distribution of embryos from all embryonic stages, whereas a short collection defines a narrow window, which can be allowed to mature to the desired stage).

The rate of embryonic development is temperature dependent. For example, at 25°C, embryos take approximately 24 hours to hatch, but at 18°C this process takes almost twice as long. Shifting from one temperature to another can be used to one's advantage in obtaining staged collections on a convenient schedule; however temperature also affects distinct biological processes differently, and often has dramatic effects on the expression of certain phenotypes. Thus, one must be careful to include controls when manipulating the rate of development with temperature. In addition, it is wise to use a variety of staging criteria when determining the precise developmental "age" of an embryo (see "Staging Embryos" under "Preparation of Samples for Microscopy.").

2. To harvest the embryos, wash embryos from collection plates with PBST (in a squirt bottle), loosening the embryos with a paintbrush (No. 1 or No. 2). Be very careful not to break up the agar, as chunks of agar will prevent efficient devitellinization later. Rinse the brush carefully between samples to avoid cross-contamination.

3. Transfer embryos (avoiding adults and hatched larvae) to one well of a multiwell plate using a disposable pipette. Care must be taken in all subsequent steps not to agitate any well too abruptly, as cross-contamination between wells can result. Once all samples are loaded into the plate, rinse the embryos with PBST (use a squirt bottle of PBST to rinse each well over a beaker).

Dechorionation and Fixation

1. Remove the embryonic eggshell (chorion) by incubating the plate in a 100 × 100-mm square Petri dish with 20 ml of 50% bleach for 5–10 minutes. The liquid volume required to cover the embryos will vary depending on the number of wells in the plate (e.g., 20 ml for 36-well, 30 ml for 25-well). Use large forceps to move the plate from dish to dish.

2. Rinse extensively with PBST, then soak in 20 ml of fresh PBST in a 100 × 100-mm dish for 5 minutes. Since embryos continue to develop until the fixation step, the previous steps should be performed quickly.

Note

The 100 × 100-mm dishes can be reused after soaking and rinsing; however, heptane will crack the plastic eventually.

3. Slowly lower the multiwell plate into a 100 × 100-mm dish with 10 ml of *n*-heptane plus 10 ml of 3.7% formaldehyde/PBS (freshly diluted) to begin fixation. Use a Pasteur pipette to create an emulsion of heptane and formaldehyde (like a

vinaigrette salad dressing) and squirt fixative from outside of the plate into each well, agitating the embryos and ensuring even access to embryos. Incubate with gentle shaking (50–100 rpm) for 25 minutes.

Notes

a. Some antibodies are very sensitive to fixatives, and may require less fixation time (see the "Troubleshooting" section at the end of this protocol for alternatives).

b. Heptane fumes are toxic, as is methanol; fixation and devitellinization should be performed in a fume hood.

4. Carefully and quickly blot away excess heptane/fix with paper towels and transfer plate to 15 ml of fresh heptane in a new 100 × 100-mm dish (do not allow the embryos to dry out). Once embryos are submerged in heptane, they are protected from dehydration; if several plates are being processed at the same time, it is safe to leave embryos under heptane for 10–15 minutes while the others are being processed.

"Cracking" Embryos out of the Vitelline Membrane

1. Although the vitelline membrane barrier is penetrated by fixative in the two-phase system, this barrier must be removed for access of antibodies and aqueous reagents to the embryo. To devitellinize or "crack" the embryos, add 5 ml of methanol to the dish containing the embryos in heptane. Squirt some methanol into each well.

Next, using a P1000-type pipetter (e.g. Rainin P1000 Pipetman) and a disposable pipette tip, plunge the embryos about 10 times through the pipette tip (set the pipetter at 300 µl); make sure the embryos are moving through the pipette tip evenly. The shear force will strip the vitelline membrane from the embryo; however, care must be taken not to use excessive force that will tear the embryonic tissue (this just takes practice). Even if the vitelline membrane is only ruptured, embryos will often stain well. Be careful not to spill embryos out into neighboring wells. Make sure the embryos are being drawn up into the pipette, and that the solvents make a nice emulsion. If the embryos clump and look like cottage cheese, be persistent to be sure that you have dispersed them.

Notes

a. Drawing up some of the solvent mixture into the pipette tip before cracking will prevent embryos from sticking to the inside of the pipette tip. If some do stick, draw up heptane and release it very slowly, dragging the embryos off of the sides of the tip.

b. Methanol itself is a very effective denaturant; many antigens (and enzymes, such as β-galactosidase) are very sensitive to methanol. Thus, it may be necessary to reduce the time that tissue is exposed to methanol, or, in some cases, one may have to avoid it completely. In such extreme cases, embryos must be manually devitellinized in PBS or Ringer's. For certain antibodies, best results are achieved by manual devitellinization and dissection prior to fixation and antibody incubation; this avoids methanol and provides much greater antibody access to internal tissues (see the "Troubleshooting" section at the end of this protocol). For those antigens that are resistant to methanol, fixed embryos can be stored under methanol at −20°C for days, if not weeks, prior to rehydration and staining.

2. After each sample has been cracked, lift the plate out of the dish and blot away solvents quickly with a paper towel. Vigorously wash away heptane in each well using a squirt bottle of methanol over a beaker (make sure the heptane is rinsed away by the methanol).

Antibody Processing

1. Rehydrate the embryos by rinsing each well with PBST from a squirt bottle, then transfer the plate to 100 ml of PBST in a baking dish on a shaker (50–100 rpm) for 15 minutes. Follow this with another rinse from the squirt bottle. It is important to remove all fixative that might nonspecifically cross-link primary antibody to tissue and create high background.

Note
If many samples are to be stained with several different antibodies, embryo samples can be transferred to small tubes (e.g., 1.5-ml centrifuge tubes), after rehydration, for separate processing. If embryos are transferred to small tubes, care must be taken during washes to ensure efficient removal of reagents (i.e., increase the number of buffer exchanges). Use of the multiwell plate for prior steps is still a very efficient way to handle multiple samples.

2. Block nonspecific background in 20 ml of PBSTS in a 100 × 100-mm dish for 30 minutes.

Notes
a. For all incubation steps in 100 × 100-mm dishes, draw the solution in the dish from outside of the plate and pipette it into each well to be sure all embryos are well bathed and mixed.
b. Additional blocking reagents can be added if nonspecific background is a problem (see the "Troubleshooting" section at the end of this protocol).

3. Replace PBSTS with 20 ml of primary antibody cocktail. Pipette cocktail into each well to mix. For most antibodies, best results are achieved with incubation at 4°C for 8–16 hours. Gentle agitation during the antibody incubation will improve the uniformity of the staining.

Primary Antibody Cocktail
The primary antibody is diluted in PBSTS. The dilution factor depends on the type and batch of antibody (e.g., low titer monoclonal hybridoma supernatants are often best from 2- to 10-fold dilution; however, high titer monoclonal ascites fluid can require 200- to 2000-fold dilution).

Notes
a. In the first primary antibody incubation, multiple antibodies can be used simultaneously. For example, to score a *lacZ*-marked balancer chromosome within a sample stained for axonal morphology, add antibody against *lacZ* and your axonal marker together. Processing multiple antibodies in this way is only a problem if one of the antibodies gives a weak signal; in such cases, it is best to process sequentially.

b. Primary and secondary antibody solutions can often be reused several times. To protect used antibodies from bacterial growth, one can add 0.02% sodium azide (but since azide blocks horseradish peroxidase activity, do not use it for peroxidase-conjugated secondary antibody), or filter-sterilize (with a 0.2-μm filter). If the solution has been used several times, you can add a "refresher" of stock antibody (less than the original amount) to restore strength, depending on the antibody titer.

4. Rinse the plate with 100 ml of PBST in a baking dish on a shaker for 10–15 minutes. Repeat three times. It is best to rinse each well from a squirt bottle of PBST between buffer changes to improve the wash efficiency.

5. Block nonspecific background in 20 ml of PBSTS in a 100 × 100-mm dish for 30 minutes. Pipette PBSTS into each well to mix.

6. Transfer the plate to a dish with secondary antibody cocktail. Incubate at 4°C for 2–16 hours. Gentle agitation during the antibody incubation will improve the uniformity of the staining.

Secondary Antibody Cocktail

The secondary antibody is diluted in PBSTS. If short incubation times are desired, increase the concentration of secondary antibody [e.g., goat antimouse conjugated to horseradish peroxidase (HRP), 1:800 in PBSTS for a 12–16-hour incubation; 1:300 for a 2-hour incubation].

Note

If the final stain is weak, signal intensity can be amplified by using multivalent reagents such as biotin-conjugated secondary plus enzyme-conjugated Avidin. Such reagents can be obtained individually or in kits (e.g., Vectastain by Vector Laboratories, Inc.).

7. Rinse as in step 4.

Development of Chromaphore and Clearing

1. Transfer the plate to a 100 × 100-mm dish with 20 ml of substrate solution (e.g., 0.5 mg/ml DAB). Equilibrate embryos in substrate for 30 minutes at room temperature.

2. Develop the reaction. For DAB, add 100 μl of 50% H_2O_2 to the DAB, mix well, and incubate the plate for 15–30 minutes to develop.

Notes

a. With HRP-conjugated antibodies, maximum signal is usually achieved within 15–20 minutes; nonspecific background may begin to develop in longer incubations. If nonspecific staining is a problem, watch the reaction proceed under a dissecting microscope and rinse the embryos with PBST to stop the reaction before background develops.

b. If samples are being prepared for high-magnification light microscopy (e.g., 1260×), signals must be strong if they are to be seen clearly. A weak detectable signal under a dissecting microscope may not be visible at high magnifications.

3. Rinse plate with PBST.

Notes
a. DAB waste is hazardous. Bleach-inactivate the DAB solution by adding 50% bleach 10 minutes prior to disposal.
b. This protocol is easily adapted for double staining for multiple antibodies. If two primary antibodies are derived from the same organism (e.g., mouse monoclonals), the staining must be done sequentially. In order for both signals to be distinguished, the first round should be reacted with a chromaphore of contrasting color [e.g., NiCl-enhanced DAB (blue-gray) in the first round, to contrast straight DAB (brown)]. After the first color reaction is complete, wash the sample in PBST several times and go back to "Antibody Processing" step 2 for the next primary antibody cycle.

4. When all stains are complete, transfer the plate to a 100 × 100-mm dish with 20 ml of 70% glycerol/PBS for several hours to clear. If glycerol is high quality, the stains are stable for a long time at room temperature.

Note
For fluorescent chromaphores, include some agent to prevent photobleaching (e.g., 2.5% DABCO from Sigma, #D-2522).

Reagents and Materials
  Egg-lay cups (100-ml Tri-pour beaker, with wings trimmed and holes made in the top with 21G needle)
  60-mm plastic Petri plates (Falcon #1007 should fit snugly into the base of the egg-lay cup) with grape juice agar medium
  100 × 100-mm square Petri plates (Falcon #1012)
  Multi-well plates (see Figure 26-1 legend for details)
  Pyrex baking dishes (8″ or 10″ square)
  Miscellaneous disposables: P-1000, Blue tips, Pasteur pipettes, etc.
Grape juice agar
(A) 333 ml Grape juice
2 g Methyl paraben (Sigma #H-5501)
33 g Dextrose
Bring to boil, cool to 60°C.
(B) 30 g Agar in 1 liter of H_2O
mix, autoclave, and cool to 60°C.
(C) Mix (A) and (B) gently with no bubbles, pour into plates, and store at 4°C (good for weeks if covered).
Yeast paste (mix Fleischmann's Active Dry Yeast #2133 and H_2O into a firm paste the consistency of peanut butter)
50% Bleach (2.63% Sodium hypochlorite)
37% Formaldehyde (a standard stock solution is adequate for the methanol cracking protocol and light microscopy; however, freshly hydrolyzed paraformaldehyde is best for live dissections or electron microscopy).
n-Heptane
Methanol
PBS = 10× stock:
  15.36 g $NaH_2PO_4 \cdot H_2O$
  71.64 g Na_2HPO_4

613.2 g NaCl (dissolve in 4 liters of H_2O)
pH to 7.2, add H_2O to 6 liters total.
PBST = 1× PBS + 0.1% Triton X-100
PBSTS = 1× PBST + 5% goat serum
Goat serum (Gibco BRL #16210-072) (heat-inactivated at 65°C for 30 minutes). This is an effective blocking reagent for secondary antibodies raised in goat.

DAB Staining Solution

0.5 mg/ml DAB (3,3' Diaminobenzidine; Sigma # D-5637) in 1× PBST. Dissolve, filter to remove any precipitate, and freeze in aliquots. DAB quality varies considerably, so test each batch of bulk reagent. *Caution*: DAB is a carcinogen! For convenience, try 10-mg pellets (Sigma # D-5905)

50% H_2O_2

70% Glycerol/PBS (For best results, use high-grade glycerol, such as Boehringer Mannheim #100 649.)

Troubleshooting

Fixative-sensitive antibodies (e.g., anti-even-skipped MAb 3C10; Patel et al. 1992)

1. Reduce fixation time. For MAb 3C10, fix for no longer than 10 minutes.
2. Use only one antibody at a time. Stain with MAb 3C10 alone in the first cycle of staining; once a good signal is obtained, a second round of processing can be added to stain for a second epitope (e.g., *lacZ* or axonal counter stain)
3. Incubate in primary antibody for 48 hours at 4°C with shaking (platform shaker works well for multiwell plate) or agitation (Adams Nutator works well for small tubes).
4. Enhance the DAB reaction product with NiCl to improve signal strength and contrast (see section "Basic Protocol" under "High-Resolution Protocol for Tissue In Situ Hybridization").

Methanol-intolerant antibodies Many antibodies perform very poorly in the methanol "cracked" protocol. Such reagents often require live dissections, with gentle fixation that avoids solvents completely. The live dissection is more difficult than the dissection of fixed tissue.

Live Dissection

1. It is difficult to stage live embryos accurately by morphological criteria under a dissection microscope. Thus, it is best to collect eggs in a short time window (1–2 hours) and then age them precisely to the stage desired.
2. Up to mid stage 16, prior to cuticle deposition, embryos stick easily to the glass surface of an untreated slide. If late-stage dissections are desired, slides should be coated; a thin layer of Sylgard (Dow Corning) works well for an adhesive surface. Dissections should be carried out under PBS or Ringer's in a dam slide. To hold the embryos in place for manual dechorionation and devitellinization, place a strip of double-stick tape over the edge of the silicone dam so that a small area of the slide surface is covered by tape (the remaining space within the dam is where you will fillet the embryos).

3. Collect and rinse embryos from collection plates on a piece of Nitex mesh. With a paintbrush, transfer embryos to the double-stick tape on the dam slide (do not get the tape too wet). Once the embryos are dry, roll them out of their eggshells with fine forceps (Dumont #5). The embryos are very delicate, so nudge them gently to break the chorion. Once out of the shell, let them stick to the area of tape inside of the silicone dam. When a number of embryos have been dechorionized, add PBS to the dam to cover them (once they are out of the eggshell, embryos will dehydrate rapidly).

4. Put the slide on a black background and select the embryos of an appropriate stage. For stages 14–17, this is most easily done by using gut morphology. The white yolk inside of the gut is easy to see. At stage 14, the gut is a thin tube; at stage 15, the gut thickens; at stage 16, one can see constrictions that break the gut into three segments; at stage 17, the posterior gut becomes a convoluted tubule. These and other stage-specific characteristics have been beautifully diagrammed by Hartenstein (1993).

5. Under PBS, gently poke open the vitelline membrane, using a tungsten needle, and lift out each desired embryo. Transfer each embryo to the glass surface, tacking it down on the anterior and posterior ends by grinding the tissue into the glass surface. Be very careful not to let the embryos touch the surface of the buffer; they will float and stick to the surface. Since the yolk will eventually coat the glass and make it difficult to stick embryos down, lay out and anchor the embryos first (start with 10 embryos per slide or fewer). Be careful to lay down each embryo in the dorsoventral orientation that is appropriate for your desired view (i.e., dorsal-side up or ventral-side up); once the embryo is stuck to the glass, it is difficult to pull it off without damage. The dorsal side of the embryo is straight compared with the bowed curve of the ventral side.

6. Once each embryo is tacked down, use two hands and a single needle to make a series of cuts along the top surface of the epidermis. Each time, hook a bit of epidermis and pull up to break it open. It is often easier to start on the end of the embryo farthest away from you and work back; otherwise, you may lift the whole embryo off of the slide. Imagine that you are cutting the laces off of a shoe (in steps), with a knife, in order to open up the shoe from the top. Once there is a cut along the entire length, then spread the epidermis to each side and tack it down on the slide. It may be easiest to remove the gut first, depending on the stage. Lift the gut up to the surface of the bath and it will stick there.

7. After all the embryos are laid out flat, remove the strip of tape and begin your fixation (e.g., 4% formaldehyde/PBS for 10–30 minutes). To exchange buffers without allowing the embryos to be exposed to air, hold a pipette in each hand and add solution at one corner of the dam while removing the old buffer with the other hand at the opposite corner of the dam. All of the processing can be done in the dam. After the stained fillets are cleared in glycerol, remove excess glycerol with a pipette, cut away the silicone dam with a razor blade, and mount a coverslip on top.

Materials
Sylgard slides
Mix up a small volume of Sylgard 184 silicone elastomer (Dow Corning Corp.) and spread a thin, even coating of the unpolymerized mixture across each slide surface, using another slide to scrape the Sylgard across the surface of the other

slide. Build silicone dams on the slide surface (as described below) and then allow the slides to cure for 48 hours prior to use.

Dam slides
Use clear silicone bathtub sealant or aquarium sealant to build a rectangular dam in the center one-third of a microscope slide, roughly 3 mm high; dry overnight.

Ringer's Solution
6.5 g NaCl
0.14 g KCl
0.2 g $NaHCO_3$
0.12 g $CaCl_2$
0.01 g NaH_2PO_4 in 1 liter ddH_2O

Dissecting tools
Use 0.005" diameter tungsten wire (Ted Pella, Inc., #27-11) mounted in a 27G × 0.5" long needle (Monoject aluminum hub #200516); the needle cover of this brand makes a convenient protective sheath for the needle after it has been made (as long as the needle is not too long). Each needle is sharpened in a bath of 1 M KOH, with electrical current passing through the needle into the bath via a DC power supply (e.g., Staco transformer, through VWR Scientific #62546-050). Note: KOH is highly corrosive and will damage equipment such as microscopes, and it can burn the skin. Current of 10 V is sufficient (be careful not to touch the needle to the bath electrode — you will hear a nasty pop that will melt the needle). Repeated dipping of the wire into the bath will sharpen the tip; the depth of the dip controls the taper; however, a long taper is sometimes too flexible to be useful. To control the taper, it is easiest to swing 1–2 mm of the tip in and out of the bath in an arc (like a pendulum). Inspect the needle under a dissecting microscope until the desired taper is attained. The tungsten dissolves quickly at 10 V. When the needle is bent or dulled, simply return it to the KOH bath; very little sculpting is required to restore a good needle. Once the needle is made, it can be mounted on a makeshift handle (e.g., a 5-ml plastic disposable pipette broken off at the broad end to a comfortable length).

A far more expensive alternative is the Micro-tool (Ted Pella, Inc., #13600 handle and #13625 Ultra Micro-needle); this can be sharpened using the same technique as for the home-made version.

Antibodies with high background Some antibodies produce enough nonspecific background signal to interfere with visualization and photography, especially in multiply stained embryos. Polyclonal sera often require affinity purification; this is particularly true of antibodies raised in rabbit. Mouse and rat sera tend to be cleaner, but are much less plentiful. Monoclonals are usually quite clean, depending on how they were originally screened. Sometimes, a modification in the staining protocol can make a difference; three different options are described.

1. Add BSA to PBST to help block nonspecific binding.
2. Preincubate the primary antibody with a small sample of embryos prior to adding the antibody to your sample. Sometimes, the nonspecific background can be depleted from your reagent without significant loss of the specific anti-

body. This works really well if a viable protein null mutant is available, avoiding any loss in specific antibody titer.

3. Boehringer Mannheim manufactures a Blocking Reagent (designed for nucleic acid hybridizations; BMB #1096176) that works well for some antibodies (e.g., anti-Sex lethal MAb 104). Use of this reagent requires a modification of the standard protocol (see "Multiwell Plate Protocol for Immunocytochemistry"), switching the buffer from phosphate to maleic acid buffer as described in the Blocking Reagent instructions. After the final washes are finished, prior to the DAB reaction, the buffer can be switched back to PBST.

Signal Enhancement Protocols

Nickel Chloride Enhancement of DAB

For improved signal intensity and color contrast, a standard DAB reaction can be modified by the addition of 150 µl of 8% $NiCl_2$ per 20 ml of 0.5 mg/ml DAB, prior to preincubation of embryos in substrate. This produces a black reaction product which contrasts nicely with the normal brown DAB precipitate in double-antibody staining experiments.

Silver Enhancement of DAB Precipitate

Silver enhancement of the DAB precipitate represents the most sensitive method for detecting HRP reaction products (e.g., Gallyas et al. 1982; Liposits et al. 1984; Kopczynski and Muskavitch 1992). The resulting opaque black reaction product provides optimal contrast for photography.

Silver Enhancement Protocol

1. Follow the above "Basic Protocol" of the "Multiwell Plate Protocol for Immunocytochemistry" immunostaining protocol, but do not clear embryos in glycerol. Do not overstain; the signal should be just visible with little or no background.

2. Rinse embryos three times with PBST, then wash twice in 2% sodium acetate (trihydrate) for 15 minutes per wash.

3. Incubate embryos for 6–14 hours at 4°C in 10% thioglycolic acid (in dH_2O).

4. Wash embryos three times in 2% sodium acetate for 10 minutes per wash.

5. Rinse embryos with dH_2O, then wash for 10 minutes in dH_2O.

6. Add developer. Incubate until color begins to darken. If no change is seen by 10 minutes, add fresh developer (it is often useful to let half of the embryos develop beyond the initial darkening stage to produce both "light" and "dark" specimens).

7. Wash embryos with 1% acetic acid for 5 minutes followed by 2% sodium acetate for 5 minutes.

8. Tone reaction product with 0.05% gold chloride (in dH_2O) for 10 minutes at 4°C.

9. Wash embryos with 2% sodium acetate for 5 minutes.

10. Wash embryos twice, for 10 minutes each, in 3% sodium thiosulfate (pentahydrate).
11. Wash embryos twice, for 5 minutes each, in 2% sodium acetate, then wash once with PBST for 10 minutes.
12. Clear embryos in 70% glycerol as above "Basic Protocol" of the "Multiwell Plate Protocol for Immunocytochemistry".

Notes
a. Silver enhancement can also be used to intensify in situ hybridization signals. In the "Hybridization and Staining" protocol below, simply substitute a 1:200 dilution of antidig-HRP for the 1:2000 dilution of antidig-AP antibody.
b. Silver intensification is particularly helpful when the antigen is membrane-bound.

Reagents and Materials
  Sodium acetate (trihydrate)
  Thioglycolic acid
  Gold chloride
  Sodium thiosulfate (pentahydrate)
Developer
Prepare developer immediately before use by slowly adding an equal volume of solution B to an equal volume of solution A (A and B made fresh in sterile disposable plastic containers).

Solution A
5% sodium carbonate (anhydrous) in dH_2O.
Solution B
Add to 50 ml of dH_2O, in the following order:
  0.1 g Ammonium nitrate
  0.1 g Silver nitrate
  0.5 g Tungstosilicic acid (Fluka)
  0.2 ml 37% formaldehyde.

High Resolution Protocol for Tissue In Situ Hybridization

Whole-embryo in situ hybridization is a powerful tool for determining the spatial and temporal regulation of gene expression during *Drosophila* development. The most popular protocols are based on the use of digoxigenin-labeled DNA probes as originally described by Tautz and Pfeifle (1989). However, the success of such protocols depends heavily on two factors that can be difficult to control: (1) the size of the DNA probe, (2) the extent of protease digestion required to allow probe penetration into the embryo. In practice, these two variables can make consistent results difficult to achieve. Furthermore, even brief protease treatment can have a dramatic effect on the integrity of tissues as viewed by Nomarski optics.

The protocol below is a modification of the Tautz and Pfeifle procedure that allows for better control of probe size and eliminates the need for protease digestion. Antisense RNA probes are used in place of DNA probes, which allows an optimal probe size of 50–150 nucleotides to be consistently produced by alkaline hydrolysis. Efficient size reduction of the RNA probe eliminates the need for protease treatment of embryos, resulting in lower background staining and greater sensitivity than is typically achieved with DNA probes. Furthermore, the lack of protease treatment allows embryos to be double-stained with antibodies following in situ hybridization. A high-throughput version of this protocol is also presented that is based on in situ hybridization in 96-well plates.

Basic Protocol

Preparation of Probe

1. Linearize 1 μg of plasmid DNA with an appropriate restriction enzyme in a 13-μl reaction. Miniprep-purified DNA works well. Restriction buffers based on K^+ salts work best; avoid restriction enzymes that require high salt (100 mM NaCl). If a high-salt buffer must be used, the DNA should be purified by ethanol-precipitation after digestion.
2. Prepare transcription mix and incubate at 37°C for 2 hours.
3. Add 10 μl of DNase mix and incubate at 37°C for 15 minutes.
4. Add 80 μl of 125 mM sodium carbonate, pH 10.2, and incubate at 60°C for 20 minutes to reduce the size of the RNA probe. If the initial probe length is less than 1 kb, the incubation time should be reduced to 10 minutes; if it is greater than 3 kb, the incubation time should be increased to 30 minutes.
5. Add 50 μl of 7.5 M ammonium acetate, mix briefly, then add 375 μl of 100% ethanol and place the tube on ice for 10 minutes.
6. Spin the tube for 15 minutes in a microfuge (13,000 rpm) to pellet the RNA. Drain well, then resuspend the damp pellet in 250 μl of 50% formamide/50% TE 7.5/0.1% Tween 20. This is a 100× probe stock for hybridization (approximately 25 μg RNA/ml).

Preparation of Tissue
Note: All volumes are approximately 1 ml unless otherwise noted.

1. Prepare an egg collection chamber and harvest embryos of the desired age as described above.
2. Transfer embryos to a 1.5-ml microfuge tube and wash three times with 0.02% Triton X-100 (Tx). *Note*: do not exceed 200 μl of settled embryos per tube.
3. Dechorionate embryos for 5 minutes in 50% bleach, then wash three times with 0.02% Tx.
4. Add 600 μl of 4% formaldehyde/PBS and 600 μl of *n*-heptane to the tube. Invert the tube several times to bring the embryos to the heptane interface. Allow the embryos to fix for 20 minutes with occasional inversion.
5. Remove the formaldehyde/PBS fixative from below the heptane and replace with 700 μl of methanol. Shake the tube for 1 minute then allow the embryos to settle.

6. Discard the heptane (upper phase) along with any embryos and vitelline membranes remaining at the interface. Rinse the settled embryos three times with methanol. The embryos can now be stored in methanol at −20°C for later use.
7. Rehydrate embryos for 2 minutes in methanol/4% paraformaldehyde (3:1)/PBS followed by 5 minutes in methanol/4% paraformaldehyde (1:3)/PBS.
8. Fix the embryos for an additional 10 minutes in 4% paraformaldehyde/PBS.
9. Wash the embryos three times with PBST to remove fixative. Transfer the embryos to a new 1.5-ml microfuge tube. The embryos are now ready for hybridization.

Hybridization and Staining
1. Add 500 µl of hybridization buffer to 50–100 µl of settled embryos. Prehybridize for 1 hour in a 52°C oven with rocking.
2. Add 5 µl of 100× probe stock and mix by inversion. Return the tube to the oven and incubate at 52°C for 12–18 hours *without* rocking (place the tube on its side).
3. Add 500 µl of wash buffer (50% formamide, 2× SSC, 0.1% Tween 20) to the hybridization mix, mix well, then place the tube upright in an oven to allow the embryos to settle. Remove the hybridization mix.
4. Rinse embryos once with wash buffer then wash the embryos at 52°C for 6–18 hours with four or more changes of wash buffer (resume rocking).
5. Rinse embryos once with PBST, then wash at room temperature for 30 minutes in PBST.
6. Replace PBST with a 1:2000 dilution of antidig-AP antibody in PBSTS. Incubate for 1.5 hours.
7. Rinse embryos once with PBST, then wash four times in PBST, for 20 minutes each wash.
8. Rinse embryos twice with AP buffer, then wash for an additional 5 minutes in AP buffer.
9. Add 300 µl of AP buffer containing 2.7 µl of NBT stock and 2.1 µl of BCIP stock. Incubate (with rocking) until desired color development is achieved (20 minutes to overnight).
10. Rinse embryos three times with PBST. Remove as much PBST as possible, then add 500 µl of 70% glycerol to clear. Embryos are ready to mount or dissect once they have settled to the bottom of the tube. Alternatively, the glycerol step can be skipped and the embryos counterstained with an antibody of choice.

Antibody Counterstain
In order to counterstain embryos after development of the AP reaction, rinse the embryos several times with PBST and then transfer to PBSTS blocking solution, beginning at the "Antibody Processing" step 2 in the "Basic Protocol" section of the "Multiwell Plate Protocol for Immunocytochemistry."

Notes
1. Protease treatment of embryos increases the rate at which the signal develops but also increases background staining. Without protease treatment, the signal develops more slowly, but the sensitivity is greatly increased since the staining

reaction can proceed for 18 hours or longer without significant background development.
2. This protocol can also be used for imaginal disks. When preparing disks for hybridization, a single fixation time of 15 minutes in 4% paraformaldehyde (omit heptane) is sufficient prior to hybridization. Proteinase digestion is not required.

Screening 96 Clones by In Situ Hybridization

Preparation of Probe
1. Digest 96 individual clones with an appropriate enzyme as described above "Basic Protocol: Preparation of Probe".
2. Place 5 ml of digested plasmid (100 mg–500 μg) into each well of a 96-well plate.
3. Using a multichannel pipetter, add 5 μl of 2× polymerase mix to each well. Cover the plate and place in an oven at 37°C for 2 hours.
4. Add 10 μl of DNase mix to each well. Incubate for another 15 minutes at 37°C.
5. Add 80 μl of 125 mM sodium carbonate, pH 10.2, to each well. Incubate at 60°C for 20 minutes.
6. Place the plate on ice, then quickly add 50 μl of 7.5 M ammonium acetate to each well.
7. Transfer the samples to 1.5-ml microfuge tubes containing 375 μl of 100% ethanol. Vortex to mix, then incubate for 10 minutes at room temperature.
8. Spin the tubes in a microfuge (13,000 rpm) for 15 minutes to pellet the RNA probe. Drain the tubes well, then resuspend each damp pellet in 150 μl of 50% formamide/50% TE 7.5/0.1% Tween 20. This is a 50× stock for hybridization in 96-well plates.

Hybridization in 96-Well Plates
1. Prepare fixed embryos as in the main protocol but scale up to obtain at least 1 ml of embryos.
2. Add 2 ml of hybridization buffer *without* dextran sulfate to 1 ml of settled embryos.
3. Rock the embryos for ≥ 1 hour at room temperature.
4. Using a multichannel pipetter with cutoff yellow tips, add 20 μl of embryo suspension to each well of a 96-well filtration plate.
5. In each well of a separate (nonfilter) 96-well plate, mix 200 μl of hybridization buffer (*without* dextran sulfate) with 4 μl of probe.
6. Using a multichannel pipetter, add probes to the corresponding wells of the filtration plate.
7. Seal the plate with electrical tape and hybridize, with rocking, at 55°C overnight.
8. Carefully remove the tape, then place the plate on a Millipore vacuum manifold. Set the manifold vacuum control to *low*, then apply vacuum (house vacuum lines work well). Press on top of the plate to form a seal, then quickly turn the vacuum off once the last bit of hybridization buffer has been removed from the wells.

Note
Be sure the vacuum is set so that it takes about 10 seconds or longer for the buffer to be removed — if the vacuum is set too high, the embryos will become flattened and stick against the membrane.

9. Rinse the embryos by adding 200 µl of wash buffer to each well, then remove the wash buffer with low vacuum. Repeat once.
10. To wash embryos, add 200 µl of wash buffer to each well and rock for 1 hour at 55°C. Repeat the wash seven times. For the final wash, seal the plate with tape and rock overnight at 55°C.
11. Rinse embryos briefly with 200 µl of PBST, then add another 200 µl of PBST and rock for 30 minutes at room temperature.
12. Remove PBST, add 200 µl of PBSTS + antidig-AP antibody (1:2000), and rock for 2 hours at room temperature.
13. Rinse embryos twice with PBST, then wash embryos with nine changes of 200 µl of PBST, for 10 minutes per wash.
14. Rinse embryos twice with 200 µl of AP buffer, then wash embryos for 5 minutes in 200 µl of AP buffer.
15. To develop, add 200 µl of AP buffer containing NBT and BCIP (prepared as in the "Hybridization and Staining" protocol above) to each well. Incubate the plate, with rocking, until desired color development is achieved (20 minutes to overnight). To stop development of individual wells, remove the staining solution from the well and add 200 µl of PBST.
16. Once development of all wells is complete, rinse the embryos three times with 200 µl of PBST. To clear embryos, add 200 µl of 70% glycerol to each well. Embryos are ready to mount or dissect once they have settled to the bottom of the well.

Reagents and Materials
Transcription Mix (10 µl)
6.5 µl Template DNA (0.5 µg)
1 µl 10× BMB transcription buffer (0.4 M Tris, pH 8.0, 20 mM spermidine, 60 mM $MgCl_2$, 100 mM DTT)
1 µl 10× digoxigenin NTP mix [10 mM each ATP, CTP, and GTP, 6.5 mM UTP, 3.5 mM dig-11-UTP (Boehringer Mannheim #1-209-256)]
0.5 µl RNase inhbitor (5 units, Boehringer Mannheim #799-017)
1 µl T3 or T7 RNA polymerase (20–40 units, Boehringer Mannheim #1-031-163/881-767)
DNase Mix (10 µl)
1 µl 10× DNase I buffer (200 mM Tris, pH 8.0, 100 mM $MgCl_2$)
8 µl dH_2O
1 µl RNase-free DNase I
(30 units, Boehringer Mannheim #776-785)
1 M Sodium Carbonate, pH 10.2 stock
1.68 g $NaCO_3$ (monobasic)
3.18 g $NaCO_3$ (dibasic)
50 ml ddH_2O
Probe suspension buffer
50% Formamide

50% TE 7.5
0.1% Tween 20

Hybridization Buffer (11 ml)
5 ml Deionized formamide (BRL #15515-018)
2 ml 20× SSC (3 M NaCl, 0.3 M Na_3 citrate)
1 ml 10× Denhardt's (0.2% Ficoll 400, 0.2% polyvinylpyrrolidone, 0.2% BSA)
0.25 ml 10 mg/ml tRNA (Sigma #R- 9001)
0.25 ml 10 mg/ml Denatured salmon sperm DNA (Sigma #D-1626) Boil for 1 hour or sonicate to shear DNA.
0.1 ml 5 mg/ml Heparin (Sigma #H- 3393)
0.1 ml 10% Tween 20 (Sigma #P- 7949)
1.3 ml dH_2O
1 ml 50% Dextran sulfate (Pharmacia #17-0340-01)
Wash buffer 50% formamide, 2× SSC, 0.1% Tween 20

Antidig-AP antibody (Boehringer Mannheim #1-093-274)
Antidig-HRP antibody (Boehringer Mannheim #1-207-733)
AP buffer = 100 mM Tris, pH 9.5, 100 mM NaCl, 50 mM $MgCl_2$, 0.1% Tween 20 (add Tween just before use)
NBT stock = 75 mg/ml NBT (Sigma #N-6876) in dimethylformamide (not all will go into solution)
BCIP stock = 50 mg/ml BCIP (Sigma #B-8503) in dimethylformamide

2× Polymerase Mix (96 reactions)

250 μl ddH_2O
100 μl 10x BMB transcription buffer
50 μl dig NTP mix
50 μl RNase inhibitor (5 units/reaction)
50 μl T3 or T7 polymerase (20 units/rxn, 1000 units total)

96-well Filtration plate, Millipore MADV N65
Vacuum manifold, Millipore MAVM 096 01

Troubleshooting

1. To determine if the transcription reaction is successful, it is often useful to include a small amount of ^{32}P-UTP (1–2 μCi) in the reaction as a tracer. The ^{32}P allows accurate quantitation of the amount of probe produced and the low specific activity of the probe prevents significant radiolysis. To determine the percentage of NTP incorporation into the probe, compare the number of Cerenkov counts present in 1/20 of the sample *prior* to EtOH precipitation (total cpm) with the number of counts obtained from 1/20 of the sample *after* EtOH precipitation (incorporated cpm). A good reaction will result in at least 30% incorporation of NTPs, which is equivalent to 4 μg of RNA probe.

2. If the use of radioactive tracer is inconvenient, a 1-μl aliquot of each probe can be checked on a standard agarose (1× TBE or 1× TAE buffered with 0.05 μg/ml ethiduim bromide) minigel just prior to probe fragmentation. Even though the molecular weight and amount of probe is not accurately determined in this gel system, an approximation is often sufficient for the purpose of determining the success of the transcription reaction, the time of fragmentation, and the dilution factor for final probe resuspension.

Preparation of Samples for Microscopy

Staging Embryos

Although more complete descriptions of embryonic development are available (e.g., Campos-Ortega 1993; Goodman and Doe 1993; Jan and Jan 1993; Campos-Ortega and Hartenstein 1997), Hartenstein (1993) provides a very useful and affordable guide to the stage-specific features of *Drosophila* development. Each major organ system is followed in both whole-mount-view and section-view illustrations in this handbook. How one approaches the task of selecting embryos of a specific stage depends largely on how the embryos are to be viewed. Staging living embryos under a dissection microscope (as described in the "Live Dissection" protocol in the "Multiwell Plate Protocol for Immunochemistry") is very different than staging them after staining and clearing. Each antibody highlights (and sometimes hides) different features. Even when embryo collections are tightly synchronized, it may be important to discriminate between minor differences in stage by following the development of specific cells or organs in each embryo. For this reason, it may be necessary to learn the intimate details of one particular organ system to stage effectively with a given antibody.

Much of embryonic nervous system development occurs within an 8–10 hour period. Since the process is rapid, precise staging is often important for accurate phenotypic analysis. For example, the axonal extension of neurons that pioneer the MP1 pathway between segments within the CNS occurs in less than 2 hours. Figure 26-2 shows sketches of the morphology of specific neurons in the CNS and PNS at three time points (early stages 13, late 13, and early 14) across roughly 90 minutes of developmental time (at 25°C) as visualized with the monoclonal antibody 22C10 (Zipursky et al. 1984).

Dissection of Stained Embryos

Dissection Single-cell resolution of anatomical features within the CNS is easiest after the dissection of stained embryos. Once embryos have been cleared in glycerol, dissections are much easier than in the case of live tissue. Simply transfer the desired embryos to a small droplet of glycerol on a slide (staged as described in "Staging Embryos" above and/or genotyped with the use of histological balancers such as CyO-P[actin5C-lacZ, w^+]; this and other marked balancers are available through the Drosophila Stock Center, Department of Biology, Indiana University, Bloomington, IN 47405-6801). Under a stereo microscope, line up the embryos in the center of the slide and blot away excess glycerol until each embryo is held close to the glass by surface tension (Figure 26-3A).

First, make sure you are comfortable, and that your hands are resting on the table in front of the dissecting microscope. If your hands are pressed into

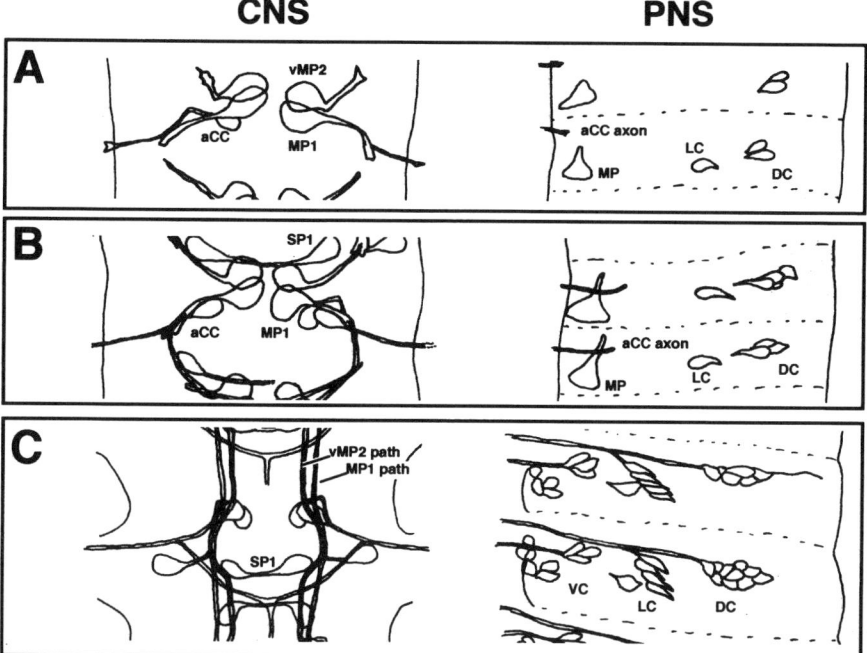

Figure 26-2 Stages of nervous system development. Drawings of central and peripheral neurons are shown as visualized with the monoclonal antibody 22C10 (Zipursky et al. 1984) in three stages that span approximately 90 minutes of development. (A) Early stage 13: Within the CNS, MP1 and vMP2 have initiated axonal growth to pioneer the MP1 longitudinal pathway. The aCC motor neuron is just extending outside of the CNS to pioneer the ISN pathway. At this stage a ventral muscle pioneer (MP) stains, in addition to the first peripheral neurons in the lateral and dorsal clusters (LC and DC). (B) Late stage 13: MP1 and vMP2 have crossed segment boundaries. SP1 axons have grown across the midline. The dorsal cluster of PNS neurons includes three or four precursors. (C) Early stage 14: Both medial (vMP2 pathway) and lateral (MP1 pathway) longitudinal axon fascicles are visible. All major clusters of the PNS are visible (VC, LC, DC); at this stage, the lateral cluster (LC) is composed of five or six chordotonal neurons. The PNS axon pathways are forming.

the table surface, vibrations will be less problematic. The fine tungsten needle (described in the "Live Dissection" protocol in the "Multiwell Plate Protocol for Immunochemistry") can be controlled with very subtle finger muscle movements in an otherwise anchored hand. If vibration from hand movement is a problem, steady the hand holding one needle with the other hand. With practice (and no caffeine intake) it is possible to dissect with two needles simultaneously; this is rapid and more precise than the one-handed method. In the one-handed mode, it is very important to remove as much glycerol as possible from the slide surface to provide friction that will

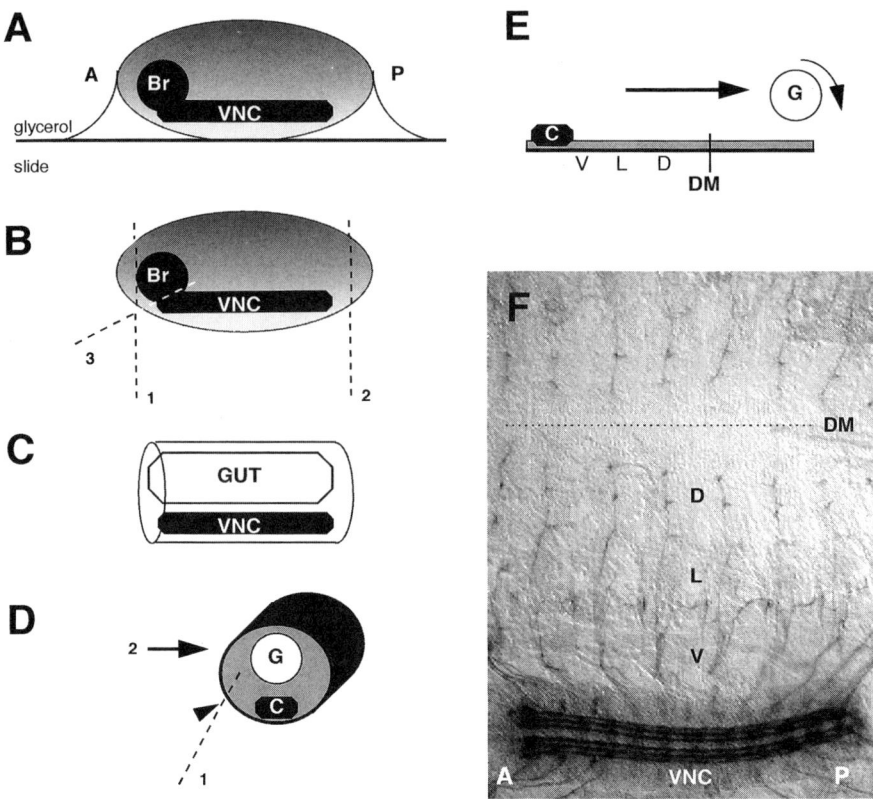

Figure 26-3 Dissection of stained embryos. (A) One stained embryo is placed on a microscope slide with a small amount of glycerol. (B) The anterior and posterior ends of each embryo are removed (dashed lines 1 & 2) prior to a cut (dashed line 3) that severs the connection between the VNC and brain. (C) The first cuts leave the embryo as a cylinder. (D) One dissection needle is inserted into the embryo to cut the embryo along its length by pressing down on the slide surface (a lateral cut is shown in order to preserve the dorsal midline; however, if ventral structures are of interest, the first cut can be made at the midline). A second needle is held parallel to the VNC and pushed sideways to unroll the embryo into a flat fillet. (E) A cross-sectional view of the fillet is shown. The gut of each embryo is rolled across the fillet surface and then removed. (F) A full fillet of a stage 17 embryo is shown stained with MAb 1D4 (Van Vactor et al. 1993). If only the VNC is to be photographed, the VNC can be lifted away from the fillet and viewed in isolation to avoid background from the epidermis. The ventral nerve cord (VNC, C), brain (Br), and gut (G) are marked. A, anterior; P, posterior; V, ventral; L, lateral; D, dorsal; DM, dorsal midline.

hold the embryos in place. With two needles, you can hold the embryo with one hand while you cut with the other.

Using a single needle, cut off the very anterior and posterior ends of the embryo (Figure 26-3B) by gently pressing the needle through the tissue towards the slide surface (you can move the needle back and forth to cut through the flesh if necessary). Next, slip one needle into the anterior end of the embryo and gently sever the connection between the brain and the ventral nerve cord; this can damage one side of the epidermis, so pick your favorite side and be careful. The embryo is now a tube (Figure 26-3C). Push one needle into this tube and press down on the slide surface to cut through the epidermis along the anterior–posterior axis (Figure 26-3D); choose whether you want the fillet boundaries to be lateral or along the ventral midline (see Figure 26-3E and F) before you insert the needle. Then, using one or two needles, peel back the epidermis on each side of the cut and push it down onto the slide surface. Since the gut is soft, it can be used as a rolling pin to spread out the fillet without damaging the fillet itself. Roll the gut out over the surface of the fillet and then remove small debris that might obscure your view of desired features (e.g., trachea, gonad, visceral mesoderm, and fat body sometimes get in the way).

Mounting Samples for Photography

After dissection, remove all unwanted tissue and debris from the slide surface. Add a tiny droplet of glycerol (~10 µl) just to one side of the fillets. Hold a clean coverslip over the fillets and drop one edge of the coverslip into the droplet of glycerol while holding the other edge elevated over the fillets, so that the glycerol wicks along the entire edge of the coverslip that is touching the slide surface. Then, gently drop the coverslip and the glycerol will sweep across the slide surface as the coverslip settles, chasing bubbles away as it fills the gap between coverslip and slide. If there is not enough glycerol to fill the gap, a small drop can be placed at the side of the coverslip and it will be drawn in. Excess glycerol is a problem because the working distance (depth of field) of high-magnification objectives (e.g., Zeiss 63×/1.40 Oil Plan-Apochromat Ph3) is very limited, making it difficult to reach deeper focal planes unless the coverslip is snug. Many objectives are designed to perform best with No. 1.5 coverslips; however, we recommend the thinner No. 1 coverslips if a full range of focal planes through the nervous system are to be reached without compressing the sample under the objective lens. After the glycerol fills the gap between slide and coverslip, gently seal the perimeter of the coverslip with fingernail polish to anchor the coverslip in place. Keep the amount of nail polish to a minimum to avoid obstructing high-magnification objectives near the edge of the coverslip.

For high-resolution color photomicrographs, Kodak Professional Ektachrome EPY 64T (color balanced for illumination with a tungsten filament) produces excellent results. For consistent color balance, it is best to saturate the tungsten light source during exposures; neutral density filters can be used to adjust light levels without changing the color temperature. As for the optics, the Zeiss 63×/1.40 Oil Plan-Apochromat Ph3 objective performs well for Nomarski (Differential Interference Contrast, DIC) optics and provides ample magnification for most photomicroscopy of the developing nervous system.

ACKNOWLEDGMENTS   We would like to thank Dr. Corey S. Goodman, and the members of his laboratory, for the exchange of ideas and many helpful technical suggestions in developing these protocols. We thank Jack Bateman and Chris Korey for helpful comments on the manuscript.

References

Campos-Ortega, J. A. (1993) Early neurogenesis in *Drosophila melanogaster*. In *The Development of Drosophila melanogaster*, Bate, M., and Martnez Arias, A., eds. Cold Spring Harbor, NY, Cold Spring Harbor Laboratory Press, pp. 1091–1129.

Campos-Ortega, J. A., and Hartenstein, V. (1997). *The Embryonic Development of Drosophila melanogaster*, 2nd ed. New York, Springer-Verlag.

Gallyas, F., Gorcs, T., and Merchenthaler, I. (1982). High-grade intensification of the end-product of the diaminobenzidine reaction for peroxidase histochemistry. J. Histol. Cytochem. 30:183–184.

Goodman, C. S., and Doe, C. Q. (1993). Embryonic development of the Drosophila central nervous system. In *The Development of Drosophila melanogaster*. Bate, M., and Martnez Arias, A., eds. Cold Spring Harbor, NY, Cold Spring Harbor Laboratory Press, pp. 1131–1206.

Hartenstein, V. (1993). *Atlas of Drosophila Development*. Plainview, NY, Cold Spring Harbor Laboratory Press.

Jan, Y. N. and Jan, L. Y. (1993). The peripheral nervous system. In *The Development of Drosophila melanogaster*. Bate, M., and Martnez Arias, A., eds. Cold Spring Harbor, NY, Cold Spring Harbor Laboratory Press, pp. 1207–1244.

Kolodziej, P. A., Jan, L. Y., Jan, Y. N. (1995). Mutations that affect the length, fasciculation, or ventral orientation of specific sensory axons in the Drosophila embryo. Neuron 15:273–286.

Kopczynski, C. C., and Muskavitch, M. A. T. (1992). Introns excised from the Delta primary transcript are localized near sites of Delta transcription. J. Cell Biol. 119:503–512.

Kopczynski, C. C., Davis, G. W., and Goodman, C. S. (1996). A neural tetraspanin, encoded by late bloomer, that facilitates synapse formation. Science 271:1867–1870.

Liposits, Z. S., Setalo, G. Y., and Lerko, B. F. (1984). Application of the silver-gold intensified 3,3-diaminobenzidine chromatin to the light and electron microscopic detection of the leutenizing hormone releasing hormone detection system of the rat brain. Neuroscience 13:513–525.

Patel, N. H. (1994). Imaging neuronal subsets and other cell types in whole-mount Drosophila embryos and larvae using antibody probes. In *Drosophila melanogaster: Practical Uses in Cell and Molecular Biology*. Goldstein, L. S. B., and Fyrberg, E. A., eds. Boston, MA, Academic Press, pp. 445–487.

Patel, N. H., Ball, E. E., and Goodman, C. S. (1992). Changing role of even-skipped during the evolution of insect pattern formation. Nature 357:339–342.
Salzberg, A., D'Evelyn, D., Schulze, K. L., Lee, J. K., Strumpf, D., Tsai, L., and Bellen, H. J. (1994). Mutations affecting the pattern of the PNS in Drosophila reveal novel aspects of neuronal development. Neuron 13:269–287.
Seeger, M. A., Tear, G., Ferres-Marco, D., and Goodman, C. S. (1993). Mutations affecting growth cone guidance in Drosophila: genes necessary for guidance towards or away from the midline. Neuron 10:409–426.
Tautz, D., and Pfeifle, C. (1989). A non-radioactive in situ hybridization method for the localization of specific RNAs in Drosophila embryos reveals translational control of the segmentation gene *hunchback*. Chromosoma 98:81–85.
Tessier-Lavigne, M., and Goodman, C. S. (1996). The molecular biology of axon guidance. Science 274:1123–1133.
Van Vactor, D., Sink, H., Fambrough, D., Tsoo, R., and Goodman, C. S. (1993). Genes that control neuromuscular specificity in Drosophila. Cell 73:1137–1153.
Zalokar, M., and Erk, I. (1977). Phase-partition fixation and staining of Drosophila eggs. Stain Technol. 52:89–92.
Zipursky, S. L., Venkatesh, T. R., Teplow, D. B., and Benzer, S. (1984). Neuronal development in the Drosophila retina: monoclonal antibodies as molecular probes. Cell 36:15–26.

Appendix: Suppliers

Boehringer Mannheim Corp., 9115 Hague Road, P.O. Box 50414, Indianapolis, Indiana 46250

Dow Corning Corp., Midland, Michigan 48640

Gibco BRL, P.O. Box 68, Grand Island, New York 14072

Pharmacia Biotech, Inc., 800 Centennial Ave., P.O. Box 1327, Piscataway, New Jersey 08855

Sigma, P.O. Box 14508, St. Louis, Missouri 63178

Ted Pella, Inc., P.O. Box 492477, Redding, California 96049

Tetko Inc., 333 S. Highland Ave., Briarcliff Manor, New York 10510

Vector Laboratories, Inc., Burlingame, California 94010

VWR Scientific Products, P.O. Box 1002, S. Plainfield, New Jersey 07080 (Distributors of Adams, Falcon, and Staco products)

Zeiss Optical Systems, Inc., Thornwood, New York 10594

Index

Acidosis, 53, 54
Acrosome reaction, 11, 20, 21
Actin labeling by phalloidin, 404
Adenylation control element (ACE), 98
Alkalosis, 49, 51
Allele-specific gene expression, 131, 135
Anesthetics, 181, 257
Animal caps, 323, 332
Antibodies
  biotinylation of, 120
  oocytes injection of, 375
  production of, 115, 386
Antisense oligonucleotides, 91, 92, 341
  mapping with, 93
  melting temperature of, 95
  modifications of, 345–347
  nonspecific effects of, 95
  optimal concentration of, 95
  optimal length of, 99
  purification of, 93
  selection for mRNA ablation, 343, 344
  stability in injected oocytes, 95
  synthesis of, 93
Antisense RNA, 259, 286
Axis formation, 357
Axon guidance, 490

β-galactosidase, 373, 406, 407 (*see also* Reporter constructs)
Bacterial fusion proteins, 386
Bacteriophage P1, 121
Blastocoel, 65, 66
  ion concentration of, 65–67
Blastocyst, 26, 68, 105
Blastomeres
  deletion of, 364–370
  dissociation of, 375–377
  explant culture of, 370–372
  preventing communication of, 377–378

516 Index

Blastomeres (*continued*)
  transplantation of, 372–374
Blastopore, 320
Blue fluorescent protein, 297
Bouin's fixative, 120

Caged compounds, use in transcription analysis, 482
Calcium ionophore, 20, 28–29, 58–61, 75, 201
  calculation of, 60
CCD camera, 44, 477
Cell cycle, 116
  determination by BrdU-labeling, 404
Cell cycle extracts
  buffers for preparation of, 222–226
  preparation of, 198, 200, 202
Cell gradients, 229, 231
Cell lineage, 355
Chloride transport, 62–63
Chorion protein, 391
Chorionic gonadotropin, 4, 17, 25, 26, 347, 348
Chromatin assembly
  analysis in egg extracts, 210–213
  analysis in oocytes, 249
  buffers for, 223–224
  preparation of extracts for, 202
  supercoiling assay, 212–214
Collagenase, 257, 399
Cordycepin (3'-deoxyadenosine), 89, 90
Cre/*lox*P, 121, 122
Culture media, 3–6
  Barth's, 182, 222, 275, 280
  Earle's salts, 16, 150
  for embryo development, 8
  for fertilization, 5, 6
  gas mixture for, 4–6, 45, 84
  KSOM, 3, 6, 7, 25, 106, 150
  M199, 16–18
  MEM, 84, 85
  MMR, 182, 223, 250, 353
  modified Ringer's, 315
  OCM, 353
  OR-2, 182, 243, 275
  Waymouth, 5
Cumulus (granulosa) cells, 4, 25, 105
Cytocholasin B, 222

Cytoplasmic polyadenylation element (CPE), 89, 103
Cytoskeleton, 33, 34, 450

Defolliculation, 348
Dejellification, 350
Dibutyryl cAMP, 84, 85
Differential display, 148, 152, 157
  of embryo RNA, 157
  of genital ridge RNA, 162
  primers used for, 166
  verification of, 167
Digitonin, 73
Dissecting tools, 500
DNA, 67
  analysis of, 191, 243
  damage, 67
  extraction from oocytes, 186
  semiconservative replication of, 216
  supercoiling of, 244
DNA repair, 177
  substrates for, 179
  triplex formation in, 179
Dominant negative mutations, 418
Dye (RIM-1), 27, 28, 34 (*see also* Vital dyes)

Egg chambers, 285, 400
  culture of, 420–421
  immunostaining of, 405
  staining of, 403
Egg extracts
  buffers used for, 222–223
  energy mix for, 222
  preparation of, 200–203
Eggs, 17
  collection from *Drosophila*, 400–401, 453–454, 474–475
  collection from mouse, 25, 28
  collection from *Xenopus*, 199
  fixation of, 28
  immobilization of, 26
  permeabilization of, 31
  storage of, 18
Eggshell proteins, 387
Embryogenesis
  Drosophila, 471
  time-lapse recording of, 474
Embryonic stem cells, 121

Embryos, 17
  collection of, 17, 150, 159, 493, 474–475
  dechorionation of, 474, 493
  dissection of, 498, 508
  fixation of, 429–431, 493
  localized gene expression in, 320
  preparation for in situ hybridization, 330
  RNase treatment of, 431
  storage of, 18
  synchronization in vivo, 105
  See also Eggs
Enhancer trapping, 408–411

Fate commitment, 363
Fate mapping, 357
Fertilization, 23
  in vitro, 69, 70, 115
  of *Xenopus* eggs, 349
Fluorescein, 45
Fluorescence indicators, 480
Fluorescent protein analogs, 479–480
Fluorophores, 40–76
  Ca^{2+}-sensitive, 41, 56–59, 62
  calibration of, 76
  Cl-sensitive, 42, 63–66
  method of detection, 43
  Na^+/K^+-sensitive, 42, 65
  pH-sensitive, 66
  pitfalls of using, 71
  sequestration of, 72
  toxicity of, 66, 69
  use in cytoplasmic localization, 73
Follicle cell clones, 414–418

GAL4/UAS, 402
GAL4-VP16, 482
Gastrulation, 322
Genital ridge, 157–167
Germ-line clones, 412
Germinal vesicle
  bulk isolation of, 187–189
  manual dissection of, 187
Green fluorescent protein, 292, 294, 296, 298, 373, 402, 405, 473

Heat shock, 245
Histones, 251

HPLC, 13, 15, 94

Image processing, 484–486
Immunoblotting (Western blotting), 115–119, 344, 387–391
Immunogold labeling, 391
Immunoprecipitation, 116–118
Immunostaining, 272–273
  artifacts and pitfalls of, 303–305
  bleaching, 303
  buffers for, 278, 314–315
  central nervous system, 492
  cryostat sections, 309–311
  detection by electron microscopy, 391–395, 442–443
  fixatives for, 301, 393–394
  multi-well plates in, 492
  ovaries and embryos, 435–438
  protocols for, 495–497
  reagents for, 497–498
  sections, 308
  signal enhancement of, 501–502
  special slide for, 300
  troubleshooting, 498
  visualization reagents for, 306
  whole mount, 120, 293, 299–303, 307–308, 491
Imprinted loci, 128, 130
  determination of, 140
  H19, 130
  Igf2r, 130
  in oocytes and embryos, 138
Imprinting, 127
  developmental regulation of, 129
In situ hybridization, 279, 327–332, 358, 453–454, 502–504
  apparatus for, 289
  buffers for, 276–278
  coupled with immunostaining, 264–266, 504–505
  detection by electron microscopy, 438–441
  double staining, 283–284
  fluorescence detection of, 447, 454–455
  large scale processing, 286–289
  nonradioactive detection of, 284–286
  optimal probe length for, 428
  to sections, 433–435

In situ hybridization (*continued*)
    troubleshooting, 433
    for use in detecting multiple transcripts, 449
    using multi-well plates, 505–507
    in whole mount, 258, 260–264, 268–271, 280–287, 427–435
Inner cell mass, 149, 153
Interspecies mating, 132
Ion exchange, 52–56, 63

Lineage tracing, fluorescent markers for, 359–363

Maternal mRNA, 102 (*see also* RNA)
Maturation promoting factor, 30
Meiosis, 175
Mesoderm induction, 317, 322
Microforge, 180
Microinjection, 83
    of *Drosophila* embryos, 475
    of germinal vesicles, 181, 183, 184, 243
    linear vs. circular DNA, 185
    of mouse oocytes, 84–96
    specifics of, 180
    of *Xenopus* oocytes and embryos, 180, 183, 243, 256, 268
Micromanipulator, 180, 345, 476
Micropipette injector, 345
Micropipette puller, 344, 476
Microscopy
    confocal, 29, 44, 271, 449, 454
    electron, 35, 311
    fluorescence, 297, 473
    preparation of samples for, 508
    time-lapse video, 477–479
Mid-blastula transition, 248, 250, 253
Mos, 118, 121
Mosaic analysis, 411
mRNA
    polyadenylation, 88
    proteins in, 228
    See also RNA
mRNP. *See* Polysomes/mRNPs
myc epitope, 294, 297

Nerve cells, 400
Nervous system, 321–327, 509
Neural induction, 333, 345
Nigericin, 48, 49
Northern blotting, 91, 109, 233–234
Nuclei
    preparation from embryos, 203–205
    preparation from sperm, 205–206
    See also Germinal vesicle
Nucleosome assembly, 244
Nycodenz, 229, 232, 234, 239

Oligo(dT), 87, 90, 91, 236, 240
Oocyte, 1, 25
    ^{32}P-labeling, 116
    culture of, 266
    identification of, 401–402
    injection of, 97, 344
    isolation of, 25, 105, 137, 181, 230, 257
    maturation of, 1, 2, 85, 111, 113
    methionine labeling, 87, 96, 117
    transfer of to recipient female, 347–349
Oogenesis
    in *Drosophila*, 400, 447
    in the mouse, 136
    in *Xenopus*, 173, 174
Ovary, 12
    dissection of, 388, 398–399
    fixation of, 399–400, 434
    homogenization buffer for, 12
    isolation of, 182, 230, 257
    isolation of nuclei from, 419
    selection of, 347
Oviduct, 26
Ovulation, 17
    superovulation, 17

Pattern formation
    anteroposterior, 321
    central nervous system, 321, 324
    dorsoventral, 317
    markers for, 318–319
    neural molecules involved in, 323
Percoll, 12, 14
pH, 40
    conversion to, 47

dependence of exchange activity, 52
indicators, 40, 43, 46, 49, 68
measurement of, 46
Phalloidin. *See* Actin labeling by phalloidin
Phenotyping, 133
Photoablation, 359
Photography, 511
Plakoglobin, 298
  depletion of mRNA for, 346
Plasmid rescue, 410
Plasminogen, 86, 87
Plasminogen activator, 84
  tPA, 84, 85, 107
  uPA, 84, 85
Poly(A) polymerase, 88
Poly(A) tail, 103, 104
  addition in vitro, 89, 107
  length determination of, 107, 110, 488
  See also Poly(A) test
Poly(A) test (PAT assay), 458–463
  of *bicoid* and *oskar* RNAs, 468
Polyacrylamide gel electrophoresis, 9, 15, 85–89, 97, 389
Polyadenylation, 89–107
  cytoplasmic, 89
  See also Poly(A) tail
Polysomes/mRNPs
  analysis of, 231
  buffers for isolation of, 239
  fractionation of, 230, 231
  gradients of, 233
Poly(U) Sepharose, 240
Post-translational modification, 113, 115
Primer extension, 243, 249
Primordial germ cells, 137
Progesterone, 249, 347, 348
Promoters
  active in oogenesis, 420
  cytomegalovirus, 244, 245, 251
  hsp70, 244, 245
  thyroid hormone receptor, 244, 246
Protease inhibitor, 28, 30, 31, 34, 35, 86, 201, 222
Protein A-Sepharose CL-4B, 118
Protein kinase, 24
  assay for, 27
  inhibitors of, 31, 32
  MAP kinase, 30, 118

PKA, 27
PKC, 27, 28, 29, 30, 34
  substrates for, 29, 30

Rapid amplification of cDNA ends (RACE), 149
Recombination
  homologous, 175, 178
  induced by FLP/FRT, 411–414
  induced by X-rays, 411
  nonhomologous, 175
  preparation of extracts for, 187, 190
  schematic of, 412
  single strand annealing, 176
  substrates for, 177
  See also Cre/*lox*P
Recordings, whole patch, 70
Replication, 196
  analysis in cell gradients, 217–218
  analysis in injected eggs, 214–217
  analysis using egg extracts, 206–210
  buffers used in analysis of, 225–226
  calculation of activity in extracts, 207–210
  identification of origins by two-dimensional gel, 218–220
  preparation of extracts for, 197
  See also DNA
Reporter constructs, 107, 109
  β-galactosidase, 109–114, 247, 293, 358
  chloramphenicol acetyltransferase, 243, 247, 248
  green fluorescent protein, 296, 358
Restriction fragment length polymorphism (RFLP), 131–133, 140
Ring canals, 400
RNA
  3′UTR, 83, 103, 109–113, 458
  analysis of, 229
  buffers for in vitro synthesis, 278
  capping of, 88
  concentration for PAT assay, 466
  density of, 232
  digoxygenin, biotin, or fluorescein labeling of, 260, 280, 329, 452
  DNase treatment of, 466
  extraction from embryos, 333

RNA (*continued*)
  in vitro synthesis of, 88, 260, 280, 329, 428, 451–452, 507
  incorporation of amino-allyl-UTP, 451
  injection of, 89, 104, 139, 150, 160 (*see also* Microinjection)
  isolation of, 466
  localization of, 256
  polyadenylation of, 88 (*see also* Poly(A) tail)
RNase
  A, 104, 235
  H, 90, 91, 94
  T1, 92, 97, 235
RT–PCR, 90, 109, 131, 134, 139, 151, 316, 332, 335–336
  primers for mesodermal markers, 319
  primers for neural ectoderm markers, 325

Single nucleotide primer extension (SnuPE), 131, 141
Single strand conformational polymorphism, 130, 132, 134
Somites, 159
Spemann's Organizer, 321
Sperm, 5
  capacitation, 16
  culture of, 16–17
  incubation with eggs, 18–19
  isolation of, 16
  preparation of demembranated nuclei of, 205
  receptor, 16
Spin, 116–119
Sry, 158

Superovulation, 159

TATA box binding proteins, 251
Three-dimensional reconstruction, 449, 450
Tissue explants, 332, 334
Transcription
  analysis in eggs and embryos, 247
  analysis in oocytes, 242, 246
  by GAL4-VP16 system, 252, 253
  See also DNA; Microinjection
Transgenic lines, 418
Translational control, 5, 83, 102–108 (*see also* Polyadenylation; Antisense oligonucleotides)
Trophectoderm, 148, 153

Uniparental disomy, 128
UV-crosslinking, 96, 97, 228, 235
  use of 5-bromouridine in, 235

Vimentin, 295
Vital dyes, 348, 354
Vitelline membrane, 494
Vitellogenin, 266–267

Western blotting. *See* Immunoblotting

Xenopus laevis, commercial suppliers of, 257, 347

Y chromosome, 159, 160

Zona pellucida, 10–15, 20, 24
Zp3 promoter, 121
Zymography, 85, 86, 94